Quantitative Fundamentals of Molecular and Cellular Bioengineering

Quantitative Fundamentals of Molecular and Cellular Bioengineering

K. Dane Wittrup, Bruce Tidor, Benjamin J. Hackel, and Casim A. Sarkar

The MIT Press
Cambridge, Massachusetts
London, England

© 2019 The Massachusetts Institute of Technology

All rights reserved. No part of this book may be reproduced in any form by any electronic or mechanical means (including photocopying, recording, or information storage and retrieval) without permission in writing from the publisher.

This book was set in Times New Roman by Westchester Publishing Services. Printed and bound in the United States of America.

Library of Congress Cataloging-in-Publication Data

Names: Wittrup, K. Dane, author. | Tidor, Bruce, author. | Hackel, Benjamin J., author. | Sarkar, Casim A., author.
Title: Quantitative fundamentals of molecular and cellular bioengineering / K. Dane Wittrup, Bruce Tidor, Benjamin J. Hackel, and Casim A. Sarkar.
Description: Cambridge, MA : The MIT Press, [2019] | Includes bibliographical references and index.
Identifiers: LCCN 2018051193 | ISBN 9780262042659 (hardcover : alk. paper)
Subjects: | MESH: Bioengineering | Biotechnology | Genetic Engineering | Cell Physiological Phenomena
Classification: LCC R857.B54 | NLM QT 36 | DDC 610.284--dc23 LC record available at
  https://lccn.loc.gov/2018051193

10 9 8 7 6 5 4 3 2 1

Contents

Preface and Acknowledgments xv

1 Introduction to Biological Rate Processes   1
  1.1 Biological Rate Processes   1
  1.2 Why Develop a Model?   9
  1.3 Mechanistic Model Formulation   10
  1.4 Model Validation   16
    1.4.1 Consistency with Experiment Is Necessary but Not Sufficient for a Model's Correctness   17
    1.4.2 Comparison to Alternative Models   19
    1.4.3 Prediction and Falsifiability   19
    1.4.4 Parameter Sensitivity Analysis   20
  1.5 Basic Themes in Rate Process Modeling   22
    1.5.1 Steady State versus Equilibrium   22
    1.5.2 Perturbations   22
    1.5.3 Rate Forms   24
    1.5.4 Characteristic Time and Length Scales   25
  Suggestions for Further Reading   27
  References   27

2 Noncovalent Binding Interactions   29
  2.1 Kinetic Rate Constants   31
    2.1.1 Typical Ranges for k_{on} and k_{off} 33
  2.2 Thermodynamics   34
    2.2.1 State Functions   34
    2.2.2 Free Energy and Standard States   36
  2.3 Energetic Contributions to Binding Affinity   43
    2.3.1 Electrostatics   43
    2.3.2 Hydrogen Bonding   45
    2.3.3 van der Waals Interactions   46
    2.3.4 Hydrophobic Effect   49
    2.3.5 Deformation Energy and Entropy   50
  2.4 Energetics of Protein Binding Interfaces   51
    2.4.1 Thermodynamic Cycle Analysis   54

　　　　　Case Study 2-1 P. J. Carter, G. Winter, A. J. Wilkinson, and A. R. Fersht.
　　　　　　　The use of double mutants to detect structural changes
　　　　　　　in the active site of the tyrosyl-tRNA synthetase
　　　　　　　(*Bacillus stearothermophilus*). *Cell* 38: 835–840 (1984).　59
　　　　　Case Study 2-2 Z. S. Hendsch and B. Tidor. Do salt bridges stabilize proteins?
　　　　　　　A continuum electrostatic analysis. *Protein Sci.* 3: 211–226 (1994).　61
　　2.5　Environmental Impacts on Binding Rate　64
　　　　　2.5.1　Arrhenius Relationship and Transition State Theory　64
　　　　　2.5.2　Electrostatic Effects on Binding Kinetics　69
　　2.6　Molecular Measurements: Light-Matter Interactions　75
　　　　　2.6.1　Scattering　76
　　　　　2.6.2　Absorbance　77
　　　　　2.6.3　Fluorescence　81
　　　　　Case Study 2-3 L. Song, E. J. Hennink, I. T. Young, and H. J. Tanke.
　　　　　　　Photobleaching kinetics of fluorescein in quantitative fluorescence
　　　　　　　microscopy. *Biophys. J.* 68: 2588–2600 (1995).　83
　　2.7　Fluorescence Applications for Biomolecular Measurements　89
　　　　　2.7.1　Biosynthetic Fluorescent Labeling　89
　　　　　Case Study 2-4 B. G. Reid and G. C. Flynn. Chromophore formation in green
　　　　　　　fluorescent protein. *Biochemistry* 36: 6786–6791 (1997).　92
　　　　　2.7.2　Common Fluorophores for In Vitro Conjugation　93
　　　　　2.7.3　Fluorescence Instrumentation　96
Suggestions for Further Reading　101
Problems　102
References　106

3　Binding Equilibria and Kinetics　111
　　3.1　Equilibrium Monovalent Protein-Ligand Binding　111
　　　　　3.1.1　Monovalent Binding Isotherm　111
　　　　　3.1.2　Graphical Representations　115
　　3.2　Binding Kinetics　115
　　3.3　Multiple Binding Sites　121
　　　　　3.3.1　Independent Sites　121
　　　　　3.3.2　Cooperativity　125
　　　　　3.3.3　Avidity and Effective Concentration　128
　　　　　3.3.4　Equilibrium Bivalent Binding at Cell Surfaces　136
　　　　　Case Study 3-1 J. D. Stone, J. R. Cochran, and L. J. Stern. T-cell activation
　　　　　　　by soluble MHC oligomers can be described by a two-parameter
　　　　　　　binding model. *Biophys. J.* 81: 2547–2557 (2001).　139
　　　　　Case Study 3-2 Y. Mazor, A. Hansen, C. Yang, P. S. Chowdhury, J. Wang,
　　　　　　　G. Stephens, H. Wu, and W. F. Dall'Acqua. Insights into the
　　　　　　　molecular basis of a bispecific antibody's target selectivity.
　　　　　　　mAbs 7: 461–469 (2015).　143
　　3.4　Fast or Complex Reaction Measurements　149
　　　　　3.4.1　Relaxation Kinetics　150
　　　　　3.4.2　Double-Jump Experiments to Detect Hidden Rate Processes　151
　　3.5　Theory and Practice of Biomolecular Measurements　152
　　3.6　General Issues in Measuring Binding Affinity and Kinetics　153

Contents

    3.7   Methods That Detect Altered Localization on Binding   159
          3.7.1   Cell-Surface Titrations and Competition   160
          Case Study 3-3 E. T. Boder and K. D. Wittrup. Optimal screening of surface-displayed polypeptide libraries. *Biotechnol. Prog.* 14: 55–62 (1998).   164
          3.7.2   Biosensors   166
          3.7.3   Microplate Immunoassays   168
          3.7.4   Microfluidic Assays   170
          3.7.5   Dialysis   171
          Case Study 3-4 T. J. Silhavy, S. Szmelcman, W. Boos, and M. Schwartz. On the significance of the retention of ligand by protein. *Proc. Natl. Acad. Sci. U.S.A.* 72: 2120–2124 (1975).   171
    3.8   Methods to Detect Changes in Intrinsic Properties on Binding   174
          3.8.1   Optical Spectroscopy   175
          3.8.2   Isothermal Titration Calorimetry   176
          3.8.3   Altered Mobility   178
    3.9   Fitting Models to Data   179
          3.9.1   Data Collection   179
          3.9.2   Nonlinear Least Squares   180
          3.9.3   Confidence Intervals on Fitted Parameters   186
          3.9.4   Global Nonlinear Least Squares Fitting   190
Suggestions for Further Reading   190
Problems   191
References   206

4   Enzyme Kinetics   209
    4.1   Enzymes Are Catalysts   209
    4.2   Enzymatic Rate Laws   212
          4.2.1   Michaelis-Menten Rate Law   212
          4.2.2   Bisubstrate Kinetics   227
          4.2.3   Substrate Inhibition   230
          4.2.4   Cooperative Enzymes   231
          4.2.5   pH Effects   232
          4.2.6   Temperature Effects   234
    4.3   Reversible Enzyme Inhibition   235
          4.3.1   Reversible Inhibitors   235
          4.3.2   Pharmacological Inhibition   243
          4.3.3   Tight-Binding Inhibitors   247
          Case Study 4-1 R. A. Spence, W. M. Kati, K. S. Anderson, and K. A. Johnson. Mechanism of inhibition of HIV-1 reverse transcriptase by nonnucleoside inhibitors. *Science* 267: 988–993 (1995).   248
    4.4   Signaling Pathways   251
          4.4.1   Tyrosine Kinases   251
          Case Study 4-2 P. A. Schwartz, P. Kuzmic, J. Solowiej, S. Bergqvist, B. Bolanos, C. Almaden, A. Nagata, K. Ryan, J. Feng, D. Dalvie, J. C. Kath, M. Xu, R. Wani, and B. W. Murray. Covalent EGFR inhibitor analysis reveals importance of reversible interactions to potency and mechanisms of drug resistance. *Proc. Natl. Acad. Sci. U.S.A.* 111: 173–178 (2014).   252

4.4.2 Trimeric G Proteins and G Protein-Coupled Receptors   256
4.5 Metabolism   257
    4.5.1 Flux-Control Coefficients   257
    4.5.2 Constraint-Based Models   258
4.6 Hydrolytic Regulatory Enzymes   260
    4.6.1 Caspases   260
    4.6.2 Blood Coagulation Cascades   261
    Case Study 4-3 K. C. Jones and K. G. Mann. A model for the tissue factor pathway to thrombin. II. A mathematical simulation. *J. Biol. Chem.* 269: 23,367–23,373 (1994).   262
    4.6.3 Extracellular Metalloproteases   269
Suggestions for Further Reading   270
Problems   270
References   277

5 Gene Expression and Protein Trafficking   281

5.1 Synthesis, Degradation, and Growth   281
    Case Study 5-1 H. C. Towle, C. N. Mariash, H. L. Schwartz, and J. H. Oppenheimer. Quantitation of rat liver messenger ribonucleic acid for malic enzyme during induction by thyroid hormone. *Biochemistry* 20: 3486–3492 (1981).   287
    Case Study 5-2 S. B. Lee. and J. E. Bailey. Analysis of growth rate effects of productivity of recombinant *Escherichia coli* populations using molecular mechanism models. *Biotechnol. Bioeng.* 26: 66–73 (1984).   295
    5.1.1 Pure Delays in Biosynthesis   296
5.2 Compartmental Models of Protein Sorting   297
    5.2.1 Receptor-Ligand Trafficking   306
    Case Study 5-3 J. M. Haugh. Mathematical model of human growth hormone (hGH)-stimulated cell proliferation explains the efficacy of hGH variants as receptor agonists or antagonists. *Biotechnol. Prog.* 20: 1337–1344 (2004).   310
    5.2.2 Vesicular Transport   313
    Case Study 5-4 K. Hirschberg, C. M. Miller, J. Ellenberg, J. F. Presley, E. D. Siggia, R. D. Phair, and J. Lippincott-Schwartz. Kinetic analysis of secretory protein traffic and characterization of Golgi to plasma membrane transport intermediates in living cells. *J. Cell Biol.* 143: 1485–1503 (1998).   315
Suggestions for Further Reading   317
Problems   317
References   320

6 Network Dynamics   323

6.1 Nonlinear Dynamics   324
6.2 Switches and Thresholds   337
    6.2.1 Ultrasensitivity through Homomultimerization   337
    6.2.2 Ultrasensitivity through Molecular Titration   340
    6.2.3 Zero-Order Ultrasensitivity   342
    6.2.4 Bistable Switches   345

Case Study 6-1 N. A. Shah and C. A. Sarkar. Robust network topologies for generating switch-like cellular responses. *PLOS Comput. Biol.* 7: e1002085 (2011); S. Palani and C. A. Sarkar. Synthetic conversion of a graded receptor signal into a tunable, reversible switch. *Mol. Syst. Biol.* 7: 480 (2011).   348

   6.3   Adaptation   351
          6.3.1   Negative Feedback   352

Case Study 6-2 N. Barkai and S. Leibler. Robustness in simple biochemical networks. *Nature* 387: 913–917 (1997); U. Alon, M. G. Surette, N. Barkai, and S. Leibler. Robustness in bacterial chemotaxis. *Nature* 397: 168–171 (1999); T.-M. Yi, Y. Huang, M. I. Simon, and J. Doyle. Robust perfect adaptation in bacterial chemotaxis through integral feedback control. *Proc. Natl. Acad. Sci. U.S.A.* 97: 4649–4653 (2000).   355

          6.3.2   Incoherent Feedforward   358
          6.3.3   Additional Adaptive Topologies   358
   6.4   Oscillations   359

Case Study 6-3 A. Hoffmann, A. Levchenko, M. L. Scott, and D. Baltimore. The IκB–NF-κB signaling module: Temporal control and selective gene activation. *Science* 298: 1241–1245 (2002).   363

          6.4.1   Cell Cycle   365
          6.4.2   Circadian Rhythms   365

Suggestions for Further Reading   367
Problems   368
References   370

7 Population Growth and Death Models   373

   7.1   Typical Growth Curves   374
          7.1.1   Exponential Growth   375
          7.1.2   Subexponential Growth Kinetics   378
   7.2   Limitations on Growth of Cell Cultures   389
          7.2.1   Consumption of Nutrients   389
          7.2.2   Growth-Factor Dose Responses   391
          7.2.3   Toxic Metabolite Accumulation   394
   7.3   Bioreactors   396
          7.3.1   Cell Growth as a Pseudochemical Reaction   396
          7.3.2   Steady-State Monod Chemostat   403
          7.3.3   Fed-Batch Fermentors   409
          7.3.4   Product-Formation Kinetics   411
   7.4   Cell and Organismal Death   413
          7.4.1   Death by Injury   413
          7.4.2   Death by Senescence   417

Case Study 7-1 B. J. Tolkamp, M. J. Haskell, F. M. Langford, D. J. Roberts, and C. A. Morgan. Are cows more likely to lie down the longer they stand? *Appl. Anim. Behav. Sci.* 124: 1–10 (2010).   424

Case Study 7-2 M. Z. Levy, R. C. Allsopp, A. B. Futcher, C. W. Greider, and C. B. Harley. Telomere end-replication problem and cell aging. *J. Mol. Biol.* 225: 951–960 (1992).   428

          7.4.3   Cell-Cell Death Signaling   431

7.5 Compartmental Growth Models 431
 7.5.1 Cell Cycle 431
 Case Study 7-3 J. A. Smith and L. Martin. Do cells cycle?
  Proc. Natl. Acad. Sci. U.S.A. 70: 1263–1267 (1973). 432
 7.5.2 In Vivo Cell Population Models 433
 Case Study 7-4 H. Mohri, A. S. Perelson, K. Tung, R. M. Ribeiro, B. Ramratnam,
  M. Markowitz, R. Kost, A. Hurley, L. Weinberger, D. Cesar,
  M. K. Hellerstein, and D. D. Ho. Increased turnover of T lymphocytes
  in HIV-1 infection and its reduction by antiretroviral therapy.
  J. Exp. Med. 194: 1277–1287 (2001). 435
 Case Study 7-5 H. Quastler and F. G. Sherman. Cell population kinetics
  in the intestinal epithelium of the mouse. *Exp. Cell Res.*
  17: 420–438 (1959). 441
 7.5.3 Epidemics and Infections 443
 Case Study 7-6 A. S. Perelson, A. U. Neumann, M. Markowitz, J. M. Leonard,
  and D. D. Ho. HIV-1 dynamics in vivo: Virion clearance rate,
  infected cell life-span, and viral generation time. *Science*
  271: 1582–1586 (1996). 447
 7.5.4 Growth of Mixed Populations 451
Suggestions for Further Reading 461
Problems 462
References 471

8 Coupled Transport and Reaction 477
8.1 Diffusion, Collision, and Binding 477
 8.1.1 Transport Effects on Binding 477
 8.1.2 Two-Step Binding Processes 484
 Case Study 8-1 A. D. Vogt and E. Di Cera. Conformational selection
  or induced fit? A critical appraisal of the kinetic mechanism.
  Biochemistry 51: 5894–5902 (2012). 486
 8.1.3 Smoluchowski Diffusion Limit 489
 8.1.4 Surfaces: Local Depletion and Recapture 495
 Case Study 8-2 S. Y. Shvartsman, H. S. Wiley, W. M. Deen,
  and D. A. Lauffenburger. Spatial range of autocrine signaling: Modeling
  and computational analysis. *Biophys. J.* 81: 1854–1867 (2001). 497
8.2 Scaling Analyses of Tissue Penetration 500
 Case Study 8-3 G. M. Thurber and K. D. Wittrup. Quantitative spatiotemporal
  analysis of antibody fragment diffusion and endocytic consumption
  in tumor spheroids. *Cancer Res.* 68: 3334–3341 (2008). 503
8.3 Compartmental Models 504
 8.3.1 The Mass-Transfer Coefficient 504
 Case Study 8-4 D. G. Myszka, X. He, M. Dembo, T. A. Morton, and B. Goldstein.
  Extending the range of rate constants available from BIACORE:
  Interpreting mass transport–influenced binding data.
  Biophys. J. 75: 583–594 (1998). 505
 8.3.2 Physiologically Based Compartmental Pharmacokinetic Models 508
 8.3.3 Two-Compartment Model 508

Contents

    8.4   Macromolecular Crowding   512
        8.4.1   Macromolecular Crowding Effects on Diffusion   514
        8.4.2   Macromolecular Crowding Effects on Binding Equilibria   516
        8.4.3   Macromolecular Crowding Effects on Rates   518
Suggestions for Further Reading   520
Problems   520
References   525

9   Discrete Stochastic Processes   527

    9.1   Poisson Statistics   528
        Case Study 9-1 S. E. Luria and M. Delbrück. Mutations of bacteria from virus sensitivity to virus resistance. *Genetics* 28: 491–511 (1943).   535
    9.2   Stochastic Simulations   538
        9.2.1   Stochastic Gillespie Simulations   539
        Case Study 9-2 T. S. Gardner, C. R. Cantor, and J. J. Collins. Construction of a genetic toggle switch in *Escherichia coli*. *Nature* 403: 339–342 (2000).   544
        9.2.2   Approximate Stochastic Methods: A Bridge between Exact Stochastic Methods and Deterministic Formalisms   548
    9.3   Stochastic Gene Expression   550
    9.4   Stochasticity in Phage Infection   551
        Case Study 9-3 A. Arkin, J. Ross, and H. H. McAdams. Stochastic kinetic analysis of developmental pathway bifurcation in phage λ-infected *Escherichia coli* cells. *Genetics* 149: 1633–1648 (1998); F. St-Pierre and D. Endy. Determination of cell fate selection during phage lambda infection. *Proc. Natl. Acad. Sci. U.S.A.* 105: 20,705–20,710 (2008).   552
    9.5   Diffusion and Random Walks   553
    9.6   Combinatorial Bioinformatics   555
        Case Study 9-4 T. A. Hopf, J. B. Ingraham, F. J. Poelwijk, C. P. I. Schärfe, M. Springer, C. Sander, and D. S. Marks. Mutation effects predicted from sequence co-variation. *Nat. Biotechnol.* 35: 128–135 (2017).   562
Suggestions for Further Reading   563
Problems   563
References   566

Index   567

For our students.

Preface and Acknowledgments

In 2005, the Massachusetts Institute of Technology established a new Biological Engineering undergraduate major, integrating molecular cell biology with quantitative engineering analysis and design. The following core curricular topics were identified: biotransport, statistical thermodynamics, molecular biology lab, instrumentation lab, and biological rate processes. In teaching the graduate and undergraduate offerings of the last subject (respectively titled *Biomolecular Kinetics and Cellular Dynamics* and *Analysis of Biomolecular and Cellular Systems*), we found there to be no single textbook that comprehensively captured the essential topics that every biological engineer should be conversant with.

This book is intended to address this need. It has been used in the MIT courses just mentioned as well as the required *Introduction to Biomolecular Engineering* course in Chemical Engineering and Materials Science at the University of Minnesota. It is intended for chemical engineering and bioengineering courses focusing on quantitative descriptions of biomolecular and cellular dynamics. Such courses are typically offered for juniors, seniors, and first-year graduate students, with calculus and life science or organic chemistry prerequisites.

The text focuses on construction and application of mechanistic models of biomolecular interaction rate processes. These dynamics underpin most biological functions: protein-ligand binding, enzymatic reactions, gene expression, receptor trafficking, biomolecular networks, and cell growth. We systematically develop the concepts necessary to understand and study complex biological phenomena by moving from the simplest elements at the smallest scale (molecular binding) and sequentially adding complexity at the cellular organizational level, focused on incisive experimental testing of mechanistic hypotheses. Such models are powerful tools for chemical and biological engineers to study biomolecular and cellular phenomena, as illustrated in more than 30 case studies from the literature. This emphasis is foundational for students and faculty interested in studying and working at the interface of life sciences and engineering.

The full contents of this book are beyond the realistic scope of a one-semester course, and so instructors should pick and choose the topics most relevant to the objectives of their own syllabus. To assist in this choice, a brief precis of each chapter follows.

Chapter 1, Introduction to Biological Rate Processes. Here we provide an overview of the key philosophical perspectives of the text, explicating the motivations for formulation of mathematical rate process models in biology. Illustrative examples are provided, and recurrent modeling themes are enumerated.

Chapter 2, Noncovalent Binding Interactions. While the reader may have taken an introductory course in the life sciences or organic chemistry, this often doesn't include essential preliminaries regarding the biophysical chemistry of protein-protein interactions. We also include a brief primer on fluorescence, because such measurements are common in biological kinetics.

Chapter 3, Binding Equilibria and Kinetics. These topics are foundational for the remainder of the book. We provide quantitative descriptions of the transient, steady-state, and equilibrium interactions of proteins and their ligands. General and specific considerations for interpretation of common measurement modalities are also provided, including best practices for parameter estimation.

Chapter 4, Enzyme Kinetics. Classic enzyme rate laws are derived and applied, with examples in pharmacology and an extended case study on the proteolytic cascade driving the blood clotting process.

Chapter 5, Gene Expression and Protein Trafficking. Extending to more complex cellular phenomena, we explore here mass-action-type models of transcription, translation, protein degradation, and protein trafficking between cellular compartments.

Chapter 6, Network Dynamics. Systems built up from multiple interacting proteins can begin to exhibit emergent dynamic behaviors, such as switchlike transitions, adaptive homeostatic control, or sustained oscillations. Examples of molecular mechanisms capable of producing such behaviors are provided in each case.

Chapter 7, Population Growth and Death Models. Biological engineers often encounter situations where a quantitative description of growth dynamics is required, such as designing bioreactors, studying tumor growth, and following epidemics. Statistical descriptions of these processes that take a mass-action kinetic structure have stood the test of time. Although they generally diverge from the molecular mechanistic character of all the models derived up to this point in the book, their high empirical utility makes them an essential tool.

Chapter 8, Coupled Transport and Reaction. The mass-action assumption made for the models of chapters 2–6 requires complete mixing; in this chapter, we address those situations where spatial concentration gradients must be accounted for, either by compartmental modeling or scaling analyses. Examples are given for cell or biosensor surface binding, tissue penetration, and pharmacokinetics.

Chapter 9, Discrete Stochastic Processes. In the final chapter, we address those situations in which the continuum hypothesis breaks down and a small number of molecular events must be considered in a stochastic framework. Applications of Poisson statistics and Gillespie simulations are presented.

A word on the case studies appearing throughout the book. In our graduate course, students were asked to implement a half dozen classic modeling papers in MATLAB, reproducing key figures and then extending the analysis in new directions by computational experimentation. Despite the expected occasional angst for students of varying coding comfort levels, these assignments are extremely valuable pedagogically. The Socratic roundtable discussion of these implementations has always been a highlight of the course for both students and teaching staff. We have included many of these papers as case studies in the book to help seed such efforts in others' classrooms.

Acknowledgments

We particularly thank the following individuals for commenting on drafts: Yash Agarwal, Ian Andrews, Cristina Coralys Torres Caban, Amanda Chen, Sean Aiden Clarke, Jennifer Cochran, Sarah Cowles, Justin DeCarlucci, Shelby Doyle, Paulina Eberts, Maxine Jonas, Byong Kang, Emi Lutz, George Markou, Christina McGovern, Naveen Mehta, Noor Momin, Joseph Palmeri, Lauren Purcell, Adrienne Rothschilds, Kyra Schwartz, Sarah Schwartz, Allie Sheen, George Sun, and Alison Tisdale.

KDW. Doug Lauffenburger, the founding department head of Biological Engineering at MIT, encouraged and materially supported our writing efforts, and for this, we offer our heartfelt gratitude. Formulating and teaching *Biomolecular Kinetics and Cellular Dynamics* for a decade with BT structured the intellectual framework for this book, and it has been a singular privilege and pleasure to work with my coauthors to realize this vision. Over the years our students have helped drive the project forward by sharing their excitement in the discovery of the power of these methods to help understand and engineer biology. And of course much love to Jean, the unwavering light of my life.

BT. This work has benefited greatly from my teachers, whose example set me on the path; my research mentors, collaborators, and students, who have shown me, over and over, the positive feedback between fundamentals and applications; KDW, who has been a colleague in the deepest sense and a leader in this project; my coauthors, with whom writing this text has been uniquely gratifying; and my family members, who sustain me.

BJH. Thanks go to my mentors: Eric Shusta, coauthor KDW, and Sam Gambhir. To the students of the classrooms and laboratories at the University of Minnesota, whose curiosity and drive to learn stimulated my contributions to this book. To my fellow educators in the Department of Chemical Engineering and Materials Science for their mentorship. To my family for their inspiration and support.

CAS. I was inspired to work on this textbook by my parents, Husain and Durriya Sarkar, who instilled in me an enduring sense of intellectual curiosity. I owe large debts of gratitude to Doug Lauffenburger, Andreas Plückthun, George Georgiou, and Jim Collins, who helped shape how I think about molecular and cellular bioengineering. Over the years my students and colleagues have made it a joy to learn and to teach, and these interactions have greatly enhanced my contributions herein. I also acknowledge the University of Minnesota for a sabbatical, during which this project was completed. Finally, I thank my coauthors for the pleasure of writing this book with them.

1

Introduction to Biological Rate Processes

The universe is full of magical things patiently waiting for our wits to grow sharper.
—Eden Phillpotts

1.1 Biological Rate Processes

Living systems are not static. Cells grow and divide; signals propagate over space and time; organisms are born, develop, grow, age, and die. This constant change is driven by networks of time-varying chemical transformations, macromolecular binding events, and changes in subcellular localization that combine to create complex functional behaviors. Evolution shapes the functions of molecules, the properties of cells, the characteristics of organisms, and the very existence of species. These dynamic rate processes propel biology at all length and time scales, from single molecules to worldwide ecosystems, from picoseconds to eons. In this text, we examine these rate processes and develop a quantitative toolkit useful for studying and engineering biological systems.

Many of the intellectual underpinnings for these approaches were originally developed in the discipline of chemical engineering, wherein understanding the networks of reactions forming and breaking covalent bonds was essential to the design and operation of petrochemical plants. Analogously, individual noncovalent binding events are the underlying building blocks of most biological functions. An understanding of these events is central to developing a quantitative and predictive description of any biological system. Those individuals with an engineering mindset are often best motivated by understanding the real-world applications of a body of knowledge. With that goal in mind, let's first take a look at some biological systems where dynamic rate processes drive overall outcomes.

Example 1-1 **Biosensors** Perhaps the most basic example consists of measuring the rate and extent of complex formation between two noncovalent binding partners, as shown in figure 1.1. Despite the apparent simplicity of

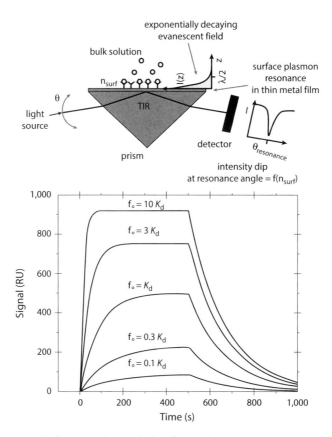

Figure 1.1 Schematic diagram and example data from a biosensor that detects complex formation via a phenomenon called surface plasmon resonance. The device shown at the top of the figure measures in real time (using optics from below the film) the extent of binding of soluble ligand at concentration f_o (small circles) to surface-immobilized protein (Y-shaped objects). The curves in the lower figure indicate how formation of macromolecular complexes on the sensor surface increases the device's signal readout as a function of time. Once free ligand is washed away continuously starting at 500 seconds, the signal decreases. Extracting quantitative information about the rate of formation and dissociation of the complex requires formulation of a mathematical description of the binding process [1]. TIR = total internal reflection.

this single binding and dissociation event, a surprising variety of complications can arise in performing and interpreting experiments with this type of device. A mathematical description is an essential tool for navigating toward a fuller understanding of the binding event being measured. We will consider biosensors in greater detail in chapter 3 (as well as in case study 8-4 in chapter 8).

Example 1-2 **Gene expression**   The dynamics of gene expression and the sensitivities of each gene's expression to changing environmental conditions are of fundamental importance in cell function and engineering. The process of expressing a given gene is the sum total of many binding and chemical transformation events: from transcription factors binding DNA to the machinelike complexity of ribosomal translation of an mRNA molecule into a polypeptide. A quantitative description of the time-varying dynamics of gene expression can be essential to understanding or engineering its behavior. In figure 1.2, one level of quantitative description is represented. We will discuss the fundamental analysis of enzymatic processes in chapter 4 and return to gene expression models in chapter 5 (with additional considerations on stochasticity in chapter 9).

In figure 1.3, the behavior of a synthetic switchlike gene network is shown, in which the transient presence of specific inducer molecules alters the pattern of genes expressed until different inducer molecules are transiently applied. Network dynamics are the focus of chapter 6. This particular con-

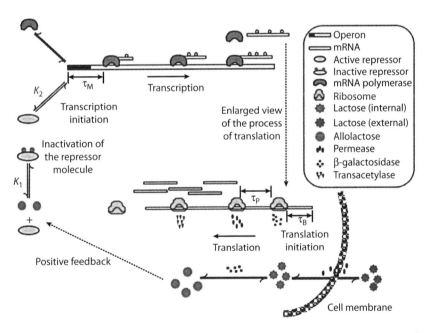

Figure 1.2 A simplified mechanistic description of the rate processes involved in gene expression. Such models can be useful for understanding the steady-state and dynamic variation in the levels of a gene product as a function of system inputs, such as lactose concentration [2].

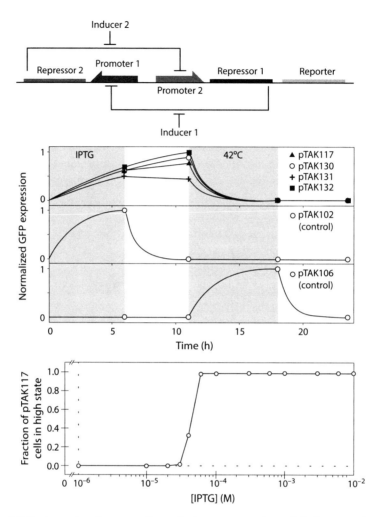

Figure 1.3 Design and performance of a synthetic gene expression switch. Incorporation of two mutual repressors leads to a stable toggle in output in response to the transient presence of input inducer. The sharp change in expression level of a reporter gene as inducer increases is an intentional feature of the system, designed with the help of a simple mathematical model of gene expression dynamics [3].

struct (further examined in chapter 9) inspired legions of biological engineers to work in the field of synthetic biology, wherein a designed genetic construct might be translated to a predictable, reliable, robust, and useful biological behavior.

Example 1-3 **Protein trafficking**   All cells transport proteins to desired subcellular locations. Prokaryotes route proteins either to the extracellular space or cytoplasmic compartment, while eukaryotes possess a more intricate transportation system for movement to numerous, distinct membrane-bound organelles. Relocalization drives various biological processes, such as the receptor endocytic trafficking pathway shown in figure 1.4. Mathematical models of protein trafficking are valuable tools for analyzing and understanding these events and will be elaborated on in chapter 5.

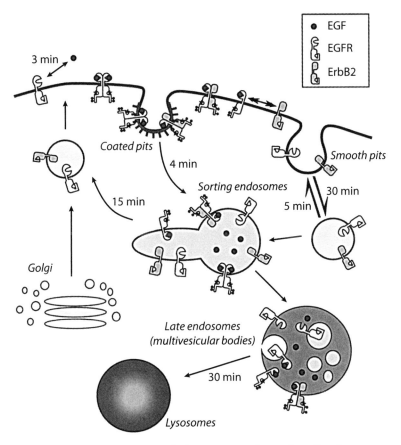

Figure 1.4 Schematic diagram of the vesicular transport pathways for cell-surface receptors [4]. Internalization and degradation following growth factor binding serves as a negative feedback function to return the system to steady state following a transient pulse of growth factor exposure. Each of the complex vesicular transport events represented by arrows in the figure, to a first approximation, can be described as a single pseudochemical reaction characterized with a single rate constant or characteristic time scale.

Example 1-4 Cell signaling Cells must respond to changes in their environment by altering gene expression. Transduction of information from the cell surface, where environmental inputs are generally sensed, to the cell's DNA, where changes in gene expression are typically effected, is often via a chain of protein binding and covalent modification events. As described in chapter 4, such events frequently involve the enzymatic addition or removal of a phosphate group from tyrosine, serine, and threonine side chains in a series of proteins. This process is termed *cell signaling*, and a schematic of the typical components is shown in figure 1.5. The outcomes of cell signaling events can be among the most dramatic biological functions, including the life or death of the cell. Mathematical models of these events, considered in chapter 6, can be useful in understanding the relationship between time-varying inputs and the resulting effects on gene expression.

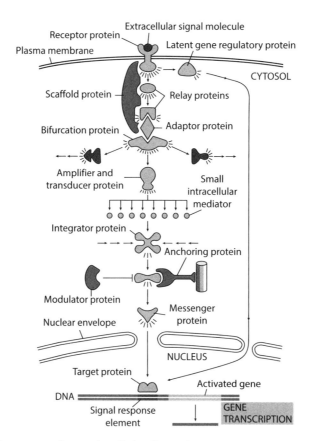

Figure 1.5 Components of a generic cell signaling pathway. Receptor-ligand complex formation at the cell surface triggers a series of binding events and covalent modifications that culminates in altering gene expression [5].

Example 1-5 **Populations** The same mathematical formalisms used to describe individual molecular events can be extended to usefully account for the dynamics of the birth, growth, and death of single cells or populations of multicellular organisms. In figure 1.6, a schematic diagram of T cell infection by the HIV virus is shown. Through a simplified representation of the complex process of infection and cell death, a dynamic accounting of circulating levels of each component can be performed. Such models are considered in chapter 7.

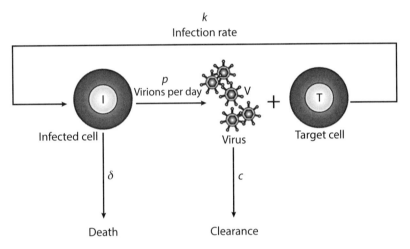

Figure 1.6 The dynamics of HIV infection and treatment have been usefully captured in simple mathematical models that treat immune cells and virions as pseudochemical species that undergo apparent chemical transformations [6]. Models such as this were critical for understanding how the slow progression of the HIV disease state over several years could arise from cellular infection and death processes that occur rapidly and continuously over the course of hours and days.

Example 1-6 **Pharmacokinetics** A central concern in pharmaceutical development is inferring the concentration of a drug at its site of action in the body as a function of the time-varying administration via a variety of possible modalities (pills, inhalers, injections, etc.). This field, called pharmacokinetics, uses highly abstract and simplified representations of complex physiological systems to explain and predict the distribution of drugs throughout multiple organ systems. These models are then useful for specifying molecular properties and administration protocols for drug delivery to the proper bodily locations at appropriate concentrations and for determining desirable time intervals. Such models are schematically represented in figure 1.7. This class of models is considered further in chapter 8.

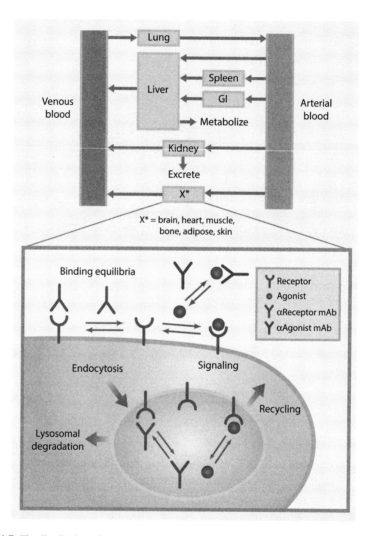

Figure 1.7 The distribution of a drug throughout the body can be represented as a network of transitions between different organs, each treated as an independent, well-mixed volume. Although clearly a significant simplification of reality, such models are nevertheless powerful tools in the development of new drugs, analysis of preclinical animal data, and design and interpretation of clinical trials [7].

1.2 Why Develop a Model?

Now that we have seen a few biological systems where mathematical modeling has contributed to improved understanding, let's step back and consider some fundamental questions regarding the motivation and practice of constructing such models. The first question to be answered when embarking on a mathematical modeling exercise in biology is: Why? There is no singular, unique, or "best" mathematical description for any biological system, so the extent and nature of detail incorporated must be appropriate to the needs at hand.

Hypothesis testing. Models serve many roles; one important role is to concisely and specifically express a scientific hypothesis. Disagreement between experimental results and model predictions is evidence that one or more of the model's mechanistic assertions is incorrect and needs revision. Agreement with the experiment supports but cannot *prove* the model.

A mathematical model provides a means to test the consistency of alternative hypothesized mechanisms with overall behavior. When trying to determine the mechanism for a biological function, equally convincing rhetorical arguments can be made for very different explanations. Rather than rely on comparisons of the investigators' persuasive skills, mathematical modeling can reduce the number of potentially correct explanations by constraining each hypothetical mechanism to comply with well-validated physical and chemical rate laws.

Developing mechanistic insight into biological systems. Experiments can measure outcomes, but it is rare for experiments to directly provide an explanation for how an outcome occurred. By contrast, model simulations contain all of the mechanistic detail needed to produce, and thus explain, the outcome in a directly accessible form. Thus, in principle, models provide a ready route to understanding the processes responsible for biological function. In practice, this is often achieved by dissection ("*How much of the outcome is due to activated receptor on the cell surface versus activated receptor in the endosomes?*") or model interrogation ("*Does electrostatic steering contribute to the outcome?*").

With the best models, quantitative calculation leads to new qualitative understanding. Rather than simply producing curves on a plot, a well-designed modeling study leads to insight expressible in words rather than numbers. To reach this ideal, modeling cannot be considered as an alternative to thinking. Rather, it is a formal language for comparing the distinguishable functional consequences of alternative mechanisms. Indeed, although this qualitative mechanistic understanding might not have been readily accessible experimentally, often the insights discovered through modeling studies can then be experimentally tested. This process can produce conclusions that are independent of the models that led to them.

Recalling and predicting system behaviors. Models can also be used as compact mathematical representations of empirical data. They are, in effect, convenient lookup tables—but are arguably better, if they contain correct mathematical functional forms for interpolating between the data used for fitting (and potentially for extrapolating beyond available data ranges). Such is the case for many pharmacokinetic models, which match a given patient's propensity to distribute and dispose of an administered drug against the well-known patterns for such events in experimental animals or other patients. Although models of this type can be useful for interpolation in the range of parameters for which experimental data are available, extrapolation of such an empirical model in the absence of a mechanistic underpinning is a risky bet. For models with a stronger mechanistic basis, moderate extrapolation is a more secure endeavor.

Engineering biological systems. Accurate and well-validated models are uniquely useful as a substrate for design. In the development of medical therapeutic approaches, where subtle interventions in human biological processes are needed to alter an outcome robustly, models are extremely valuable in the design of strategies and agents for intervention. Even though the ability of models to make predictions may allow one to exhaustively test "all" avenues and select the most viable, often much more efficient approaches are available. By using approaches from mathematics and computer science, such as optimization and constraint propagation, one can more directly compute the best interventions directly, without exhaustively testing all possibilities.

1.3 Mechanistic Model Formulation

A generalized protocol for biological mathematical model formulation is shown in figure 1.8. The first step is to hypothesize a mechanism that captures the essential features of the process. Representation for such a mechanism is often most convenient in "cartoon" form, as shown for the example systems referenced earlier. Hypothesizing a mechanism consists of identifying which molecules to take into account, which interactions among them to include, how many distinguishable compartments there are for these molecules to reside in, and what intercompartmental transport steps occur.

Let's illustrate the process of model formulation with an example of cell surface receptor-ligand interaction and internalization. In particular, we will consider receptor downregulation and attempt to predict the decrease in surface receptors by a process of accelerated endocytosis of a receptor-ligand complex. Despite the dauntingly complex cell biology of receptor gene expression, trafficking, and ligand binding, we will formulate a simple model intended to help

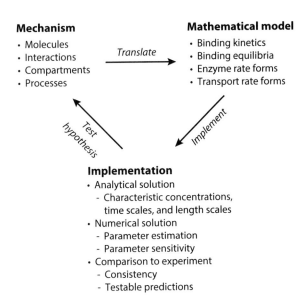

Figure 1.8 A typical flowsheet for construction and use of a mechanistic mathematical model of a biological process. The hypothesized mechanism is often in cartoon form, as is the case for the process examples in this chapter (figures 1.1–1.7, 1.9). Once written out, this mechanism is translated into mathematical form using either mechanistic mass-action kinetic rate forms or empirical first- or second-order rate expressions. These equations can then be analytically solved to identify key behaviors (or numerically solved if the model is more complex). The final step in this process is to compare the model outputs to previously observed experiments to parameterize and validate the model, or more powerfully, to predict new experiments that can then be tested for consistency with the originally hypothesized mechanism, closing the loop.

understand how ligand binding kinetics interact with endocytic processes to determine steady-state levels of both free receptor and receptor-ligand complex at the cell surface.

The mechanism we hypothesize is shown in figure 1.9. Several simplifying choices, or assumptions, have been made.

1. Only three molecular species will be tracked: free ligand, unbound receptor, and receptor-ligand complex.
2. Endocytosis is irreversible, so receptor recycling is not included (an assumption that would require independent experimental verification).
3. Only the cell surface will be considered, as a well-mixed compartment, so the dynamics in the endoplasmic reticulum, Golgi apparatus, and sorting endosomes will not be included in this description.
4. Five rate processes will be considered: receptor synthesis, receptor endocytosis, receptor-ligand binding, complex dissociation, and complex internalization.

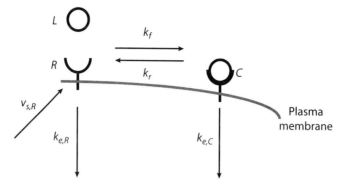

Figure 1.9 A simplified mechanistic description of the rate processes involved in receptor-ligand interactions at the cell surface. This cartoon is a typical graphical representation of a hypothesis regarding the significant molecular species, their interactions, and the compartmental organization of the system. L is free ligand, R is free receptor, and C is the receptor-ligand complex.

5. Ligand is in excess in the culture volume, so its concentration in solution is not measurably decreased by binding to the cell surface.

The next step (see figure 1.8) in formulating the model is to translate the hypothesis into mathematics, turning the cartoon into equations. This involves a process of molecular accounting that, in the discipline of chemical engineering, is termed *material balances*. For each species being considered, a single differential equation will be written that includes a rate expression for each arrow in the cartoon that leads to or away from the species:

$$Accumulation = In - Out + \Sigma(Produced\ by\ reaction)$$
$$- \Sigma(Consumed\ by\ reaction) \quad (1.1)$$

It is often very useful to first write the material balance equations in words so as to understand the meaning of each resultant mathematical term. For the receptor downregulation problem, this corresponds to:

$$Receptor\ accumulation = Expression - Receptor\ endocytosis$$
$$+ Complex\ dissociation - Ligand\ binding \quad (1.2)$$
$$Complex\ accumulation = -Complex\ endocytosis + Ligand\ binding$$
$$- Complex\ dissociation \quad (1.3)$$

Note that we have restricted our attention to the cell-surface compartment, which is reflected in our cartoon (see figure 1.9). These equation descriptions are specific to this compartment. Had we separately considered the accumulation of receptors and complexes in an intracellular compartment composed of endosomes, the process of endocytosis would actually lead to a positive influx of species due to their movement into vesicles from the cell surface. Thus, the signs associated with specific processes in these equations depend on the definition of the compartment.

Now let us substitute mathematical expressions for each of the terms in the word equations. The *Accumulation* terms in the equations above are each time derivatives of the concentration of the species being accounted for in that equation:

$$\frac{d[R]}{dt} = Expression - Receptor\ endocytosis + Complex\ dissociation$$
$$- Ligand\ binding \qquad (1.4)$$

$$\frac{d[C]}{dt} = -Complex\ endocytosis + Ligand\ binding - Complex\ dissociation \qquad (1.5)$$

In this text, we will use square brackets to denote the concentration of the named molecular species. We must now choose particular mathematical rate forms to represent the binding kinetics, equilibria, and transport steps. Each arrow in figure 1.9 is a rate process that will be described by a mathematical formula.

Moving from left to right in equation 1.4, we begin with the receptor expression term. We choose to make no assumptions with respect to how it might depend on time or on any other variables in the problem; hence, we will simply name it as a variable: $Expression = v_{s,R}$. The letter v is often used to indicate the *velocity* of a reaction, the subscript s stands for *synthesis*, and subscript R for *receptor*. Because this rate does not depend on receptor concentration, it is termed a *zeroth-order rate form*, as it depends on concentration to the zeroth power: $Expression = v_{s,R}[R]^0$.

The default rate law for a reaction involving only one molecular species is generally a first-order form; in other words, we assume that the rate of the process is proportional to the concentration of the species in question. Therefore, we define our next term as: $Receptor\ endocytosis = k_{e,R}[R]$. We will consider this simplified representation of protein trafficking kinetics in chapter 5.

We again use a first-order expression for the complex dissociation rate. In other words, we assume that the rate of dissociation is proportional to present concentration of complex and has no memory of previous conditions other than the value of $[C]$, such that $Complex\ dissociation = k_r[C]$.

We will often require mathematical expressions for the formation of a complex. The simplest such rate law is called mass-action kinetics, wherein the rate of complex formation is proportional to the product of the concentrations of the unbound partners. We will explore the physical-chemical basis of this assumption in more depth in chapters 3 and 8; it is generally a good approximation when the binding reaction occurs in a well-mixed compartment. To finish the differential equation describing receptor levels on the surface, we have *Ligand binding* $= k_f[R][L]$.

Going through the analogous steps for the equation describing concentration of the receptor-ligand complex (equation 1.5) leaves us with the following set of two differential equations:

$$\frac{d[R]}{dt} = v_{s,R} - k_{e,R}[R] + k_r[C] - k_f[R][L] \tag{1.6}$$

$$\frac{d[C]}{dt} = -k_{e,C}[C] + k_f[R][L] - k_r[C] \tag{1.7}$$

To obtain a fully specified set of equations, we must define the initial conditions: $[L] = [L]_o$ and $[C] = 0$ at time zero. Before time zero, the ligand has not yet been added and the system is assumed to be at steady state. Therefore, the initial receptor level, just prior to time zero, can be found by setting the time derivative to zero (the steady-state condition) and solving for $[R]_o$ in equation 1.6, yielding $[R]_o = v_{s,R}/k_{e,R}$ at time zero. Stating this result in words, before ligand is present, the steady-state surface receptor level is given by the ratio of the protein synthesis rate to the endocytosis rate constant.

This completes the formulation steps; we have a well-posed mathematical model. (This is not to say that the biology is necessarily correct, simply that the hypothesis represented by figure 1.9 has been unambiguously translated into solvable mathematical terms.) Of course, so far we haven't learned anything about receptor trafficking dynamics. To do so, we must implement the model—which is to say, we must use it for something.

Whenever possible, attempting even a partial or approximate analytical solution can often provide the necessary insights sought by the modeling study. In this case, we set out to model receptor downregulation to determine the extent to which a receptor will be reduced on the cell surface given the rate processes we hypothesized. To obtain a steady-state solution, we first substitute 0 for all time derivatives in the equations and add a subscript "ss," denoting steady state, to each concentration variable. Note that we also assume excess ligand such that it is not depleted by binding and internalization ($[L] = [L]_o =$ constant):

$$0 = v_{s,R} - k_{e,R}[R]_{ss} + k_r[C]_{ss} - k_f[R]_{ss}[L]_o \tag{1.8}$$
$$0 = -k_{e,C}[C]_{ss} + k_f[R]_{ss}[L]_o - k_r[C]_{ss} \tag{1.9}$$

Solving equation 1.9 for $[C]_{ss}$, we have:

$$[C]_{ss} = \frac{k_f[R]_{ss}[L]_o}{k_r + k_{e,C}} \tag{1.10}$$

Then solving equation 1.8 for $[R]_{ss}$ gives:

$$[R]_{ss} = \frac{v_{s,R} + k_r[C]_{ss}}{k_{e,R} + k_f[L]_o} \qquad (1.11)$$

Solving equations 1.10 and 1.11 along with the aforementioned relationship of $[R]_o = v_{s,R}/k_{e,R}$ yields the following expression for the fractional remaining total receptor on the surface at steady state:

$$\frac{[R]_{ss} + [C]_{ss}}{[R]_o} = \frac{1 + \kappa}{\delta + \kappa} \qquad (1.12)$$

where $\delta \equiv k_{e,C}/k_{e,R}$ expresses the relative increase in endocytotic rate for receptor-ligand complexes, and $\kappa \equiv \frac{k_r + k_{e,C}}{k_f[L]_o}$ is a dimensionless effective equilibrium dissociation constant that can be conceptualized as the ratio of first-order rate constants that lead to removal of cell-surface complexes (i.e., $k_r + k_{e,C}$) to the effective first-order rate constant that leads to production of cell-surface complexes (i.e., $k_f[L]_o$).

Thus, what this analysis enables is a prediction of the degree of receptor down-regulation as a function of the relevant binding and endocytic rate constants. We can now predict the decrease in surface receptors by a process of accelerated endocytosis of a receptor-ligand complex. We can immediately see that if $\delta = 1$, as is the case for some transport receptors, then the fraction is 1, regardless of the value of κ. For $\delta > 1$, as is the case for most signaling receptors, the fraction still approaches 1 for large κ values, but it can be very low for systems with large δ and small κ values. (In chapter 5, we further analyze related models to help understand how receptor trafficking can alter the apparent potency of growth factors.)

Of the various steps involved in this example, the hardest is the choice of a succinct mechanism for a complicated cell biological process. The simplified schematic in figure 1.9 leaves out many molecules, structures, and rate processes that are known to occur but have been hypothesized to not contribute to the overall phenomenon of interest at the moment. Once the cartoon has been drawn, translation to mathematical form is essentially automatic and straightforward; implementation requires us to look at the steady-state solution and drive the algebra toward the particular simplified form shown. At this point, we can close the loop by using trustworthy experimental data or performing experiments to determine whether we made good choices in constructing figure 1.9. Is our hypothesis consistent with experiment, or does it need to be altered?

1.4 Model Validation

Essentially, all models are wrong, but some are useful.
—George Box

The best material model for a cat is another, or preferably the same cat.
—Arturo Rosenblueth and Norbert Wiener

What is the "best" model—the one that is most accurate or the one that gives the greatest insight? We would argue the latter. The reductio ad absurdum of the cat quote above illustrates how seeking the greatest accuracy would only lead one back to the original system. One can never prove a model to be correct, in the sense that no model describes a system in perfect detail. A nonscientific illustration from Lewis Carroll's story *Sylvie and Bruno*, about two young children visiting the man in the moon, illustrates the point concisely:

"What a useful thing a pocket-map is!" I remarked.
"That's another thing we've learned from your Nation," said Mein Herr, "map-making. But we've carried it much further than you. What do you consider the largest map that would be really useful?"
"About six inches to the mile."
"Only six inches!" exclaimed Mein Herr. "We very soon got to six yards to the mile. Then we tried a hundred yards to the mile. And then came the grandest idea of all! We actually made a map of the country, on the scale of a mile to the mile!"
"Have you used it much?" I enquired.
"It has never been spread out, yet," said Mein Herr: "the farmers objected: they said it would cover the whole country, and shut out the sunlight! So we now use the country itself, as its own map, and I assure you it does nearly as well."

For any model smaller than the system itself, it is important to take steps to confirm that the most essential features for the chosen purpose have been correctly and adequately captured. As with a map, such a simplified representation can be very useful despite being formally incorrect, due to omission of some aspects of the system. One can use the general steps outlined here to test the validity of a given model.

It is vain to do with more what can be done with fewer.
—William of Occam

Everything should be made as simple as it can be, but not simpler.
—Common paraphrase of Albert Einstein[1]

---

1. Ironically, the actual quote is less simple: "It can scarcely be denied that the supreme goal of all theory is to make the irreducible basic elements as simple and as few as possible without having to surrender the adequate representation of a single datum of experience."

Science may be described as the art of systematic oversimplification.
—Karl Popper

Paring away unnecessary details is fundamental to the process of model formulation in general. Is the model no more complex than necessary to describe the system? This question is particularly pressing for biological systems, given the presence of a potentially bewildering profusion of irrelevant components and activities in a biological system that exert no significant influence on the main outcome of interest. The concept of model simplification will run throughout the approaches and examples discussed in this text.

The availability of large biological data sets has motivated attempts to discern underlying statistical relationships among variables. Machine learning and statistical data mining are extremely valuable tools for generating mechanistic hypotheses from such reams of data, but these topics are beyond the scope of this text [8].

We emphasize these points at the outset of this text, because it is often assumed by the starting student that greater accuracy is the unquestioned goal of any modeling project. In truth, some sacrifice of *correctness* is effectively required to extract simple and *useful* insights about a system.

1.4.1 Consistency with Experiment Is Necessary but Not Sufficient for a Model's Correctness

Literally an infinite number of mathematical models of equivalent statistical consistency can be constructed for a given set of data. Statistical consistency means that the model's inaccuracies are of the same magnitude as the variation in data due to random noise. Consistency is a *necessary*, but by no means *sufficient*, condition for a particular model to be valid. We must always hold this point in mind, because there is a tremendous temptation to declare victory in a modeling exercise when the model curve intersects all the data symbols—yet such a condition represents only a single, not very difficult, step toward confirming the utility of a model.

As a demonstration of this principle, consider the yeast cell culture growth data and models represented in figure 1.10. Cell growth is a rate process of considerable interest, and many alternative models have been developed to describe growth kinetics. We will consider these models in detail in chapter 7; here, three of them are represented in the figure. They are:

Gompertz equation:

$$X = X_f \left(\frac{X_o}{X_f}\right)^{e^{-at}} \qquad (1.13)$$

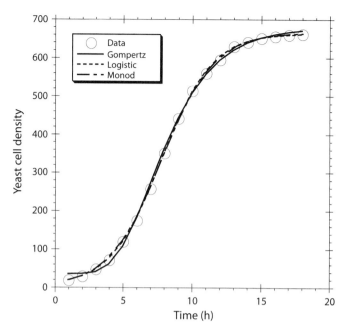

Figure 1.10 Consistency of three alternative growth models with a given set of data. Yeast were grown in shake-flask cultures [9] and exhibited a sigmoidal increase in cell mass versus time. This behavior is well reproduced by three different mathematical models that have all been used frequently to describe growing populations: the Gompertz, Logistic (Verhulst-Pearl-Reed), and Monod models. Clearly, no meaningful discrimination among these three different models can be made simply on the basis of their consistency with the data, because all three models exhibit an excellent fit.

Logistic (Verhulst-Pearl-Reed) equation:

$$X = \frac{X_f}{1 + \left(\frac{X_f}{X_\circ} - 1\right)e^{-at}} \quad (1.14)$$

Monod equation (integrated and accounting for depletion of the growth substrate):

$$at = \left(1 + \frac{b}{X_f}\right)\ln\left(\frac{X}{X_\circ}\right) - \frac{b}{X_f}\ln\left(\frac{X_f - X}{X_f - X_\circ}\right) \quad (1.15)$$

where $X \equiv$ yeast cell density in mass per volume; $t =$ time; a, $b =$ fit parameters.

The parameters in each of these models were fit to the data by nonlinear least squares, and in each case, the fit is excellent. Which, if any, of these models is *correct*? All three provide a compact representation of the data, and in fact are often used for this purpose in practical bioprocess applications. However, additional information beyond that presented in figure 1.10 is required to learn more about the underlying mechanisms at play in yeast growth. As it happens, the Gompertz and Logistic equations make no pretense of arising from a mechanistic derivation—both were originally presented as curve-fit formulas. The Monod equation shrinks the black box a bit by incorporating an observed relationship between sugar concentration and the specific growth rate into its derivation. As such, one would have greater confidence that the Monod model could describe other data sets under conditions of varying sugar concentration, whereas the Gompertz and Logistic forms must be refit to the new data sets under each new set of conditions.

This example provides a direct illustration of the central concept that consistency with data cannot uniquely identify any single model as valid or correct. The three very different growth models tested here all look correct, because each curve touches all the data symbols. Demonstration of consistency with experiment is only the first of several steps required to validate a model.

1.4.2 Comparison to Alternative Models

Whenever a theory appears to you as the only possible one, take this as a sign that you have neither understood the theory nor the problem which it was intended to solve.
—Karl Popper

Far too often, a single model is considered correct without also considering alternatives. In fact, the data from figure 1.10 were presented long ago as evidence for the value of the Verhulst-Pearl-Reed logistic model, without consideration of alternative models [10]. It is essential to examine plausible alternative models consistent with the current mechanistic understanding of the system for equivalent consistency with the data, before preferring one model over the universe of alternatives.

1.4.3 Prediction and Falsifiability

Predictions are hard to make, especially about the future.
—Danish proverb, subsequently paraphrased by many others, including Niels Bohr, Mark Twain, and Yogi Berra

Good tests kill flawed theories; we remain alive to guess again.
—Karl Popper

If a model does not literally predict the outcome of new experiments, then it cannot be meaningfully tested against the set of alternative models that are equally consistent with whatever empirical evidence is available a priori. A model that represents one data set will, in general, not extrapolate to systems under different conditions unless it captures essential mechanistic features of the process.

The word prediction is often regrettably used to describe a retrospective comparison of model and data. It requires powerful intellectual discipline to actually formulate the mathematical model for a process of interest *before* performing experiments. In this fashion, one is better prepared to learn something new about the world when things (almost inevitably) don't go as planned.

There are two possible outcomes: if the result confirms the hypothesis, then you've made a measurement. If the result is contrary to the hypothesis, then you've made a discovery.
—Enrico Fermi

If a mathematical model survives all attempts to prove it incorrect (i.e., *falsify* it), confidence is raised that the mechanistic assumptions incorporated into the model's development are an adequate representation of the system's actual inner workings. In this way, mechanistic model building closely follows the classic scientific method, in that no hypothesis can ever be proven correct—only shown to be consistent with the available evidence.

1.4.4 Parameter Sensitivity Analysis

In any simulation with many parameter values that have not been independently measured (i.e., the great majority of kinetic models), a question that naturally arises is: How significantly does each parameter affect the model outputs of most interest? Local parameter sensitivity analysis provides a useful tool for addressing this question.

Let's say that the objective output Y of interest is some arbitrary function of the outputs, for example, a χ^2 goodness of fit between the model and data, the maximum concentration of some component, or the time required to reach some threshold concentration. The question of a parameter's significance can then be rephrased in mathematical form as an evaluation of the magnitude of the partial derivative of the output variable of interest with respect to each parameter, k_i:

$$s(Y; k_i) \equiv \frac{\partial Y}{\partial k_i} \quad (1.16)$$

where $s(Y; k_i)$ is termed the *local objective sensitivity*. The sensitivity is local, because the partial derivative is evaluated at just one point in the parameter space.

However, different parameters often have different units and magnitudes, making direct comparisons among the partial derivatives impossible. It is therefore convenient to normalize the partial derivatives as follows:

$$S(Y; k_i) \equiv \frac{k_i}{Y} \frac{\partial Y}{\partial k_i} \quad (1.17)$$

$$= \frac{\partial \ln Y}{\partial \ln k_i} \qquad (1.18)$$

where $S(Y; k_i)$ is the normalized local objective sensitivity, a parameter that is very useful for comparing the significance of all the parameters in a model. In some cases, these partial derivatives can be evaluated analytically. However, often the output Y is a complicated function of the parameters, and the partial derivatives must be evaluated numerically by a central finite difference in the parameter k_i:

$$S(Y; k_i) \approx \left(\frac{k_i}{Y(k_i)}\right) \frac{Y(k_i \cdot (1 + \frac{\delta}{2})) - Y(k_i \cdot (1 - \frac{\delta}{2}))}{\delta \cdot k_i} \qquad (1.19)$$

$$= \frac{Y(k_i \cdot (1 + \frac{\delta}{2})) - Y(k_i \cdot (1 - \frac{\delta}{2}))}{\delta \cdot Y(k_i)} \qquad (1.20)$$

where δ is a dimensionless fraction by which the parameter is perturbed to approximate the partial derivative as a finite difference. What value should be used for δ? On one hand, the estimate of the derivative will improve as $\delta \to 0$. On the other hand, every simulation has some uncertainty, noise, or error involved, and too small a step in the parameter value k_i runs the risk of differentiating noise. If the model error is ε_Y and the desired uncertainty in the sensitivity estimate is ε_S, then the parameter step should satisfy the following inequality:

$$\delta k_i \geq \frac{\varepsilon_Y Y(k_i)}{\varepsilon_S \frac{\partial Y}{\partial k_i}} \qquad (1.21)$$

Substituting equation 1.17 into the denominator and recognizing that the smallest value of δ is desirable, yields:

$$\delta = \frac{\varepsilon_Y}{\varepsilon_S S(Y; k_i)} \qquad (1.22)$$

Because each parameter has a different sensitivity $S(Y; k_i)$, in principle a different differential step should be used in the estimation of each parameter's sensitivity. This is a nicety often not observed, although an iterative check that none of the differentials δ used were too small by the criterion of equation 1.22 is recommended.

In practice, the list of normalized local objective sensitivities for the parameters in a model paints a qualitative picture of some variables significantly driving the model, while others exert little effect. This helps identify the key reactions and mechanisms in the model, and conversely it shows which parameters or reactions might actually be trimmed from the model equations with little effect. The topic of sensitivity analysis is actually far richer than discussed here, and the interested reader can find greater detail elsewhere [11, 12].

1.5 Basic Themes in Rate Process Modeling

1.5.1 Steady State versus Equilibrium

When observing an unperturbed biological system, the concentrations of many of the molecular species may not vary as a function of time, giving a potentially misleading impression of a static condition. In fact, most biomolecules in a living system are undergoing numerous dynamic processes: binding, trafficking, not to mention synthesis and degradation. However, when the environment in which the system is immersed remains constant over time, all rates eventually balance out such that the pools of each molecule reach a *steady state*. By contrast, in a closed system, a state of dynamic equilibrium can be attained that is then amenable to analysis by the powerful tools of classical and statistical thermodynamics. A first step in formulating a mathematical model is determining whether the system is open or closed (figure 1.11).

1.5.2 Perturbations

When a biological system is at steady state, we cannot measure the various underlying rate processes that create this condition. This is because by definition, nothing changes with time at steady state! It is therefore necessary to somehow introduce a time-varying component to the system, to essentially push it away from its current steady-state position. If the push is transient—say, just a brief pulse—then the system may shrug off the input and return to the same steady state, but the time

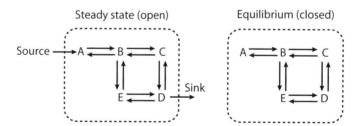

Figure 1.11 Open and closed systems reach different types of time-invariant states. In an open system, new mass enters and exits the system, so a time-invariant state is reached only once the sum of the input rates equals the sum of the output rates. In a closed system, dynamic equilibrium is attained when all internal rate processes balance out such that the concentrations of each component do not change with time. Biological organisms are open systems, so steady state rather than equilibrium is generally the relevant condition. Nevertheless, if the source and sink terms vary much more slowly than the internal dynamics, conditions approaching equilibrium may often be attained.

Introduction to Biological Rate Processes

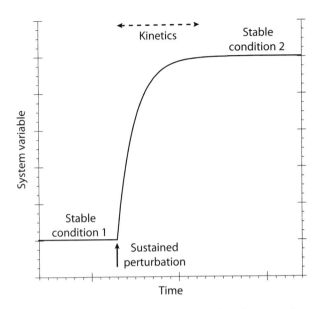

Figure 1.12 A typical step response curve. Before the perturbation, the system is at steady state and is time invariant. At some time point, a variable in the system's environment is suddenly changed to a different value and maintained at that value until a new steady state is reached.

dependence of that return will be informative. If the push is sustained as a step change in one of the environmental variables of the system, the concentrations of species of interest may move to a new steady state, and information can be obtained about both the nature of the new steady state and the rate of approach (figure 1.12).

For biomolecular and cellular systems, the step change of interest is often the addition of a new molecule. For instance, imagine at time zero a growth factor is added to a cell culture, or a stream of ligand begins flowing across a biosensor surface. Although such inputs can often be accomplished very rapidly and therefore approximate a perfect step function, the biological response is generally more sluggish—a measurable time is required for the various rates to rebalance to a new steady-state condition. Careful measurement of the time-varying response to such abrupt step changes can quantitatively discriminate between alternative hypothesized mechanisms for how the overall system works. The development of this indirect means for obtaining a window into the rapid internal reactions of a system has been recognized by more than one Nobel Prize—once for Chemistry in 1967 to Eigen, Norrish, and Porter "for their studies of extremely fast chemical reactions, effected by disturbing the equilibrium by means of very short pulses of energy," and once to George Palade in 1974 for Physiology & Medicine, for his use

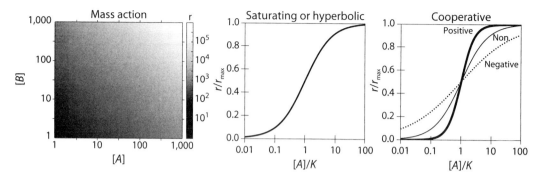

Figure 1.13 Graphical representation of various rate forms for biological processes.

of pulse-chase metabolic radiolabeling to show how "secretory proteins, produced by the ribosomes on the outside of the reticulum enter the space between its membranes, migrate to a special organelle, the Golgi complex, where they are changed to a form suitable for secretion."

1.5.3 Rate Forms

A few rate form types are frequently observed for formation of biomolecular complexes and are briefly summarized here (figure 1.13).

Mass-action kinetics When mechanistic descriptions of biological processes are available, the process rate can often be described by mass-action kinetics, in which the observed rate is proportional to the concentration of each relevant component:

$$\text{rate} = k[A][B] \tag{1.23}$$

Saturating or hyperbolic In numerous instances, when the concentration of one component is low or modest, biological processes are observed to have a first-order dependence on the concentration. However, when the concentration increases enough, the rate becomes independent of concentration (i.e., zeroth order). These saturating dependencies appear when evaluating enzyme kinetics (as a function of substrate, in the Michaelis-Menten model) and cell growth kinetics (as a function of a key nutrient, in the Monod model):

$$\frac{\text{rate}}{\text{rate}_{\max}} = \frac{[A]/K}{1 + [A]/K} \tag{1.24}$$

Cooperative Cooperativity, in which a molecular binding event impacts the likelihood of additional binding, can alter the sensitivity of response. Strong positive cooperativity can yield switchlike behavior in molecular binding, cellular signal transduction, and enzyme kinetics. By contrast, negative cooperativity can dull responses:

$$\frac{\text{rate}}{\text{rate}_{\max}} = \frac{\left(\frac{[A]}{K}\right)^n}{1 + \left(\frac{[A]}{K}\right)^n} \qquad (1.25)$$

1.5.4 Characteristic Time and Length Scales

The speeds of biological processes span many orders of magnitude, from subsecond enzymatic catalysis to day-long human cell doubling (and beyond for such processes as human development; figure 1.14). Quantitative mathematical modeling of these processes, detailed in this text, requires quantitative parameterization with variable impact of each parameter's precision and accuracy (as detailed above). It is useful, both for model implementation and evaluation, to understand the relevant time scales for various processes. In many cases, an order-of-magnitude scaling analysis can provide substantial insight to differentiate the primary effectors within a system.

An understanding of typical length scales is similarly beneficial for consideration of biological systems, which encompass a broad array of length scales, from angstrom-long covalent bonds to industrial bioreactors tens of meters in diameter (figure 1.15).

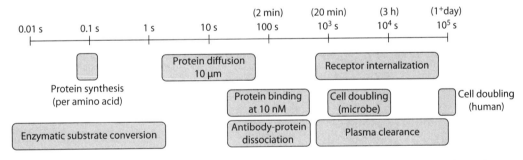

Figure 1.14 Typical time scales for various biological processes. Limits are not absolute but rather representative of typical ranges.

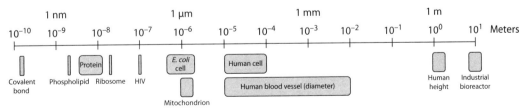

Figure 1.15 Typical length scales for various biological systems. Limits are not absolute but are representative of typical ranges.

Example 1-7 Estimate the amount of time to translate all proteins and replicate all DNA in an *Escherichia coli* bacterium. Compare these results to the time needed for *E. coli* cell growth.

Solution The amount of protein in a cell can be approximated as the product of the density, the volume, and the fraction of mass that is protein:

$$m_{\text{protein}} = \rho_{\text{cell}} \cdot V_{\text{cell}} \cdot f_{\text{dry}} \cdot f_{\text{protein}} \tag{1.26}$$

$$= \left(1 \frac{\text{g}}{\text{cm}^3} \cdot \frac{1 \text{ cm}^3}{10^{12} \, \mu\text{m}^3}\right) (0.4 \, \mu\text{m}^3) \left(1 - 0.7 \frac{\text{g water}}{\text{g cell}}\right) (0.60) \tag{1.27}$$

$$= 7.2 \times 10^{-14} \frac{\text{g}}{\text{cell}} \tag{1.28}$$

The protein synthesis rate can be calculated assuming 20,000 ribosomes per bacterium and 50 ms per amino acid ("AA" below) translated per ribosome ("rib" below):

$$r = \left(\frac{1 \text{ AA}}{0.05 \text{ s} \cdot \text{rib}}\right)\left(119 \frac{\text{g}}{\text{mol} \cdot \text{AA}}\right)\left(\frac{1 \text{ mol AA}}{6.02 \times 10^{23} \text{ AA}}\right)\left(2 \times 10^4 \frac{\text{rib}}{\text{cell}}\right)$$
$$\tag{1.29}$$

$$= 7.9 \times 10^{-17} \frac{\text{g}}{\text{s} \cdot \text{cell}} \tag{1.30}$$

Thus, the time to produce all the protein is the ratio of the mass and rate:

$$t = \frac{m_{\text{protein}}}{r} \tag{1.31}$$

$$= \left(\frac{7.2 \times 10^{-14} \text{ g/cell}}{7.9 \times 10^{-17} \text{ g/(s} \cdot \text{cell)}}\right)\left(\frac{1 \text{ min}}{60 \text{ s}}\right) = 15 \text{ min} \tag{1.32}$$

For DNA replication, consider bidirectional addition to a circular chromosome, so that there are two replication forks each with two strands, corresponding to four active polymerases. The DNA polymerase addition rate can be estimated to be 1,000 nucleotides (nt) per second:

$$t = \frac{\text{number of bases}}{\text{base addition rate}} \tag{1.33}$$

$$= \left(\frac{4.6 \times 10^6 \text{ bp/cell}}{1000 \text{ nt/(s} \cdot \text{polymerase)} \times 4 \text{ polymerases/cell}}\right) \left(\frac{2 \text{ nt}}{\text{bp}}\right) \left(\frac{\text{min}}{60 \text{ s}}\right) \tag{1.34}$$

$$= 38 \text{ min} \tag{1.35}$$

These times scales are comparable to (within twofold of) the doubling time for *E. coli*. Notably, *E. coli* can actually double faster than this estimated time for DNA replication, because the daughter genomes begin replicating before cell division is completed.

Suggestions for Further Reading

R. Milo and S. Phillips. *Cell Biology by the Numbers*. Garland Science, 2015.

W. H. Press, S. A. Teukolsky, W. T. Vetterling, and B. P. Flannery. *Numerical Recipes in C++: The Art of Scientific Computing*. Cambridge University Press, 2002.

References

1. P. Schuck. Use of surface plasmon resonance to probe the equilibrium and dynamic aspects of interactions between biological macromolecules. *Annu. Rev. Biophys. Biomol. Struct.* 26: 541–566 (1997).

2. N. Yildirim and M. C. Mackey. Feedback regulation in the lactose operon: A mathematical modeling study and comparison with experimental data. *Biophys. J.* 84: 2841–2851 (2003).

3. T. S. Gardner, C. R. Cantor, and J. J. Collins. Construction of a genetic toggle switch in *Escherichia coli*. *Nature* 403: 339–342 (2000).

4. H. S. Wiley. Trafficking of the ErbB receptors and its influence on signaling. *Exp. Cell. Res.* 284: 78–88 (2003).

5. B. Alberts, A. D. Johnson, J. Lewis, D. Morgan, M. Raff, K. Roberts, and P. Walter. *Molecular Biology of the Cell*, sixth edition. W. W. Norton & Company (2014).

6. A. S. Perelson. Modelling viral and immune system dynamics. *Nat. Rev. Immunol.* 2: 28–36 (2002).

7. B. M. Rao, D. A. Lauffenburger, and K. D. Wittrup. Integrating cell-level kinetic modeling into the design of engineered protein therapeutics. *Nat. Biotechnol.* 23: 191–194 (2005).

8. K. A. Janes and D. A. Lauffenburger. A biological approach to computational models of proteomic networks. *Curr. Opin. Chem. Biol.* 10: 73–80 (2006).

9. T. Carlson. Über Geschwindigkeit und Grösse der Hefevermehrung in Würze. *Biochemische Zeitschrift* 57: 313–334 (1913).

10. R. Pearl. The growth of populations. *Q. Rev. Biol.* 2: 532–548 (1927).

11. H. Rabitz, M. Kramer, and D. Dacol. Sensitivity analysis in chemical kinetics. *Annu. Rev. Phys. Chem.* 34: 419–461 (1983).

12. A. Varma, M. Morbidelli, and H. Wu. *Parametric Sensitivity in Chemical Systems*. Cambridge University Press (1999).

2

Noncovalent Binding Interactions

Teach us the nature of the ties which bind them together.
—Socrates, quoted by Plato, in *The Republic*

What causes one biomolecule to bind specifically to another? In this chapter, we will examine the structures of protein complexes, their rates of formation, and the energies and forces that drive biological molecules to bind together. These binding interactions are generally noncovalent and reversible. The primary contributions to complex formation come from hydrogen bonding, electrostatics, van der Waals interactions, and the hydrophobic effect. Intermolecular contacts are visible in high-resolution, atomic-level structures of biomolecular complexes. However, intramolecular energies and forces are also important, because changes in shape and energetics within a partner may also accompany binding. This set of energetics and forces determines two fundamental characteristics of binding that we return to throughout this book: the extent of binding (affinity) and the rates of binding and unbinding (kinetics). Together, these two properties generate much of the dynamics of biological systems.

A significant portion of biological function is mediated by protein-protein contacts, which are generally very specific. To get a feeling for the specificity of such interactions, consider that a cell's cytoplasm is typically 20–30% macromolecular solute by volume, which has been described as a condition of *macromolecular crowding* [1]. These solute molecules consist of thousands of different proteins, nucleic acids, and small organic metabolites.

The images in figure 2.1 provide vivid graphic representations of the crowding encountered by proteins, both inside and outside cells. As we draw highly simplified schematics of rate processes throughout this book, it bears remembering that the actual physical situation is significantly more complex. It is the power and beauty of some simple mechanistic kinetic models that they can quantitatively predict phenomena in such a complex system.

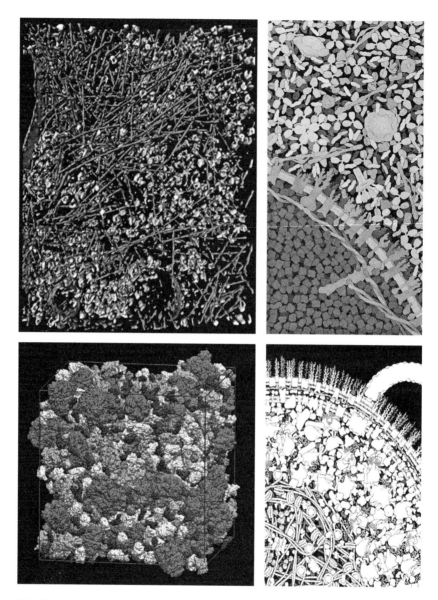

Figure 2.1 Macromolecular crowding in different biological environments. Upper-left panel: a cryogenic electron microscopic image of the interior of a eukaryotic cell [2]. The long strings are actin filaments, and the spheroid particles are predominantly ribosomes. Smaller proteins, nucleic acids, and organic molecules are not represented in this image. Upper-right panel: a representation of the environment inside and outside a red blood cell in whole blood. The Y-shaped molecules outside the cell are IgG antibodies; the chunks inside the cell are hemoglobin [3]. Lower-left panel: a molecular dynamics simulation of the environment inside an *Escherichia coli* cell at 300 mg/mL based on [4]. For scale, the cube is 100 nm on each side; the largest objects are ribosomes. Lower-right panel: a representation of macromolecular crowding in an *E. coli* cell [5].

2.1 Kinetic Rate Constants

Fast bind, fast find
—Shylock in William Shakespeare's *The Merchant of Venice*

Each protein in a cell collides with a tremendous variety of other molecules every second, but the great majority of these interactions are fleeting and have no functional consequences. However, in a small minority of these collisions, a more persistent interaction occurs between the protein and its ligand, docking the two into a precise relative orientation. This complex may stay assembled anywhere from seconds to days and perform a variety of functions: catalysis of a chemical reaction, construction of an intracellular structure or compartment, or transmission of signaling information from the environment or between intracellular compartments.

How does one quantify the rate and extent of noncovalent binding between a protein P and ligand L? When mixed in solution, some molecules of P and L bind to form the complex C, and some complexes of C dissociate to reform P and L:

$$P + L \underset{k_{\text{off}}}{\overset{k_{\text{on}}}{\rightleftharpoons}} C \qquad (2.1)$$

After sufficient time, a dynamic equilibrium will be reached where the net concentrations of P, L, and C remain constant. This equilibrium is dynamic in that the binding and dissociation reactions have not stopped. Rather, their rates are equal, and their effects are opposite, resulting in no net change in the species concentrations.

For protein and ligand molecules homogeneously dispersed in solution, the rate of complex formation is proportional to the current concentrations of the components, and the proportionality constant is termed the association rate constant k_{on},

$$\text{Rate of complex formation} = k_{\text{on}}[P][L] \qquad (2.2)$$

where square brackets denote the concentration of that component in mol L^{-1} ($\equiv$ M). The rate of complex formation has units of M s^{-1}, so k_{on} has units of M^{-1} s^{-1}. We will examine the basis of this broadly applicable relationship—known as mass-action kinetics—more closely in chapter 8. A key assumption of this rate form is that the two species are well mixed, without spatial heterogeneity, in the volume of interest.

The rate of complex dissociation is proportional to $[C]$ with a dissociation rate constant k_{off}, which therefore has units of s^{-1},

$$\text{Rate of complex dissociation} = k_{\text{off}}[C] \qquad (2.3)$$

The central assumption of such first-order kinetics is history independence: The rate constant k_{off} is not itself a function of time. Such processes follow what are known as Poisson statistics, which are considered in chapter 9.

By definition, at equilibrium, these rates are equal,

$$k_{on}[P]_{eq}[L]_{eq} = k_{off}[C]_{eq} \tag{2.4}$$

or, combining concentration terms on the left and rate constants on the right,

$$\frac{[P]_{eq}[L]_{eq}}{[C]_{eq}} = \frac{k_{off}}{k_{on}} \equiv K_d \tag{2.5}$$

K_d is an equilibrium dissociation constant, sometimes called an affinity constant, and has units of molar concentration. When the free ligand concentration $[L]_{eq} = K_d$, half of the protein is complexed with ligand (i.e., $[P]_{eq} = [C]_{eq}$). As an equilibrium dissociation constant, a smaller value of K_d corresponds to a stronger, higher-affinity interaction, in which only a small concentration of ligand is needed to yield significant complex formation.

Example 2-1 A buffered solution of protein and ligand is at equilibrium and contains 10 nM free ligand, 10 nM free protein, and 10 nM protein-ligand complex. If the volume is doubled by adding pure buffer (containing no ligand or protein), what is the new concentration of complex at equilibrium?

Solution From equation 2.5,

$$K_d = \frac{(10 \text{ nM})(10 \text{ nM})}{10 \text{ nM}} = 10 \text{ nM} \tag{2.6}$$

From conservation of mass ($[P]_{total} = [P]_{eq} + [C]_{eq}$ and $[L]_{total} = [L]_{eq} + [C]_{eq}$), we know that the total concentrations of protein and ligand are each 20 nM before the dilution step.

Thus, after dilution, the total concentrations of protein and ligand are each 10 nM. Using equation 2.5 and the mass balances for protein and ligand,

$$K_d = 10 \text{ nM} = \frac{(10 \text{ nM} - [C]_{eq})^2}{[C]_{eq}} \tag{2.7}$$

Solving the resulting quadratic equation, $[C]_{eq} = 3.8$ nM. Note that a twofold increase in the volume does not simply result in a twofold decrease in $[C]_{eq}$, due to the nonlinear equilibrium relationship dictated by equation 2.7.

2.1.1 Typical Ranges for k_{on} and k_{off}

One way to get a sense for the strength of a protein-ligand interaction is to determine how long, on average, a newly formed protein-ligand complex pair persists before dissociating. To estimate typical values of k_{off} for proteins of varying K_d, we can take advantage of a very rough empirical rule of thumb that 10^5 M^{-1} s^{-1} < k_{on} < 10^6 M^{-1} s^{-1} for protein-ligand binding, with protein ligands typically having k_{on} values at the lower end of this range and small-molecule ligands at the upper end [6]. (We will examine this approximation in chapter 8.) To estimate dissociation half-life, the time for 50% completion of any rate process with a first-order rate constant k_{off} is $(\ln 2)/k_{off}$.

The persistence of protein-protein complexes spans a broad range of times, from milliseconds to months (table 2.1). Most proteins are weakly sticky, forming clumps of nonspecific short-lived clusters at millimolar concentrations. Most functionally meaningful binding interactions have submillimolar K_d values, distinguishing them from such random contacts. Cell-surface proteins involved in adhesion and migration typically exhibit micromolar affinities for their ligands, with the fleeting nature of these binding interactions functionally compensated by the multivalency of many such complexes per cell (to be considered in chapter 3). An important component of cellular immunology, the T cell receptor–peptide MHC binding interaction, is also in the micromolar affinity range. This interaction is responsible for the adaptive immune system's successful recognition of intracellular pathogens via interactions lasting only seconds at most. More stable interactions with nanomolar K_d values are exemplified by antibody-antigen interactions, which label extracellular pathogens for the minutes to hours necessary to ensure destruction by phagocytic immune cells. Growth factors and their receptors often possess affinities in the picomolar range, such that at equilibrium, a significant fraction of receptors is complexed by growth factors, which are often only present at picomolar concentrations in plasma and tissues. Some of the tightest noncovalent complexes in nature are hydrolase-inhibitor complexes, which bind with femtomolar affinity constants and ensure essentially irreversible inhibition of degradative enzymes, such as trypsin or RNase.

Table 2.1 Typical protein-protein complex half-lives.

K_d	Dissociation half-life	Examples
mM	milliseconds	nonspecific stickiness
μM	milliseconds–seconds	multivalent cell surface interactions
nM	minutes–hours	antibodies, enzymes
pM	hours–weeks	growth factors and their receptors
fM	months	hydrolase inhibitors

> **Example 2-2** Show that a protein-protein complex with a K_d value in the nanomolar range does indeed have a dissociation half-life in the minutes-to-hours range, as stated in table 2.1.
>
> **Solution** From equation 2.5, $k_{off} = k_{on} K_d$. Estimating a k_{on} of 10^5 M^{-1} s^{-1}, the range of k_{off} values for nanomolar binders ($K_d = 1$–100 nM) is 0.0001–0.01 s^{-1}.
> Therefore, the range of half-lives, $(\ln 2)/k_{off}$, is 69–6900 s, or ≈ 1 min to ≈ 2 h.

2.2 Thermodynamics

Thermodynamics provides a description of biological systems at equilibrium. One property that we accept readily—but is really quite remarkable—is that the equilibrium reached does not depend on the system's history. For example, if we add purified DNA containing a transcription factor binding site to a solution containing the corresponding transcription factor, an equilibrium will be reached with defined concentrations of free transcription factor, free DNA, and bound complex. If we instead prepare the same amount of material as purified bound complex (isolated from a gel, for instance) and dissolve it in the same amount of buffer solution, equilibrium will be reached with the same concentrations of free transcription factor, free DNA, and bound protein found earlier. That is, the same equilibrium state is reached, no matter from which direction it is approached; this independence is a fundamental prerequisite to be able to write such expressions as equation 2.5. Thus, it is appropriate to describe equilibrium as a *state*, because its properties do not depend on the history of the system or the pathway followed to reach it.

2.2.1 State Functions

Much can be gained by using the tools of thermodynamics to study the driving forces for biochemical reactions. There are special thermodynamic quantities that describe the properties of a system and depend only on its state, called state functions. For biochemical systems at constant pressure and temperature, the most convenient of these are Gibbs free energy (G), enthalpy (H), and entropy (S). The change between any two states can be characterized by the corresponding changes ΔG, ΔH, and ΔS, which are related by the expression

$$\Delta G = \Delta H - T \Delta S \tag{2.8}$$

where T is the temperature.

At constant pressure and temperature, the Gibbs free energy of the system is the overall driving force for the reaction. A ΔG equal to zero corresponds to no driving force, and the reaction is at equilibrium. A ΔG less than zero means that the reaction proceeds spontaneously in the forward direction, and a ΔG greater than zero means that the reverse reaction is spontaneous (because ΔG for the reverse reaction is equal to $-\Delta G$ for the forward reaction). Note that although the free energy change indicates that a reaction is favorable and is termed "spontaneous," the reaction still may not happen to an appreciable extent. If there is a sufficient activation energy that serves as a kinetic barrier to the reaction, it may need an input of energy for initiation.

What is it about a chemical change that causes it to be spontaneous with a negative ΔG? This is a direct consequence of the second law of thermodynamics, which states that the entropy of the universe (system + surroundings) always increases over time. Under conditions of constant pressure and temperature, which is often the case in biological systems, it can be shown that a negative ΔG for the system alone corresponds to an increase in the entropy of the universe and can therefore be used to assess spontaneity.

Two general trends lead to increased favorability of biochemical reactions. Natural systems tend to a state of lower energy and to a state of higher disorder. Enthalpy is one useful measure related to the energy of a system, and entropy is a measure of disorder. Thus, these two quantities represent the two components of the driving force for spontaneous change, as shown in equation 2.8.

A simple example is the addition of a drop of a hydrophilic dye to a cup of water. The dye will spontaneously disperse throughout the water without performing any work (e.g., agitation) on this system. Why does this occur? If the dye molecules and water molecules all interact with comparable energies, the change in the enthalpy is minimal. The system will therefore evolve to maximum disorder, which corresponds to uniformly dispersed dye molecules in the solution. By contrast, consider the addition of a drop of hydrophobic oil to a cup of water. Why don't the oil molecules also spontaneously disperse throughout the water? Although this would appear to be entropically favorable, the oil-oil and water-water energetic interactions are much more favorable than oil-water interactions, so the enthalpic term in equation 2.8 dictates that the oil and water molecules remain phase separated, even though the entropic term is seemingly unfavorable. (This is a bit of a simplification for the sake of example. In reality, the oil and water molecules are also more mobile when they do not have to interact with the other species.)

The folding of a protein generally involves these opposing effects. The unfolded state of a protein is highly disordered, and so it has high entropy compared to the relative order of the folded state. Like the dispersed hydrophilic dye, many arrangements or conformations are equally available in that state. This unfavorable entropic contribution to protein folding is more than balanced at sufficiently low temperatures by the stronger interactions, due to packing and other effects, available in the folded state (a favorable enthalpic contribution). At high enough temperature, however, equation 2.8 indicates that the unfavorable entropic term becomes more strongly weighted, which helps explain why proteins unfold at elevated temperatures. This simplified analysis neglects solvation effects, which make particularly strong contributions to the entropy, as well as the temperature dependence of the enthalpic and entropic changes.

2.2.2 Free Energy and Standard States

The Gibbs free energy of a system is the sum of an enthalpic term and an entropic term; if there is a path for the system to move between two states, it will tend to move to the state with the lower free energy via changes in its enthalpy and entropy. A system is at equilibrium between two states when their free energies are equal. For the following reaction,

$$P + L \rightleftharpoons C \qquad (2.9)$$

the net change in free energy for the forward (binding) reaction is

$$\Delta G = \Delta G^\circ + RT \ln \frac{[C]}{[P][L]} \qquad (2.10)$$

where ΔG is the change in free energy on complex formation, R is the gas constant (1.987 cal/(mol·K)), T is the temperature, and ΔG° is the standard-state free energy at arbitrary (standard) reference conditions. For biomolecular interactions in water, by convention, the standard reference state is: 25°C, pH = 7, pressure = 1 atmosphere, [NaCl] = 150 mM, and concentration referenced to a standard state of 1 M. Of course, such a concentration is so high as to actually exceed the solubility limits of many proteins; however, this fictional reference state need not be physically realized to serve as a reference point. At equilibrium, the free energy of the bound and unbound states is equal ($\Delta G = 0$), and so:

$$\Delta G^\circ = -RT \ln \frac{[C]_{eq}}{[P]_{eq}[L]_{eq}} \qquad (2.11)$$

And from equation 2.5,

$$\Delta G^\circ = -RT \ln \left(\frac{1}{K_d} \right) = RT \ln K_d \qquad (2.12)$$

Note that the dissociation constant K_d has units of concentration, which corresponds to the concentration in the standard-state free energy change ΔG°. That is, K_d must be defined with respect to the particular concentration and units originally chosen for the reference state. Comparing the values in table 2.1 to equation 2.12, we see that the typical range for standard-state free energy change $\Delta G°$ of biomolecular interactions is in the range of -5 to -20 kcal/mol. A decrease in standard-state binding free energy $\Delta\Delta G° = -1$ kcal/mol corresponds to an approximately fivefold decrease in the affinity constant K_d at room or body temperature, from equation 2.12.

The temperature dependence of the binding constant K_d is given by the van't Hoff relationship. This can readily be derived from the Gibbs-Helmholtz equation under standard-state reference conditions, which is given by

$$\Delta H° = -T^2 \left(\frac{\partial \left(\frac{\Delta G°}{T}\right)}{\partial T}\right)_P \tag{2.13}$$

where the subscript P indicates that the partial derivative is taken under constant pressure conditions. Enthalpy is thus especially useful for estimating how the free energy change or position of equilibrium will be altered as the temperature is varied. Combining equations 2.12 and 2.13, we obtain

$$\left(\frac{\partial \ln K_d}{\partial T}\right)_P = \frac{-\Delta H°}{RT^2} \tag{2.14}$$

where we have followed the convention that $\Delta H°$ is for the binding reaction and K_d corresponds to the dissociation reaction. If $\Delta H°$ were invariant over some temperature range (an uncommon situation in biochemical systems), then K_d at a new temperature T could be determined by integration from its value at a given temperature T_o, to give the van't Hoff formula:

$$\ln\left(\frac{K_d(T)}{K_d(T_o)}\right) = \frac{\Delta H°}{R}\left(\frac{1}{T} - \frac{1}{T_o}\right) \tag{2.15}$$

where $K_d(T)$ denotes the K_d at temperature T.

However, it has been found that $\Delta H°$ often does vary with temperature, because the temperature derivative of enthalpy (called the heat capacity C_p) is different for the complex versus the unbound protein and ligand:

$$\Delta C_p \equiv \left(\frac{\partial \Delta H}{\partial T}\right)_P \tag{2.16}$$

Consequently, the temperature dependence of enthalpy and entropy changes upon binding can be determined from the following integrals:

$$\Delta H = \int_{T_\circ}^{T} \Delta C_p dT + \Delta H(T_\circ) \qquad (2.17)$$

$$\Delta S = \int_{T_\circ}^{T} \frac{\Delta C_p}{T} dT + \Delta S(T_\circ) \qquad (2.18)$$

in which the assumption that ΔH (or ΔS) is independent of temperature over the relevant range has been relaxed.

Under standard-state conditions, it has been found that ΔC_p° remains approximately constant across physiological temperature ranges for most protein-ligand interactions, simplifying integration of equations 2.14 and 2.17:

$$\ln\left(\frac{K_d(T)}{K_d(T_\circ)}\right) = \frac{\Delta H^\circ(T_\circ)}{R}\left(\frac{1}{T} - \frac{1}{T_\circ}\right) - \frac{\Delta C_p^\circ}{R}\left(\ln\frac{T}{T_\circ} + \frac{T_\circ}{T} - 1\right) \qquad (2.19)$$

As it happens, $\Delta C_p^\circ < 0$ for most protein-ligand interactions, largely because of the decrease in water-accessible macromolecular surface area when a protein-ligand complex forms. In fact, ΔC_p° for protein-ligand complex formation is well correlated with the changes in solvent-accessible polar and nonpolar surface areas by the following expression: $\Delta C_p^\circ(\text{cal/mol} \cdot \text{K}) = \alpha \Delta ASA_{ap} + \beta \Delta ASA_{pol}$, where ΔASA_{ap} and ΔASA_{pol} are the changes in solvent-accessible apolar and polar surface areas, respectively, in units of Å^2, $\alpha = 0.36$ to 0.45, and $\beta = -0.25$ to -0.26 [7].

Experimental isothermal titration calorimetry data for ΔH°, $-T\Delta S^\circ$, and ΔG° as functions of temperature for six different protein-ligand interactions are shown in figure 2.2. Several features are apparent in each data set:

1. ΔH° is negative, so the enthalpy contribution to ΔG° is favorable at all temperatures tested.
2. ΔG° is essentially independent of temperature in the range measured.
3. ΔH° decreases with temperature approximately linearly; hence, ΔC_p° is negative and is approximately constant in this range.
4. Decreases in ΔH° appear to be quantitatively canceled by increases in $-T\Delta S^\circ$ for each system, resulting in the observed temperature independence of ΔG°.

As for the six systems represented in figure 2.2, it has been found that a common feature of protein-ligand interaction energetics is that ΔH° and $-T\Delta S^\circ$ change with temperature in opposite directions and at roughly equivalent magnitudes, resulting in enthalpy-entropy compensation and a much smaller effect on ΔG° [11, 12].

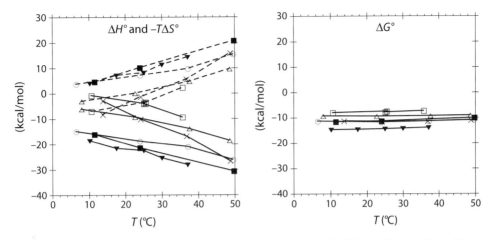

Figure 2.2 Experimentally determined $\Delta H°$ (solid lines, left panel), $-T\Delta S°$ (dashed lines, left panel), and $\Delta G°$ (right panel). Calorimetric data for six protein-ligand systems are shown (five antibody-protein antigen interactions [8, 9] and the Met repressor interaction with its DNA operator binding site [10]).

The observed enthalpy-entropy compensation is a straightforward consequence of the temperature dependencies represented by equations 2.17 and 2.18. For a $\Delta C_p°$ constant with respect to temperature, these relationships can be integrated to give

$$\Delta H°(T) = \Delta H°(T_\circ) + \Delta C_p°(T - T_\circ) \tag{2.20}$$

and

$$\Delta S°(T) = \Delta S°(T_\circ) + \Delta C_p° \ln\left(\frac{T}{T_\circ}\right) \tag{2.21}$$

Measurements of protein-ligand energetics are often made in the physiological temperature range 10–40°C. Because the temperature term in equation 2.21 is in units of absolute temperature (kelvin), this range corresponds only to an approximately 10% change (i.e., from 283 to 313 K). Over this small relative change in temperature, the approximation $\ln(1 + x) \approx x$ can be used to simplify equation 2.21, as follows:

$$\Delta S°(T) = \Delta S°(T_\circ) + \Delta C_p° \ln\left(1 + \frac{T - T_\circ}{T_\circ}\right) \tag{2.22}$$

$$\approx \Delta S°(T_\circ) + \Delta C_p° \frac{T - T_\circ}{T_\circ} \tag{2.23}$$

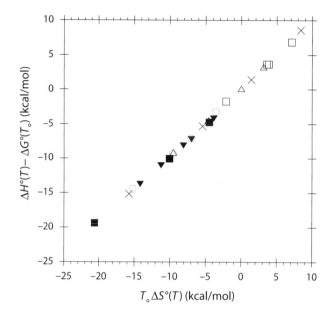

Figure 2.3 Apparent enthalpy-entropy compensation in the systems represented in figure 2.2, for arbitrarily selected $T_\circ = 25°C$. As predicted by equation 2.25, there is a linear relationship between $\Delta H°(T)$ and $\Delta S°(T)$, largely because the variation in temperature is small relative to the magnitude of the absolute temperature in units of kelvins.

Equations 2.23 and 2.20 can be combined to yield

$$\Delta H°(T) \approx \Delta H°(T_\circ) + T_\circ \left(\Delta S°(T) - \Delta S°(T_\circ) \right) \tag{2.24}$$

$$= \Delta G°(T_\circ) + T_\circ \Delta S°(T) \tag{2.25}$$

Taking the data for the six systems represented in figure 2.2, one finds the expected unit slope and zero intercept for a plot of $\Delta H° - \Delta G°(T_\circ)$ versus $T_\circ \Delta S°$ (figure 2.3), arbitrarily taking $T_\circ = 25°C$.

Although figure 2.2 indicates that the binding constant K_d can be expected to be relatively independent of temperature, this relationship can be predicted if $\Delta C_p°$ is known. By the definition of free energy ($G \equiv H - TS$),

$$\Delta G° = \Delta H° - T \Delta S° \tag{2.26}$$

Substituting in equations 2.20 and 2.23 yields

$$\Delta G°(T) \approx \Delta H°(T_\circ) - T \Delta S°(T_\circ) - \Delta C_p° \frac{(T - T_\circ)^2}{T_\circ} \tag{2.27}$$

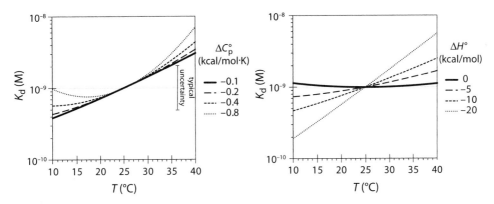

Figure 2.4 Predicted dependence of the equilibrium dissociation constant K_d on temperature, assuming $K_d = 1$ nM at 25°C. (Left panel) Using equation 2.28, four values of ΔC_p° spanning the typical experimentally observed range are shown. For most experimental measurements of K_d, an error within twofold is typical and is represented on the right side of the plot to demonstrate that the predicted temperature effects on affinity are often within experimental error. (Right panel) Using equation 2.19, four values of ΔH° spanning the typical experimentally observed range are shown.

$$\approx \Delta G^\circ(T_\circ) - \Delta C_p^\circ \frac{(T - T_\circ)^2}{T_\circ} \quad (2.28)$$

ΔC_p° has been measured for a broad range of protein-ligand interactions and is generally found to be in the range of -0.1 to -0.6 kcal/(mol·K) [13–15]. This estimate excludes dimer and other multimer formation, which often exhibits more negative ΔC_p° due to burial of larger hydrophobic areas. The temperature dependence of affinity predicted by equation 2.28 is plotted in figure 2.4. Binding affinity is a fairly weak function of temperature; hence measurements made at cooler temperatures can often be extrapolated to physiological temperature without gross error.

Example 2-3 Thymidine kinase has affinity constant $K_d = 5.2$ μM for thymidine at 25°C. $\Delta H^\circ = -19.1$ kcal/mol at 25°C, and $\Delta C_p^\circ = -0.36$ kcal/(mol·K). Predict the thermodynamic parameters ΔG°, ΔH°, and $T\Delta S^\circ$ for the thymidine/kinase interaction in the temperature range 5–40°C.

Solution From equation 2.20,

$$\Delta H^\circ(\text{kcal/mol}) = -19.1 - 0.36(T(°C) - 25) \quad (2.29)$$

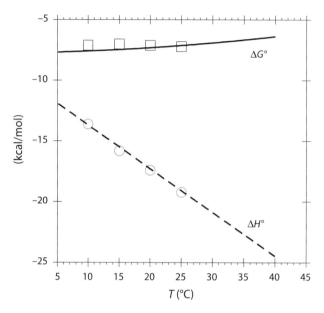

Figure 2.5 Relationships for $\Delta H°(T)$ from equation 2.29 and $\Delta G°(T)$ from equations 2.26, 2.29, and 2.33. The symbols are experimental data [15]. Note that free energy (and therefore affinity) changes very little in this temperature range.

At $25°C$, $\Delta G° = RT \ln K_d = \left(1.987 \times 10^{-3} \text{kcal/(mol·K)}\right) \cdot (298\text{K}) \ln(5.2 \times 0^{-6})$ $= -7.2$ kcal/mol. The entropy change is calculated from the free energy and enthalpy as follows:

$$\Delta S°(25°C) = \frac{\Delta H°(25°C) - \Delta G°(25°C)}{298 \text{ K}} \quad (2.30)$$

$$= \frac{(-19.1) - (-7.2) \text{ kcal/mol}}{298 \text{ K}} \quad (2.31)$$

$$= -0.040 \text{ kcal/(mol·K)} \quad (2.32)$$

From equation 2.21,

$$\Delta S°(\text{kcal/mol}) = -0.040 - 0.36 \ln \frac{T(\text{K})}{298} \quad (2.33)$$

Free energy is calculated from equations 2.26, 2.29, and 2.33. These relationships are plotted in figure 2.5.

2.3 Energetic Contributions to Binding Affinity

Noncovalent complexes are held together by four different types of interaction—electrostatic attraction between oppositely charged groups, hydrogen bonds, van der Waals contacts, and the hydrophobic effect—as well as energetic and entropic contributions from conformational changes upon binding. The balance of these interactions between the protein and its ligand versus the unbound pair and solvent molecules determines the affinity of the protein for its ligand and the rate of complex association. Each affinity-determining interaction can be thought of as an energy (U) or equivalently as a force (F), because force is the negative derivative of energy with respect to distance ($F = -dU/dr$).

2.3.1 Electrostatics

Interactions between electrically charged groups must be balanced to stably bring two protein surfaces together in a complex. Opposite charges attract, such as the positively charged side chain of arginine interacting with the negatively charged side chain of aspartic acid in figure 2.6, which produces a favorable contribution toward binding. By contrast, atoms sharing the same positive or negative charge repel each other. The strength of interaction between opposite charges is proportional to the magnitudes of the charge of each of the two groups and inversely proportional to the distance between them. The actual strength of the interaction depends greatly on the spatial distribution of charges and properties of the surrounding medium, although for two isolated charges in a uniform medium, the potential energy U_{elec} of the interaction is given by Coulomb's law,

$$U_{elec} = \frac{q_1 q_2}{\varepsilon r} \tag{2.34}$$

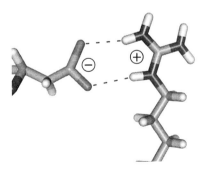

Figure 2.6 Oppositely charged chemical groups attract and produce a favorable contribution toward binding. Here the positively charged side chain of the amino acid arginine interacts with the negatively charged side chain of the amino acid aspartic acid.

```
          H₃C
            \
             N—H
   H₃C      /
      \δ⁻ δ⁺   δ⁻ δ⁺
       N—H -----O=C
      /          \
   O=C            CH₃
      \
       CH₃
```

Figure 2.7 A hydrogen bond involves the short-range sharing of a hydrogen atom between a donor and an acceptor. The dashed line represents a hydrogen bond. The preceding N is the hydrogen donor, and the following O is the hydrogen acceptor. The pattern of partial charges, indicated by δ^+ and δ^-, indicates that the electropositive hydrogen atom is shared between two electronegative atoms.

where q_1 and q_2 are the strengths of the two charges, ε is the dielectric constant of the medium, and r is the distance between the charges. The dielectric constant is a measure of the extent to which the medium filling the volume between two charges screens their interaction. This screening is due to the polarizability of the medium— water has a dielectric constant of about 80; in the interior of a protein, $\varepsilon \approx 2-4$; and in a vacuum, $\varepsilon = 1$. This dielectric screening can qualitatively alter the nature of charge interactions. For example, sodium and chloride ions are dispersed when dissolved in water yet tightly bind together in a low dielectric crystalline environment. Some chemical groups bear a formal charge (a nonzero integer charge, such as $+1, +2, +3, -1, -2, -3$). Others are neutral but carry a strong dipole moment due to the relative electronegativity of adjacent atoms (such as the amide groups shown in figure 2.7). The interaction falls off more quickly over distance for charge-dipole and dipole-dipole interactions, which decrease proportionally to $1/r^2$ and $1/r^3$, respectively.

Proteins, nucleic acids, and other biological macromolecules are large and composed of many chemical groups. Although it is sometimes advantageous to consider the individual interactions among small groups of atoms in each binding partner, often it is the overall net interaction between binding partners that matters.

The total electrostatic energy of a collection of atoms, such as a bound complex, is a Coulombic sum over all atom pairs. For a protein complex, it is convenient to construct sums for atoms with each of the two binding partners (first two terms on the right side of equation 2.35) and a third sum over pairs with one atom in the ligand and the other in the protein (third term on the right side of equation 2.35). This third term is the electrostatic interaction energy between a protein and a ligand.

$$U_{\text{elec}} = \sum_{i=1}^{N_P-1} \sum_{j=i+1}^{N_P} \frac{q_i q_j}{\varepsilon r_{ij}} + \sum_{i=1}^{N_L-1} \sum_{j=i+1}^{N_L} \frac{q_i q_j}{\varepsilon r_{ij}} + \sum_{i=1}^{N_P} \sum_{j=1}^{N_L} \frac{q_i q_j}{\varepsilon r_{ij}} \quad (2.35)$$

Here, N_P and N_L are the number of atoms in the protein and ligand, respectively, q_i and q_j are partial atomic charges representing the molecular charge distribution, ε is again the dielectric constant of the medium, and r_{ij} is the distance between the atoms. Strong electrostatic interactions between protein and ligand often replace strong solvation interactions between the unbound protein and water, as well as the unbound ligand and water. The overall effect accounts for the replacement of solvation interactions with protein-ligand electrostatic interactions, which is often a much smaller effect than either interaction alone. Protein and ligand may also change conformation upon binding, resulting in additional electrostatic binding contributions. Finally, because the environment surrounding individual atoms is generally not a uniform medium, other approaches have been usefully applied to study electrostatic interactions in proteins. Continuum methods, such as solutions to the Poisson-Boltzmann equation, explicitly account for different dielectric properties distributed in the environment but are beyond the scope of this discussion [16, 17].

2.3.2 Hydrogen Bonding

A particularly common interaction involves the partial sharing of a hydrogen atom between two electronegative atoms. The hydrogen atom is chemically bonded to one of the electronegative atoms, called the hydrogen donor, with a bond length of about 1 Å. The hydrogen points toward the other electronegative atom, called the hydrogen acceptor, at a distance of 1.8–2.6 Å (figure 2.7). In proteins, the hydrogen donor is often a nitrogen atom, and the acceptor is often an oxygen or other atom bearing a lone pair of electrons. A hydrogen bond can be between a pair of neutral polar groups, or it can involve a formal charge on one or both participants.

The hydrogen bond is a quantum mechanical effect that generally involves mutual polarization of dipolar groups, which makes dipole magnitudes larger than they would be in the absence of the hydrogen bond. The hydrogen bond also involves interaction between molecular orbitals, particularly those for lone pair electrons of the hydrogen acceptor. Hydrogen bonds are very short range (beyond which, any force between a donor and acceptor atom is considered to be an ordinary electrostatic effect). The strength has very little dependence on the H–O–C angle in figure 2.7 (generally termed *hydrogen-acceptor-antecedent*) and is relatively constant between values of 100° and 180°. There is stronger dependence on the N–H–O angle (donor-hydrogen-acceptor), whose preferred geometry is linear (180°). The strength of hydrogen bonds between uncharged partners in the gas phase is on the order of −6 kcal/mol (that is, favorable by 6 kcal/mol). Groups that form hydrogen bonds across the interface in a bound complex often form hydrogen bonds with solvent in the unbound state for aqueous reactions. Because bound-state hydrogen bonds essentially replace those with solvent in the unbound state, the net

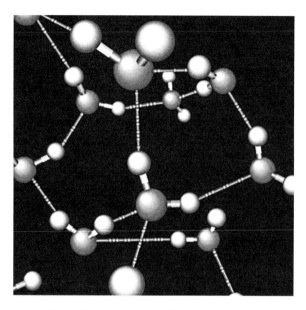

Figure 2.8 Water in the liquid state forms an extensive, fluctuating network of hydrogen bonds, with each molecule participating in up to four such bonds (one through each hydrogen and two through oxygen) [18].

energetic contribution to binding is generally much smaller in magnitude than either the bound- or the unbound-state hydrogen bonding energy and is estimated to be approximately -1.5 kcal/mol.

Water is the solvent for most biological interactions. Each water molecule can donate up to two hydrogen bonds (one through each of its hydrogen atoms), and each water molecule can accept up to two hydrogen bonds, one through each electron lone pair (figure 2.8). In the solid state (ice), each water molecule is involved in four simultaneous hydrogen bonds with its neighbors. In the vapor phase, as steam, water molecules make essentially no hydrogen bonds with their neighbors, because they are in constant, independent motion. Liquid water is an intermediate state, in which each water molecule makes roughly two hydrogen bonds to its neighbors as part of a fluctuating network structure that constantly changes. It is this fluctuating structure, coupled through hydrogen bonds, that is responsible for some of water's most interesting properties.

2.3.3 van der Waals Interactions

In interplanetary physics, one of the most important and ubiquitous forces is gravity. The objects one typically considers—stars, asteroids, and planets—exert gravitational forces on one another, and the strength of this force depends on the mass of the objects and the distance between them. In the physical chemistry of protein interactions, the analogously important and ubiquitous force is the van der Waals interaction (named after Dutch physicist Johannes Diderik van der Waals). All atoms attract one another with a relatively weak van der Waals force at long distances that becomes stronger at shorter distances. At very short distances, the interaction becomes a repulsion that keeps atoms from occupying the same space. The source of these forces is in the underlying structure of each atom, with a strong, dense, positively charged nucleus surrounded by an electron cloud. At any instant in time, the electron cloud surrounding a particular nucleus may be off center, creating an asymmetric charge distribution with a dipole moment. The van der Waals attraction is due to instantaneous, temporary, induced alignments between the atomic dipoles of atom pairs. When nonbonded atoms approach very closely, however, a strong repulsion develops because of the overlap of their electron clouds.

Mathematically, the van der Waals interaction energy between a pair of nonbonded atoms is reasonably well approximated by the empirical Lennard-Jones potential (figure 2.9, top),

$$U_{\text{vdW}} = \frac{A}{r^{12}} - \frac{B}{r^6} \qquad (2.36)$$

where A and B are parameters describing the interaction, and r is the distance between the atom centers. The $1/r^6$ term corresponds to the strength of induced-dipole–induced-dipole interactions that are the source of this term. By contrast, the $1/r^{12}$ term is a mathematically convenient and appropriate form. It is computationally efficient, once one has computed $1/r^6$, to multiply the term by itself to get $1/r^{12}$. The total van der Waals energy of a collection of atoms is a sum over all atom pairs. For a protein-ligand complex, it is convenient to construct sums for atoms in each of the two binding partners (first two terms on the right-hand side of equation 2.37) and a third sum over atom pairs spanning the protein and ligand (third term of equation 2.37),

$$U_{\text{vdW}} = \sum_{i=1}^{N_P-1} \sum_{j=i+1}^{N_P} \left(\frac{A_{ij}}{r_{ij}^{12}} - \frac{B_{ij}}{r_{ij}^6} \right) + \sum_{i=1}^{N_L-1} \sum_{j=i+1}^{N_L} \left(\frac{A_{ij}}{r_{ij}^{12}} - \frac{B_{ij}}{r_{ij}^6} \right) + \sum_{i=1}^{N_P} \sum_{j=i}^{N_L} \left(\frac{A_{ij}}{r_{ij}^{12}} - \frac{B_{ij}}{r_{ij}^6} \right)$$

(2.37)

In very large systems, computing the pairwise van der Waals and electrostatic interactions dominates the computational cost of molecular dynamics simulations. Some approximations have been developed to make this portion of the calculation more efficient. The simplest is to recognize that the van der Waals interactions fall away as $1/r^6$ and can be truncated directly or smoothed to zero at sufficiently long distances. The distance dependence of electrostatic and van der Waals interactions also

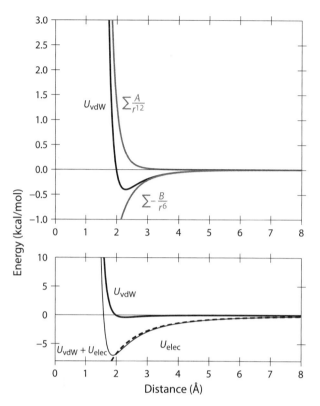

Figure 2.9 The top portion of the figure shows the Lennard-Jones potential of equations 2.36 and 2.37, which is a good approximation to the van der Waals interaction. The dark line is the total van der Waals potential, and the lighter lines represent the attractive and repulsive components separately. The bottom portion shows the same van der Waals potential together with an electrostatic potential and the sum of both. The potentials are for the colinear approach of a protein backbone carbonyl and N–H group in hydrogen bonding configuration, using geometric and van der Waals parameters from reference 19 and partial atomic charges from reference 20. Notice that the electrostatic potential is much stronger than the van der Waals at intermediate distances. The total nonbonded potential is similar in shape to the van der Waals potential, but the minimum is much deeper and somewhat shorter than the pure van der Waals potential, due to this attractive electrostatic effect.

directly impacts the rate constants of binding: k_{on}, which characterizes the rapidity of association between two molecules that are initially far apart, is often dominated by long-range electrostatics; k_{off}, which describes how quickly a complex dissolves, is largely driven by short-range van der Waals interaction energies.

The van der Waals interaction energies are also responsible for general properties of biological molecule solutions in the laboratory or in living organisms. Proteins and solvent fill space, resulting in neither large voids nor strongly overlapping atoms. When proteins bind to one another, they form a complementary interface from which bulk solvent is generally expelled but in which individual water molecules may be included. The binding process involves a continuous squeezing out of water molecules as the proteins come together, with the minimization of voids or steric clashes.

2.3.4 Hydrophobic Effect

The burial of surface area—particularly hydrophobic surface area—is a driving force in the folding of proteins during their synthesis and processing and also in the binding of proteins to their ligands. The magnitude of the energetic contribution is generally taken as proportional to the surface area buried, although it is understood that this is an approximation (figure 2.10). The physical basis for this interaction is not fully understood, but a significant portion results from the reduced degrees of freedom and interactions available to water molecules lining a hydrophobic surface (and therefore decreased entropy), compared to water molecules in bulk solution.

At nonpolar surfaces, liquid water converts from a rapidly fluctuating network of hydrogen bonds to a more persistently structured hydrogen bonding network. The net result of the hydrophobic effect is that hydrophobic solute molecules in aqueous solvent tend to associate. The relative strengths of the solute-solute

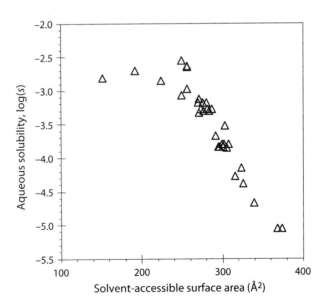

Figure 2.10 A strong correlation exists between solubility and solvent-accessible surface area for nonpolar, nonaromatic alkane molecules. (Free energy of solvation is directly proportional to $-\log s$.) Data from reference 21.

and solute-solvent van der Waals attractions can be closely balanced, but the hydrophobic effect produces an apparent force due to the relative favorability of the solvent when hydrophobic surface area exposure to the bulk solvent is minimized. Because the "force" driving the solute molecules together is indirect, acting through the solvent, the hydrophobic effect is referred to as a *potential of mean force*.

When two molecules associate in solution, it is useful to consider the reaction as an exchange in which each binding partner trades interactions with solvent in the unbound state for interactions with each other in the bound state. This concept is particularly important for polar and charged groups at the protein surface that become desolvated and buried at the binding interface. This desolvation is an unfavorable contribution to the binding free energy, because water reorients to make favorable interactions with polar and charged functional groups. Complementary interactions are generally formed at binding interfaces, including hydrogen bonds and salt bridges, which form solvent-screened interactions in solution. Thus, an electrostatic trade-off takes place between unfavorable desolvation and favorable protein-ligand interactions that together contribute to binding affinity.

Likewise, the interaction of solvent with nonpolar functional groups is responsible for the hydrophobic effect. However, whereas the electrostatic contributions are trade-offs that can be net favorable or unfavorable, the association of hydrophobic groups in aqueous solution is essentially always favorable.

2.3.5 Deformation Energy and Entropy

When a protein and ligand bind, if the binding partners are thought of as rigid entities, then binding can be viewed as the joining of three-dimensional jigsaw-puzzle pieces. A more accurate picture views the binding partners as flexible molecules with substantial internal motion in both the bound and unbound states. The average conformation and the distribution of conformations can be different between the bound and unbound states. Differences in average conformation result in a conformational change upon binding; differences in the relative amount of internal motion result in an internal entropy change upon binding.

Most proteins and ligands undergo some change in conformation and internal motion upon binding, though often these changes are modest. In exceptional circumstances large conformational changes accompany binding. Sometimes these changes prevent side reactions from occurring by bringing reactive groups into proximity only when all reactants are bound. Some enzymes have an active site cleft that permits hingelike bending motions in the absence of bound ligand; the binding of ligand fills the active-site cleft and appears to reduce the freedom of motion. The result is an unfavorable contribution to binding affinity, which must be more than counterbalanced by other, favorable enthalpic or entropic contributions, such as the hydrophobic effect. In a similar manner, the binding of peptides that are unstructured when free in solution, but become ordered when bound, results in an unfavorable entropic binding contribution. The mere introduction of interactions that preorganize the unbound peptide to its bound-state conformation leads to enhanced binding affinity. Concepts such as these are of critical importance in the redesign or de novo design of binding partners.

2.4 Energetics of Protein Binding Interfaces

Protein-protein interface amino acid composition is similar to nonbound surfaces. One might expect that the portions of molecular surfaces that bind partner molecules should have chemical features that distinguish them from nonbinding surfaces. In some chemical sense, perhaps they should be "stickier." Interestingly, this idea appears not to be true, and one cannot generally identify the binding surface of a protein simply on the basis of its composition. In a survey of 75 reversible protein interaction sites (excluding stable interfaces, such as in multimeric proteins and viral capsids), characteristics of surface residues that contacted a binding partner are compared to noncontacting surface residues (table 2.2; [22]). It is noteworthy that protein-protein contact residues are no more likely on average to be hydrophobic or charged than noncontact surface residues. The surveyed complex structures had, on average, 10 bridging hydrogen bonds and 3–50 (average, 18) water molecules trapped at the interface. Additionally, 19% of the contact atoms were from the polypeptide backbone as opposed to side-chain atoms. One notable difference between the two groups is that contact residues are, on average, more aromatic: Tyrosine, tryptophan, phenylalanine, and histidine comprise 21% of contact residues, as opposed to 8% elsewhere on the surface. Despite our poor mechanistic understanding of what comprises a binding patch on a protein surface, data-driven computational approaches, such as machine learning, have been used to predict amino acids implicated in binding [23], indicating a molecular signature unique to such interfaces.

Alanine scanning identifies energetically important contact residues. Structures determined by X-ray crystallography or nuclear magnetic resonance (NMR) spectroscopy clearly identify where two proteins make contact. An example of such

Table 2.2 Relative surface compositions at interfaces versus elsewhere on protein surfaces [22].

Residue character	Interface residue	Noninterface surface
Nonpolar	56%	57%
Neutral polar	29%	24%
Charged polar	15%	19%

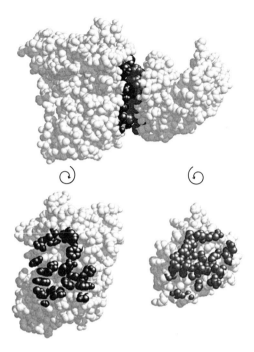

Figure 2.11 The structure of an antibody Fv (left) in complex with an antigen (right), as determined by X-ray crystallography [26]. The antibody in this case is the Fv fragment of mouse monoclonal antibody D1.3, and the antigen is hen egg-white lysozyme.

a structure is shown in figure 2.11. What is not clear from the structure alone, however, is the positive or negative contribution that any of the observed contacts makes to the overall strength of the interaction. A mosaic of interactions at the binding interface includes hydrophobic, polar, and charged residues making complementary interactions. One measure of the energetic contribution of a particular amino acid residue is the loss in binding affinity that results from its removal. This loss is often approximated by mutating the residue of interest to alanine, whose side chain is simply a methyl group [24]. Each binding partner of the antibody-antigen complex shown in figure 2.11 has been *alanine scanned* in this fashion, with the result shown in figure 2.12. Note that many of the contact residues shown in figure 2.11 are readily exchanged to alanine without loss of binding affinity, while a few contribute critically to stabilizing the complex. A large collection of studies of this type indicates that many residues involved in structural contacts at the binding interface can be replaced by alanine with little or no loss in affinity. Only a few contact residues cause large decreases in affinity when replaced with alanine [25]. Typically, a *hot spot* of several side chains, located near the center of the

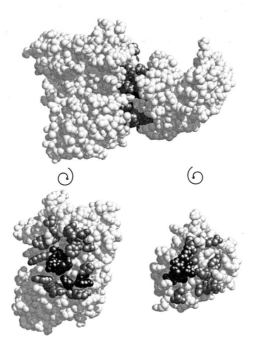

Figure 2.12 The results of an alanine scan performed on the antibody-antigen complex shown in figure 2.11. Individual residues were changed to alanine by site-directed mutagenesis, and the change in binding free energy was determined [27,28]. The darkest gray shade represents a loss of more than 2 kcal/mol of binding free energy; the next lightest shade a loss of 1–2 kcal/mol; and the next lightest shade a loss of less than 1 kcal/mol. All atoms within 5.0 Å of the binding partner are shaded darker, and in the bottom portion of the figure, the antibody and antigen are rotated to display the contact surface.

protein-protein interface, causes loss of much of the binding strength when each residue is individually substituted with alanine. Mutation to amino acids other than alanine can provide additional insight into the nature of the interaction.

Thus, structural examination of protein binding interfaces reveals that surfaces engaged in binding appear remarkably similar in character to nonbinding surfaces. Moreover, systematic trimming of amino acid side-chain atoms through alanine scanning reveals significant heterogeneity in the magnitude of the resulting energetic effects; many side chains contribute small binding effects, with only a few yielding very large effects. However, it should also be noted that affinity is not the be-all, end-all of protein function in the crowded intracellular milieu. Binding selectivity is also essential to ensure faithful cellular processes, so some amino acid residues that do not contribute significantly to the desired target affinity may nevertheless contribute to selectivity by minimizing undesired interactions with the thousands of other biomolecules present in the cell.

2.4.1 Thermodynamic Cycle Analysis

A useful tool for the analysis of the effect of structural changes on binding interactions employs thermodynamic cycles. These representations of relationships between chemical reactions are valuable, because they properly account for contributions to reactivity; other approaches are prone to omitting some effects and double-counting others.

Figure 2.13 depicts a thermodynamic cycle for the analysis of binding free energy changes due to mutating phenylalanine to alanine. Alanine scanning mutagenesis is applied in an attempt to find the primary determinants of binding by systematically performing a series of such mutations to alanine. The horizontal reactions in the figure represent the two binding events—the top reaction for wild type and the bottom for the mutant, with corresponding free energy changes $\Delta G_{bind}^{o,wt}$ and $\Delta G_{bind}^{o,mut}$. It is useful to consider the vertical reactions, in which a phenylalanine side chain is changed into an alanine, in either the bound ($\Delta G_{mutate}^{o,bnd}$) or unbound ($\Delta G_{mutate}^{o,unb}$) state. These reactions are unusual in that they do not even represent balanced chemical

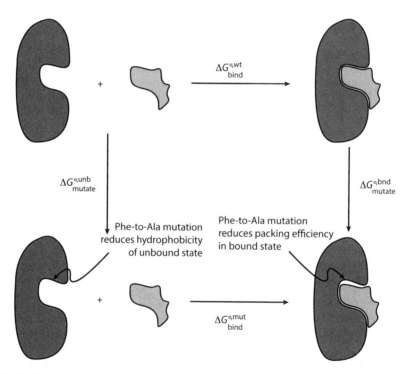

Figure 2.13 Thermodynamic cycle for studying the binding affinity difference for wild-type and a mutant protein associating with ligand. The two horizontal reactions represent the binding for wild-type (top) and mutant (bottom) complexes. The vertical arrows represent the free energy for mutating the protein from wild type to mutant in the unbound (left) and bound (right) states.

changes and thus could be termed alchemical. One consequence of free energy being a thermodynamic function of state is that the free energy for a sequence of steps that returns back to the starting state must be zero. For the current case, the total free energy change for traversing around the cycle must be zero:

$$\Delta G_{\text{bind}}^{\circ,\text{wt}} + \Delta G_{\text{mutate}}^{\circ,\text{bnd}} - \Delta G_{\text{bind}}^{\circ,\text{mut}} - \Delta G_{\text{mutate}}^{\circ,\text{unb}} = 0 \quad (2.38)$$

or, equivalently,

$$\Delta\Delta G^{\circ} = \Delta G_{\text{bind}}^{\circ,\text{mut}} - \Delta G_{\text{bind}}^{\circ,\text{wt}} = \Delta G_{\text{mutate}}^{\circ,\text{bnd}} - \Delta G_{\text{mutate}}^{\circ,\text{unb}} \quad (2.39)$$

where $\Delta\Delta G^{\circ}$ represents the difference in binding free energy due to the mutation to alanine, and equation 2.39 indicates that the free energy difference for the two horizontal reactions equals that for the vertical reactions in figure 2.13. Conceptual analysis of the horizontal reactions, which are the actual experimental observables, is difficult, because each represents tremendous complexity, including stripping solvent molecules from the binding interface, possible titration changes and conformational changes of binding partners, and intermolecular packing interactions introduced in the bound state. In many cases, nearly all of these complex changes are essentially the same for wild type and mutant—they virtually cancel and don't contribute to the overall $\Delta\Delta G^{\circ}$.

The vertical reactions—which equation 2.39 demonstrates are equally relevant for analyzing the contributions to the binding free energy change—focus attention on changes due to the phenylalanine-to-alanine mutation in the bound and unbound states, which automatically ignores many of the interactions that cancel in the overall effect. In this case, the analysis points to less efficient packing interactions for alanine versus phenylalanine at the binding interface in the bound state and greater exposure of hydrophobic surface area in the unbound state for phenylalanine over alanine. Thus, one expects a decreased binding affinity for the mutant due to smaller hydrophobic and packing contributions for alanine relative to phenylalanine. The use of a thermodynamic cycle is an efficient way to focus attention on the key contributions to binding. Here, the cycle was used to compare binding affinity of wild-type and mutant proteins. A very similar cycle is used in structure-based drug design efforts to study the effect of variations in ligand structure on binding affinity.

Thermodynamic cycles are a useful tool to analyze a wide variety of problems in biochemical equilibria. Protonation of functional groups can have significant effects on protein binding. The thermodynamic cycle in figure 2.14 provides a framework for studying pH-dependent binding events. The horizontal reactions correspond to binding of a neutral ($\Delta G_{\text{bind}}^{\circ,L}$) and protonated ($\Delta G_{\text{bind}}^{\circ,LH^+}$) ligand to its receptor,

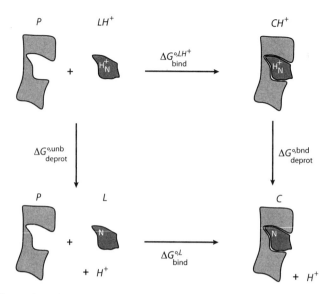

Figure 2.14 Thermodynamic cycle for studying the binding affinity for different protonation states of the same ligand. The two horizontal reactions represent the binding for protonated (top) and unprotonated (bottom) ligand. The vertical arrows represent the free energy for titrating the ligand in the unbound (left) and bound (right) states. The symbols L and LH^+ represent the unprotonated and protonated ligand, respectively; C and CH^+ represent unprotonated and protonated complex, respectively.

whereas the vertical reactions correspond to deprotonation of the ligand in the unbound ($\Delta G^{\mathrm{o,unb}}_{\mathrm{deprot}}$) and bound ($\Delta G^{\mathrm{o,bnd}}_{\mathrm{deprot}}$) states. Both types of reactions are balanced chemical reactions; neither is alchemical.

Thermodynamic cycle analysis requires

$$\Delta\Delta G^{\circ} = \Delta G^{\mathrm{o},LH^+}_{\mathrm{bind}} - \Delta G^{\mathrm{o},L}_{\mathrm{bind}} = \Delta G^{\mathrm{o,unb}}_{\mathrm{deprot}} - \Delta G^{\mathrm{o,bnd}}_{\mathrm{deprot}} \qquad (2.40)$$

One important insight that results directly from equation 2.40 is that differential affinity of the protonated and unprotonated forms requires a different $\Delta G^{\circ}_{\mathrm{deprot}}$ in the bound and unbound states. Another way of stating this, shown further below, is that the pK_a for the ligand must be different in the bound and unbound states. Moreover, equation 2.40 states that the magnitude of the binding affinity difference for the protonated and unprotonated forms of ligand is directly related to the pK_a difference.

In some situations, it is appropriate to consider the binding of one pure protonation state of ligand to form one pure protonation state of complex. If the bound and unbound ligand are both protonated (or both unprotonated), the binding reaction and free energy change are given by one of the horizontal arrows in figure 2.14. If the bound and unbound ligands are of different titration states, then two legs of

Noncovalent Binding Interactions

the thermodynamic cycle can be used to study the binding reaction. Imagine that the unbound ligand is protonated but the bound ligand is unprotonated. The free energy change corresponds to that between the upper-left and lower-right corners of the figure. This can be obtained by summing the left vertical reaction with the lower horizontal reaction (or by summing the other two reactions, because they form a thermodynamic cycle).

For intermediate pH values, the reactant ligand and product complex may be a mixture of protonated and unprotonated forms, and the thermodynamic cycle enables analysis of this case as well. It is useful to consider the apparent equilibrium constant, which resembles the usual binding equilibrium constant, except that for free ligand and bound complex, we consider the sum of unprotonated and protonated forms:

$$K_{bind}^{apparent} = \frac{[C] + [CH^+]}{[P]\left([L] + [LH^+]\right)} \tag{2.41}$$

Binding equilibria can be described by equilibrium association or dissociation constants and their corresponding $\Delta G°$ values. For consistency with convention for this topic, we consider equilibrium association constants. This expression can be expanded and re-arranged:

$$K_{bind}^{apparent} = \left(\frac{[C]}{[P]} + \frac{[CH^+]}{[P]}\right)\left(\frac{1}{[L] + [LH^+]}\right) \tag{2.42}$$

$$= \left(K_{bind}^{L}[L] + K_{bind}^{LH^+}[LH^+]\right)\left(\frac{1}{[L] + [LH^+]}\right) \tag{2.43}$$

$$= K_{bind}^{L}\left(1 + \frac{K_{bind}^{LH^+}[LH^+]}{K_{bind}^{L}[L]}\right)\left(\frac{[L]}{[L] + [LH^+]}\right) \tag{2.44}$$

where the equilibrium constants in equation 2.43 have been substituted for the equilibrium concentrations of equation 2.42 and correspond to the equilibria indicated in figure 2.14. Defining α as the fraction of free ligand that is protonated (equation 2.45) and using the corresponding equilibrium from the figure (where K_{deprot}^{unb} is an equilibrium *dissociation* constant) yields

$$\alpha = \frac{[LH^+]}{[L] + [LH^+]} \tag{2.45}$$

$$= \frac{[H^+]}{K_{deprot}^{unb} + [H^+]} \tag{2.46}$$

Equation 2.44 can be simplified:

$$K_{\text{bind}}^{\text{apparent}} = K_{\text{bind}}^{L}\left[1 + \left(\frac{K_{\text{bind}}^{LH^+}}{K_{\text{bind}}^{L}}\right)\left(\frac{\alpha}{1-\alpha}\right)\right](1-\alpha) \qquad (2.47)$$

$$= K_{\text{bind}}^{L}\left[1 + \alpha\left(\frac{K_{\text{bind}}^{LH^+}}{K_{\text{bind}}^{L}} - 1\right)\right] \qquad (2.48)$$

The pH dependence of the binding can be made explicit through α using equation 2.46,

$$K_{\text{bind}}^{\text{apparent}} = K_{\text{bind}}^{L}\left[1 + \left(\frac{[H^+]}{K_{\text{deprot}}^{\text{unb}} + [H^+]}\right)\left(\frac{K_{\text{bind}}^{LH^+}}{K_{\text{bind}}^{L}} - 1\right)\right] \qquad (2.49)$$

and noting the relationship to the pK_a of the unbound ligand and the pH:

$$\text{p}K_a^{\text{unb}} = -\log_{10} K_{\text{deprot}}^{\text{unb}} \qquad (2.50)$$

$$\text{pH} = -\log_{10}[H^+] \qquad (2.51)$$

Equation 2.49 is a useful representation of the pH dependence of binding for the case of a single site that titrates. It has the proper limits observed in the previous qualitative analysis. If the pH is high (corresponding to $[H^+] \ll K_{\text{deprot}}^{\text{unb}}$), then the unbound and bound states will be unprotonated, $\alpha \approx 0$, and $K_{\text{bind}}^{\text{apparent}} \approx K_{\text{bind}}^{L}$, the equilibrium constant for unprotonated ligand binding protein to form unprotonated complex. At the other extreme, if the pH is low ($[H^+] \gg K_{\text{deprot}}^{\text{unb}}$), then the unbound and bound states will be protonated; $\alpha \approx 1$; and $K_{\text{bind}}^{\text{apparent}} \approx K_{\text{bind}}^{LH^+}$, the equilibrium constant for protonated ligand binding protein to form protonated complex.

To examine the case of intermediate protonation states, it is useful to rearrange equation 2.49:

$$K_{\text{bind}}^{\text{apparent}} = K_{\text{bind}}^{L}\left(\frac{1 + \frac{[H^+]}{K_{\text{deprot}}^{\text{bnd}}}}{1 + \frac{[H^+]}{K_{\text{deprot}}^{\text{unb}}}}\right) = K_{\text{bind}}^{LH^+}\left(\frac{K_{\text{deprot}}^{\text{bnd}} + [H^+]}{K_{\text{deprot}}^{\text{unb}} + [H^+]}\right) \qquad (2.52)$$

where $K_{\text{deprot}}^{\text{bnd}} = K_{\text{deprot}}^{\text{unb}} \cdot K_{\text{bind}}^{L}/K_{\text{bind}}^{LH^+}$, which is a constraint imposed by the thermodynamic cycle (combining equations 2.40 and 2.12).

If $[H^+]$ is between the K_{deprot} values for the bound and unbound states, then the result is pH-dependent binding. For example, if the unbound ligand is protonated but the bound complex is unprotonated, then $K_{\text{deprot}}^{\text{unb}} \ll [H^+] \ll K_{\text{deprot}}^{\text{bnd}}$ and $K_{\text{bind}}^{\text{apparent}} \approx \frac{K_{\text{bind}}^{L} K_{\text{deprot}}^{\text{unb}}}{[H^+]} = \frac{K_{\text{bind}}^{LH^+} K_{\text{deprot}}^{\text{bnd}}}{[H^+]}$. This equates the apparent binding constant with either traversing the left vertical reaction followed by the bottom horizontal reaction in figure 2.14 $\left(\frac{K_{\text{bind}}^{L} K_{\text{deprot}}^{\text{unb}}}{[H^+]} \right)$, or equivalently, with traversing the top horizontal reaction followed by the right vertical reaction $\left(\frac{K_{\text{bind}}^{LH^+} K_{\text{deprot}}^{\text{bnd}}}{[H^+]} \right)$.

Case Study 2-1   P. J. Carter, G. Winter, A. J. Wilkinson, and A. R. Fersht. The use of double mutants to detect structural changes in the active site of the tyrosyl-tRNA synthetase (*Bacillus stearothermophilus*). *Cell* 38: 835–840 (1984) [29].

Thermodynamic cycles have been applied to understanding relationships between protein structure and reactivity, including binding, folding, and catalysis. Although single mutants can indicate the relative importance of individual protein residues, an understanding of *how* significant residues exert their role can be more elusive. More complex thermodynamic cycles involving multiple mutants can be used for this purpose. The most popular version involves pairs of mutations and is termed *double mutant cycle analysis*.

The main idea for using double mutant cycles is illustrated in figure 2.15 (upper panel) for the case of protein stability. The individual effect of two single mutants can be measured by their effect on the unfolding free energy. The figure indicates a three-dimensional thermodynamic cube whose top represents the thermodynamic cycle for the mutation a → A, and whose back represents the thermodynamic cycle for b → B (lowercase letters indicate wild-type residues; uppercase letters indicate mutant). The cycle on the bottom of the cube gives the effect of introducing the a → A mutation into the background with the B mutation, and the difference between the $\Delta\Delta G°$ for the top and bottom cycles gives the effect of one mutation on the other. If the two mutations are completely independent, the effect of one will be unchanged by the presence of the other, and the two $\Delta\Delta G°$ values will be the same. Likewise, the cycle on the front face of the cube examines the effect of introducing the b → B mutation into the A background. The requirement that the free energy change for traversing a cycle that returns to its start is zero guarantees that the differences between the $\Delta\Delta G°$ values are the same for any pair of parallel faces of the cube. Thus, the effect of the b → B mutation on a → A is the same as the effect of A on B and is the $\Delta\Delta\Delta G°$ for the cycle.

Because the cube representing the cycle can be unwieldy, a two-dimensional projection can be used to convey the same information more compactly. The lower panel of the figure indicates such a projection (viewed from beyond the cube's left face). Each corner of the projected cycle represents the unfolding free energy for the wild type, one of the two single mutants, or the double mutant. Each arrow represents a $\Delta\Delta G°$, and the $\Delta\Delta\Delta G°$ for the cycle is indicated by the diagonal arrow.

The power of double mutant cycles is their ability to detect mutual effects between positions in a protein or complex. The existence of a significant $\Delta\Delta\Delta G°$ indicates an energetic interaction between the positions observed through the protein unfolding reaction, which

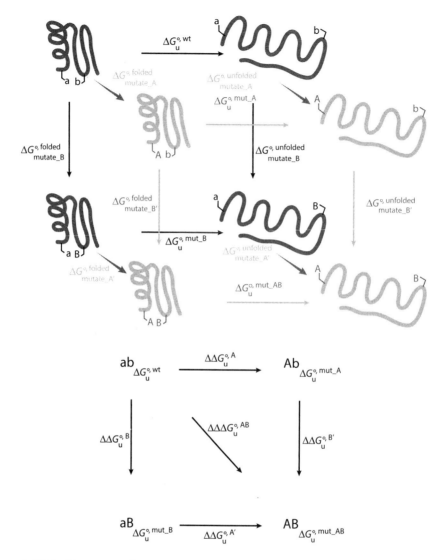

Figure 2.15 Double mutant cycle analysis illustrating the interaction of two mutations on protein stability. The upper panel represents six thermodynamic cycles that together form the six faces of a cube. The top face probes the role of the mutation at position a (the mutation a→A) on protein stability in the context of the wild type; the back face examines the corresponding mutation at position b. The bottom face examines the role of a in the context of mutant B, while the front face examines the role of b in the context of mutant A. The lower panel shows a projection, looking at the left face of the cube. Each of the four corners represents the stability of the wild type or one of the mutants; the horizontal and vertical arrows represent the $\Delta\Delta G°$ values due to mutation; and the diagonal arrow represents the coupling energy, a $\Delta\Delta\Delta G°$.

is in addition to the effect of each mutation individually on protein unfolding. Sometimes, the two residues can be seen to interact directly in protein structures. Other times the two residues contribute to a third factor, such as stabilizing a conformational change, and exert their cooperative effect without direct interaction.

Case Study 2-2   Z. S. Hendsch and B. Tidor. Do salt bridges stabilize proteins? A continuum electrostatic analysis. *Protein Sci.* **3: 211–226 (1994) [30].**

A survey of high-resolution protein structures reveals a significant number of *ion pairs*—a positively and a negatively charged side chain nearby in the folded structure. Often the charged portions are close enough to make at least one hydrogen bond, in which case the interaction is called a salt bridge. On average, about one buried or partially buried salt bridge occurs per medium-sized protein domain [31]. In the unfolded protein, ionic side chains are expected to be exposed to solvent, where they may make extremely strong interactions. In the folded protein, their burial at least partially removes the side chains from solvent. Strong interactions are made between the bridging side chains, as well as to other polar and possibly charged functional groups in the region. Thus, in the process of protein folding, the ion-pairing side chains lose solvent interactions but gain protein interactions. One approximate measure of the upper bound on the strength of solvent interactions lost is the free energy of transfer from water to the gas phase of amino acid side-chain analogs (see table 2.3). These values are on the order of 65–75 kcal/mol for each member of an ion pair.

An important question regarding the fundamental energetics and forces of protein interactions is whether salt bridges stabilize proteins electrostatically. To address this question, Hendsch and Tidor used computational modeling to quantify three electrostatic contributions of salt-bridging side chains toward folding [30]. Using a database of 21 salt bridges, the average cost of desolvating the pair of ionic side chains was computed to be +11.7 kcal/mol. This is significantly less than the ≈140 kcal/mol that would result from individually desolvating the side chains into vacuum, because the side chains, even when completely buried, can interact favorably with solvent through long-range electrostatic effects and because the interior of a protein is not as low a dielectric environment ($\varepsilon \approx 2$–4) as vacuum ($\varepsilon = 1$). The favorable interaction between the salt-bridging pair was computed to average 5.0 kcal/mol, and the "bystander" interaction with other polar and charged groups was computed to contribute an additional favorable 3.2 kcal/mol (see figure 2.16). The unfolded protein is in the upper left corner, and the folded protein is in the lower right. A salt bridge is formed in the folded protein by the two charged side chains indicated. $\Delta\Delta G^\circ_{total}$ is the continuum electrostatic contribution of the two charged side chains shown to the free energy change. It (as well as the other thermodynamic quantities in the cycle) is termed a $\Delta\Delta G^\circ$, because its connection to experiments is through the difference of two measurements: one with the charged side chains as they are, and the other in which they are replaced by their hydrophobic isosteres (but all structures remain the same). $\Delta\Delta G^\circ_{total}$ is the sum of $\Delta\Delta G^\circ_{solv}$, $\Delta\Delta G^\circ_{bridge}$, and $\Delta\Delta G^\circ_{prot}$.

Although the results vary greatly (see table 2.3), on average, a significant portion (roughly 40%) of the recovered interaction free energy came from "bystander" interactions. Moreover, of the 21 salt bridges, 17 were computed to destabilize the protein

Table 2.3 Electrostatic contribution of salt bridges to protein stability.[a]

Salt bridge	% Burial	$\Delta\Delta G°_{total}$	$\Delta\Delta G°_{solv}$	$\Delta\Delta G°_{bridge}$	$\Delta\Delta G°_{prot}$
T4 Lysozyme [2lzm]					
E11–R145	88	10.13 (1.03)	18.66 (1.47)	−7.72 (0.82)	−0.81 (0.15)
D10–R148	88	6.24 (1.01)	18.64 (1.01)	−8.84 (0.16)	−3.57 (0.18)
H31–D70	75	3.46 (0.26)	11.62 (0.35)	−7.91 (0.21)	−0.26 (0.09)
Lambda repressor (dimer; N-terminal domain)					
D14c–R17c	83	3.79 (0.51)	15.26 (0.72)	−6.24 (0.26)	−5.23 (0.39)
D14d–R17d	86	2.86 (0.80)	15.53 (1.06)	−7.17 (0.62)	−5.50 (0.20)
K67c–E89d	65	2.89 (1.14)	9.50 (1.22)	−4.36 (0.23)	−2.26 (0.19)
K67d–E89c	66	5.14 (1.52)	11.38 (1.45)	−2.64 (0.07)	−3.60 (0.22)
Ubiquitin [1ubq]					
K11–E34	66	4.09 (0.46)	7.63 (0.56)	−3.07 (0.13)	−0.46 (0.02)
Uteroglobin [1utg] (dimer)					
E22a–K35a	37	−1.10 (0.43)	3.31 (0.50)	−3.37 (0.37)	−1.03 (0.02)
E22b–K35b	37	−1.03 (0.68)	3.41 (0.75)	−3.42 (0.30)	−1.03 (0.02)
K42a–D46a	93	4.72 (0.35)	23.79 (0.98)	−10.52 (0.48)	−8.54 (0.38)
K42b–D46b	93	4.48 (0.57)	23.28 (1.39)	−10.40 (0.69)	−8.41 (0.50)
GCN4 leucine zipper					
K15b–E20a	52	2.57 (0.50)	4.21 (0.59)	−1.73 (0.12)	0.08 (0.05)
E22b–K27a	66	3.61 (0.72)	5.62 (0.79)	−1.75 (0.08)	−0.26 (0.05)
E22a–K27b	69	4.73 (0.68)	8.60 (0.65)	−1.87 (0.07)	−2.00 (0.20)
E22a–R25a	64	5.76 (0.47)	9.18 (0.49)	−3.79 (0.11)	0.36 (0.11)
K8a–E11a	49	3.67 (0.67)	6.83 (0.63)	−2.90 (0.07)	−0.26 (0.05)
Hen egg white lysozyme [6lyz]					
D48–R61	58	0.09 (0.49)	8.42 (0.80)	−1.29 (0.11)	−7.04 (0.35)
Erabutoxin [5ebx]					
K27–E38	58	4.12 (0.70)	7.91 (0.72)	−2.33 (0.07)	−1.46 (0.10)
434 Repressor [1r69] (R1–69)					
R10–E35	92	−0.43 (0.65)	19.68 (0.56)	−10.66 (0.16)	−9.45 (0.12)
Barnase					
D12–R110	68	3.03 (0.70)	13.39 (0.93)	−4.15 (0.15)	−6.21 (0.29)
Average	69	3.47 (0.68)	11.71 (0.84)	−5.05 (0.25)	−3.19 (0.18)

[a] All free energy values are in kcal/mol. A positive value indicates that the salt bridge destabilizes the folding protein electrostatically. Values in parentheses are standard deviations of ten runs translated relative to one another on the finite-differences grid. The line marked "Average" contains the average of each column, even though five pairs of bridges are related by symmetry or pseudo-symmetry. This table reproduced from reference 30.

electrostatically, because the desolvation penalty was not sufficiently compensated by interactions made in the folded state. Thus, even though the interaction between salt-bridging groups may be attractive, these calculations suggest that the net effect on protein stability need not be favorable.

One prediction from these results is that the replacement of salt-bridging side chains with hydrophobic groups of roughly the same size and shape could lead to more stable

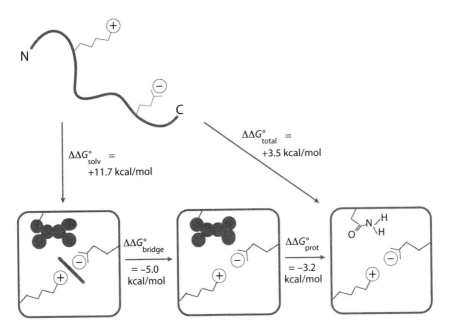

Figure 2.16 Thermodynamic cycle used to analyze salt bridges. In the lower left of the figure is the folded molecule, in which all polar and charged groups in the protein (except for the two charged side chains) have had their partial atomic charges turned off (indicated by shading of a representative side chain). The two charged side chains do not interact electrostatically (indicated by the diagonal bar). In the lower center is a similar representation, but the interaction between the two charged side chains is restored (no diagonal bar). In the lower right (folded state), the interactions with other charged and polar groups in the protein are also restored (no shading). Not shown below the plane of the figure is an identical cycle, but with the salt-bridge side chains replaced by hydrophobic isosteres. Each $\Delta\Delta G°$ in the current figure represents the difference of two $\Delta G°$ values (one in this plane and the corresponding one in the hydrophobic plane below) [30].

proteins. Experiments in this vein have now been carried out and are consistent with this prediction. One of the earliest was reported by R. T. Sauer and co-workers at MIT, who carried out an elegant combinatorial experiment in which each member of a salt-bridge triad (Arg31, Glu36, and Arg40) in Arc repressor was replaced with all 20 amino acids. A selection was performed to identify variants (of the 8,000 possibilities) that were as stable, or more stable, than wild type. The only fully active isolates recovered were wild type and mutants in which all three positions were replaced by hydrophobes. When some of these mutant isolates were purified, they were found to be 1–2 kcal/mol per monomer more stable than wild type [32]. S. H. White and co-workers at the University of California, Irvine, used a peptide model system to measure the free energy of salt-bridge formation in solution. Their findings, that "salt-bridged charge pairs do not, and probably cannot, provide as much stabilization to proteins as a pair of hydrophobic residues of about the same volume," further support the notion that salt bridges do not necessarily contribute to the electrostatic stability of proteins [33].

The observation that many salt bridges don't appear to contribute to the stability of proteins raises two interesting questions: (1) Why do proteins have salt bridges and what role do they play? (2) Although salt bridges aren't necessarily automatically stabilizing, might they sometimes be? Although the desolvation penalty is often not recovered in salt bridges, the penalty for burying a pair of charges and not satisfying them is much worse. Thus, the need to either leave all charged groups on the surface or bury them in pairs with hydrogen-bonding partners adds strong geometric constraints on the folding. Some lines of evidence suggest that these constraints can have an important role in contributing specificity and uniqueness to the folded protein. Proteins from organisms that live at elevated temperature (thermophiles) are under selective pressure to improve the stability of their proteins over that of their mesophilic counterparts. Comparison of homologous proteins indicates some cases where the thermophile has more salt bridges. In a few cases, calculations have shown that these salt bridges contribute favorably to the thermodynamic stability of the protein [34].

2.5 Environmental Impacts on Binding Rate

2.5.1 Arrhenius Relationship and Transition State Theory

In warm-blooded animals (and their parasites, such as *E. coli*), the great majority of chemical reactions occur at 37°C. However, often measurements are made in vitro at room temperature (20–25°C) or on ice (0–4°C). How do biochemical rates vary as a function of temperature?

It has been found empirically that most simple chemical reactions depend on temperature by the following equation, called an Arrhenius relationship:

$$k = Ae^{\frac{-E_a}{k_B T}} \qquad (2.53)$$

where k is a rate constant, A is a pre-exponential factor, E_a is the activation energy, T is absolute temperature (often in kelvins), and k_B is the Boltzmann constant (1.38×10^{-23} J/K). For the range of temperatures in which a protein is stably folded, its catalytic and binding reactions are generally found to follow Arrhenius dependence on temperature (figure 2.17). Single-step catalytic and binding processes are shown to follow equation 2.53 in the top two panels. Perhaps more surprisingly, the multistep process of ribosomal protein synthesis, and even overall cell growth rate, are shown to exhibit Arrhenius temperature dependence in the bottom two panels. At temperatures above 37°C, bacterial growth rate increases no further, possibly due in part to the destabilization of enzymes important for cell growth; however, bacteria also mount a specific response to the stress of high temperatures, a component of which is to decrease biosynthesis. Furthermore, even the behaviors of whole organisms have often been found to follow Arrhenius temperature relationships (figure 2.18), an observation that in the early part of the twentieth century helped support the concept of continuity of chemical principles between physical and biological systems.

Noncovalent Binding Interactions

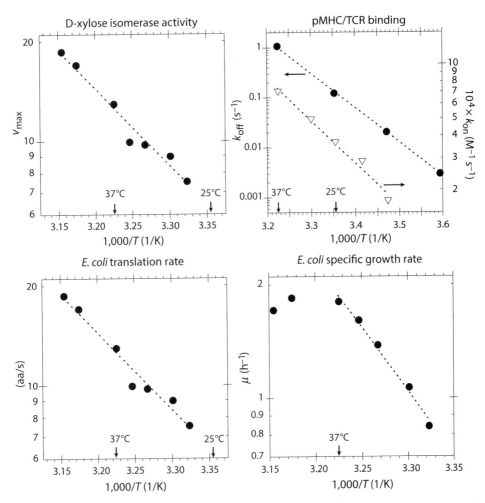

Figure 2.17 Biological reactions generally exhibit a dependence on temperature well described by equation 2.53. The upper-left plot shows the enzyme-catalyzed rate of glucose isomerization to fructose versus temperature [36]. This enzyme is used industrially for high-fructose corn syrup production. The upper-right plot shows the binding and dissociation rate constants for the interaction of soluble T-cell receptors and peptide/MHC complexes [35]. The lower-left plot shows the rate of ribosomal peptide elongation in *E. coli* as a function of temperature [37]. The lower-right plot shows the specific growth rate of *E. coli* as a function of temperature [37]. 37°C and 25°C are marked on the x-axes for reference.

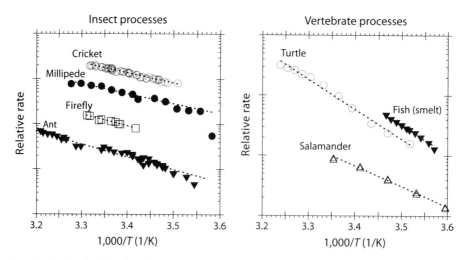

Figure 2.18 Complex behaviors in cold-blooded animals, both invertebrate (left panel) and vertebrate (right panel), often follow an Arrhenius dependence of rates on temperature. On the left are plotted the rates of cricket chirping (chirps/min; $E_a = 10.2$ kcal/mol) [38]; millipede crawling (cm/min; $E_a = 10.0$ kcal/mol) [39]; firefly flashing (flashes/min; $E_a = 11.5$ kcal/mol) [40]; and ant crawling (cm/s; $E_a = 12.9$ kcal/mol) [41]. On the right are a variety of vertebrate rate processes found to obey Arrhenius kinetics: growth rate prior to hatching of a particular fish species ($E_a = 21.4$ kcal/mol) [42]; heartbeat frequency of a salamander (beats/s; $E_a = 15.5$ kcal/mol) [43]; and eye flicker response rate of a turtle (reciprocal of critical light flash intensity, $E_a = 22.7$ kcal/mol) [44].

An interesting rule of thumb is illustrated by the following data: A temperature shift from 25°C to 35°C increases the *E. coli* specific growth rate 2.3-fold, translation rate 1.8-fold, D-xylose isomerase activity 2.2-fold, pMHC/TCR association 1.8-fold, and pMHC/TCR dissociation 5.6-fold (this latter reaction was noteworthy for its strong temperature dependence [35]). In other words, an increase in temperature of 10°C approximately doubles the rate of many reactions of biological interest in this relevant temperature range.

There is some theoretical justification for the particular mathematical form of the Arrhenius dependence on temperature. If the reaction of interest consists of first forming a high-energy intermediary, or *transition state*, which is readily converted to either products or reactants, the progress of the reaction can be schematically represented by an energy diagram as shown in figure 2.19.

The reaction in figure 2.19 is imagined to occur along a one-dimensional path characterized by a reaction coordinate that increases as the reaction progresses. Whereas reactants $L + P$ and product C correspond to low-energy states at the start and end of the reaction coordinate axis, respectively, at some intermediate value of the reaction coordinate, the energy reaches a peak. The chemical species corresponding to the energy peak is an unstable entity with sufficient energy to

Noncovalent Binding Interactions

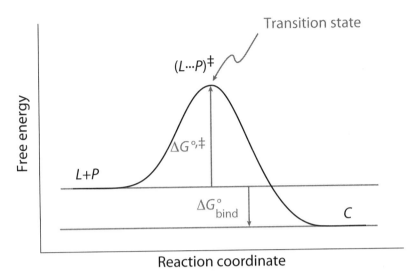

Figure 2.19 The free energy reaction profile for ligand (L) and binding protein (P) to pass through a transition state $(L\cdots P)^\ddagger$ and go on to form a complex (C) is illustrated schematically. The highest free energy intermediate along the reaction coordinate is called the transition state. The free energy difference between reactants and transition state is indicated by $\Delta G^{\circ,\ddagger}$, and the overall binding free energy difference is indicated by ΔG°_{bind}.

react, called an activated complex or a transition state. The resulting transition state theory that is built on this diagram is a useful approximation for describing reaction dynamics, particularly in solution. The key assumption in this treatment is that the transition state is modeled as participating in an equilibrium with the reactants. It is typical to indicate the transition-state species and its properties with a superscript double dagger symbol $\ddagger$. Thus, the transition state itself is $(L\cdots P)^\ddagger$, and its free energy above the ground state reactants is $\Delta G^{\circ,\ddagger}$.

The concentration of the transition state can be expressed in terms of the reactant concentrations and the free energy of activation using the pseudo-equilibrium expression,

$$[(L\cdots P)^\ddagger] = \frac{[L][P]}{K_d^\ddagger} \quad (2.54)$$

$$= [L][P]e^{\frac{-\Delta G^{\circ,\ddagger}}{k_B T}} \quad (2.55)$$

The overall reaction rate is related to the rate at which the activated complex decays to the product C. This is typically expressed as the product of two terms. The first is the rate of decay of the activated complex, whose derivation is beyond the scope of the current treatment but is generally expressed as $k_B T/h$, where h is Planck's constant; the second is the fraction of decays that continue forward over the reaction barrier to form products, as opposed to re-creating reactants, which is expressed as a transmission coefficient κ ($0 \leq \kappa \leq 1$) and is often approximated as the integer 1. Thus, the overall reaction rate is

$$\frac{d[C]}{dt} = \kappa \left(\frac{k_B T}{h}\right) e^{\frac{-\Delta G^{\circ,\ddagger}}{k_B T}} [L][P] \qquad (2.56)$$

and allows the association of the prefactor with the forward rate constant:

$$k_{on} = \kappa \left(\frac{k_B T}{h}\right) e^{\frac{-\Delta G^{\circ,\ddagger}}{k_B T}} \qquad (2.57)$$

Equation 2.57 is often referred to as the *Eyring equation*, though it was also separately and almost simultaneously developed by Evans and Polanyi. (When developing his theory, Eyring intended to use the then-traditional asterisk to denote the activated complex, but his assistant mistyped a double dagger instead, and that has been used ever since [45].)

It is particularly informative to compare the Arrhenius expression of equation 2.53 with the transition state theory result of equation 2.57. To do this, we make the substitution $\Delta G^{\circ,\ddagger} = \Delta H^{\circ,\ddagger} - T\Delta S^{\circ,\ddagger}$ in the latter, which gives

$$k_{on} = \kappa \left(\frac{k_B T}{h}\right) e^{\frac{\Delta S^{\circ,\ddagger}}{k_B}} e^{\frac{-\Delta H^{\circ,\ddagger}}{k_B T}} \qquad (2.58)$$

Taking the temperature derivative of the natural logarithm of both sides yields

$$\frac{d \ln(k_{on})}{dT} = \frac{1}{T} + \frac{\Delta H^{\circ,\ddagger}}{k_B T^2} \qquad (2.59)$$

$$= \frac{\Delta H^{\circ,\ddagger} + k_B T}{k_B T^2} \qquad (2.60)$$

under the assumption that κ, $\Delta H^{\circ,\ddagger}$, and $\Delta S^{\circ,\ddagger}$ are independent of temperature. This assumption varies in quality but enables a useful analytical evaluation of the activation energy.

Comparing with the temperature derivative of the natural logarithm of equation 2.53 permits the association,

$$E_a = \Delta H^{\circ,\ddagger} + k_B T \qquad (2.61)$$

Thus, the phenomenological observations of the Arrhenius equation and the associated activation energy E_a can be related to thermodynamic quantities describing the enthalpy difference between the reactants and the transition state in activated state theory, which is a notable confluence of observation and simplified theory.

2.5.2 Electrostatic Effects on Binding Kinetics

Net attractive electrostatic interactions can accelerate binding interactions by steering molecules toward binding configurations more readily than would occur purely by collisions through random diffusive motions. This steering can continue at close range, drawing two binding interfaces closer into the correct orientation to form a stable interaction. It is also possible, though less common, for electrostatic steering effects to be repulsive and result in slower binding. Because the strength of electrostatic interactions between approaching binding partners can be screened by salt ions in solution, one method of detecting and quantifying electrostatic steering effects is to study binding kinetics as a function of salt concentration. In this section, we develop some of the theory and present key results framing the effects of long-range electrostatic interactions and their diminution through ionic screening.

The presence of salt ions in solution can significantly affect interactions between macromolecules in two general ways:

1. Nonspecific ionic screening diminishes the strength of electrostatic interactions between charges, be they attractive or repulsive. This occurs because the intervening ions in solution arrange themselves to interact with the fixed charge distribution of the macromolecules, which effectively results in a weakening of the electrostatic interactions between fixed charges.
2. Specific binding ions may bind persistently at particular sites on proteins and act essentially as an extension or modification of the protein itself.

The theory for nonspecific effects has been well worked out and will be described here. Specific-ion effects can be modeled together with protein structure where they are known to exist. Generally, nonspecific effects are independent of ionic character (that is, NaCl behaves the same as KCl and NaBr), whereas specific-binding effects due either to sodium or chloride would be modified by substituting another 1:1 salt to replace the specific-binding ion.

Ionic strength Nonspecific ionic screening can be described by an analysis first developed by Debye and Hückel in a classic 1923 paper [46]. In this model, nonspecific ions in solution are assumed to have zero volume and lack correlation among their positions. Ions fixed at particular sites are not accounted for. A detailed derivation of this model is beyond the scope of this text, but we will sketch part of the argument and highlight key results.

Imagine an ion of radius a fixed at the origin of coordinates. The fundamental electrostatic law describing the potential of the system outside the ion is the Poisson-Boltzmann equation, which for this case can be written,

$$\nabla^2 \phi(\vec{r}) = \kappa^2 \phi(\vec{r}) \quad \text{for } |\vec{r}| \geq a \tag{2.62}$$

where ϕ is the electrostatic potential; κ is the inverse Debye length given by

$$\kappa^2 = \frac{e^2}{\varepsilon k_B T} I \tag{2.63}$$

where e is the magnitude of the charge of an electron; ε is the dielectric constant of the medium in the absence of salt; k_B is the Boltzmann constant; T is the absolute temperature; and I is the ionic strength, defined as

$$I \equiv \frac{1}{2} \sum_i m_i z_i^2 \tag{2.64}$$

The summation i in equation 2.64 runs over all the ionic species (for example, if the only salt in the solution is NaCl, then i runs over Na^+ and Cl^-), m_i is the molality of ionic species i, and z_i is the formal charge on species i.

Because the system is spherically symmetric, equation 2.62 can be written

$$\frac{1}{r} \frac{d^2(r\phi)}{dr^2} = \kappa^2 \phi \quad \text{for } r \geq a \tag{2.65}$$

whose solution can readily be confirmed to be

$$\phi = \frac{A \exp(-\kappa r)}{r} + \frac{B \exp(+\kappa r)}{r} \quad \text{for } r \geq a \tag{2.66}$$

with arbitrary constants A and B set through the boundary conditions. Because the potential approaches zero at infinite distance, $B = 0$. A more detailed analysis produces a value for A, such that the complete solution is

$$\phi = \left(\frac{ze}{4\pi\varepsilon}\right) \left(\frac{\exp(\kappa a)}{1+\kappa a}\right) \left(\frac{\exp(-\kappa r)}{r}\right) \quad \text{for } r \geq a \tag{2.67}$$

where z is the charge on the central fixed ion.

The properties of equation 2.67 reveal interesting characteristics of the electrostatic environment of the fixed ion due to the nonspecific sea of ions surrounding it. First, if there were no ions in the surrounding solution ($I = 0$), the potential would be given by Coulomb's law, which also corresponds to equation 2.67 with $\kappa = 0$.

Noncovalent Binding Interactions

This effectively removes the term in the middle set of parentheses and the $\exp(-\kappa r)$ term. Thus, the ionic atmosphere causes the potential to fall off faster with distance than $1/r$ by $\exp(-\kappa r)$, as well as scaling the potential by the distance-independent factor $\frac{\exp(\kappa a)}{1+\kappa a}$.

It can also be shown that the charge density at some point in solution is given by $\rho(r) = -\kappa^2 \varepsilon \phi$, which permits the calculation of the net ionic solution charge within a given distance of the fixed ion:

$$q_{r \leq R} = \int_a^R 4\pi r^2 \rho(r) dr \tag{2.68}$$

$$= -\int_a^R 4\pi r^2 \kappa^2 \varepsilon \phi \, dr \tag{2.69}$$

$$= -\kappa^2 z e \left(\frac{\exp(\kappa a)}{1+\kappa a}\right) \int_a^R r \exp(-\kappa r) dr \tag{2.70}$$

$$= -ze \left(1 - \left(\frac{1+\kappa R}{1+\kappa a}\right) \exp(-\kappa(R-a))\right) \tag{2.71}$$

Examining the near limit, $R = a$, we find that $q_{r \leq R} = 0$. That is, there is no net charge due to the ionic environment inside the fixed ion, because it does not permit ions to penetrate its radius a. At the other limit, $R \to \infty$, $q_{r \leq R} \to -ze$. That is, at large distances the total charge of the ionic atmosphere just counterbalances the fixed charge at the origin, due to the system's charge neutrality. Viewed from this long distance, the fixed charge at the origin is completely canceled by the counteracting effect of the ionic atmosphere. At all intermediate distances, the ionic atmosphere presents a partial but incomplete compensating charge, leading to screening of the central charge. Figure 2.20 shows this function plotted for several values of κ, the screening parameter. At low ionic strength (small κ), a very large region of space must be included to neutralize the central ion, whereas at high ionic strength (high κ), the central ion is essentially neutralized over a very short distance, and farther away, the effect of the central ion is essentially screened.

Equation 2.67 corresponds to the superposition of the potential due to the central ion (which we imagine as a biological macromolecule) and the potential due to the sea of ions surrounding it (which we refer to as the *atmosphere*). The potential due to the ion atmosphere alone is given by

$$\phi_{atm} = \left(\frac{ze}{4\pi\varepsilon}\right)\left[\left(\frac{\exp(\kappa a)}{1+\kappa a}\right)\left(\frac{\exp(-\kappa r)}{r}\right) - \frac{1}{r}\right] \quad \text{for } r \geq a \tag{2.72}$$

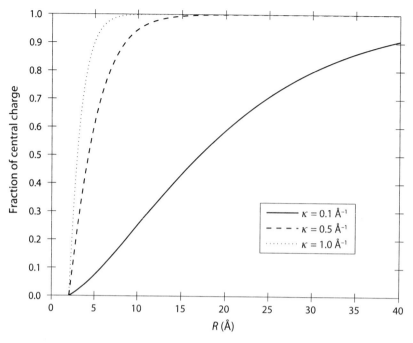

Figure 2.20 The ionic atmosphere of a central charge in the Debye-Hückel model. For a central charge at the origin of coordinates, the y-axis gives the total charge of the ion atmosphere within a sphere whose radius is given by the x-axis value. Lines are plotted for $\kappa = 0.1$, 0.5, and 1.0 Å^{-1}, using equation 2.71. A physiologically relevant value for κ, at 150 mM NaCl and 37°C, is 0.14 Å^{-1}.

whose value at the surface of the central ion is

$$\phi_{\text{atm}}(r = a) = \left(\frac{-ze}{4\pi\varepsilon}\right)\left(\frac{\kappa}{1+\kappa a}\right) \quad (2.73)$$

In the limit of low ionic strength, this simplifies to

$$\lim_{\kappa a \to 0} \phi_{\text{atm}}(r = a) = \left(\frac{-ze}{4\pi\varepsilon}\right)\kappa \quad (2.74)$$

If we imagine that the charge density of the central ion resides entirely at the surface, then the electrostatic interaction energy between the central ion and the sea of ions forming the atmosphere is given by the product of the potential in equation 2.73 and the total charge on the central ion, ze. The interaction energy has a negative sign because it is favorable, and its magnitude decreases with increasing size.

The presence of the ionic atmosphere surrounding a central ion has an effect on the free energy of the system. The magnitude is different from just the interaction energy given above, because the introduction of the central ion changes the distribution of ions, with a consequent energy change. To treat the total effect, we integrate the free energy change for "turning on" the charge of the central ion, using the switching parameter λ to go from "off" ($\lambda = 0$) to "on" ($\lambda = 1$):

$$\Delta G_{\text{ionic}} = \int_0^1 (\lambda z e) \left(\frac{-ze}{4\pi\varepsilon}\right) \left(\frac{\kappa}{1+\kappa a}\right) d\lambda \tag{2.75}$$

$$= \left(\frac{-z^2 e^2}{4\pi\varepsilon}\right) \left(\frac{\kappa}{1+\kappa a}\right) \int_0^1 \lambda d\lambda \tag{2.76}$$

$$= \left(\frac{-z^2 e^2}{8\pi\varepsilon}\right) \left(\frac{\kappa}{1+\kappa a}\right) \tag{2.77}$$

In the limit of low ionic strength, this simplifies to

$$\lim_{\kappa a \to 0} \Delta G_{\text{ionic}} = \left(\frac{-z^2 e^2}{8\pi\varepsilon}\right) \kappa \tag{2.78}$$

For the association of two molecules, P and L, we can construct a transition state theory estimate for the ionic strength dependence of the reaction rate using these results from the Debye-Hückel ionic screening model. The presentation here is a simplified version of that given by Zhou and coworkers [47]. Modeling protein and ligand as spherical molecules of radius R_P and R_L, respectively, and the transition state as contacting molecules, the electrostatic contribution to the activation energy can be estimated as the interaction energy between the molecules (although this neglects their mutual desolvation, which a more refined theory would include). One approximation of this quantity has been given by Debye [48]:

$$\Delta G_{\text{int,ionic}} = \left(\frac{z_P z_L e^2}{2\varepsilon}\right) \left(\frac{\exp(\kappa R_P)}{1+\kappa R_P} + \frac{\exp(\kappa R_L)}{1+\kappa R_L}\right) \frac{\exp(-\kappa(R_P + R_L))}{(R_P + R_L)} \tag{2.79}$$

$$= \left(\frac{z_P z_L e^2}{2\varepsilon(R_P + R_L)}\right) \left(\frac{\exp(-\kappa R_L)}{1+\kappa R_P} + \frac{\exp(-\kappa R_P)}{1+\kappa R_L}\right) \tag{2.80}$$

Applying transition state theory, the reaction rate is proportional to $\exp(-\Delta G^{\circ,\ddagger}/k_B T)$, which leads to

$$k_{\text{on}}(\kappa) = k_{\text{on}}(\kappa = \infty) \exp\left[\left(\frac{-z_P z_L e^2}{2\varepsilon k_B T (R_P + R_L)}\right) \left(\frac{\exp(-\kappa R_L)}{1+\kappa R_P} + \frac{\exp(-\kappa R_P)}{1+\kappa R_L}\right)\right] \tag{2.81}$$

where $k_{on}(\kappa = \infty)$ corresponds to the rate at infinite inverse Debye length, so that all electrostatic interactions are completely screened. Further manipulation leads to

$$\ln k_{on}(\kappa) = \ln k_{on}(\kappa = \infty) - \left(\frac{z_P z_L e^2}{2\varepsilon k_B T (R_P + R_L)}\right) \left(\frac{\exp(-\kappa R_L)}{1 + \kappa R_P} + \frac{\exp(-\kappa R_P)}{1 + \kappa R_L}\right) \quad (2.82)$$

$$= \ln k_{on}(\kappa = 0) + \left(\frac{z_P z_L e^2}{2\varepsilon k_B T (R_P + R_L)}\right) \left(2 - \frac{\exp(-\kappa R_L)}{1 + \kappa R_P} - \frac{\exp(-\kappa R_P)}{1 + \kappa R_L}\right) \quad (2.83)$$

$$\approx \ln k_{on}(\kappa = 0) + \left(\frac{z_P z_L e^2}{\varepsilon k_B T}\right) \left(\frac{\kappa}{1 + \kappa (R_P + R_L)}\right) \quad (2.84)$$

where equation 2.83 results from solving equation 2.82 for $\kappa = 0$ and substituting, and where equation 2.84 results from expansion of the exponentials.

Detailed consideration of the source of these expressions suggests that z_P and z_L might more accurately refer not to the total charge of protein and ligand but rather to the charge of the patches on each protein, respectively, that are involved in strong intermolecular interactions in the transition state model. Likewise, $R_P + R_L$ might more accurately refer to the distance between these charge clusters rather than the center-to-center distance between the protein and ligand [47].

Abstraction of this expression together with the relationship between ionic strength and the inverse Debye screening length given in equation 2.63 gives the following expression for the predicted dependence of association rate constants on salt concentration:

$$\ln \frac{k_{on}(I)}{k_{on}(I = 0)} = \frac{A z_P z_L \sqrt{I}}{1 + Ba\sqrt{I}} \quad (2.85)$$

At the limit of low ionic strength, this can be simplified further to

$$\ln \frac{k_{on}(I)}{k_{on}(I = 0)} = A z_P z_L \sqrt{I} \quad (2.86)$$

In these equations, $k_{on}(I = 0)$ is the association rate constant in a solution with zero salt, A and B are constants, $a = (R_P + R_L)$, z_P and z_L are respectively the net charges of the interacting molecules P and L, and I is the ionic strength of the solution.

Equation 2.85 has been tested and validated using experimental measurements on the kinetic binding rate for the complementarily charged, tight-binding, and rapidly associating complex between the proteins barnase and barstar [47,49]. Binding rates were measured at a range of ionic strengths for the wild-type protein complex and mutants with altered total charge by G. Schreiber and A. R. Fersht [49]. The results agree quite well with the theory outlined here.

2.6 Molecular Measurements: Light-Matter Interactions

Building mechanistic models of biological phenomena, particularly at the level of cells and tissues, requires the ability to measure amounts, rates, and activities of a variety of species in a range of contexts. Mechanistic understanding of experimental methodology is essential for proper interpretation of thermodynamic and kinetic data. Often, dynamic aspects of the measurement process itself can obscure the underlying phenomena of interest. We now consider the interaction of light with matter, which is the basis for many biological measurements. Additionally, it is important to consider the location at which reactions occur, the identity and disposition of binding partners, and potential rearrangements of molecular and cellular structure. Short-range probes producing a unique short-range signal that depends on relative geometry can be particularly useful for detecting binding events and conformational changes.

The spectrum of electromagnetic radiation runs from short-wavelength gamma rays, through X-ray, ultraviolet, visible, infrared, and microwave radiation, to long-wavelength radio waves. These are all electromagnetic waves with the same form but differing wavelength, so for simplicity, we refer to them all as *light* (figure 2.21). Light has special interest in biological research because it interacts with matter (molecules, cells, and tissues) in ways that facilitate making measurements and observations of biological samples.

Light consists of oscillating electric and magnetic fields oriented perpendicular to each other and to the direction of propagation of the light wave. The oscillating electromagnetic fields interact with electrons and protons in molecules and

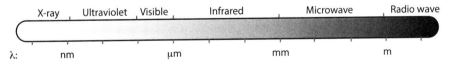

Figure 2.21 Electromagnetic spectrum. Wavelengths of electromagnetic radiation most relevant to biophysical research are indicated, together with the names of the regions. Interestingly, the region of visible light is relatively narrow compared to others—this results from the particular range of energies that the retina's photoreceptors absorb and respond to.

alter their states. Often, light transfers energy to molecules; energy is transferred in discrete quanta of magnitude $E = h\nu$, where h is Planck's constant ($h = 6.626 \times 10^{-34}$ J s) and ν, the frequency of light in cycles per second, is inversely proportional to the wavelength of light. Thus, higher frequency (shorter wavelength) light has larger quanta of energy. All wavelengths of light travel at the same speed in a given medium ($c =$ the speed of light in a vacuum), which provides a simple way to relate wavelength (λ; the distance between peaks in successive waves) and frequency (ν; the number of waves that pass per unit time):

$$\lambda \cdot \nu = c = 2.998 \times 10^8 \text{ m s}^{-1} \tag{2.87}$$

Light is slowed when it passes through any type of matter; the index of refraction gives the ratio of the speed of light in a vacuum to the speed in that particular matter. For air, light is barely slowed at all; but for aqueous solutions, the index of refraction is roughly 1.3, causing a significant slowing of light. A quantum of light (a photon) can be absorbed by a molecule only if the corresponding energy matches an allowed transition between molecular energy levels (states). The resulting absorption will coincide with a change in state from lower to higher energy. Likewise, a quantum of energy emitted by a molecule coincides with a transition from a higher energy to a lower energy state, with the energy difference corresponding to the energy of the light emitted. Thus, light from different parts of the electromagnetic spectrum not only has different characteristic energies but also promotes transitions of different energies and types in molecules. Light from the visible and near-ultraviolet ranges, for example, has energies corresponding to transitions of electronic energy levels and is particularly convenient for studying some properties of biological molecules.

Three mechanisms of interaction are especially important and will be discussed here: scattering, absorbance, and fluorescence. For the purposes of performing biological kinetic measurements, we consider absorbance and fluorescence more extensively than scattering.

2.6.1 Scattering

Scattering produces oscillations of molecular constituents, whereas absorbance and fluorescence are due to the movement of electrons between occupied and unoccupied orbitals. Scattering results from interactions of radiation with matter. Biological scattering experiments are most frequently used to learn about the size and shape of biological objects—molecules and cells. The general effect is that applied radiation produces oscillations, which then reradiate to give scattered intensity. X-rays incident on molecules cause oscillations of the electrons in their orbitals; incident neutrons cause oscillations of atomic nuclei. If the sample is pure and highly ordered in a crystalline array, then the scattered light produces a diffraction pattern that can be used to determine the three-dimensional atomic protein structure. On a more macroscopic scale, light scattering can also be used to infer the size and complexity of cells or particles (e.g., in flow cytometry).

2.6.2 Absorbance

When light is directed at a molecular sample, some frequencies are at least partially absorbed, and others pass through undiminished. This behavior can be understood by considering that a collection of molecules is capable of absorbing only electromagnetic radiation corresponding to the energies of allowed electron transitions, and the amount of energy absorbed from an incident beam depends on the frequency. In a typical absorbance spectrum, the incident radiation is scanned across a wide range of individual, narrow wavelength bands. The amount of light absorbed depends on the existence of a transition at that frequency, the number of molecules in the transition's starting state (called the ground state), and the efficiency of the transition. If the light is in the ultraviolet ($\lambda = 200-400$ nm) or visible range ($\lambda = 400-750$ nm), the transition activated generally moves an electron from an occupied to an unoccupied orbital, and efficient transfer requires a dipole-moment difference between the ground and excited states. Chemical groups that possess such a transition frequently have conjugated aromatic systems of pi-orbitals. Typical biological examples include the aromatic amino acids, nucleic acid bases, NADH, heme, some polyunsaturated systems, and certain transition metal ions.

The structures of some biological chromophores and their absorption spectra are shown in figure 2.22. Conjugated double bond systems and aromatic ring systems are frequently key components of chromophores (also called dyes and pigments). An interesting feature of absorption spectra for complex molecules (like proteins) or molecules in condensed phase (like aqueous solvent) is that wide peaks or broad bands appear in the spectra, which indicates a range of frequencies for which essentially the same electronic transition occurs. This is in contrast to the very sharp peaks and narrow lines of spectra for simple molecules in the gas phase. Narrow lines for molecules with essentially only one conformation and no neighbors indicate that each transition occurs for only a very narrow band of energies. Linewidth broadening results from a population of molecules, each of which has a narrow transition but not at precisely the same frequency. The distribution of frequencies is due to differences in environment from multiple conformations, neighbors, or arrangements of solvent.

It is observed that the absorbance is often time independent, which implies that mechanisms exist to return molecules to their ground state. If the absorbance

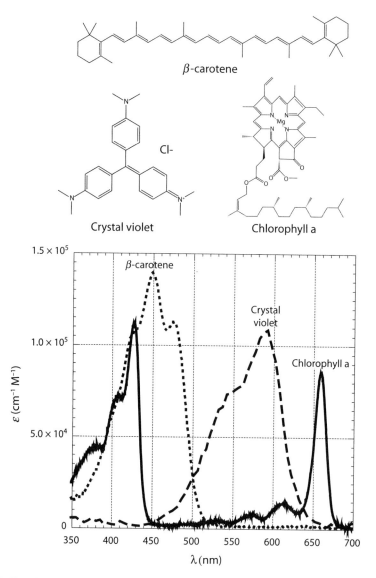

Figure 2.22 Chemical structures and absorbance spectra of three natural chromophore molecules with significant absorption in the visible spectrum. The presence of conjugated series of double bonds, often in ring structures, is a common feature of chromophores. The extinction coefficient ε is a frequency-dependent measure of the efficiency of light absorption (equation 2.88). Each spectrum is a trace that extends across visible wavelengths (400–700 nm) and consists of a few dominant peaks or bands. For example, β-carotene absorbs in the low wavelength (blue) region and tends to transmit or reflect yellow, orange, and red wavelengths, giving substances that contain it an orange appearance. Crystal violet dye absorbs in the green and red spectrum and therefore appears violet; and chlorophyll appears green due to absorbance in the red region. Data from reference 50.

of light progressively promoted molecules from the ground to the excited state without some mechanism for their return, the amount of absorbance would eventually decrease with time. Mechanisms that allow repopulation of the ground state include collisions and internal transitions—some of which emit light, while others do not.

The amount of light absorbed depends on intrinsic properties of the chromophore, its concentration, and the distance along which the light interacts with the molecular solution (called the path length). The relationship between absorbance and concentration is given by the Beer-Lambert law,

$$A = \log_{10}\left(\frac{I_o}{I_t}\right) = \varepsilon \cdot [C] \cdot l \qquad (2.88)$$

$$I_t = I_o 10^{-\varepsilon \cdot [C] \cdot l} \qquad (2.89)$$

where A is absorbance at a particular frequency of light and is defined as a log ratio of the incident (I_o) and transmitted (I_t) intensities. The extinction coefficient, or molecular absorption coefficient ε, gives a measure of the intrinsic strength of the absorbance at the given frequency. [C] is the concentration, and l is the path length. Equation 2.89 shows that the intensity of transmitted light leaving a sample is diminished below the incident intensity by more efficient chemical mechanisms of absorption (i.e., greater ε) and a larger number of absorbers in the beam (through a combination of greater concentration [C] and longer path length l). The absorbance due to mixtures of chromophores is given by the sum of the absorbance due to each chromophore.

Absorbance in the ultraviolet wavelength range is commonly used to measure the concentration and purity of proteins and nucleic acids (figure 2.23). Because the majority of a protein's absorbance at 280 nm is due to tryptophan, tyrosine, and disulfide bonds (figure 2.23), it has been shown that it is possible to predict a protein's extinction coefficient with a root mean square accuracy of 5% from the following formula [51]:

$$\varepsilon(280 \text{ nm})(\text{M}^{-1}\text{ cm}^{-1}) = 5{,}500(\# \text{ Trp}) + 1{,}490(\# \text{ Tyr}) + 125(\# \text{ Disulfides}) \qquad (2.90)$$

It should be noted that the estimation of the extinction coefficient and the actual absorption measurement typically introduce an uncertainty of $\pm 10\%$ to the concentration of protein in any biochemical assay. This error propagates together with other measurement errors to create uncertainty in estimated equilibrium and rate constants.

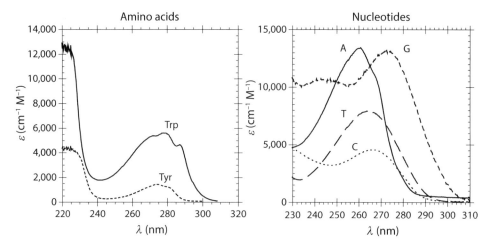

Figure 2.23 Absorbance spectra in the ultraviolet spectrum of the amino acids tryptophan and tyrosine (left panel) and individual nucleotides (right panel). Other amino acids (with the exception of disulfide bonds) negligibly absorb light in the 250–300 nm range. Data from reference 50.

Example 2-4 What absorbance value at 280 nm would you predict for a tenfold dilution of a 1 mg/mL stock solution of bovine serum albumin (BSA)?

Solution In the polypeptide sequence of BSA from a variety of databases (e.g., Swiss-Prot [52], GenBank [53], NCBI Entrez [54]), there are two tryptophan residues, 20 tyrosine residues, and 35 cysteine residues. Note that the tryptophan at residue 3 and the tyrosine at residue 17 are processed and removed as a signal peptide, and are therefore absent from the mature protein, represented by residues 25–607. (This nicety is sometimes overlooked by automated extinction coefficient prediction programs!) Of the cysteine residues, the database entries note that 17 disulfides are formed (and one cysteine is unpaired). From equation 2.90,

$$\varepsilon_{BSA}(280 \text{ nm})(M^{-1}\text{ cm}^{-1}) = 5{,}500(2) + 1{,}490(20) + 125(17) \quad (2.91)$$
$$= 42{,}925 \text{ M}^{-1}\text{ cm}^{-1} \quad (2.92)$$

The predicted extinction coefficient of 42,925 M^{-1} cm^{-1} compares favorably with the reported experimental value of 43,820 ± 530 M^{-1} cm^{-1} [51]. At a molecular weight of 66,463 g mol^{-1}, a 1 mg mL^{-1} solution will be at a molar concentration of 15.0 μM, and a tenfold dilution will be at 1.50 μM. From the Beer-Lambert law (equation 2.88), assuming a 1 cm pathlength, as is typical in most commercial instruments, the absorbance is

$$A = \varepsilon \cdot [C] \cdot l \tag{2.93}$$
$$= (42{,}925 \text{ M}^{-1} \text{ cm}^{-1})(1.50 \times 10^{-6} \text{ M})(1 \text{ cm}) \tag{2.94}$$
$$= 0.065 \tag{2.95}$$

Note that absorbance is a dimensionless quantity. The particular reliable linear range for a given spectrophotometer should be directly determined by measuring the absorbance of a series of dilutions of a standard solution, but it is typically in the approximate range $0.01 < A < 0.5$. Consequently, a tenfold dilution of the BSA stock solution in this example is necessary to obtain an absorbance signal in the linear range.

2.6.3 Fluorescence

Fluorescence refers to the emission of light almost immediately following its absorption. Not all molecules fluoresce; those that do are termed *fluorophores*. The process generally involves four steps (figure 2.24). First, the absorption of a photon promotes a molecule from ground to excited state. Next, the molecule

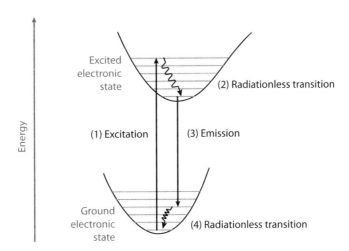

Figure 2.24 Fluorescence spectroscopy. In a four-step process, first a quantum of light (photon) is absorbed, and the molecule is excited from a low vibrational ground electronic state to a high vibrational excited electronic state. Second, through a radiationless transition, the molecule relaxes to a lower vibrational state in the excited electronic state. Third, a photon is emitted as the molecule drops to a high vibrational state in the ground electronic state. Finally, through a second radiationless transition, the molecule relaxes to a low vibrational state in the ground electronic state. The horizontal lines in each electronic state represent vibrational energy levels.

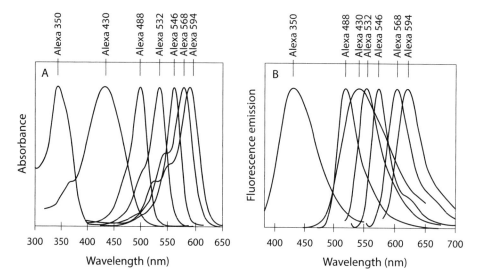

Figure 2.25 Excitation (A) and emission (B) spectra of a series of Alexa Fluor dyes commonly used to label proteins [55]. Note that the emission from each fluorophore is red-shifted relative to its absorbance maximum (e.g., Alexa 350 $\lambda_{ex,opt} \approx 350$ nm and $\lambda_{em,opt} \approx 430$ nm). Judicious selection of excitation (ex) light sources and emission (em) interference filters allows multiple signals (from two to as many as a dozen) to be detected simultaneously in a single biological sample.

relaxes in its excited electronic state, often to a lower energy vibrational state, without emission (a so-called radiationless transition). Third, the molecule drops from the excited electronic state to the ground electronic state, while simultaneously emitting a photon. Finally, the molecule relaxes in the ground electronic state, also without emission. Because of the radiationless relaxations, the energy difference is smaller for the emission than for the absorption transition, and the emitted photon is of correspondingly lower energy. Thus, emitted fluorescence is shifted toward red (longer) wavelengths relative to the incident beam that produces the absorption. This shift of emission to lower energy than the excitation light is called the Stokes shift (figure 2.25).

The physical processes involved in fluorescence are extremely fast relative to most biological processes. The time scales for excitation and emission are on the order of 10^{-15} s, and the lifetime of the excited state is generally 10^{-9}–10^{-6} s (usually the lower end of this range). Not all excitation events lead to a subsequent emission in the measurement range. The excess energy can be released by other mechanisms, including collisions and chemical reactions. Some mechanisms restore the molecule to a similar ground state, where it can participate in further fluorescence events; others chemically modify the chromophore so that it is no longer fluorescent, at least not in the same wavelength ranges. The quantum yield is the fraction of excitation events that lead to an observable emission. Both radiationless transitions and destructive processes can lead to a decrease of quantum yield below its ideal value of 1. Fluorophores with high quantum yields are desirable, because they permit sensitive measurements to be made on small and dilute samples; high quantum yields are especially important for spectroscopic studies in which single molecules are monitored individually, as opposed to measuring a population average. The effect of covalent processes that lead to degradation and a continuous loss of fluorescent probe over time is called photobleaching, because over time, the application of light has reduced the level of fluorescence emission from the sample.

Case Study 2-3   L. Song, E. J. Hennink, I. T. Young, and H. J. Tanke. Photobleaching kinetics of fluorescein in quantitative fluorescence microscopy. *Biophys. J.* 68: 2588–2600 (1995) [56].

When a fluorophore has absorbed a photon and is in an excited state, the excess energy is dissipated in three general ways: emission of a photon (resulting in fluorescence); nonradiative emission by intermolecular collisions; or a covalent reaction that creates a new, nonfluorescent product. This third process, photobleaching, is important for practical fluorescence measurements, because ongoing photobleaching must be either minimized or accounted for when attempting to quantify the number of fluorophores in a sample via fluorescence emission intensity. Rapid photobleaching by high-intensity illumination is also used intentionally to remove the signal from a limited sample area, followed by recovery of the signal by diffusion of fluorophores into the bleached area. This method is called fluorescence recovery after photobleaching (FRAP) (section 2.7.2).

There are two general chemical mechanisms for photobleaching, involving the collision of an excited-state fluorophore either with oxygen or another fluorophore molecule [57]. The different energetic states of interest are shown in figure 2.26.

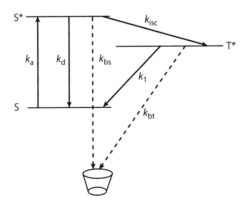

Figure 2.26 Interconversions among electronic states in a fluorophore.

The asterisked species are excited states, S is the singlet state, and T is the triplet electronic energy state. Of particular note is that the activated singlet S* is very short lived (half-life of approximately 10 nanoseconds), whereas the triplet excited state T* persists much longer (half-life of microseconds to hundreds of milliseconds). Because the photobleaching reactions require molecular collisions of an excited state with another molecule, the triplet state rather than the excited singlet state is the main source of quenching. A brief analysis of diffusive lifetimes supports this conclusion. The diffusivity D of fluorescein in water is approximately 4.3×10^{-6} cm^2 s^{-1} and for oxygen is approximately 2.1×10^{-5} cm^2 s^{-1}. As we shall see in chapter 8, a characteristic distance for diffusion is $\sqrt{Dt}$; in 10 nanoseconds, a fluorescein molecule will diffuse approximately 20 Å and an oxygen molecule will diffuse about 50 Å. In a 1 mM fluorescein solution, the average fluorophore intermolecular distance is 120 Å, whereas in a saturated oxygen solution (250 μM), the average distance between oxygen molecules is 190 Å. Comparing these lengths, it's reasonable to conclude that, on average, an S* will emit a photon and become S before colliding with either a fluorophore or an oxygen molecule. By contrast, T* will undergo many intermolecular collisions prior to relaxation to the S state.

Song et al. [56] modeled this process to determine whether the observed kinetics of photobleaching were consistent with the proposed mechanisms and also to determine whether oxygen-mediated photobleaching reactions were significant under conditions of fluorescence microscopy. It had been observed that although dilute solutions of fluorescein exhibit photobleaching decays consistent with a single exponential time constant, stained biological samples exhibited a more complex decay curve that was not adequately described by a single exponential time constant, as shown in figure 2.27.

The reactions considered and parameter estimates for this model are shown in the table below:

Reaction	Reaction number	Description	Rate constants
$S + h\nu \to S^*$		Absorption	$k_a = 3.8 \times 10^8$ s^{-1}
$S^* \to S + h\nu'$		Fluorescence emission	$k_d = 2.134 \times 10^8$ s^{-1}
$S^* \to S$		Internal conversion	
$S^* \to T^*$		Intersystem crossing	$k_{isc} = 6.6 \times 10^6$ s^{-1}
$T^* \to S$	1	Radiationless deactivation	$k_1 = 50$ s^{-1}
$T^* + T^* \to T^* + S$	2	Triplet quenching	$k_2 = 5 \times 10^8$ M^{-1} s^{-1}
$T^* + S \to S + S$	3		$k_3 = 5 \times 10^7$ M^{-1} s^{-1}
$T^* + T^* \to R + X$	4	Electron transfer	$k_4 = 6 \times 10^8$ M^{-1} s^{-1}
$T^* + S \to R + X$	5	Electron transfer	$k_5 = 5 \times 10^7$ M^{-1} s^{-1}
$T^* + X \to S + X$	6	T* quenching by X	$k_6 + k_7 = 1 \times 10^9$ M^{-1} s^{-1}
$T^* + R \to S + R$	7	T* quenching by R	k_7
$T^* + O_2 \to S + O_2$	8	Physical quenching by O_2	$k_8 = 1.56 \times 10^9$ M^{-1} s^{-1}
$T^* + O_2 \to X + HO_2$ (or O_2^-)	9	Chemical quenching by O_2	$k_9 = 1.4 \times 10^8$ M^{-1} s^{-1}

Notes: S, ground state dye; S*, singlet excited state dye; T*, triplet excited state dye; R, semi-reduced form of the dye; X, semi-oxidized form of the dye; O_2, oxygen.

Translating these reactions into the ordinary differential equations describing the material balance on each species yields:

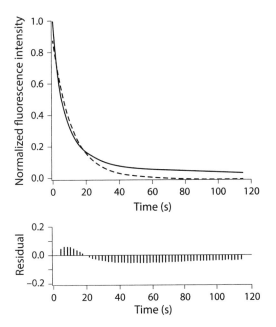

Figure 2.27 The nuclear DNA of human lymphocytes was labeled with fluorescein-conjugated nucleic acid probes (a technique known as fluorescence in situ hybridization, or FISH) and imaged continuously in a fluorescence microscope with a mercury arc lamp light source. The solid line is the experimental data, and the dashed line is the best fit with a single-exponential decay curve. The lower panel shows the residuals (curve fit minus data), and clearly displays a nonrandom pattern, indicating that the single exponential fit is inadequate.

$$\frac{d[S]}{dt} = k_d[S^*] + k_1[T^*] + k_2[T^*]^2 + k_3[T^*][S] + (k_6[X] + k_7[R])[T^*]$$
$$+ k_8[T^*][O_2] - k_a[S] - k_5[T^*][S]$$

$$\frac{d[S^*]}{dt} = k_a[S] - k_d[S^*] - k_{isc}[S^*]$$

$$\frac{d[T^*]}{dt} = k_{isc}[S^*] - \Big(k_1[T^*] + (k_2 + 2k_4)[T^*]^2 + (k_3 + k_5)[T^*][S]$$
$$+ (k_6[X] + k_7[R])[T^*] + (k_8 + k_9)[T^*][O_2]\Big)$$

$$\frac{d[X]}{dt} = k_4[T^*]^2 + k_5[T^*][S] + k_9[T^*][O_2]$$

$$\frac{d[R]}{dt} = k_4[T^*]^2 + k_5[T^*][S]$$

Although oxygen consumption by reaction 9 was initially included in the model, the authors later determined that this was unnecessary, because oxygen will rapidly be repleted by diffusion under the experimental conditions. Note that in the table of rate constants, k_d is a lumped rate constant for fluorescence emission and internal conversion. Additionally,

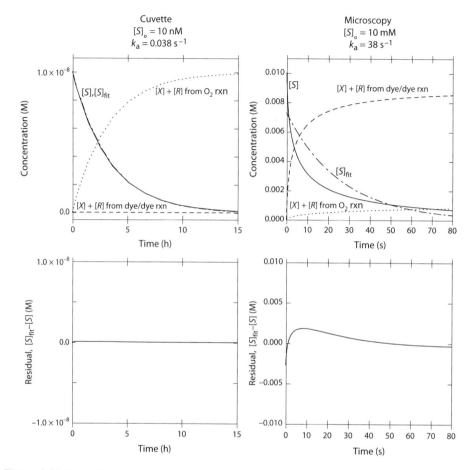

Figure 2.28 Two different experiments were simulated: a low concentration of fluorescein ($[S]_o = $ 10 nM) under low illumination in a spectrofluorometer ($k_a = 0.038$ s^{-1}), and fluorophore densely constrained to binding DNA as in the in situ hybridization experiment ($[S]_o = 10$ mM) as well as under more intense arc lamp illumination ($k_a = 38$ s^{-1}). (Note that at these high concentrations, the assumption that all photobleaching occurs from the triplet state rather than the singlet state may become questionable.)

the values for k_6 and k_7 are a bit ambiguous; however, each of these was intended to be estimated as 10^9 M^{-1} s^{-1}, an approximate value for a diffusion-limited reaction such as collisional quenching (to be discussed in depth in chapter 8).

The ordinary differential equations were numerically simulated using a *stiff* differential equation solver, due to the wide span of time scales involved (from nanoseconds to hours). The results are shown in figure 2.28.

The key take-home messages from the simulation results are (1) under the high fluorophore concentrations and excitation intensities of fluorescence microscopy, the rate of decay of signal due to photobleaching may not be well described by a single exponential function; and (2) the mechanism of photobleaching under fluorescence microscopy conditions may not depend sensitively on oxygen concentration, by contrast with experiments performed at low concentrations and low intensity. The latter conclusion would be significant if one were to seek to minimize photobleaching in microscopy by reducing soluble oxygen concentration, which this analysis suggests would be rather ineffective.

Fluorescence resonance energy transfer A mechanistic understanding of biological phenomena requires identifying key binding events and conformational changes that transduce chemical and mechanical signals. Fluorescence resonance energy transfer (also known as Förster resonance energy transfer, both abbreviated FRET) is a mechanism used to spectroscopically probe binding events and conformational change through short-range interaction between two fluorophores in a distance-dependent manner.

Ordinarily the presence of two different fluorophores, even in close proximity, produces spectra that are the sum of what would be observed with either fluorophore alone. However, if the fluorophores are chosen such that the emission spectrum of one overlaps with the absorption spectrum of the other, then a form of intermolecular energy transfer can occur from the former (called the *donor*) to the latter (*acceptor*). This basic scheme is illustrated in figure 2.29.

The emission spectrum for each fluorophore is shifted to lower energy relative to its absorption spectrum, as discussed previously (illustrated in figure 2.24). The overlap between the donor emission and acceptor absorption is the key new feature. When this situation occurs, and donor and acceptor are close enough in space, FRET is observed. Light of a wavelength under the donor absorption peak is input, and emission is observed not only in the region of the donor emission band but also in the acceptor emission band. The amount of donor-emitted light is less than would be observed in the absence of acceptor, and the amount of acceptor-emitted light is greater than would be observed without the donor. Although it is tempting to imagine that the acceptor captures some of the donor emitted light before it can reach the detector, absorbs it, and reemits it, the physics of the situation is somewhat different: A concerted mechanism of energy transfer instead occurs that does not involve sequential emission and reabsorption of a photon. The efficiency of Förster transfer falls off as the sixth power of the distance (due to the dipole-dipole coupling mechanism, analogous to the van der Waals relation seen earlier in this chapter) and also depends on the relative orientation and quantum yields of donor and acceptor. The distance dependence can be represented as

$$E_\mathrm{F} = \frac{1}{1 + \left(\frac{r}{r_\circ}\right)^6} \tag{2.96}$$

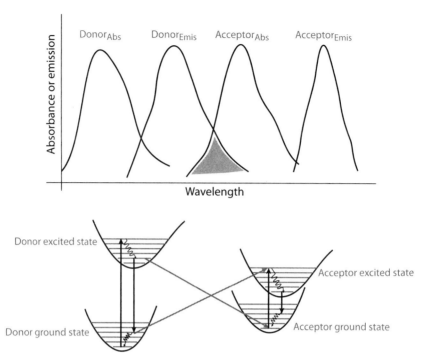

Figure 2.29 Fluorescence resonance energy transfer (FRET). (Top) To observe FRET, the emission spectrum of the donor must overlap the absorption spectrum of the acceptor, which is indicated by the shaded area. When the donor is excited, a portion of the light that would otherwise be emitted from the donor is instead transferred to the acceptor through the Förster mechanism when donor and acceptor are in close proximity. The result is diminished donor emission and enhanced acceptor emission, which can be detected experimentally. (Bottom) In FRET, the donor is excited and then returns to the ground state, and the acceptor is excited and then returns to the ground state, which is similar to what happens for a pair of radiative processes. However, as indicated by the crossed gray arrows, the donor drops to the ground state through a concerted process that simultaneously promotes the acceptor to the excited state without radiation or release of a photon. This internal Förster mechanism operates over short distances with a distance dependence that goes as the inverse sixth power (equation 2.96).

where E_F is the efficiency, r is the distance between donor and acceptor, and r_o is the distance at which an efficiency of 50% is achieved. For most fluorophores of interest, r_o is in the range of 10–100 Å.

Detection and quantification of fluorescence is important for ascertaining the presence and concentration of molecules that possess intrinsic fluorescence or, more commonly, have fluorescent labels attached. In localization studies, fluorescence imaging can be used to observe the position, amounts, and distribution of molecules in tissues and cells. In more complex arrangements, such as FRET, conformational changes and interactions between molecules can be observed and quantified. Technologies have been developed to take full advantage of these physical phenomena and are discussed throughout the rest of this chapter. These technologies include genetic constructs that produce fluorescent protein molecules without the need for external introduction of fluorophores and matched sets of such proteins for FRET experiments.

2.7 Fluorescence Applications for Biomolecular Measurements

The theory and principles described in the previous section provide a framework for understanding many molecular measurements. For biological applications, labeling technology considerations include whether they can be used in the context of live cells, propagating cultures, or tissues; access (whether and how the label enters the cell) and distribution (which cellular locations the label enters); the kinetics for label attachment and activation as well as for its degradation, including control over expression and activation of the label and whether it is propagated to daughter cells; and interference (whether attachment of the label to a protein target interferes with the activity of that protein target in vivo). Detection issues include whether data can be collected continuously in real time or must be taken periodically by processing aliquots of cells from a batch experiment; whether the methodology requires a label or is label free; whether the experiment will be done in high throughput, whereby many samples are processed in parallel; whether the technology is amenable to measurement in an imaging apparatus with subcellular resolution; and whether the detection technique damages the cells. A further issue is whether techniques move beyond simple measurement to permitting high-throughput isolation and manipulation of cells with specific fluorescent properties.

2.7.1 Biosynthetic Fluorescent Labeling

To track proteins in living cells, it is desirable to encode a detection tag in the form of a polypeptide fusion to the protein of interest. When the additional polypeptide is intrinsically fluorescent no further reagents need be added to detect the fusion protein. Other methods involve the addition of a second exogenous reagent to bind specifically to the polypeptide tag.

Intrinsically fluorescent proteins, first cloned from jellyfish, have proven extremely valuable for quantifying gene expression, determining protein localization and trafficking rates, and measuring protein-protein interactions in vivo. The primary advantage of fluorescent proteins is that the fluorescence signal is entirely genetically encoded, so living cells can be assayed without the addition of any exogenous chemicals.

The paradigmatic fluorescent protein is *green fluorescent protein* (GFP), a generic term generally inclusive of some closely related proteins with mutations to improve expression, solubility, brightness, and the rate of acquisition of fluorescence. GFP fluoresces due to the presence of a fluorophore that forms spontaneously in the center of the protein (figure 2.30) in the presence of molecular oxygen [58]. Unfortunately,

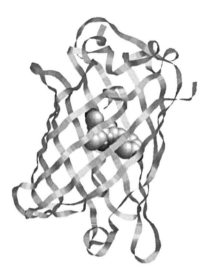

Figure 2.30 Crystal structure of GFP. The fluorophore (solid spheres) is buried in a cage formed by the β-barrel structure (ribbons) of the protein. The surrounding structure of the protein positions the constituents of the chromophore, which is formed by covalent cyclization of the polypeptide backbone and oxidation. This also shields the chromophore from the surrounding solvent to create an environment for brighter fluorescence emission [60].

this reaction is rather slow, requiring an hour or so to generate the fluorescent chromophore (case study 2-4). A veritable rainbow of fluorescent proteins (figure 2.31) has been engineered through laboratory evolution of GFP and bioprospecting from luminescent sea life for desirable spectral properties [59].

GFP can be used as a *reporter protein* to signal the onset or cessation of transcription from a given promoter to which its coding sequence is fused [61]. GFP is widely used as a reporter to study promoter regulation or the behavior of synthetically designed gene expression circuits. Its high stability can lead to accumulation of GFP in cells as a result of *leaky* expression, a problem that has been addressed by intentional destabilization via the addition of polypeptide sequences that target the protein to the proteasomal degradation apparatus. Construction of transgenic animals containing GFP reporter fusion genes allows noninvasive tracking of gene expression in various tissues in response to different environments.

It has been found, somewhat fortuitously, that fusion of GFP to either the N or C terminus of various proteins often does not substantially alter either the biological function or localization of the parent protein. Consequently, expression of GFP gene fusions can enable identification of a given protein's subcellular localization. Additionally, the rate of movement of GFP fusion proteins across the cell surface or between organellar compartments can be quantified by monitoring the recovery

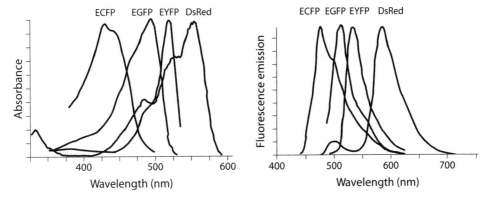

Figure 2.31 Fluorescent proteins with cyan, green, yellow, and red emissions are in common use. EGFP possesses the mutations F64L and S65T; ECFP has the mutations F64L, S65T, Y66W, N146I, M153T, and V163A; and EYFP has the mutations S65G, V68L, S72A, and T203Y. DsRed is an unrelated protein whose gene was isolated from a coral species. EGFP, ECFP, and EYFP stand for green, cyan, and yellow fluorescent proteins, respectively; the "E" signifies *enhanced*. Adapted from the Clontech Living Colors user manual PT2040-1.

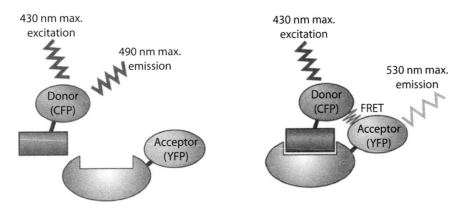

Figure 2.32 GFP mutants with shifted excitation and emission spectra (cyan fluorescent protein, CFP; yellow fluorescent protein, YFP) can signal proximity of their attached fusion partners via the presence of a FRET signal [61].

of fluorescence after intentional photobleaching, as discussed in more detail in section 2.7.2.

Formation of complexes between two proteins in vivo can be detected by expressing one protein as a fusion to a *donor* fluorescent protein and the other to an *acceptor* fluorescent protein, and detecting the proximity of the two fluorophores by FRET (figure 2.32; see section 2.6.3). Common FRET pairs for this purpose are CFP and YFP, given the overlap in their spectra (figure 2.31).

Case Study 2-4 B. G. Reid and G. C. Flynn. Chromophore formation in green fluorescent protein. *Biochemistry* 36: 6786–6791 (1997) [62].

An often overlooked shortcoming of GFP as a gene expression reporter is slow formation of the fluorescent chromophore following translation and folding of the protein. In this paper, Reid and Flynn dissect the separate kinetic steps of folding, backbone cyclization, and oxidation, determining rate constants for each step at room temperature (figure 2.33). The folding rate was measured by urea-mediated unfolding of GFP lacking the mature chromophore, followed by refolding on dilution, leading to protection from protease digestion; this provides only a crude estimate for the rate of the in vivo folding process, which is mediated by chaperones and occurs simultaneously with emergence of the polypeptide from the ribosome during translation.

This mechanistic schematic translates to the following set of differential equations for the material balances for the four forms of GFP (U, unfolded; F, folded; FC, folded and backbone cyclized; FO, chromophore fully oxidized):

$$\frac{d[U]}{dt} = -k_1[U]$$

$$\frac{d[F]}{dt} = k_1[U] - k_2[F]$$

$$\frac{d[FC]}{dt} = k_2[F] - k_3[FC]$$

$$\frac{d[FO]}{dt} = k_3[FC]$$

with the initial conditions $[U] = [U]_\circ$, $[F] = [FC] = [FO] = 0$ at $t = 0$.

This set of differential equations can be solved analytically to yield the concentration of fluorescent GFP $[FO]$:

$$\frac{[FO]}{[U]_\circ} = 1 - k_1 k_2 k_3 \left(\frac{e^{-k_1 t}}{\alpha_1} + \frac{e^{-k_2 t}}{\alpha_2} + \frac{e^{-k_3 t}}{\alpha_3} \right) \quad (2.97)$$

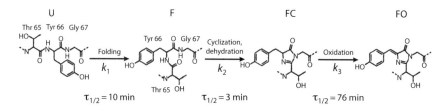

Figure 2.33 The mechanism for autocatalytic formation of the GFP chromophore is shown, together with the experimentally determined rate constants. Only the atoms involved in fluorophore formation are shown; the remainder of the protein is omitted for clarity. Adapted from reference 62.

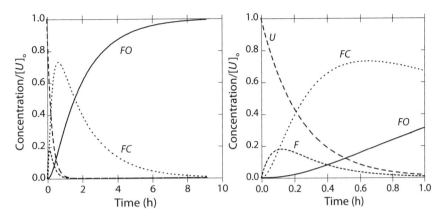

Figure 2.34 It is noteworthy that the fluorescent GFP form (FO) does not reach full fluorescence for several hours (left panel); this delay complicates interpretation of rapid dynamic gene expression experiments. Also, the FO concentration has an inflection point when plotted as a function of time. Note that in the first 0.2 h (right panel), little FO has been formed. Even after 1 h, the yield of fluorescent protein is still less than 40%.

where

$$\alpha_1 = k_1(k_2 - k_1)(k_3 - k_1)$$
$$\alpha_2 = k_2(k_1 - k_2)(k_3 - k_2)$$
$$\alpha_3 = k_3(k_1 - k_3)(k_2 - k_3)$$

The modeled concentrations of fluorescent GFP and each of the intermediates are shown in figure 2.34.

2.7.2 Common Fluorophores for In Vitro Conjugation

Proteins can be covalently labeled with fluorophores in vitro for subsequent use in labeling cells. Common points of attachment include the primary amines of lysines and the N-terminus, the thiol of cysteines not involved in a disulfide bond, or sugars present in glycoproteins.

Fluorescein and Alexa Fluor dyes For many years, fluorescein was the most commonly used chemical fluorophore in biological labeling applications; its excitation peak is near the commonly used 488-nm line of an argon laser, and its emissions are in the green wavelength range, at which the human eye is most sensitive. However, two particular shortcomings limit the utility of fluorescein: fairly rapid photobleaching (see case study 2-3), and strongly pH-sensitive fluorescence (lower intensity at acidic pH). A series of related compounds termed *Alexa Fluor dyes* has been

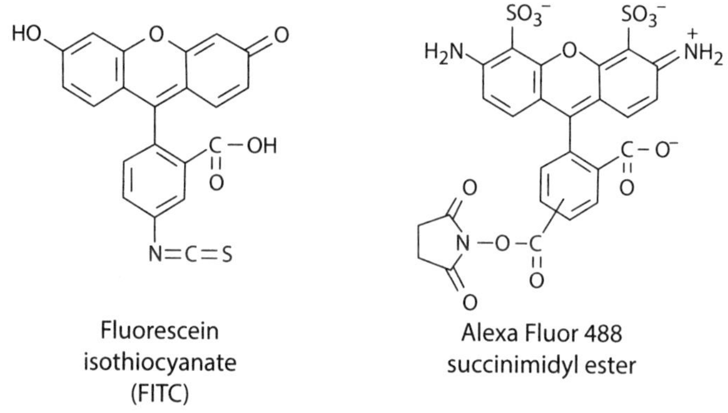

Figure 2.35 Chemical structures of fluorescein and Alexa Fluor 488. The absorbance and emission spectra of these two fluorophores are nearly superimposable, but the substitutions in Alexa Fluor 488 significantly decrease the rate of photobleaching and pH sensitivity of fluorescence emissions. The isothiocyanate and succinimidyl ester portions at the bottom of these structures are amine-reactive groups used for labeling proteins and do not significantly alter the optical properties of the dyes.

developed with different substituents attached to the 3-ring structure of fluorescein, giving them more desirable properties in applications (figures 2.35 and 2.25) [55].

Phycobiliproteins Proteins present in the light-harvesting antennae of blue-green and red algae are often exploited as fluorophores, given their very high extinction coefficients and consequent bright fluorescence emissions. This class of proteins, called phycobiliproteins, consists of three different subtypes: allophycocyanins, phycoerythrins, and phycocyanins. The most commonly exploited of these is R-phycoerythrin (R-PE), due to its convenient high absorption at the 488-nm wavelength of argon lasers and large Stokes shift to red emissions (figure 2.36, top). Allophycocyanin absorbs and emits in the more red region (absorbance maximum 650 nm, emission maximum 660 nm). A crystal structure of a portion of R-PE is shown in figure 2.36 (bottom). In the light harvesting complex, called a phycobilisome, there are 10 hexamers of the R-PE heterodimer, as well as 12 hexamers of the phycocyanin heterodimer and 34 copies of the allophycocyanin heterodimer [63]. These photosynthetic organisms' efforts to efficiently capture light with phycobiliprotein complexes is well aligned with the technological needs for fluorescent labeling.

Pulse-activated and inactivated fluorophores The kinetics of labeling processes, whether biosynthetic or exogenous, can often interfere with direct measurement of the underlying biological rates of interest. The delay in GFP chromophore activation can obscure the timing of gene expression events it is intended to

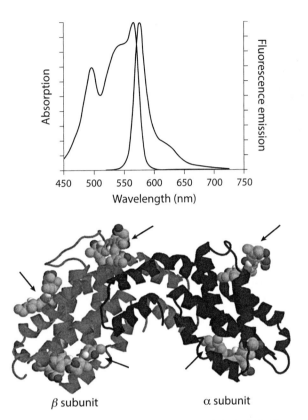

Figure 2.36 (Top) Excitation and emission spectra of R-phycoerythrin (R-PE). (Bottom) Crystal structure of the $\alpha\beta$ heterodimer subunit of R-PE from the red alga *Griffithsa monilis* (PDB: 1B8D) [64]. Two fluorophore molecules are bound to the α subunit, and three to the β subunit (indicated by black arrows pointing to space-filling representation of fluorophore; protein is in ribbon representation). This heterodimer assembles to a hexamer that contains 30 fluorophore molecules, resulting in an extinction coefficient of 2×10^6 M^{-1} s^{-1} at 566 nm. For comparison, the extinction coefficient of Alexa Fluor 488 is 71,000 M^{-1} s^{-1} at its absorption maximum wavelength. Consequently, a single R-PE complex produces a 20- to 30-fold brighter signal than a correspondingly labeled small-molecule dye.

report. The rate of a fluorescently labeled antibody binding to its target can be delayed by slow mixing or permeation through tissues or cell membranes. To obtain more rapid labeling, two complementary approaches have been developed using light as the trigger: photoactivatable fluorophores and fluorescence recovery after photobleaching (FRAP).

1. *Photoactivatable fluorophores.* Various mutant fluorescent proteins have been developed that increase their fluorescence more than 100-fold within 1 second of intense ultraviolet laser illumination [65]. Once irreversibly and effectively instantaneously activated, these fluorescent proteins can be used to track the

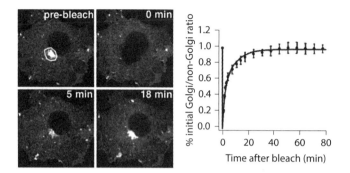

Figure 2.37 An example of FRAP. A glycosylphosphatidylinositol-anchored GFP was expressed in COS-7 cells (left panel). An area corresponding to the Golgi compartment was bleached by high-intensity laser illumination, and the recovery of fluorescence in that region was quantified as a function of time (right panel) [68].

motion of individual proteins, organelles, or cells for many hours. This same approach can also be applied to binding probes (e.g., antibodies, ligands, oligodeoxynucleotides) conjugated to *caged* fluorophores. Attachment of a photolabile group, such as 2-nitrobenzyl, quenches the fluorescence emissions of fluorescein and other fluorophores. Brief pulses of ultraviolet illumination lead to the photolysis of the caging group and effectively instantaneous appearance of fluorescence emission [66]. These photoactivatable fluorophore methods essentially allow the investigator to unambiguously define time zero in the experiment: Before the activating light pulse, no signal is present; following the pulse, fluorescent probes can be followed in space and time by their emissions.

2. *Fluorescence recovery after photobleaching (FRAP).* A complementary approach to fluorophore photoactivation is localized photodestruction by bleaching. In this approach, the label is first allowed to distribute throughout the volume of interest, either by biosynthesis or exogenous labeling. Then at time zero, a particular patch of the organelle, cell, or tissue of interest is bleached by high-intensity laser illumination. Recovery of fluorescence due to diffusion of unbleached fluorophores in the surrounding space can then be measured with time, allowing quantitative measurement of diffusivities in biological samples [67]. Example FRAP data are shown in figure 2.37.

2.7.3 Fluorescence Instrumentation

Kinetic data are often obtained by quantifying fluorescence emission as a function of time, whether in solution, at particular locations in cells or tissues, or from individual cells in heterogeneous populations. The following is an overview of some of the most common devices used to make such measurements.

Stopped-flow spectrometry When the initiation of a kinetic experiment occurs by mixing two solutions together, there is an inevitable delay of a few seconds when the mixing is accomplished manually, while the two solutions intermingle due to shaking or stirring. High-speed *stopped-flow* mixing devices can accelerate the process such that two solutions are effectively well mixed within 0.1–5 milliseconds, allowing one to measure spectroscopic signals in the millisecond–second time window (figure 2.38).

Fluorescence microscopy Fluorescence microscopy is often used to quantify the abundance of proteins or their rate of transport between cellular compartments. These compartments often have features smaller than the wavelength of the light used to image them, however, and consequently, their spatial details cannot be resolved using conventional light microscopy. As light passes through a small hole, it forms a diffraction pattern due to constructive and destructive wave interference. In the plane of focus, this pattern consists of a bright central disk (called the Airy disk after its discoverer, George Biddell Airy) with weaker rings around it. The radius r_{xy} of the Airy disk in the xy plane for an infinitesimally small light source is

$$r_{xy} = \frac{0.61\lambda}{\text{NA}} = \frac{0.61\lambda}{n \sin \theta} \qquad (2.98)$$

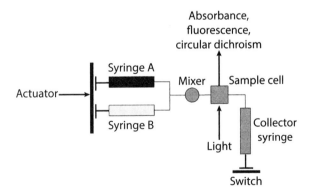

Figure 2.38 Schematic diagram of a stopped-flow apparatus. A mechanical actuator simultaneously depresses the plunger on two syringes, mixing their contents in a sample cell and triggering data collection by expansion of a plunger in a collection syringe. These devices have been engineered to have mixing dead times of 0.1–5 milliseconds.

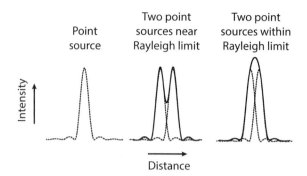

Figure 2.39 Graphical representation of the Rayleigh imaging limit (equation 2.98). An infinitesimally small point source of light creates a larger image due to the finite wavelength of the light—this image is described by the equation for an Airy disk, for which a one-dimensional intensity histogram is shown. When two points approach sufficiently close for their Airy disks to overlap, it is not possible to distinguish the presence of two distinct points.

where λ is the wavelength of light being imaged, and NA is the numerical aperture of the microscope objective, with $NA = n \sin \theta$, where n is the index of refraction of the medium between the sample and the objective, and θ is the angle of the cone of light collected from the sample by the objective. When the gap between sample and objective is air, $NA < 1$, which is why immersion oil objectives are often used to increase NA to values as high as 1.4. The length scale given by equation 2.98 is the smallest image that can be formed by a traditional light microscope, regardless of how small the light-emitting object is. For example, fluorescein-labeled ($\lambda = 520$ nm) polystyrene beads of radius 50, 100, or 200 nm all produce a fuzzy Airy disk image of radius 400 nm when imaged by a microscope with $NA = 0.8$. Consequently, two such small objects look like a brighter single object when they are within this distance of each other, as represented in figure 2.39. To quantify the resolution limit, the British physicist Lord Rayleigh proposed the following convention: Two identical point sources are just resolvable if the principal diffraction maximum of one point source coincides with the first minimum of the second. Note that this distance is identical to the Airy disk radius, so equation 2.98 is also called the Rayleigh limit. Thus, even with favorable experimental conditions of low-wavelength visible light ($\lambda = 450$ nm) and a high NA (1.4), the Rayleigh limit is ≈ 200 nm. For comparison, the tubulovesicular structures of the endoplasmic reticulum and the Golgi apparatus have thicknesses on the order of 50–200 nm, necessitating the use of electron microscopy to directly discern their fine spatial features. An additional limitation arises from the inherent randomness of photon statistics when very dim samples are being imaged.

Axial resolution (along the vertical z axis) is generally worse than the xy Rayleigh limit for two reasons. One is that the diffraction pattern is longer in this dimension, with the resolution in z given by the following expression:

$$r_z = \frac{2\lambda}{NA^2} \tag{2.99}$$

(The inverse-squared dependence on numerical aperture indicates significantly greater improvements in vertical resolution when using oil-immersion objectives.) A second major contributor to loss of vertical resolution is the fluorescence from out-of-focus emitters: As the excitation light travels through a sample, fluorophores not in the plane of focus absorb and emit light that is then captured in the image. There are four general approaches for minimizing this effect:

1. *Confocal microscopy.* In confocal microscopy, out-of-focus light is physically excluded from detection by placing pinhole apertures at both the light source and the detector. The detector used for laser scanning confocal microscopy is generally a photodiode, which is less sensitive than the CCD imaging camera used for deconvolution microscopy. As a result, photodamage from laser scanning confocal microscopes prevents their more general application to living samples.

2. *Deconvolution microscopy.* In this method, the response characteristics of the microscope's optical train are precisely measured with nonbiological standard reference samples and used to back-calculate light contributions from different depths in the sample after taking a series of images across a range of vertical distances through the sample. This method can provide excellent quantitative precision and linearity [69], but it doesn't work well for thicker samples with multiple cell layers ($z > 20$ μm), for which confocal or multiphoton microscopy are preferred [70].

3. *Multiphoton microscopy.* The central concept of multiphoton (most commonly two-photon) microscopy is that if a fluorophore absorbs more than one photon in a sufficiently short period of time, the absorbed energy is equivalent to a single photon of half the wavelength (i.e., twice the energy). For example, simultaneous absorption of two infrared photons can excite a fluorescein molecule in the same fashion as a single blue photon. The most common light sources for this method are Ti:sapphire lasers emitting high-intensity pulses of infrared light at wavelengths tunable from 700 to 1,000 nm, with each pulse several hundred femtoseconds in duration at the sample plane and about 100 million pulses per second. Because infrared light is not well absorbed by many naturally occurring biological chromophores, this intense illumination causes surprisingly negligible photodamage to living cells and tissues, and significantly less photobleaching of the fluorophore of interest outside the focal plane [71]. This lesser absorption

of infrared light can also enable imaging as deep as several hundred microns into living tissue.

4. *Super-resolution microscopy.* To study structural details at length scales below the Rayleigh limit, fluorescence microscopy can exploit diffraction-structured illumination, single-molecule excitation, or near-field illumination point sources to image at resolutions one or even two orders of magnitude smaller than the illuminating light [72]. The far-field illumination approaches were the subject of the 2014 Nobel Prize in Chemistry for super-resolved light microscopy. Structured illumination methods, such as stimulated emission depletion (STED), use diffraction-pattern illumination to suppress fluorescence emissions from a ring surrounding a central illuminating spot, achieving resolutions of 10–50 nm with organic dyes and fluorescent proteins. In a fundamentally different approach, single-molecule fluorophore point sources are randomly activated in the photoactivated localization microscopy (PALM) and stochastic optical reconstruction microscopy (STORM) techniques. PALM and STORM can achieve spatial resolutions of about 10 nm.

Flow cytometry A flow cytometer is an instrument for measuring the multicolor fluorescence intensity of single cells at high speeds (up to 50,000 cells per second with current instruments). In some instruments, individual cells can be electrostatically isolated using their fluorescence profiles as a criterion (figure 2.40).

Why might it be important to measure the properties of individual cells when techniques such as Western blots and mass spectrometry can measure the average properties of the mass of cells in a population? In many instances, population

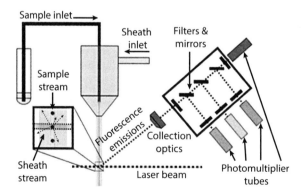

Figure 2.40 Schematic diagram of a flow cytometer. If an instrument additionally has the capability to sort individual cells, it is called a fluorescence-activated cell sorter (FACS). Multicolor fluorescence emissions from single cells can be quantified as fast as 50,000 cells per second. In FACS instruments, the exiting stream is broken into droplets, and cells matching defined fluorescence criteria can be electrostatically sorted into separate volumes.

Noncovalent Binding Interactions

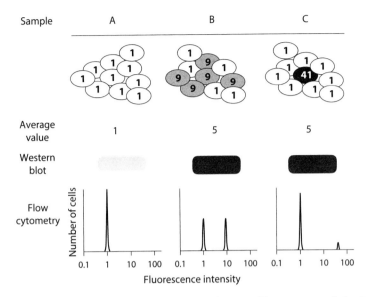

Figure 2.41 A schematic example of how flow cytometry data reveal important population heterogeneity characteristics [73]. The two subpopulations on the right have an average value of 5-fold higher levels of some factor, as revealed by Western blot (middle row). However, by staining each cell individually for this factor and measuring the distribution of single-cell fluorescence by flow cytometry, it becomes clear that half the cells are 9-fold amplified in one case, while only 10% of cells have a 41-fold amplification in the other case.

averages mask biologically significant variation of subpopulations of cells (e.g., the variability in the population may not be normally distributed), and thus they do not accurately capture the underlying phenomena, as illustrated schematically in figure 2.41. Such situations occur frequently when studying the immune system, because it is made up of many different, highly specialized cells [73]. In fact, flow cytometry was initially developed by Herzenberg et al. to study lymphocyte subpopulations labeled with fluorescent antibodies [74]. However, this method is now widely used to study and isolate cells in heterogeneous populations—from bone marrow, to ocean plankton, to combinatorial microbial libraries used in directed evolution.

Suggestions for Further Reading

C. R. Cantor and P. R. Schimmel. *Biophysical Chemistry: Part I: The Conformation of Biological Macromolecules*. W. H. Freeman & Company, 1980.

T. E. Creighton. *Proteins: Structures and Molecular Properties*, second edition. W. H. Freeman & Company, 1992.

A. R. Fersht. *Structure and Mechanism in Protein Science: A Guide to Enzyme Catalysis and Protein Folding*, third edition. Worth Publishing, 1998.

G. G. Hammes, *Thermodynamics and Kinetics for the Biological Sciences*. John Wiley & Sons, 2000.

J. Kuriyan, B. Konforti, and D. Wemmer. *The Molecules of Life: Physical and Chemical Principles*. Garland Science, 2013.

J. R. Lakowicz. *Principles of Fluorescence Spectroscopy*, third edition. Springer, 2011.

H. M. Shapiro. *Practical Flow Cytometry*, fourth edition. Wiley-Liss, 2003.

Problems

2-1 The van der Waals potential between two atoms has a minimum, which represents the most favorable distance of approach, R_{min}, with favorable interaction energy, $-\varepsilon$. Derive relationships for R_{min} and ε in terms of A and B in equation 2.36. The parameter σ gives the interatomic separation at which the van der Waals energy is zero. Derive an expression for σ in terms of R_{min}. For a pair of atoms whose Lennard-Jones parameters are $A = 3{,}251$ $\text{Å}^{12}\,\text{kcal mol}^{-1}$ and $B = 34.021$ $\text{Å}^{6}\,\text{kcal mol}^{-1}$, what is the most favorable interaction distance?

2-2 Plot the electrostatic interaction as a function of distance for charged and polar groups. For the charged groups, use point charges of $+1e$ and $-1e$; for the polar groups, use a pair of point charges of magnitude $+0.5e$ and $-0.5e$ separated by a distance of 1 Å. Plot charge–charge, charge–polar, and polar–polar interactions across a distance range of 2–20 Å with $\varepsilon = 2$, aligning groups in their most favorable orientation.

2-3 What is the magnitude of the Coulombic interaction energy between an oxygen atom and a hydrogen atom participating in a hydrogen bond in water, assuming that the partial charge on the oxygen is $-0.8e$, and the partial charge on the hydrogen is $+0.4e$? What is the interaction energy of this hydrogen bond in the interior of a protein? (Inspired by Tinoco et al. [75].)

2-4 The association rate constant for a protein-ligand pair is 1.5×10^5 $\text{M}^{-1}\,\text{s}^{-1}$ at 150 mM NaCl, and 7.1×10^4 $\text{M}^{-1}\,\text{s}^{-1}$ at 500 mM NaCl. What value for k_{on} would you predict in neat water (i.e., zero ionic strength)? What value for k_{on} would you predict in 10 mM $MgCl_2$?

2-5 Can ionic strength affect the equilibrium binding constant for a reaction, or only the kinetic constants? Briefly justify your answer.

2-6 Imagine a pair of wild-type proteins that bind to each other with an association rate constant of 5.78×10^5 M^{-1} s^{-1} at zero ionic strength. One binding partner (the receptor) has a net charge of $+2$, and the other (the ligand) has a net charge of -2. Imagine further that a mutant of the ligand is created that has a net charge of -4 (rather than -2) and is found to bind with an association rate constant of 6.21×10^5 M^{-1} s^{-1} at zero ionic strength. At what ionic strength is the association rate for the mutant equal to the association rate for the wild type at an ionic strength of 0.1 molal?

2-7 What is the standard-state free energy change $\Delta G°$ for a protein-ligand complex that binds with an equilibrium constant $K_d = 0.1$ nM? If the affinity of the complex is increased such that $K_d = 0.01$ nM, what is the value of $\Delta \Delta G°$?

2-8 A protein and a ligand mixed in a solution can bind to form a complex. After waiting a sufficiently long time for equilibrium to be reached, the concentrations of the protein, the ligand, and the complex in the 1 mL volume are measured to be 0.3 µM, 0.8 µM, and 6.5 µM, respectively. The association rate constant is 3×10^6 M^{-1}s^{-1}. (Contributed by L. Jeng.)

   a. What is the rate of complex formation?
   b. What is the rate of complex dissociation?
   c. What is the dissociation rate constant?
   d. What is the equilibrium dissociation constant?
   e. In a negligible additional volume, 100 picomoles of ligand are spiked into the mixture. What will be the new equilibrium complex concentration?

2-9 You are studying the binding equilibria between two proteins, A and B, in solution. You have determined the K_d for the pair at 37°C to be 25 nM. However, you are now doing an experiment at room temperature (25°C), and you need to know K_d for your experiment at this temperature. (Contributed by D. Liu.)

   a. What additional information or physicochemical properties do you need to know to calculate K_d at room temperature?
   b. Using the following values (at 37°C and pH = 7, unless otherwise noted), perform the calculation. State all assumptions used.

$$\Delta H° = -17.3 \text{ kcal/mol}$$
$$\Delta S° = -0.02 \text{ kcal/(mol} \cdot \text{K)}$$
$$\Delta C_p° = -0.25 \text{ kcal/(mol} \cdot \text{K), constant between 25°C and 37°C}$$
$$\text{MW of protein } A = 37 \text{ kDa}$$
$$\text{MW of protein } B = 37 \text{ kDa}$$

$$\text{pI of protein } A = 5.3$$
$$\text{pI of protein } B = 6.2$$
$$k_{on} = 2 \times 10^6 \text{ M}^{-1}\text{s}^{-1} \text{ at } 25°\text{C}$$

2-10 If water were a linear molecule, would it still be polar? Would water molecules hydrogen bond together? (Inspired by Chang [76].)

2-11 A water molecule fixed in an inorganic salt crystal loses approximately 10 cal mol^{-1} K^{-1} of entropy compared to a water molecule in solution [77]. Assuming this to be an approximate maximum value, by how much can K_d change and in which direction at 25°C when an additional water molecule is fixed in a protein-ligand complex?

2-12 A given protein is 50% unfolded at its $T_m = 75°$C, and the enthalpy change of unfolding is $\Delta H(T_m) = 100$ kcal/mol. The heat capacity change on unfolding is $\Delta C_p = 1{,}000$ cal mol^{-1} K^{-1} and is approximately independent of temperature. Plot the protein stability curve ($\Delta G(T)$ versus T) from 0 to 100°C.

2-13 The enthalpy of a protein-ligand binding interaction is measured at two temperatures: $\Delta H°(25°\text{C}) = -10$ kcal/mol, and $\Delta H°(50°\text{C}) = -20$ kcal/mol. At 25°C, the affinity constant K_d is 4 nM. What is the value of K_d at 10°C?

2-14 What would be the absorbance of a 2 mg/mL hen egg lysozyme solution?

2-15 The following binding affinities were measured for complex formation of an antibody and antigen at 298 K, as well as mutants of each as denoted:

Antibody	Antigen	K_d (nM)
Wild type	Wild type	12.5
Y32A	Wild type	227
Wild type	Q121A	1,610
Y32A	Q121A	1,000
Wild type	I124A	100
Y32A	I124A	1,700

Use thermodynamic analysis of a double-mutant cycle to determine the interaction free energy between Y32 and Q121. Also perform such an analysis for Y32 and I124. Based on these results, is Y32 more likely to make contact with residue Q121 or I124 [28]?

2-16 Show that the differences between $\Delta\Delta G°$ for any pair of parallel faces on a double-mutant-cycle cube (figure 2.15) are the same. (Contributed by A. Gai.)

2-17 A cell-surface receptor ligand binding affinity was measured to be $K_d = 0.5$ nM at 4°C to prevent receptor endocytic trafficking. Using the approximate relationship represented in figure 2.4, estimate the range of likely values for K_d at 37°C.

2-18 Data for the thermal unfolding of the oxidized form of the protein plastocyanin from the cyanobacteria *Synechocystis* is as follows [78]:

Temperature (°C)	Equilibrium folded fraction
62.5	0.87
65.0	0.67
67.5	0.40
70.0	0.20
72.5	0.08
75.0	0.05

Assuming that $\Delta H°$ is constant over the given temperature range, determine $\Delta H°$ for the unfolding process. (Contributed by R. Krishnan.)

2-19 To investigate the properties of protein-ligand binding at various solution pH values, a biophysical experiment is set up to measure the free energy of binding of these biomolecules. The four states of the protein and ligand are as shown in figure 2.14. Suppose that $\Delta G_{bind}^{o,L}$ is found to be –8.0 kcal/mol, and $\Delta G_{bind}^{o,LH^+}$ is found to be –10 kcal/mol at a temperature of 298 K. At a pH of 7.5, the apparent equilibrium dissociation constant of the protein-ligand complex is determined to be 1.96×10^{-7} M. Given this information, determine the fraction of unbound ligand that is deprotonated at pH 7.5. (Contributed by J. Spangler.)

2-20 You wish to study the interaction between a protein and a ligand, and you have developed a special technique, with which you can directly detect the number of complexes in solution. You start with a solution that initially contains 1,000 nM protein and 1,000 nM ligand, and you wait until equilibrium is reached. At 37°C, the concentration of complexes at equilibrium is 729.8 nM; at 25°C, this concentration is 758.2 nM; and at 20°C, this concentration is 763.8 nM.

   a. What are the K_d values for this interaction at 37°C, 25°C, and 20°C?

   b. Given the result in part a and without invoking additional assumptions, calculate $\Delta H°$ at 30°C.

2-21 Tyrosine is frequently found at protein-protein interfaces. Briefly describe at least three interaction energies or forces that help explain this observation.

2-22 A protein and ligand, each with a molecular weight of 20,000 g/mol, bind reversibly in solution to form a noncovalent complex. Kinetic and equilibrium binding experiments were performed at different pH values at 25°C, and the results are shown in the table below. These experiments were performed in very low ionic strength buffers. The reported k_{obs} is for the association phase of a pseudo-first-order binding reaction where the concentration of the excess species is 10 nM.

pH	k_{obs} (s^{-1})	$\Delta G°$ (kcal/mol)
2	9.8	−12.3
7	870.0	−15.0
11	10.4	−11.9

a. Estimate the theoretical maximum association rate constant for these proteins in the limit of very high ionic strength (i.e., high salt concentration).

b. Determine the association rate constant for formation of the complex at each of the pH values in the table above.

c. Propose an explanation for the trend observed in part b above. Compare the values for the association rate constant determined in part b to the theoretical maximum value calculated in part a; is there a discrepancy? Explain.

References

1. R. J. Ellis. Macromolecular crowding: Obvious but underappreciated. *Trends Biochem. Sci.* 26: 597–603 (2001).

2. O. Medalia, I. Weber, A. S. Frangakis, D. Nicastro, G. Gerisch, and W. Baumeister. Macromolecular architecture in eukaryotic cells visualized by cryoelectron tomography. *Science* 298: 1209–1213 (2002).

3. D. S. Goodsell. Illustrations for public use. http://mgl.scripps.edu/people/goodsell/illustration/public.

4. T. Ando and J. Skolnick. Crowding and hydrodynamic interactions likely dominate in vivo macromolecular motion. *Proc. Natl. Acad. Sci. U.S.A.* 107: 18,457–18,462 (2010).

5. D. S. Goodsell. Inside a living cell. *Trends Biochem. Sci.* 16: 203–206 (1991).

6. G. Schreiber, G. Haran, and H.-X. Zhou. Fundamental aspects of protein–protein association kinetics. *Chem. Rev.* 109: 839–860 (2009).

7. B. M. Baker and K. P. Murphy. Prediction of binding energetics from structure using empirical parameterization. *Methods Enzymol.* 295: 294–315 (1998).

8. F. P. Schwarz, D. Tello, F. A. Goldbaum, R. A. Mariuzza, and R. J. Poljak. Thermodynamics of antigen-antibody binding using specific anti-lysozyme antibodies. *Eur. J. Biochem.* 228: 388–394 (1995).

9. D. E. Hyre and L. D. Spicer. Thermodynamic evaluation of binding interactions in the methionine repressor system of *Escherichia coli* using isothermal titration calorimetry. *Biochemistry* 34: 3212–3221 (1995).

10. K. A. Hibbits, D. S. Gill, and R. C. Willson. Isothermal titration calorimetric study of the association of hen egg lysozyme and the anti-lysozyme antibody HyHEL-5. *Biochemistry* 33: 3584–3590 (1994).

11. A. Cooper, C. M. Johnson, J. H. Lakey, and M. Nöllmann. Heat does not come in different colours: Entropy-enthalpy compensation, free energy windows, quantum confinement, pressure perturbation calorimetry, solvation and the multiple causes of heat capacity effects in biomolecular interactions. *Biophys. Chem.* 93: 215–230 (2001).

12. E. Gallicchio, M. M. Kubo, and R. M. Levy. Entropy-enthalpy compensation in solvation and ligand binding revisited. *J. Am. Chem. Soc.* 120: 4526–4527 (1998).

13. R. S. Spolar and M. T. Record, Jr. Coupling of local folding to site-specific binding of proteins to DNA. *Science* 263: 777–784 (1994).

14. J. M. Sturtevant. Heat capacity and entropy changes in processes involving proteins. *Proc. Natl. Acad. Sci. U.S.A.* 74: 2236–2240 (1977).

15. R. Perozzo, G. Folkers, and L. Scapozza. Thermodynamics of protein-ligand interactions: History, presence, and future aspects. *J. Recept. Signal Transduct. Res.* 24: 1–52 (2004).

16. K. A. Sharp and B. Honig. Electrostatic interactions in macromolecules: Theory and applications. *Annu. Rev. Biophys. Biophys. Chem.* 19: 301–332 (1990).

17. M. E. Davis and J. A. McCammon. Electrostatics in biomolecular structure and dynamics. *Chem. Rev.* 90: 509–521 (1990).

18. D. Chandler. Hydrophobicity: Two faces of water. *Nature* 417: 491 (2002).

19. B. R. Brooks, R. E. Bruccoleri, B. D. Olafson, D. J. States, S. Swaminathan, and M. Karplus. CHARMM: A program for macromolecular energy, minimization, and dynamics calculations. *J. Comput. Chem.* 4: 187–217 (1983).

20. D. Sitkoff, K. A. Sharp, and B. Honig. Accurate calculation of hydration free energies using macroscopic solvent models. *J. Phys. Chem.* 98: 1978–1988 (1994).

21. R. B. Hermann. Theory of hydrophobic bonding. II. The correlation of hydrocarbon solubility in water with solvent cavity surface area. *J. Phys. Chem.* 76: 2754–2759 (1972).

22. L. Lo Conte, C. Chothia, and J. Janin. The atomic structure of protein-protein recognition sites. *J. Mol. Biol.* 285: 2177–2198 (1999).

23. J. R. Bradford and D. R. Westhead. Improved prediction of protein-protein binding sites using a support vector machines approach. *Bioinformatics* 21: 1487–1494 (2005).

24. B. C. Cunningham and J. A. Wells. High-resolution epitope mapping of hGH–receptor interactions by alanine scanning mutagenesis. *Science* 244: 1081–1085 (1989).

25. A. A. Bogan and K. S. Thorn. Anatomy of hot spots in protein interfaces. *J. Mol. Biol.* 280: 1–9 (1998).

26. T. N. Bhat, G. A. Bentley, G. Boulot, M. I. Greene, D. Tello, W. Dall'Acqua, H. Souchon, F. P. Schwarz, R. A. Mariuzza, and R. J. Poljak. Bound water molecules and conformational stabilization help mediate an antigen-antibody association. *Proc. Natl. Acad. Sci. U.S.A.* 91: 1089–1093 (1994).

27. W. Dall'Acqua, E. R. Goldman, E. Eisenstein, and R. A. Mariuzza. A mutational analysis of the binding of two different proteins to the same antibody. *Biochemistry* 35: 9667–9676 (1996).

28. W. Dall'Acqua, E. R. Goldman, W. Lin, C. Teng, D. Tsuchiya, H. Li, X. Ysern, B. C. Braden, Y. Li, S. J. Smith-Gill, and R. A. Mariuzza. A mutational analysis of binding interactions in an antigen-antibody protein-protein complex. *Biochemistry* 37: 7981–7991 (1998).

29. P. J. Carter, G. Winter, A. J. Wilkinson, and A. R. Fersht. The use of double mutants to detect structural changes in the active site of the tyrosyl-tRNA synthetase (*Bacillus stearothermophilus*). *Cell* 38: 835–840 (1984).

30. Z. S. Hendsch and B. Tidor. Do salt bridges stabilize proteins? A continuum electrostatic analysis. *Protein Sci.* 3: 211–226 (1994).

31. D. J. Barlow and J. M. Thornton. Ion-pairs in proteins. *J. Mol. Biol.* 168: 867–885 (1983).

32. C. D. Waldburger, J. F. Schildbach, and R. T. Sauer. Are buried salt bridges important for protein stability and conformational specificity? *Nat. Struct. Biol.* 2: 122–128 (1995).

33. W. C. Wimley, K. Gawrisch, T. P. Creamer, and S. H. White. Direct measurement of salt-bridge solvation energies using a peptide model system: Implications for protein stability. *Proc. Natl. Acad. Sci. U.S.A.* 93: 2985–2990 (1996).

34. L. Xiao and B. Honig. Electrostatic contributions to the stability of hyperthermophilic proteins. *J. Mol. Biol.* 289: 1435–1444 (1999).

35. B. E. Willcox, G. F. Gao, J. R. Wyer, J. E. Ladbury, J. I. Bell, B. K. Jakobsen, and P. A. van der Merwe. TCR binding to peptide-MHC stabilizes a flexible recognition interface. *Immunity* 10: 357–365 (1999).

36. K. N. Allen, A. Lavie, G. K. Farber, A. Glasfeld, G. A. Petsko, and D. Ringe. Isotopic exchange plus substrate and inhibition kinetics of D-xylose isomerase do not support a proton-transfer mechanism. *Biochemistry* 33: 1481–1487 (1994).

37. A. Farewell and F. C. Neidhardt. Effect of temperature on in vivo protein synthetic capacity in *Escherichia coli*. *J. Bacteriol.* 180: 4704–4710 (1998).

38. C. A. Bessey and E. A. Bessey. Further notes on thermometer crickets. *Am. Nat.* 32: 263–264 (1898).

39. W. J. Crozier. On the critical thermal increment for the locomotion of a diplopod. *J. Gen. Physiol.* 7: 123–136 (1924).

40. C. A. Aleida and V. H. Snyder. The flashing interval of fireflies—its temperature coefficient—an explanation of synchronous flashing. *Am. J. Physiol.* 51: 536–542 (1920).

41. H. Shapley. Thermokinetics of *Liometopum apiculatum* Mayr. *Proc. Natl. Acad. Sci. U.S.A.* 6: 204–211 (1920).

42. W. J. Crozier. On curves of growth, especially in relation to temperature. *J. Gen. Physiol.* 10: 53–73 (1926).

43. H. Laurens. The influence of temperature on the rate of the heart beat in *Amblystoma* embryos. *Am. J. Physiol.* 35: 199–210 (1914).

44. W. J. Crozier, E. Wolf, and G. Zerrahn-Wolf. Temperature and the critical intensity for response to visual flicker. *Proc. Natl. Acad. Sci. U.S.A.* 24: 216–221 (1938).

45. M. J. Nye. Working tools for theoretical chemistry: Polanyi, Eyring, and debates over the semiempirical method. *J. Comput. Chem.* 28: 98–108 (2007).

46. P. Debye and E. Hückel. The theory of electrolytes. I. Lowering of freezing point and related phenomena. *Physik. Z.* 24: 185–206 (1923).

47. M. Vijayakumar, K.-Y. Wong, G. Schreiber, A. R. Fersht, A. Szabo, and H.-X. Zhou. Electrostatic enhancement of diffusion-controlled protein-protein association: Comparison of theory and experiment on barnase and barstar. *J. Mol. Biol.* 278: 1015–1024 (1998).

48. P. Debye. Reaction rates in ionic solutions. *Trans. Electrochem. Soc.* 82: 265–272 (1943).

49. G. Schreiber and A. R. Fersht. Rapid, electrostatically assisted association of proteins. *Nat. Struct. Biol.* 3: 427–431 (1996).

50. H. Du, R.-C. A. Fuh, J. Li, L. A. Corkan, and J. S. Lindsey. PhotochemCAD: A computer-aided design and research tool in photochemistry. *Photochem. Photobiol.* 68: 141–142 (1998).

51. C. N. Pace, F. Vajdos, L. Fee, G. Grimsley, and T. Gray. How to measure and predict the molar absorption coefficient of a protein. *Protein Sci.* 4: 2411–2423 (1995).

52. UniProtKB/Swiss-Prot. http://www.expasy.org/sprot/.

53. GenBank. http://www.ncbi.nlm.nih.gov/Genbank/.

54. NCBI PubMed. http://www.ncbi.nlm.nih.gov/Entrez/.

55. N. Panchuk-Voloshina, R. P. Haugland, J. Bishop-Stewart, M. K. Bhalgat, P. J. Millard, F. Mao, W. Y. Leung, and R. P. Haugland. Alexa dyes, a series of new fluorescent dyes that yield exceptionally bright, photostable conjugates. *J. Histochem. Cytochem.* 47: 1179–1188 (1999).

56. L. Song, E. J. Hennink, I. T. Young, and H. J. Tanke. Photobleaching kinetics of fluorescein in quantitative fluorescence microscopy. *Biophys. J.* 68: 2588–2600 (1995).

57. V. Kasche and L. Lindqvist. Reactions between the triplet state of fluorescein and oxygen. *J. Phys. Chem.* 68: 817–823 (1964).

58. M. Zimmer. Green fluorescent protein (GFP): Applications, structure, and related photophysical behavior. *Chem. Rev.* 102: 759–781 (2002).

59. N. C. Shaner, P. A. Steinbach, and R. Y. Tsien. A guide to choosing fluorescent proteins. *Nat. Methods* 2: 905–909 (2005).

60. M.-A. Elsliger, R. M.Wachter, G. T. Hanson, K. Kallio, and S. J. Remington. Structural and spectral response of green fluorescent protein variants to changes in pH. *Biochemistry* 38: 5296–5301 (1999).

61. P. van Roessel and A. H. Brand. Imaging into the future: Visualizing gene expression and protein interactions with fluorescent proteins. *Nat. Cell Biol.* 4: E15–E20 (2002).

62. B. G. Reid and G. C. Flynn. Chromophore formation in green fluorescent protein. *Biochemistry* 36: 6786–6791 (1997).

63. A. N. Glazer. Light guides. Directional energy transfer in a photosynthetic antenna. *J. Biol. Chem.* 264: 1–4 (1989).

64. S. Ritter, R. G. Hiller, P. M. Wrench, W. Welte, and K. Diederichs. Crystal structure of a phycourobilin-containing phycoerythrin at 1.90-Å resolution. *J. Struct. Biol.* 126: 86–97 (1999).

65. K. A. Lukyanov, D. M. Chudakov, S. Lukyanov, and V. V. Verkhusha. Innovation: Photoactivatable fluorescent proteins. *Nat. Rev. Mol. Cell Biol.* 6: 885–891 (2005).

66. S. R. Adams and R. Y. Tsien. Controlling cell chemistry with caged compounds. *Annu. Rev. Physiol.* 55: 755–784 (1993).

67. R. D. Goldman, J. R. Swedlow, and D. L. Spector (eds.) *Live Cell Imaging: A Laboratory Manual*, second edition. Cold Spring Harbor Laboratory Press (2010).

68. B. J. Nichols, A. K. Kenworthy, R. Polishchuk, R. Lodge, T. H. Roberts, K. Hirschberg, R. D. Phair, and J. Lippincott-Schwartz. Rapid cycling of lipid raft markers between the cell surface and Golgi complex. *J. Cell Biol.* 153: 529–542 (2001).

69. J. R. Swedlow, K. Hu, P. D. Andrews, D. S. Roos, and J. M. Murray. Measuring tubulin content in *Toxoplasma gondii*: A comparison of laser-scanning confocal and wide-field fluorescence microscopy. *Proc. Natl. Acad. Sci. U.S.A.* 99: 2014–2019 (2002).

70. P. Periasamy A. Skoglund, C. Noakes, and R. Keller. An evaluation of two-photon excitation versus confocal and digital deconvolution fluorescence microscopy imaging in *Xenopus* morphogenesis. *Microsc. Res. Tech.* 47: 172–181 (1999).

71. P. T. So, C. Y. Dong, B. R. Masters, and K. M. Berland. Two-photon excitation fluorescence microscopy. *Annu. Rev. Biomed. Eng.* 2: 399–429 (2000).

72. B. Huang, M. Bates, and X. Zhuang. Super-resolution fluorescence microscopy. *Annu. Rev. Biochem.* 78: 993–1016 (2009).

73. P. O. Krutzik, J. M. Irish, G. P. Nolan, and O. D. Pérez. Analysis of protein phosphorylation and cellular signaling events by flow cytometry: Techniques and clinical applications. *Clin. Immunol.* 110: 206–221 (2004).

74. L. A. Herzenberg, D. Parks, B. Sahaf, O. Pérez, M. Roederer, and L. A. Herzenberg. The history and future of the fluorescence activated cell sorter and flow cytometry: A view from Stanford. *Clin. Chem.* 48: 1819–1827 (2002).

75. I. Tinoco, K. Sauer, J. C. Wang, J. D. Puglisi, G. Harbison, and D. Rovnyak. *Physical Chemistry: Principles and Applications in Biological Sciences*, fifth edition. Pearson (2013).

76. R. Chang. *Chemistry*. McGraw-Hill (2010).

77. J. D. Dunitz. The entropic cost of bound water in crystals and biomolecules. *Science* 264: 670 (1994).

78. M. J. Feio, A. Diaz-Quintana, J. A. Navarro, and M. A. De La Rosa. Thermal unfolding of plastocyanin from the mesophilic cyanobacterium *Synechocystis* sp. PCC 6803 and comparison with its thermophilic counterpart from *Phormidium laminosum*. *Biochemistry* 45: 4900–4906 (2006).

3

Binding Equilibria and Kinetics

There exists everywhere a medium in things, determined by equilibrium.
—Dmitri Mendeleev

Nobody, I suppose, could devote many years to the study of chemical kinetics without being deeply conscious of the fascination of time and change: this is something that goes outside science into poetry; but science, subject to the rigid necessity of always seeking closer approximations to the truth, itself contains many poetical elements.
—Sir Cyril Hinshelwood, Nobel Lecture, 1956

In this chapter, we develop basic mathematical relationships to describe the rates of biomolecular interactions underpinning biological function. We first consider monovalent protein-ligand interactions, then the effects of multivalent interactions and energetic communication between binding sites. We also examine a broad variety of approaches for measuring biomolecular interactions. Finally, we briefly outline the methods for numerically extracting parameter estimates—and the uncertainties of those estimates—from data.

3.1 Equilibrium Monovalent Protein-Ligand Binding

3.1.1 Monovalent Binding Isotherm

For given initial concentrations of protein and ligand, what fraction of the molecules is bound in complex at equilibrium? The simplest case to consider is when a single protein binds a single ligand to form a complex (figure 3.1), where $P =$ protein, $L =$ ligand, and $C =$ complex, with the equilibrium dissociation constant K_d defined as

$$K_d \equiv \frac{[P]_{eq}[L]_{eq}}{[C]_{eq}} \qquad (3.1)$$

in which the brackets and subscript $[\ ]_{eq}$ indicate a concentration at equilibrium. The subscript "eq" will generally be dropped for convenience in situations where equilibrium conditions are clearly under consideration.

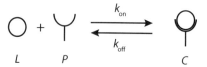

Figure 3.1 One molecule of ligand L can bind to one molecule of protein P to form one molecular complex C. Reciprocally, one C can dissociate to yield one L and one P. When the rates of binding and dissociation are equal, the system is at equilibrium.

The equilibrium dissociation constant K_d is experimentally determined by taking measurements of $[C]$ as a function of the initial concentrations $[P]_o$ and $[L]_o$. Such an experiment is called a titration. The resulting curve of $[C]$ versus $[L]_o$ is referred to as a "binding isotherm," because the experiment is generally performed at constant temperature and is phenomenologically analogous to Langmuir's classic isothermal adsorption experiments. The usual experimental strategy is to add increasing amounts of ligand $[L]_o$ to a fixed initial amount of protein $[P]_o$ and then indirectly measure the concentration of the complex $[C]$. We must wait until equilibrium is attained (i.e., $[C]$ is time invariant), and we'll shortly consider how long that takes (i.e., the kinetics of approach to equilibrium). Generally, the measured experimental variable is proportional to the fractional site saturation y, which is the number of occupied binding sites divided by the total number of sites:

$$y \equiv \frac{[C]}{[P]_o} = \frac{[C]}{[P]+[C]} \tag{3.2}$$

$$0 \leq y \leq 1 \tag{3.3}$$

A spectroscopic signal proportional to y, such as fluorescence or light absorption, is often measured; alternatively, the complex is detected by its altered physical location due to immobilization on a solid phase. Substituting equation 3.1 into equation 3.2, the following relationship is obtained:

$$y_{eq} = \frac{[L]_{eq}}{[L]_{eq}+K_d} \tag{3.4}$$

Although equation 3.4 is always correct for a system at equilibrium, it should be noted that in general, the free ligand concentration $[L]_{eq} \neq [L]_o$, because some of the initial free ligand is depleted by inclusion in complexes. But measuring the concentration of free ligand, while excluding from the measurement any ligand bound in complex, is often problematic. To simplify the analysis, one often designs experimental conditions such that $[L]_o \gg [P]_o$, which leads to the simplification $[L]_{eq} = [L]_o - [C]_{eq} \approx [L]_o$. That is, the ligand is in such great excess over the protein that even if all binding sites are occupied, the ligand concentration will decrease by a neglible amount. (Nonintuitively, it is also possible for ligand depletion to be small even if $[L]_o \ll [P]_o$, as long as $[P]_o \ll K_d$. This proof is left to the reader in homework problem 3–4.) Therefore, under conditions of excess ligand,

$$y_{eq} \approx \frac{[L]_o}{[L]_o + K_d} \tag{3.5}$$

Under this approximation, additional meaning can be directly extracted from the K_d value, which is in units of concentration. When $[L]_o = K_d$, half of the protein molecules are in complex with ligand. Thus, the K_d value represents the tipping point in initial ligand concentration between a majority unbound protein state and a majority bound protein state. (Note that the same logic can be applied to the general result, equation 3.4, but the relevant ligand concentration in that case is $[L]_{eq}$, which is harder to measure and relate to the initial ligand input.)

Sometimes ligand cannot be added in great excess to protein, perhaps because the ligand is too expensive, solubility limits are reached, signal sensitivity is insufficient to lower protein concentration $[P]_o$ further, or such a setup would not be faithful to the biology being studied. In this case, ligand depletion must be accounted for explicitly. To do this, $[L] = [L]_o - [C] = [L]_o - y[P]_o$ is substituted into equation 3.4 to obtain

$$y = \frac{[L]_o - y[P]_o}{[L]_o - y[P]_o + K_d} \tag{3.6}$$

Solving for y by the quadratic equation produces the following rather ungainly result:

$$y = \frac{K_d + [L]_o + [P]_o - \sqrt{(K_d + [L]_o + [P]_o)^2 - 4[P]_o[L]_o}}{2[P]_o} \tag{3.7}$$

(The positive root of the quadratic equation leads to nonsensical y values, as evidenced when evaluating it at $[L]_o = 0$.)

From an experimentally obtained binding isotherm (i.e., a plot of a signal proportional to y versus $[L]_o$), a value for K_d can be obtained by nonlinear least squares regression (as discussed in section 3.9) of equation 3.5 or 3.7, whichever is appropriate. Specific experimental details to bear in mind when performing such a fit are: fitting a proportionality constant α between the measured signal and y; the contribution of nonzero signal baselines; and contributions to the signal from unbound ligand and protein. It should be emphasized that equation 3.7 is always valid for equilibrium monovalent protein-ligand binding. If ligand depletion is negligible, equations 3.5 and 3.7 converge to the same result; however, if ligand depletion is significant, only equation 3.7 will yield the correct K_d when applied to experimental data.

Often, the protein in the reaction shown in figure 3.1 is bound to a solid phase, either a cell surface or a device. The description of this system is very similar but will be developed more fully in chapter 8, where we will account for the potential impact of diffusion on observed rate processes.

Example 3-1   How much monoclonal antibody must be added to quantitatively immunoprecipitate 95% of a particular protein that is present at 1 pmol in a 100 µL protein extract?

Solution   To ensure that most of the particular protein is bound by antibody, one must use a concentration of antibody that is both greater than K_d and also greater than the concentration of the protein itself. The protein's concentration is (1 pmol)/(100 µL) = 10 nM. For most monoclonal and polyclonal antibodies used in research, a value for K_d is not known. However, many such antibodies have a binding constant $K_d <$ 10 nM if they are of general use [1].

With 95% saturation as our goal and using equation 3.5,

$$0.95 = \frac{[Antibody\ binding\ sites]}{[Antibody\ binding\ sites] + 10\ nM} \tag{3.8}$$

so [Antibody binding sites] = 190 nM. A monoclonal IgG antibody has two binding sites per molecule. A common stock concentration for such an antibody is 50 µg/mL, which corresponds to a $\approx$ 667 nM concentration of antibody binding sites. However, this stock must be added to the 100 µL of protein extract, so it will be diluted in the assay. The addition of 40 µL of 50 µg/mL antibody to the 100 µL of protein extract will yield 190 nM concentration of antibody binding sites, which results in fractional complexation of

$$y = \frac{190\ nM}{190\ nM + 10\ nM} = 0.95 \tag{3.9}$$

We anticipated that the amount of antibody required would far exceed the amount of protein in the sample, so we chose to use equation 3.5, which assumes no ligand depletion. As verification that this was a reasonable assumption, we can compare $[L]_o$ (190 nM) to $[P]_o$ (10 nM). Note that even if every single protein were bound by a ligand, the ligand would be depleted by only 10 nM, which corresponds to a reduction of 5%. Thus, our assumption of no ligand depletion is a reasonable one.

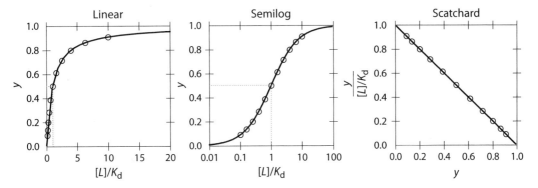

Figure 3.2 Common graphical representations for binding-isotherm data. The semilog plot most readily allows visual estimation of the half-maximum point. Also note that choosing ligand concentrations to be evenly spaced on a logarithmic coordinate places the data points evenly across the curve. In the left and middle panels, the half-saturation of binding ($y = 0.5$) is indicated by a dashed gray line, and the corresponding x-axis value of $[L]/K_d$ is unity.

3.1.2 Graphical Representations

Binding isotherms are plotted in a variety of ways, as shown in figure 3.2. The straightforward linear representation is perhaps the least useful, as it collapses the portion of the curve where the complex fraction y is changing most rapidly versus $[L]$ to a small portion of the plot. Also, note that it is difficult to accurately estimate the maximum of the curve from the ten data points alone, because even at ligand concentrations 10 times greater than K_d, binding is only 91% saturated. The semilogarithmic plot usefully spreads the data points, providing a better visual estimate of the half-maximal saturation point, where $[L] = K_d$. Another format, the Scatchard plot, was initially developed to linearize data for easier manual curve fitting, prior to the development of facile computational curve fitting. Rearrangement of equation 3.4 gives the following relationship:

$$\frac{y}{[L]} = \frac{-1}{K_d}y + \frac{1}{K_d} \qquad (3.10)$$

A Scatchard plot is often used to visually demonstrate that binding behavior differs from a simple monovalent isotherm: If a plot of $y/[L]$ versus y is not a straight line, the data are inconsistent with monovalent binding as described by equation 3.4.

3.2 Binding Kinetics

How long must one wait until a binding reaction, as in figure 3.1, effectively reaches equilibrium? To answer this question, we mathematically model the kinetics of ligand binding and complex dissociation. We first consider each of the rate processes as follows:

Change in complex = Complex formation − Complex dissociation
$$\frac{d[C]}{dt} = k_{on}[L][P] - k_{off}[C]$$

Note that this equation contains two key mathematical relationships. The rate of association of ligand and protein to form complex is equal to $k_{on}[L][P]$. This second-order (bilinear) term represents the bimolecular association of two separate molecules to form a third, noncovalent complex. The rate of dissociation of complex to reform free ligand and protein is $k_{off}[C]$. This first-order (linear) term represents the unimolecular event of a single molecular complex dissociating into its constituent parts. Throughout most of this text, expressions of this type will be used without regard to whether a more subtle set of steps occurs at the molecular level. These expressions were originally developed for gas-phase kinetics, where the chemical bond-making and bond-breaking steps tend to occur orders of magnitude faster than the time between molecular collisions. In aqueous solution, the bimolecular association involves the two molecules approaching each other by diffusion, merging their solvation shells, and squeezing water molecules out of the interface. The rate of a reaction tends to be determined by the slowest step in the process, and any of these events could be rate limiting. In certain instances, the details of the rate process become particularly important, and one must delve deeper into the reaction mechanism. However, as the simplest first approximation, the *mass-action* kinetic term $k_{on}[L][P]$ is often a very useful quantitative description of the rate of biomolecular association.

In the most general case, depletion of both protein and ligand by inclusion in complexes is accounted for; however, this analytical solution is of little practical use, because it gives $[C]$ as a complicated implicit function of time. Notably, it is often feasible to carry out a kinetic experiment under conditions where $[L]_o \gg [P]_o$, such that $[L] \approx [L]_o$:

$$\frac{d[C]}{dt} = k_{on}[L]_o[P] - k_{off}[C] \quad (3.11)$$

This is known as the pseudo-first-order approximation, with the rate of complex formation proportional to $[P]$ (i.e., first order in $[P]$) by the effective rate constant $k_{on}[L]_o$. However, we must still account for depletion of protein by complex formation, because $[P] \napprox [P]_o$. Using conservation of mass, we have

$$[P] = [P]_o - [C] \quad (3.12)$$

Binding Equilibria and Kinetics

We then substitute this relationship into equation 3.11 to give

$$\frac{d[C]}{dt} = k_{on}[L]_o([P]_o - [C]) - k_{off}[C] \tag{3.13}$$

$$= k_{on}[P]_o[L]_o - (k_{on}[L]_o + k_{off})[C] \tag{3.14}$$

To solve this equation, we must specify the initial condition, which we set as $[C] = [C]_o$ at $t = 0$. Integrating, we obtain

$$[C] = [P]_o \frac{[L]_o}{[L]_o + K_d} + \left([C]_o - [P]_o \frac{[L]_o}{[L]_o + K_d}\right) e^{-(k_{on}[L]_o + k_{off})t} \tag{3.15}$$

Two commonly used simplifications of this equation are: (1) to consider protein-ligand association with no complexes at the outset (i.e., $[C]_o = 0$ and $[L]_o > 0$ at $t = 0$); and (2) to consider protein-ligand dissociation with free ligand continuously removed from the system (i.e., $[C]_o > 0$ and $[L]_o = 0$ at $t = 0$).

In the first case, which describes association kinetics (and the subsequent potential dissociation of newly formed complex), equation 3.15 reduces to

$$[C] = [P]_o \frac{[L]_o}{[L]_o + K_d} \left(1 - e^{-k_{obs}t}\right) \tag{3.16}$$

where

$$k_{obs} \equiv k_{on}[L]_o + k_{off} \tag{3.17}$$

and

$$\tau_{\frac{1}{2}} \equiv \frac{\ln 2}{k_{on}[L]_o + k_{off}} \tag{3.18}$$

is the time for $[C]$ to reach half its equilibrium value of $[P]_o \frac{[L]_o}{[L]_o + K_d}$.

In the second case, which describes pure dissociation kinetics, equation 3.15 reduces to

$$[C] = [C]_o e^{-k_{off}t} \tag{3.19}$$

Equations 3.16 and 3.19 are used to extract association and dissociation rate constants from experimental measurements of binding kinetics. As noted in chapter 2, association rate constants are generally in the range 10^5–10^6 M^{-1}s^{-1} [2], whereas dissociation rate constants can span many orders of magnitude. Thus, the dissociation rate constant is typically the driver of affinity.

Example 3-2 For the previous example, how long should the antibody and protein extract be mixed to reach 95% of the eventual maximum complex formation?

Solution Recall that the solution in this case was to add antibody to a final concentration of 190 nM, which was predicted to complex 95% of the target protein at equilibrium. In the absence of a specific measured value, an association rate constant $k_{on} \approx 10^5$ M^{-1}s^{-1} is a reasonable estimate for a protein-protein interaction. Thus, for the given equilibrium affinity $K_d = 10$ nM, $k_{off} = k_{on}K_d = (10^5$ M^{-1}s$^{-1})(10^{-8}$ M$) = 10^{-3}$ s^{-1}. From equation 3.17, we have

$$k_{obs} = (10^5 \text{ M}^{-1}\text{s}^{-1})(1.90 \times 10^{-7} \text{ M}) + 10^{-3} \text{ s}^{-1} = 0.02 \text{ s}^{-1} \quad (3.20)$$

To reach 95% of equilibrium, $0.95 = 1 - e^{-k_{obs}t}$. Thus, $t_{95\%} = -\ln(0.05)/k_{obs} = 150$ seconds.

Example 3-3 Radiolabeled, small-molecule ligand is added to a solution containing cells at a final concentration of 10 pM. The cells have a surface receptor that binds the ligand with affinity $K_d = 10$ pM. How long must the cells be incubated in the ligand solution for the surface receptor–ligand complex to reach 95% of its equilibrium value?

Solution In the absence of known kinetic parameters, one can generally estimate that a protein–small-molecule binding reaction exhibits $k_{on} \approx 10^6$ M^{-1}s^{-1} (note that this estimate is higher than that for protein-protein binding), and hence $k_{off} = (10^6$ M^{-1}s$^{-1})(10^{-11}$ M$) = 10^{-5}$ s^{-1}. From equation 3.17, $k_{obs} = (10^6$ M^{-1}s$^{-1})(10^{-11}$ M$) + 10^{-5}$ s$^{-1} = 2 \times 10^{-5}$ s^{-1}. To reach 95% of equilibrium, $0.95 = 1 - e^{-k_{obs}t}$. Thus, $t_{95\%} = -\ln(0.05)/k_{obs} = 1.5 \times 10^5$ s, or 42 hours. Note the dramatic difference from the previous example, sufficient to cause practical experimental difficulties when working with dilute samples for high-affinity interactions.

How accurate is the pseudo-first-order approximation? Let us consider the solution to the material balance for $[C]$ without this approximation:

$$\frac{d[C]}{dt} = k_{on}[L][P] - k_{off}[C] \quad (3.21)$$

By conservation of mass, $[L] = [L]_\circ - [C]$, and $[P] = [P]_\circ - [C]$, allowing us to write the differential equation entirely in terms of $[C]$:

$$\frac{d[C]}{dt} = k_{on}[C]^2 - \{k_{on}([L]_\circ + [P]_\circ) + k_{off}\}[C] + k_{on}[L]_\circ[P]_\circ \qquad (3.22)$$

Insights into how particular rates and concentrations affect dynamic behavior can often be obtained by appropriate nondimensionalization of the variables in an equation. One generally seeks to nondimensionalize so that the dimensionless variables are approximately in the range from zero to one. By this criterion, nondimensionalizing complex concentration by the equilibrium value (with excess $[L]_\circ$) will give a variable $\hat{C}$ that ranges from zero to one:

$$\hat{C} \equiv \frac{[C]}{[C]_{eq}} \qquad (3.23)$$

$$\equiv \frac{[C]}{[P]_\circ \frac{[L]_\circ}{[L]_\circ + K_d}} \qquad (3.24)$$

To nondimensionalize the time variable, let's take the time scale specified by the pseudo-first-order approximation (equation 3.17) as a base case:

$$\hat{t} = \frac{t}{1/k_{obs}} = \frac{t}{1/(k_{on}[L]_\circ + k_{off})} = (k_{on}[L]_\circ + k_{off})t \qquad (3.25)$$

Making these substitutions into equation 3.22, following some algebra, gives

$$\frac{d\hat{C}}{d\hat{t}} = \frac{\hat{P}_\circ \hat{L}_\circ}{(1+\hat{L}_\circ)^2}\hat{C}^2 - \frac{1+\hat{L}_\circ + \hat{P}_\circ}{1+\hat{L}_\circ}\hat{C} + 1 \qquad (3.26)$$

where $\hat{P}_\circ = [P]_\circ/K_d$, and $\hat{L}_\circ = [L]_\circ/K_d$. As a reality check, equation 3.26 should reduce to the pseudo-first-order case when $[L]_\circ \gg [P]_\circ$ (or $\hat{L}_\circ \gg \hat{P}_\circ$). In this limit, we have

$$\lim_{\frac{[L]_\circ}{[P]_\circ} \to \infty} \frac{d\hat{C}}{d\hat{t}} = 1 - \hat{C} \qquad (3.27)$$

For the initial condition $\hat{C} = 0$, the solution to this equation is $\hat{C} = 1 - e^{-\hat{t}}$. This is the nondimensionalized form of equation 3.16, confirming that the solutions agree in the limit of excess ligand. More broadly, examination of equation 3.26 reveals that departure from the pseudo-first-order solution depends on the relative magnitude

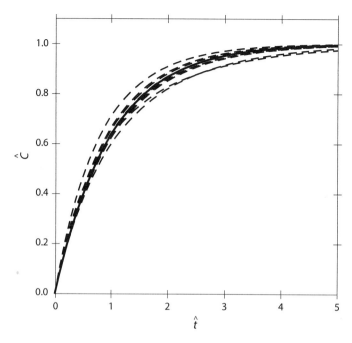

Figure 3.3 Pseudo-first-order kinetics provides an unexpectedly good approximation for cases in which ligand is not actually in excess over protein. The dimensionless complex concentration $\hat{C}$ is defined in equation 3.24, and the dimensionless time $\hat{t}$ is defined in equation 3.25. The solid curve is the theoretical solution given by the equation $\hat{C} = 1 - e^{-\hat{t}}$; the dashed lines are nine trajectories numerically calculated from equation 3.26 with $[P]_o/K_d = 0.1$, 1, or 10 and $[L]_o/[P]_o = 1$, 3, or 10.

of the initial protein and ligand concentrations ($\hat{L}_o$ versus $\hat{P}_o$) and the magnitude of these concentrations relative to the binding constant K_d. More directly, the pseudo-first-order case is achieved for $([L]_o + K_d) \gg [P]_o$. Note, however, that the validity of the pseudo-first-order approximation does not depend on the absolute values of the kinetic parameters k_{on} and k_{off}, only on their ratio K_d. This insight is not readily apparent from inspection of equation 3.22 alone, providing an example of the benefits of the nondimensionalization exercise.

Equation 3.26 was solved numerically for a wide range of $\hat{L}_o$ and $\hat{P}_o$, and the results are shown in figure 3.3. It is clear that the pseudo-first-order solution provides a surprisingly satisfactory estimate to the exact solution, even when ligand and protein are initially equimolar. Thus, the half-time calculated with equation 3.18 provides a useful estimate for the time to reach equilibrium in most cases, regardless of the relative magnitude of $[L]_o$ versus $[P]_o$.

3.3 Multiple Binding Sites

To this point, we have considered proteins P that have a single binding site for ligand L. Many problems of biological interest, however, involve proteins with multiple binding sites: Monoclonal antibodies possess two binding sites that can each interact with ligands tethered to a cell or device surface; many cell-cell interactions are mediated by weak multivalent binding events between cells and other cells or tissue structures; enzymes are often multimeric; and hemoglobin has four binding sites for oxygen.

3.3.1 Independent Sites

We begin by considering a protein with n identical binding sites. If the individual ligand binding strengths are invariant with respect to site occupancy, then the binding sites are *independent*, and in such a case, we can use statistical mechanics to examine the protein states. Let's consider individual binding equilibria for the progressive binding of each additional ligand to the complex. For notational simplicity, protein with no ligand bound is written P_0, protein with one ligand molecule bound is P_1, and so on, such that protein with i ligands bound is P_i. The relevant equilibria are written below, and the binding constants are defined for each subsequent ligand binding reaction:

$$P_0 + L \rightleftharpoons P_1 \qquad K_{d,1} = \frac{[P_0][L]}{[P_1]}$$

$$P_1 + L \rightleftharpoons P_2 \qquad K_{d,2} = \frac{[P_1][L]}{[P_2]}$$

$$P_2 + L \rightleftharpoons P_3 \qquad K_{d,3} = \frac{[P_2][L]}{[P_3]}$$

$$\vdots$$

$$P_{i-1} + L \rightleftharpoons P_i \qquad K_{d,i} = \frac{[P_{i-1}][L]}{[P_i]}$$

$$\vdots$$

$$P_{n-1} + L \rightleftharpoons P_n \qquad K_{d,n} = \frac{[P_{n-1}][L]}{[P_n]}$$

(It is understood that the concentrations are taken at equilibrium, although the subscript has been dropped from each concentration $[\ ]_{eq}$ to simplify the notation.)

Because each of the binding sites is identical and independent of all others, one might expect that the affinity for binding the second ligand to a protein should be the same as the first, in some sense. Does that imply that $K_{d,1} = K_{d,2}$? We will see that this is not, in fact, true. To understand this counterintuitive result, imagine what happens at the molecular level when a ligand binds. For the case of $n = 4$ binding

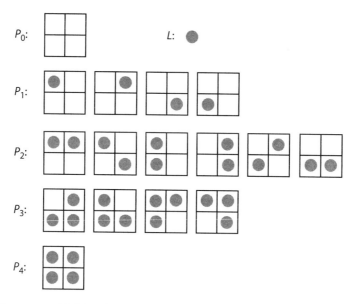

Figure 3.4 There are a multiplicity of ways to bind one, two, or three ligands to a tetrameric binding protein. Because the sites are energetically identical and independent, all species in each grouping (e.g., the six P_2 species) are energetically indistinguishable.

sites, there is only one way for a protein to have no ligand bound; however, there are four energetically indistinguishable ways of binding the first ligand, as illustrated in figure 3.4.

The way to think about this is to imagine that the binding status of each of the four binding sites could be determined individually (e.g., figure 3.4), although this would be difficult in practice. There are six ways to bind two ligands among four sites, four ways to bind three ligands, and only one way to bind four ligands. In fact, the general formula for combination (the number of ways to distribute i ligands among n sites without regard for order) is given by

$$\binom{n}{i} = \frac{n!}{i!(n-i)!} \tag{3.28}$$

where $\binom{n}{i}$ is read "n choose i," and $i! = i \times (i-1) \times (i-2) \times \cdots \times 2 \times 1$. Recall that $0! = 1$, so $\binom{n}{0} = \binom{n}{n} = 1$.

How does the macroscopic picture connect with what is happening at the molecular level? Imagine the binding of ligand to just one of the four binding sites, and let us call the equilibrium for this chemical reaction K_d^μ, where the superscript μ indicates that this is an equilibrium dissociation constant in the microscopic sense

Binding Equilibria and Kinetics

of applying to only one of the possible singly bound forms:

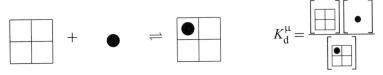

Comparing this to the macroscopic case of all possible singly bound forms,

$$P_0 + L \rightleftharpoons P_1$$

produces a relationship between K_d^μ and $K_{d,1}$:

$$K_{d,1} = \frac{[P_0][L]}{[P_1]} = \frac{[\square][\bullet]}{[\square]+[\square]+[\square]+[\square]} = \frac{1 \times [\square][\bullet]}{4 \times [\square]} = \frac{1}{4} K_d^\mu \tag{3.29}$$

On average, each of the singly bound forms will occur at the same abundance, because they are energetically identical, allowing collection of the denominator into $4\times$ one of the forms. Thus, the macroscopic equilibrium dissociation constant is one-fourth of the microscopic constant. This results directly from there being one version of P_0 and four equally stable versions of P_1, formed with equal probability. In general, we have

$$K_{d,i} = \frac{\binom{n}{i-1}}{\binom{n}{i}} K_d^\mu \tag{3.30}$$

If one is interested only in the ratio of total bound ligand to total protein, defined as

$$\nu = \frac{\sum_{i=1}^{n} i[P_i]}{\sum_{i=0}^{n} [P_i]} \tag{3.31}$$

it can be shown that

$$\nu = \frac{n[L]}{K_d^\mu + [L]} \tag{3.32}$$

which makes intuitive sense, because all binding sites act independently. A related quantity, the fractional saturation y, represents the fraction of binding sites that are occupied:

$$y = \frac{\text{Number of occupied binding sites}}{\text{Total number of binding sites}} = \frac{\nu}{n} = \frac{[L]}{K_d^\mu + [L]} \quad (3.33)$$

with $0 \leq y \leq 1$.

Example 3-4 IgM is a form of antibody found in the blood that consists of five identical IgG-like binding molecules, creating a large complex with ten identical and independent binding sites. Calculate the distribution of numbers of antigen molecules bound per IgM complex when the antigen concentration is in excess and equal to the binding constant K_d^μ for one of the binding sites in isolation.

Solution The macroscopic equilibrium dissociation constants $K_{d,i}$ can be calculated from equation 3.30 as shown in the second column below.

i	$K_{d,i}/K_d^\mu$	$[P_i]/[P_0]$	y_i
0	—	1	9.8×10^{-4}
1	0.10	10	9.8×10^{-3}
2	0.22	45	0.044
3	0.38	120	0.12
4	0.57	210	0.21
5	0.83	252	0.25
6	1.20	210	0.21
7	1.75	120	0.12
8	2.67	45	0.044
9	4.50	10	9.8×10^{-3}
10	10.00	1	9.8×10^{-4}

The relationship $K_{d,i} = \frac{[P_{i-1}][L]}{[P_i]}$ can be rearranged for this specific case, because $[L]_\circ \gg [P]_\circ$ and $[L] = K_d^\mu$, to yield: $\frac{[P_{i-1}]}{[P_i]} = \frac{K_{d,i}}{K_d^\mu}$. Thus, $\frac{[P_i]}{[P_0]} = \prod_{j=1}^{i} \frac{K_d^\mu}{K_{d,j}}$. Consequently, the third column above gives the relative proportions of each molecular species, which can be normalized by the sum of that column to give the fractional binding values y_i, listed in the fourth column.

3.3.2 Cooperativity

When the first binding site of certain multivalent proteins is occupied by ligand, the remaining sites change their shape or dynamics in such a way as to alter their ligand-binding affinity. This communication between binding sites is called cooperativity—positive cooperativity if subsequent binding events are higher affinity, and negative cooperativity if subsequent binding events are lower affinity. Let's examine a case of perfect positive cooperativity. Consider a protein with n binding sites. After the first binding site is occupied by a ligand molecule, the remaining binding sites increase their ligand-binding affinities to such an extent that they are all immediately occupied by ligand. In this situation, the protein is either unbound (P_0) or every site is occupied (P_n), and there are no intermediate states. (This is not, of course, a realistic situation, but rather an upper bound on positive cooperativity for the sake of setting a limit.)

$$P_0 + nL \rightleftharpoons P_n \quad (3.34)$$

$$K = \frac{[P_0][L]^n}{[P_n]} \quad (3.35)$$

or

$$y = \frac{[P_n]}{[P_0]+[P_n]} = \frac{[L]^n}{[L]^n + K} \quad (3.36)$$

The ligand concentration at which half of the protein is in the fully bound state ($y = 0.5$) is therefore $K^{\frac{1}{n}}$, which we define as the EC_{50}. Therefore, we can write equation 3.36 as

$$y = \frac{[L]^n}{[L]^n + EC_{50}^n} \quad (3.37)$$

It is often convenient to use this form of the equation, because EC_{50} always has units of concentration (unlike K, whose units depend on n).

Rearranging equation 3.35 yields

$$\frac{[L]^n}{EC_{50}^n} = \frac{[P_n]}{[P_0]}$$

$$= \frac{y}{1-y}$$

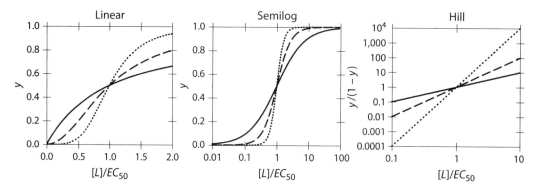

Figure 3.5 Three graphical representations for binding isotherms exhibiting positive cooperativity. The solid curve represents a noncooperative isotherm ($n=1$), the dashed curve perfect positive cooperativity with $n=2$, and the dotted curve perfect positive cooperativity with $n=4$. A definitive feature of a cooperative binding isotherm is the presence of an inflection point in the linear plot; noncooperative interactions do not exhibit this switchlike behavior, which is a functionally critical attribute of cooperativity.

or

$$\ln\left(\frac{y}{1-y}\right) = n\ln[L] - n\ln EC_{50} \qquad (3.38)$$

A plot of $\ln\left(\frac{y}{1-y}\right)$ versus $\ln[L]$ is called a Hill plot, and as shown in figure 3.5 (right panel), is a straight line with slope n. The slope of a Hill plot of experimental or simulated data is called the Hill coefficient (often symbolically represented as n_H), and from examination of equation 3.38, it must be no greater than the ligand-binding valency n of the protein, because this result represents the idealized upper limit for positive cooperativity. At very low ligand concentrations, the slope approaches 1 because of the scarcity of ligand. In other words, the transition from zero to one bound ligand will look noncooperative, similar to that for ligand binding to a monovalent protein. Similarly, a slope of unity is observed at very high ligand concentrations, because the only remaining possible transition is from $n-1$ to n ligands, which again will mirror the binding isotherm of a monovalent protein. As it happens, the switchlike behavior characteristic of positive cooperativity is of considerable interest and significance in signal transduction and cellular responses, and effective Hill coefficients are often empirically curve-fit to sigmoidal curves without any intended mechanistic interpretation for the value of the effective Hill coefficient so obtained.

Example 3-5 The following data for activation of MAP kinase (MAPK) were taken as a function of the concentration of the activating protein malE-Mos [3]. What is the apparent Hill coefficient n_H for this response?

[malE-Mos] (µM)	MAPK activity
0.010	0.019
0.013	0.0
0.017	0.0
0.020	0.012
0.025	0.21
0.033	0.40
0.051	0.78
0.098	1.0
0.20	0.93
0.25	0.96
0.48	1.0
1.0	0.99

Solution These data were fit by nonlinear least squares to a simpler non-cooperative isotherm (dashed curve in figure 3.6; equation 3.5, fitted with $K_d = 0.05$ µM) and to a perfect positively cooperative isotherm (solid curve; equation 3.37, fitted with $n_H = n = 2.7$ and $EC_{50} = 0.04$ µM). The cooperative fit better matches the observed behavior (issues of model selection and regression are discussed later in section 3.9). A Hill coefficient n_H significantly greater than 1 (in this case, 2.7) provides empirical evidence for switchlike, or "ultrasensitive," behavior (to be further discussed in chapter 6).

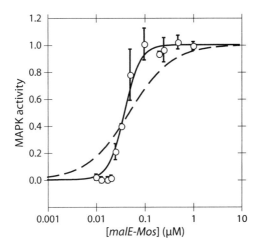

Figure 3.6 Plot of MAPK activity as a function of the concentration of malE-Mos. The dashed curve corresponds to a fit of the data to a noncooperative isotherm, whereas the solid curve corresponds to a perfectly cooperative one.

3.3.3 Avidity and Effective Concentration

A particularly important class of binding reactions involves molecules that make multiple interactions with one another. Some examples of multivalent-multivalent interactions are shown in figure 3.7. Often, these are essentially identical interactions repeated across an interface due to multivalency of both binding partners. For example, antibodies used to label the surface of a cell present two identical binding sites that can engage two identical antigen molecules on the cell surface. Likewise, a virus might present hundreds of copies of a binding site that engage a large number (perhaps dozens) of cell-surface receptors. Cell adhesion to a surface involves multiple interactions. The term *avidity* is used to describe the affinity of multivalent interactions, which often persist in complex far longer than their corresponding individual interactions. The basic framework for thinking about the possible additivity of these repeated interactions has been shown previously [4].

Consider the simplest example, a bipartite ligand—one with two separate parts connected by a linker—binding to a bipartite protein (figure 3.8). The binding of the bipartite ligand AB is clearly some function of the two interactions of A and B binding separately, so there should be some relationship between K_d^{AB} and the binding constants for the individual portions, K_d^A and K_d^B. The corresponding binding free energy for each reaction can be obtained from the dissociation constant:

$$\Delta G_{AB}^\circ = RT \ln K_d^{AB} \tag{3.39}$$

$$\Delta G_A^\circ = RT \ln K_d^A \tag{3.40}$$

$$\Delta G_B^\circ = RT \ln K_d^B \tag{3.41}$$

If the effect of combining the ligand to include A and B were equivalent to adding the interactions represented by each, then ΔG_{AB}° would be equal to $\Delta G_A^\circ + \Delta G_B^\circ$. The fundamental flaw in this thinking is that ΔG_A° and ΔG_B° both account for more than the interactions A and B make with their respective binding sites. They also account for the fundamental loss of entropy in any two-particle $\rightarrow$ one-particle reaction, due to a ligand surrendering three translational and three rotational degrees of freedom upon binding. The incorrect analysis double counts this entropy loss.

A more correct framework is illustrated in figure 3.9, in which the binding of AB is considered sequentially. Along the upper path, A binds first with B tethered; then the B portion of the ligand binds. The lower path recapitulates the same outcome, but reverses the order of binding. Notice that a key difference between figures 3.8 and 3.9 is that the latter provides a sequence of reactions, a

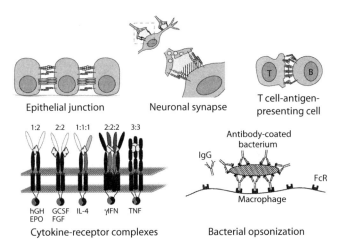

Figure 3.7 Protein-protein interactions across cell-cell interfaces are highly multivalent: In the top row, epithelial cell junctions, neuronal synapses, and immunological synapses are schematically represented [5]. At bottom left, examples of multivalent cytokine-receptor complexes are shown [6]. At bottom right, a bacterial cell is coated with antibodies, which when bound by Fc receptors on a macrophage, triggers phagocytosis of the cell [7].

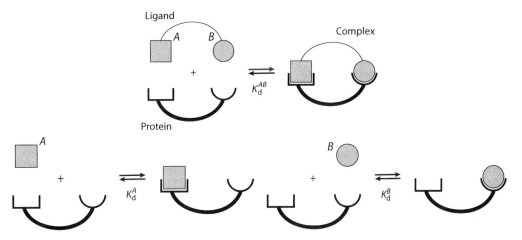

Figure 3.8 The binding of a bipartite ligand to a bipartite receptor involving a linker region connecting two parts, A and B. The dissociation constant K_d^{AB} describes the affinity for the bipartite system. One way of thinking about the contributions to the overall system is to separately consider the binding of the A and B portions of the ligand to the protein, with individual dissociation constants K_d^A and K_d^B, respectively.

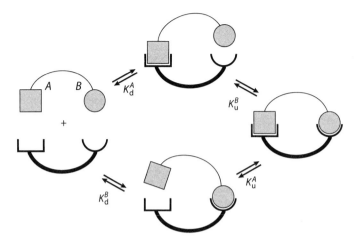

Figure 3.9 The upper and lower pathways depict two ways of dissecting the overall binding of ligand AB to protein. In the upper pathway, first the A portion of the ligand binds while B remains tethered to it but doesn't make specific protein interactions; then the B portion binds. In the lower pathway, the B portion binds first with A tethered, and then A binds.

thermodynamic pathway linking the overall reactants and products of the reaction of interest. In contrast, the former contains reactions that seem to include all the right pieces, but a pathway connecting overall reactants and products is not presented. The construction of thermodynamic pathways and cycles that include reactions of interest is a fundamental tool for understanding free energy changes and molecular interactions, as discussed in chapter 2.

In figure 3.9, the binding of AB is composed of two reactions. The first is a two-particle $\to$ one-particle reaction whose energetics are represented by a dissociation constant K_d^A or K_d^B with units of concentration (M). We assume that the presence of the linker and other binding moiety has a negligible effect on this step. The second is a one-particle $\to$ one-particle reaction with energetics given by a unimolecular unbinding constant, K_u^A or K_u^B, which is unitless. This framing of the situation satisfies dimensional analysis. Following the upper pathway gives

$$K_d^{AB} = K_d^A K_u^B \tag{3.42}$$

and following the lower pathway gives

$$K_d^{AB} = K_d^B K_u^A \tag{3.43}$$

In either case, the units agree (M on both sides). Moreover, the overall binding free energy is the sum of two terms. One is due to intermolecular binding and the other to intramolecular binding,

$$\Delta G_{AB}^\circ = RT \ln K_d^{AB} \tag{3.44}$$

$$= RT \ln K_d^A + RT \ln K_u^B \tag{3.45}$$

$$= \Delta G_A^\circ + RT \ln K_u^B \tag{3.46}$$

$$= \Delta G_B^\circ + RT \ln K_u^A \tag{3.47}$$

where equations 3.45 and 3.46 come from the upper pathway in figure 3.9, and equation 3.47 results from carrying out the same analysis on the lower pathway.

Effective concentration In $2 \to 1$ binding events, the extent of protein bound in complex at equilibrium depends on the ligand concentration. In $1 \to 1$ equilibration events, such as the two reactions to the right in figure 3.9, there is no dependence on the ligand concentration, as the change is effectively between two conformations of a complex. Essentially, the binding of B tethers A in the neighborhood of its binding site such that A is present at some *effective concentration*. A longer tether gives A more freedom to be away from its binding site, so the effective concentration experienced from the perspective of the A-binding site is lower than with a tether just long enough to place A in its site. We can also imagine a tether too short to allow A to reach its binding site, in which case the effective concentration is zero. We can explore this concept further by comparing the binding affinity of A between $2 \to 1$ and $1 \to 1$ reactions. By the definition of K_d for bimolecular reactions, we have

$$K_d^A = [A]\frac{[P]}{[C]} \tag{3.48}$$

The dissociation constant is the unbound ligand concentration times the ratio of unbound to bound protein. By analogy, let's replace $[A]$ with the concentration of *unbound ligand* in our $1 \to 1$ reaction, the effective concentration of ligand, $[A]_\text{eff}$. We also replace the ratio $\frac{[P]}{[C]}$ with K_u^A, which is the ratio of unimolecular unbound to bound complex, and rearrange to provide this relationship for effective concentration

$$[A]_\text{eff} = \frac{K_d^A}{K_u^A} \tag{3.49}$$

which has the proper units of M. A vivid colloquial analogy is helpful in understanding the concept of $[A]_\text{eff}$. Consider an angry dog to be the ligand A and one's leg to be the receptor. The effective concentration is the number of unbound dogs in a room that would result in the same leg-biting rate as a single tethered dog. Clearly, the effective concentration of dogs increases as the leash shortens.

Combining equations 3.49 and 3.43, the relationship between the overall binding constant K_d^{AB} and the individual binding constants K_d^A and K_d^B is given by

$$K_d^{AB} = \frac{K_d^A K_d^B}{[A]_{\text{eff}}} \tag{3.50}$$

Returning to figure 3.9, the four reactions form a thermodynamic cycle. If one starts with one species and carries out all four reactions in sequence, one returns to the original species. Because free energy is a state function that depends only on the state and not on the pathway taken to arrive there, the overall free energy change for a cycle that returns to its starting point must be zero. Enforcing this condition results in an interesting relationship:

$$RT \ln K_d^A + RT \ln K_u^B - RT \ln K_u^A - RT \ln K_d^B = 0 \tag{3.51}$$

$$K_d^A \cdot K_u^B = K_d^B \cdot K_u^A \tag{3.52}$$

$$\frac{K_d^A}{K_u^A} = \frac{K_d^B}{K_u^B} \tag{3.53}$$

$$[A]_{\text{eff}} = [B]_{\text{eff}} \tag{3.54}$$

That is, the effective concentrations of A and B are the same. Tethering B near its binding site through binding A is symmetric to tethering A by binding B, and this symmetry is reflected in the corresponding effective concentrations.

There is one practical difficulty with the concept of effective concentration, which is the measurement of unimolecular dissociation constants, such as K_u^B. Very demanding measurement techniques are necessary to distinguish protein with AB bound from that with just A bound. Moreover, the effective concentration is specific to a given complex, so it cannot be readily varied to create the type of titration used to fit bimolecular dissociation constants. One convenient indirect approach is to synthesize fragments of the ligand, or cleave the intact ligand, to isolate A and B separately (as in figure 3.8). One can then measure the bimolecular dissociation constants K_d^{AB}, K_d^A, and K_d^B individually and then compute the unimolecular ones:

$$K_u^B = \frac{K_d^{AB}}{K_d^A} \quad \text{and} \quad K_u^A = \frac{K_d^{AB}}{K_d^B} \tag{3.55}$$

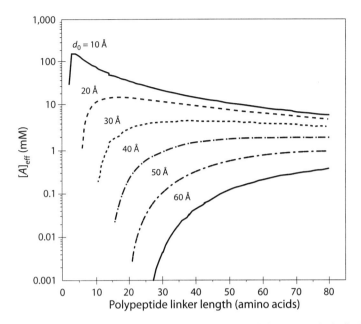

Figure 3.10 The predicted effective concentration $[A]_{\text{eff}}$ at a distance d_o from one end of a flexible polypeptide linker of length L amino acids. The linker was modeled as a "wormlike chain" polymer. For example, the effective concentration of one end of a 5-amino acid linker at a distance 10 Å from the other end is a little over 100 mM. For context, the Fv/antigen complex depicted in figure 2.11 is approximately 50 Å wide and 75 Å long. In practice, synthetic polypeptide linkers are generally 5–25 amino acids in length [8].

The assumption that K_d^A and K_d^B do not change whether the ligand is a fragment or intact is generally a good approximation, as long as binding domains are tethered together by flexible linkers to create constructs that possess independent multiple binding activities. The assumption may be less appropriate if the two binding surfaces A and B are physically attached in such a way that binding at A significantly affects the structure or dynamics at binding site B.

Useful approximations for effective concentration have been developed for two common situations: bivalent antibodies (section 3.3.4) and two protein-binding domains connected by a flexible polypeptide linker. It has been shown that the end-to-end distance for a flexible polypeptide is well described by a *wormlike chain* polymer model [8]. This simple representation allows estimation of the effective concentration $[A]_{\text{eff}}$ at a distance d_o away from one end of a linker, when a binding site is tethered at the other end of the linker (figure 3.10). Note that $[A]_{\text{eff}}$ is generally in the micromolar to millimolar range. The empirically derived algebraic relationship for $[A]_{\text{eff}}$ as a function of d_o (distance in Å) and L (polypeptide length

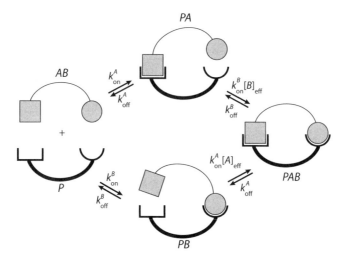

Figure 3.11 The system of figure 3.9 with rate constants labeled.

in amino acids) is as follows:

$$[A]_{\text{eff}} \, (\text{M}) = 5.03 L^{-\frac{3}{2}} e^{\frac{-6.58 \times 10^{-2} d_o^2}{L}} \left[1 - \frac{0.987}{L} + 0.139 \left(\frac{d_o^2}{L^2} \right) \right.$$

$$\left. -2.51 \times 10^{-3} \left(\frac{d_o^4}{L^3} \right) - \frac{0.308}{L^2} - 0.150 \left(\frac{d_o^2}{L^3} \right) + 0.0204 \left(\frac{d_o^4}{L^4} \right) \right.$$

$$\left. -5.17 \times 10^{-4} \left(\frac{d_o^6}{L^5} \right) + 3.14 \times 10^{-6} \left(\frac{d_o^8}{L^6} \right) \right] \quad (3.56)$$

Can we predict the apparent dissociation rate for systems such as that shown in figure 3.9? Let's expand the schematic for this bipartite interaction to include the individual rate constants, as shown in figure 3.11. Two assumptions have been made in constructing this schematic: (1) the dissociation rate constants for each ligand moiety are independent of the bound state of the other (e.g., k_{off}^B is the same from PB or PAB); and (2) the association rate constant for the second site of singly bound complexes (PA and PB) is equal to the solution phase association rate constant k_{on} times the effective concentration of the second ligand $[A]_{\text{eff}}$ or $[B]_{\text{eff}}$.

Writing the dynamic material balance for the species P, PA, PB, and PAB gives:

$$\frac{d[P]}{dt} = -k_{on}^A[AB][P] - k_{on}^B[AB][P] + k_{off}^A[PA] + k_{off}^B[PB] \tag{3.57}$$

$$\frac{d[PA]}{dt} = k_{on}^A[AB][P] - k_{on}^B[B]_{eff}[PA] - k_{off}^A[PA] + k_{off}^B[PAB] \tag{3.58}$$

$$\frac{d[PB]}{dt} = k_{on}^B[AB][P] - k_{on}^A[A]_{eff}[PB] - k_{off}^B[PB] + k_{off}^A[PAB] \tag{3.59}$$

$$\frac{d[PAB]}{dt} = k_{on}^A[A]_{eff}[PB] + k_{on}^B[B]_{eff}[PA] - k_{off}^A[PAB] - k_{off}^B[PAB] \tag{3.60}$$

Rather than attempt an analytical solution of this full set of equations, we use an approximation that is often useful in analyzing complex kinetic systems (e.g., in enzyme kinetics), known as the quasi-steady-state approximation (QSSA). To apply the QSSA, we assume that the concentrations of intermediates, such as *PA* and *PB*, quickly reach a value that adjusts rapidly to slower changes in the values of *P* and *PAB*. Consequently, we assume that $\frac{d[PA]}{dt} = \frac{d[PB]}{dt} = 0$ and solve for the QSSA concentrations for these intermediates:

$$[PA]_{QSSA} = \frac{k_{on}^A}{k_{on}^B[B]_{eff} + k_{off}^A}[AB][P] + \frac{k_{off}^B}{k_{on}^B[B]_{eff} + k_{off}^A}[PAB] \tag{3.61}$$

$$[PB]_{QSSA} = \frac{k_{on}^B}{k_{on}^A[A]_{eff} + k_{off}^B}[AB][P] + \frac{k_{off}^A}{k_{on}^A[A]_{eff} + k_{off}^B}[PAB] \tag{3.62}$$

To simplify the expression further, we can narrow our considerations to interactions for which $k_{on}^A[A]_{eff} \gg k_{off}^B$ and $k_{on}^B[B]_{eff} \gg k_{off}^A$. This means a singly bound complex (*PA* or *PB*) is much more likely to proceed to doubly bound (*PAB*) than to dissociate to fully unbound (*P* + *AB*). Typical values of k_{on}, k_{off}, and $[A]_{eff} = [B]_{eff}$ (often in the micromolar to millimolar concentration range; see figure 3.10), indicate that these approximations are often satisfied. This simplifies the QSSA intermediate concentration expressions:

$$[PA]_{QSSA} = \frac{k_{on}^A}{k_{on}^B[B]_{eff}}[AB][P] + \frac{k_{off}^B}{k_{on}^B[B]_{eff}}[PAB] \tag{3.63}$$

$$[PB]_{QSSA} = \frac{k_{on}^B}{k_{on}^A[A]_{eff}}[AB][P] + \frac{k_{off}^A}{k_{on}^A[A]_{eff}}[PAB] \tag{3.64}$$

Substituting these QSSA concentrations into equation 3.57, employing the previous assumptions again, collecting terms, and recalling that $[A]_{\text{eff}} = [B]_{\text{eff}}$, we obtain:

$$\frac{d[P]}{dt} = -\left\{k_{\text{on}}^B + k_{\text{on}}^A\right\}[AB][P] + \left\{\frac{k_{\text{off}}^B K_d^A + k_{\text{off}}^A K_d^B}{[A]_{\text{eff}}}\right\}[PAB] \qquad (3.65)$$

$$\frac{d[P]}{dt} = -k_{\text{on,app}}[AB][P] + k_{\text{off,app}}[PAB] \qquad (3.66)$$

where the apparent overall association and dissociation rate constants are

$$k_{\text{on,app}} \equiv k_{\text{on}}^B + k_{\text{on}}^A \qquad (3.67)$$

$$k_{\text{off,app}} \equiv \frac{k_{\text{off}}^B K_d^A + k_{\text{off}}^A K_d^B}{[A]_{\text{eff}}} \qquad (3.68)$$

Note that the apparent dissociation rate constant is reduced by a factor of $\approx K_d^A/[A]_{\text{eff}}$ or $\approx K_d^B/[B]_{\text{eff}}$, a very small ratio that will result in dramatic decreases in the apparent dissociation rate for a bivalent interaction.

3.3.4 Equilibrium Bivalent Binding at Cell Surfaces

A commonly encountered situation where multivalency affects binding affinity is in the use of monoclonal antibodies to label cell surfaces or targets immobilized on a solid phase. Antibodies possess two identical binding sites, connected with a certain degree of flexibility by the IgG framework. A schematic diagram showing singly and doubly bound antibody molecules is shown in figure 3.12.

Let's write down the equations of conservation for each molecular species of interest—in this case, P, R, $C1$, and $C2$. To simplify the analysis, we also assume that the soluble antibody is present in excess—that is, $[P] \gg n[R]_\text{o}/N_A$, where n = number of cells/volume, $[R]_\text{o}$ = initial number of receptors/cell, and N_A is Avogadro's number (6.02×10^{23} molecules/mol). We need only consider

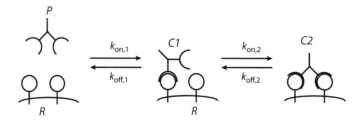

Figure 3.12 An antibody (P) can bind to a receptor (R) on a solid surface by one ($C1$) or both ($C2$) of its identical and independent binding sites.

Binding Equilibria and Kinetics

balances on $C1$ and $C2$, because $[R]$ can be calculated from $[R]_o = [R] + [C1] + 2[C2]$.

To properly describe the second binding step from $C1$ to $C2$, we must determine whether the receptor molecules are mobile on the surface. Different descriptions are necessary if the receptors are randomly distributed at fixed points (as is the case when they are immobilized on a solid surface) or if they are free to diffuse across the surface (as is often the case in a cell's plasma membrane). Let's first consider the case where receptor diffuses freely on the surface. In this case, by mass-action kinetics, the rate of formation of $C2$ will be proportional to the concentrations (in units of receptors/cell) of $C1$ and R:

$$\frac{d[C1]}{dt} = 2k_{on,1}[P]_o[R] - k_{off,1}[C1] - k_{on,2}[C1][R] + 2k_{off,2}[C2] \qquad (3.69)$$

$$\frac{d[C2]}{dt} = k_{on,2}[C1][R] - 2k_{off,2}[C2] \qquad (3.70)$$

Note the factors of 2 in these conservation equations. As with the analysis of binding to multiple equivalent sites in section 3.3.1, the initial binding reaction rate is $2k_{on,1}[P]_o[R]$, because two binding sites in the antibody are capable of binding. Similarly, dissociation may occur at two sites in $C2$, so the dissociation rate is $2k_{off,2}[C2]$.

Let's consider the equilibrium solution to these conservation equations. To do so, set each time derivative to zero and solve for the concentrations $[C1]$ and $[C2]$:

$$[C1]_{eq} = [R]_o \beta \left(\frac{-1 + \sqrt{1 + 4\delta}}{2\delta} \right) \qquad (3.71)$$

$$[C2]_{eq} = \frac{[R]_o}{2} \left(\frac{1 + 2\delta - \sqrt{1 + 4\delta}}{2\delta} \right) \qquad (3.72)$$

$$[R]_{eq} = [R]_o - [C1] - 2[C2] \qquad (3.73)$$

where

$$\beta = \frac{[P]_o}{[P]_o + \frac{K_{d,1}}{2}} \qquad (3.74)$$

and

$$\delta = \beta(1 - \beta)\frac{[R]_o}{K_{d,2}} \qquad (3.75)$$

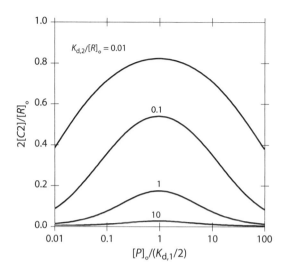

Figure 3.13 Equilibrium solution of crosslinking material balances (equation 3.72). The number of crosslinked receptors is maximum when the soluble antibody concentration $[P]_o = \frac{K_{d,1}}{2}$, and is a strong function of the crosslinking equilibrium constant $K_{d,2}$. At high $[P]_o/(K_{d,1}/2)$, $C1$ predominates over $C2$.

The affinity constants $K_{d,1}$ and $K_{d,2}$ are defined as follows:

$$K_{d,1} = \frac{k_{\text{off},1}}{k_{\text{on},1}}, \quad \text{and} \quad K_{d,2} = \frac{k_{\text{off},2}}{k_{\text{on},2}}.$$

Note that the units for $K_{d,2}$ are number of receptors/cell, as for $[C1]$, $[C2]$, and $[R]$. Somewhat unexpectedly, the concentration of crosslinked receptors $[C2]$ goes through a maximum, as shown in figure 3.13. This is because at high concentrations of antibody P, monovalently bound receptor $C1$ dominates. Growth factors that act by crosslinking receptors (e.g., human growth hormone and others in figure 3.7), also exhibit the paradoxical phenomenon of inhibition at high concentrations (figure 3.14).

Predicting avidity of antibodies The bivalency of monoclonal antibodies might be expected to increase their apparent affinity, because the doubly bound form $C2$ dissociates more slowly than singly bound $C1$. A simplified geometric analysis of this situation provides an estimate for the apparent affinity of a bivalent antibody. This approach follows one first developed by Crothers and Metzger [11]. As we shall see, the predominant effect is to substantially decrease dissociation rates rather than shift the binding isotherm leftward to lower concentrations.

Binding Equilibria and Kinetics

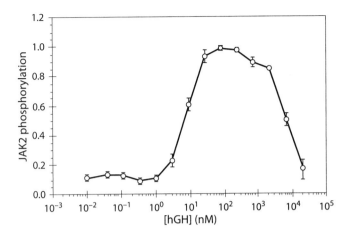

Figure 3.14 Activation of the human growth hormone (hGH) receptor exhibits the bell-shaped response as a function of [hGH] that is characteristic of a crosslinking binding isotherm (figure 3.13). hGH has been shown to form a 1:2 complex with its receptor, bridging two receptor molecules to initiate signaling [9].

Case Study 3-1 J. D. Stone, J. R. Cochran, and L. J. Stern. T-cell activation by soluble MHC oligomers can be described by a two-parameter binding model. *Biophys. J.* 81: 2547–2557 (2001) [10].

T cells are activated when their T cell receptors (TCRs) are bound by multivalent ligands, either solubly or on the surface of antigen presenting cells. Soluble ligands with higher multivalency are generally found to be more potent, meaning that the degree of cell response at a given ligand concentration is higher. What is not immediately clear, however, is whether highly multivalent ligands are more potent because: (1) clustering greater

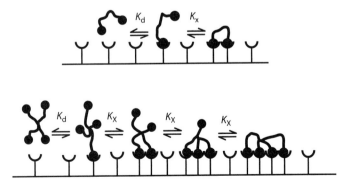

Figure 3.15 TCR activation by a dimeric or tetrameric ligand. Figure from reference 10.

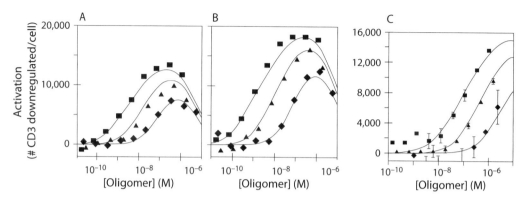

Figure 3.16 Square symbols represent tetravalent ligands; triangles, trivalent; and diamonds, bivalent. Panels A and B are responses of the HA1.7 T cell line at 12 and 27 hours, respectively, and panel C represents the C1-1 T cell line. The solid curves are fits to a model where activation is proportional to the density of pairwise crosslinks. The three curve fits (one for each panel) gave similar values for $K_x \approx 3 \times 10^{-4}$ cell, but the monovalent equilibrium binding constants are quite different for the HA1.7 T cell line ($K_d = 1.7$ µM) and the C1-1 T cell line ($K_d = 60$ µM). Figure from reference 10.

numbers of TCR molecules sends a more intense intracellular signal or (2) more highly multivalent ligands have higher apparent affinity TCR crosslinks. This latter explanation is shown to be quantitatively consistent with the data of Stone et al. [10], who modeled TCR activation as shown in figure 3.15:

$$[C_i] = \binom{v}{i}(K_x)^{i-1}\frac{[L]_o}{K_d}[R]^i \quad (3.76)$$

where $[C_i]$ is the concentration of multivalent ligand bound at i sites, v is the maximum valency of the ligand, K_x is the crosslinking equilibrium constant, K_d is the monovalent soluble equilibrium binding constant, and $[R]$ is the concentration of available receptors at equilibrium.

Experimental data for T cell responses to multivalent ligands are shown in figure 3.16 and indicate strong consistency with pairwise crosslink formation driving activation.

The proper way to consider the second binding step in bivalent interactions is in terms of the effective concentration (section 3.3.3) of the second binding site. One would expect this increase in *avidity* to be a function of the receptor density on the surface [R] (in units of receptor number per cell), as well as the distance r_{IgG} between the antibody's binding sites. Let's consider the volume swept out within a distance r_{IgG} of the receptor-bound antibody arm of $C1$. The number of receptor molecules within a disc of radius r_{IgG} on the cell surface is

Binding Equilibria and Kinetics

$$\text{Number of accessible receptors} = \pi r_{\text{IgG}}^2 \left(\frac{[R]}{4\pi r_{\text{cell}}^2} \right) \quad (3.77)$$

$$= [R] \frac{r_{\text{IgG}}^2}{4 r_{\text{cell}}^2} \quad (3.78)$$

The hemispheric volume swept out by the unbound IgG binding site is $\frac{2}{3}\pi r_{\text{IgG}}^3$. So the molarity of accessible receptors in this region is

$$[R]_{\text{accessible}} = [R] \frac{3}{8\pi r_{\text{cell}}^2 r_{\text{IgG}} N_A} \quad (3.79)$$

Can we make the simplifying assumption that the rate of second-site binding is describable by mass-action kinetics; in other words, is proportional to $[R]_{\text{accessible}}$? This assumption is clearly not quite right, because the surface-bound receptor is not uniformly well mixed in the sampled volume; however, recall that we've already made the somewhat questionable assumption that the second IgG site uniformly accesses a hemispherical space above the membrane (which would require that the antibody itself occupy no space that would block the transit of its free binding site.) All mathematical models are necessarily incomplete descriptions of physical reality, but they can be useful nevertheless when they produce straightforward mathematical expressions that adequately capture essential processes. This benefit is only justifiable if the experimental data are consistent with model predictions.

Let us then press on and assume that the rate of second-site binding is describable by the following expression:

$$\text{Second-site binding rate} = k_{\text{on},1}[C1][R]_{\text{accessible}} \quad (3.80)$$

$$= \frac{3 k_{\text{on},1}}{8\pi r_{\text{cell}}^2 r_{\text{IgG}} N_A} [C1][R] \quad (3.81)$$

$$= k_{\text{on},2}[C1][R] \quad (3.82)$$

Substituting the above relationship for $k_{\text{on},2}$ into $K_{d,2} = k_{\text{off},2}/k_{\text{on},2}$ and assuming that $k_{\text{off},2} = k_{\text{off},1}$, we obtain:

$$K_{d,2} = \frac{k_{\text{off},1}}{\frac{3 k_{\text{on},1}}{8\pi r_{\text{cell}}^2 r_{\text{IgG}} N_A}} \quad (3.83)$$

$$= \frac{8}{3}\pi r_{\text{cell}}^2 r_{\text{IgG}} N_A K_{d,1} \quad (3.84)$$

Kaufman and Jain [12] have used this relationship to correlate bivalent and monovalent antibody labeling data in fluorescence recovery after photobleaching experiments.

Let's examine typical numerical values for the parameters in equation 3.84. r_{IgG} is approximately 125 Å for most IgGs, although this distance fluctuates with time due to the flexibility of the hinge region [13]. A typical mammalian cell in suspension will have $r_{cell} \approx 7.5$ μm, allowing the following general estimate:

$$K_{d,2}(\text{number/cell}) = 3.5 \times 10^9 K_{d,1}(\text{mol/L}) \tag{3.85}$$

The effect of bivalent binding on the shape of the titration curve is shown in figure 3.17. Curiously, the concentration $[P]_o$ for half-maximal surface binding is unaltered by the avidity effect of bivalent binding. Instead the shape of the binding isotherm is shifted. At low IgG concentrations, stronger crosslinking increases the bound fraction by the avidity benefit of reduced dissociation from the doubly bound state. But at elevated IgG concentrations, stronger crosslinking populates the doubly bound ($C2$) state, which occupies two receptors with a single ligand, thereby decreasing ligand binding. Contrary to the ligand perspective, receptor occupancy is solely aided by enhanced crosslinking, as seen in the right panel of figure 3.17. It

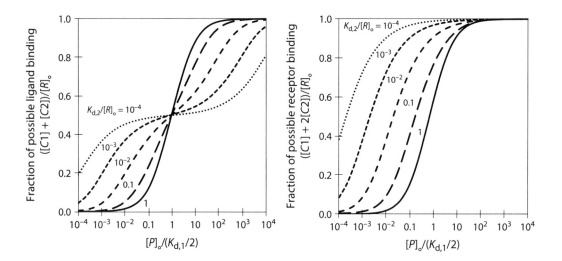

Figure 3.17 The presence of bivalent binding distorts the typical isotherm seen in monovalent binding (e.g., figure 3.2). The left panel shows measurable levels of cell-surface antibody (or ligand), and the right panel shows saturation levels of cell-surface receptor. These perspectives differ significantly, because each incremental molecule [C1] or [C2] contributes only one additional bound molecule in the left panel, but it blocks one or two surface receptors, respectively, in the right panel. Attempting to fit curves in the left panel with small values of $K_{d,2}/[R]_o$ to equation 3.5 would not produce a good fit; furthermore, the antibody concentration $[P]_o$ at which half-maximal binding is obtained would in all cases still equal $K_{d,1}/2$.

is interesting to note how fundamentally different the binding isotherm shapes are when plotted as: numbers of crosslinked receptors (figure 3.13); bound antibodies (figure 3.17, left panel); or occupied receptors (figure 3.17, right panel). It is therefore critical to specify what is being measured, or what is driving biological function, in any experiment involving titration of bivalent antibody on cell-surface receptors.

The effect of bivalency on the overall dissociation rate can be estimated by applying equation 3.68 to an antibody with identical binding sites, using the relationship for $[R]_{accessible}$ (equation 3.79) for the ligand concentration $[A]_{eff}$:

$$k_{off,app} \equiv 2 \frac{K_{d,1}}{[R]_{accessible}} k_{off,1} \tag{3.86}$$

$$= \frac{16}{3}\pi r^2{}_{cell} r_{IgG} N_A \frac{K_{d,1}}{[R]} k_{off,1} \tag{3.87}$$

$$= \frac{2K_{d,2}}{[R]} k_{off,1} \tag{3.88}$$

$$\tag{3.89}$$

Using the same numerical estimates as for equation 3.85 yields

$$k_{off,app} = 7 \times 10^9 \frac{K_{d,1}(mol/L)}{[R](\#/cell)} k_{off,1} \tag{3.90}$$

It should be noted that because these results use equation 3.68, which was derived under the assumption that $[R]_{accessible} \gg K_{d,1}$, we have $k_{off,app} \ll k_{off,1}$.

Case Study 3-2 Y. Mazor, A. Hansen, C. Yang, P. S. Chowdhury, J. Wang, G. Stephens, H. Wu, and W. F. Dall'Acqua. Insights into the molecular basis of a bispecific antibody's target selectivity. *mAbs* 7: 461–469 (2015) [14].

Mazor and colleagues evaluate the impact of avidity on the cellular specificity of a bispecific antibody. Their experimental system consists of a heterobivalent antibody with one arm binding CD4 and the other binding CD70, as well as a trio of cell types with dual expression (CD4$^+$/CD70$^+$) or either form of singular expression (CD4$^+$/CD70$^-$ or CD4$^-$/CD70$^+$). The original antibody binds the targets with 0.9 nM and 25 nM affinity, respectively, which results in strong binding to the double-positive cells and the CD4$^+$/CD70$^-$ cells, yet weaker binding to the CD4$^-$/CD70$^+$ cells. This behavior highlights two of the potential functions of a bispecific ligand that are noted by Mazor et al.: (1) the ability to bind one of the single-positive cells demonstrates the expanded breadth of binding relative to monospecific ligands, yet the weakness of binding to CD4⁻/CD70⁺ cells renders this agent an incomplete OR gate; and (2) the preferential binding of double-positive cells provides selectivity, yet the presence of relatively strong binding to CD4⁺/CD70⁻ cells renders this agent a poorly selective AND gate.

To explore the latter phenotype (i.e., selectivity for only double-positive cells), Mazor et al. impose a series of point mutations on the CD4-binding arm of the antibody to reduce its monovalent affinity for CD4 and thereby diminish binding to single-positive CD4⁺/CD70⁻ cells. Despite the reduced monovalent affinity, the bispecific antibody exhibits effectively no change in binding to double-positive cells. This result can be understood and quantitatively described as an extension of the cellular homobivalent binding model (figure 3.12) to consider heterobivalent binding. Balances can be written for the antibody ligand, the unbound and singly bound forms of each receptor, and the crosslinked receptors:

$$\frac{d[Ab]}{dt} = -k_{\text{on,CD4}}[Ab][CD4] + k_{\text{off,CD4}}[Ab{:}CD4] +$$
$$-k_{\text{on,CD70}}[Ab][CD70] + k_{\text{off,CD70}}[Ab{:}CD70] \qquad (3.91)$$

$$\frac{d[CD4]}{dt} = -k_{\text{on,CD4}}[Ab][CD4] + k_{\text{off,CD4}}[Ab{:}CD4] +$$
$$-\kappa k_{\text{on,CD4}}[Ab{:}CD70][CD4] + k_{\text{off,CD4}}[CD4{:}Ab{:}CD70] \qquad (3.92)$$

$$\frac{d[CD70]}{dt} = -k_{\text{on,CD70}}[Ab][CD70] + k_{\text{off,CD70}}[Ab{:}CD70] +$$
$$-\kappa k_{\text{on,CD70}}[Ab{:}CD4][CD70] + k_{\text{off,CD70}}[CD4{:}Ab{:}CD70] \qquad (3.93)$$

$$\frac{d[Ab{:}CD4]}{dt} = k_{\text{on,CD4}}[Ab][CD4] - k_{\text{off,CD4}}[Ab{:}CD4] +$$
$$-\kappa k_{\text{on,CD70}}[Ab{:}CD4][CD70] + k_{\text{off,CD70}}[CD4{:}Ab{:}CD70] \qquad (3.94)$$

$$\frac{d[Ab{:}CD70]}{dt} = k_{\text{on,CD70}}[Ab][CD70] - k_{\text{off,CD70}}[Ab{:}CD70] +$$
$$-\kappa k_{\text{on,CD4}}[Ab{:}CD70][CD4] + k_{\text{off,CD4}}[CD4{:}Ab{:}CD70] \qquad (3.95)$$

$$\frac{d[CD4{:}Ab{:}CD70]}{dt} = \kappa k_{\text{on,CD4}}[Ab{:}CD70][CD4] - k_{\text{off,CD4}}[CD4{:}Ab{:}CD70] +$$
$$\kappa k_{\text{on,CD70}}[Ab{:}CD4][CD70] - k_{\text{off,CD70}}[CD4{:}Ab{:}CD70] \qquad (3.96)$$

In this case, we have written the second association rate constant as the product of the intrinsic monovalent association rate constant and a dimensionless avidity parameter, κ, which is the product of the ratio of the equilibrium constants and the volume of extracellular liquid per mole of cells:

$$\kappa = \frac{K_{d,1}}{K_{d,2}} N_A V_{\text{ecf}} \qquad (3.97)$$

The monovalent association and dissociation rate constants, as well as the receptor densities on each cell type, were measured experimentally. The κ parameter can be related to

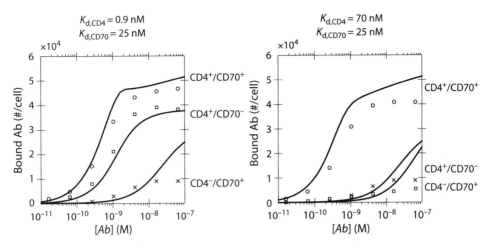

Figure 3.18 Mechanistic mathematical model with zero fitted parameters is consistent with experimental data exhibiting enhanced selectivity of a bispecific antibody on decrease of its monovalent affinity. (Left panel) Antibody with affinities of 0.9 nM (CD4) and 25 nM (CD70). (Right panel) Antibody with affinities of 70 nM (CD4) and 25 nM (CD70). Experimental results [14]: Bispecific antibody was added at the indicated concentration to CD4$^+$/CD70$^+$ cells (circles), CD4$^+$/CD70$^-$ cells (squares), or CD4$^-$/CD70$^+$ cells (crosses) for 1 h; binding was evaluated by flow cytometry. Theoretical results (solid lines) were computed from the system of differential equations (equations 3.91–3.96) using experimentally determined parameters, including kinetic rate constants, receptor densities, and reaction volume.

the geometrically derived ratio of equilibrium dissociation constants for bivalent antibody binding (equation 3.85). The conditions used in the paper (V_{ecf} = 0.001 L/300,000 cells, as well as an assumed cell diameter of 8 μm), result in a theoretical value of $\kappa = 2 \times 10^6$. Thus, the system of equations can be numerically solved without any fitted parameters. The analysis (figure 3.18) reveals a relatively good match between theory and experiment, including the enhanced selectivity achieved by reduced binding to single-positive CD4 cells while maintaining binding to double-positive cells. Although the avidity parameter could be fit, rather than rationally calculated from the geometric model, doing so yields a minimal reduction in the difference between model and experiment. Furthermore, substantially lower values of κ fail to capture the strong crosslinked binding in the lower affinity case. A numerical fit of κ requires no mechanistic insight into its value, whereas the relative strength of agreement without any fitted parameters is consistent with an understanding of the molecular system. The primary shortcoming of the model is that it indicates higher antibody binding than is experimentally observed to single-positive cells in the low-affinity monovalent cases (CD4$^-$/CD70$^+$ cells with both bispecific molecules as well as CD4$^+$/CD70$^-$ cells with the low-affinity bispecific). This may result from dissociation during the washing and analysis steps. In fact, inclusion of a brief dissociation period after the 1 h incubation (to mimic the washing steps) yields an improved fit between experiment and theory, because singly bound antibodies with weak monovalent affinity dissociate—reducing their binding as observed experimentally—whereas doubly bound antibodies and the high-affinity CD4 monovalent antibody exhibit minimal dissociation, maintaining their match to experiment.

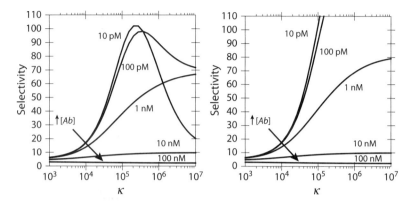

Figure 3.19 Low concentrations of bispecific antibody with a high avidity constant (κ) can provide high selectivity (left panel). Theoretical selectivities, defined as the ratio of antibody binding to double-positive (CD4$^+$/CD70$^+$) versus single-positive (CD4$^+$/CD70$^-$) cells, were computed from the system of differential equations (equations 3.91–3.96) using the parameters in the experimental system from Mazor et al. [14]. Notably, the reduced selectivity at high κ values emerges from the presence of very low densities of CD70 (100 per cell) on the CD4$^+$/CD70$^-$ cells, which enables bivalent binding at very high κ values. These low density receptors were included in the model as a conservative estimate, because they represent the lower limit of detection. Neglecting these 100 receptors yields greater selectivity (right panel).

The maintained binding to double-positive cells despite the intrinsic affinity reduction results from a multivalent avidity effect. This result is supported by evaluation of the impact of the avidity parameter on selectivity (figure 3.19). Moreover, modest intrinsic affinity is needed to diminish binding to single-positive cells, which is supported by both theory and experiment (figure 3.20). The experimental results match the model's qualitative trend of increased selectivity at lower affinity and lower antibody concentration, though the quantitative match is poor at 1 nM antibody (resulting from slightly higher experimental binding to single-positive cells) and 64 nM concentration of low-affinity antibody (perhaps resulting from the aforementioned experimental dissociation during washing).

Notably, the bispecific antibody exhibited selectivity for double-positive cells over single-positive cells despite allowing only 1 h for binding, which is insufficient to approach equilibrium. With additional time, increased antibody binding will be achieved on double-positive cells, predominantly as crosslinked CD4:Ab:CD70 ternary complex, whereas single-positive cells will not appreciably change (figure 3.21). Care should be taken to differentiate between dynamics and equilibrium when evaluating binding data. The implications of incubation time on affinity measurements will be discussed in greater depth in section 3.6 in the context of monovalent binding. Bivalent binding adds additional complexity: At early times, the low-concentration portion of a binding titration curve can appear similar to a monovalent binding curve (figure 3.21, Total bound Ab, CD4$^+$/CD70$^+$). Yet the concentration at which this transition occurs is dramatically lower than the intrinsic monovalent affinity. As previously discussed in conjunction with figure 3.17, this transition primarily represents the titration of doubly bound antibodies. A second transition with singly bound antibody occurs above the monovalent K_d. Although these phenomena can be readily handled mathematically via the analytical tools of this

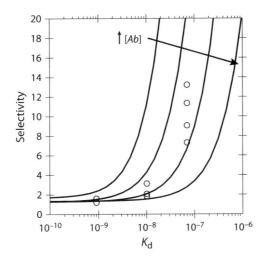

Figure 3.20 Reduced monovalent affinity improves selectivity. Theoretical selectivities, defined as the ratio of antibody binding to double-positive ($CD4^+/CD70^+$) versus single-positive ($CD4^+/CD70^-$) cells, were computed from the system of differential equations (equations 3.91–3.96) using the parameters in the experimental system from Mazor et al. [14]. Experimental data are shown for the bispecific antibody at concentrations of 1, 4, 16, and 64 nM.

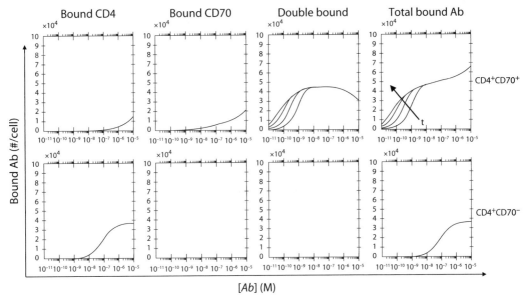

Figure 3.21 Theoretical binding profiles after 0.3, 1, 3, and 10 hours for a bispecific antibody with 25 nM affinity for CD70 and 70 nM affinity for CD4 on cells expressing both targets (top) or predominantly CD4 (bottom). Each of the three modes of antibody binding is shown as well as the cumulative total.

chapter, the form of titration curves with an incomplete concentration range and/or an insufficient approach to equilibrium can unfortunately prompt erroneous analysis using a more simplified model.

Example 3-6 What is the half-life for dissociation of a typical antibody from a cell with 10^5 binding sites on its surface?

Solution As before, it is reasonable to assume the following approximate parameters for any given antibody in the absence of specific data: $K_d = 10$ nM, $k_{on} = 10^5$ M^{-1}s^{-1}, $k_{off} = 10^{-3}$ s^{-1}. Plugging these values into equation 3.90, one obtains:

$$k_{off,app} = 7 \times 10^9 \frac{10^{-8}(\text{mol/L})}{10^5(\frac{\text{number}}{\text{cell}})} 10^{-3} \text{ s}^{-1}$$

$$= 7 \times 10^{-7} \text{ s}^{-1}$$

To find the dissociation half-life, consider a reaction without rebinding, for which the surface-bound antibody number will drop as $e^{-k_{off}t}$ (equation 3.19). It will drop by half when $t_{1/2} = (\ln 2)/k_{off,app}$; in this case, $t_{1/2} = 9.9 \times 10^5$ s ≈ 275 h! This is far longer than the time it would take for most experiments to be performed. More to the point, cellular endocytic trafficking processes would alter localization of the surface ligand on the minutes-to-hours time scale first.

Extremely slowly dissociating antibody fractions on tumor cells have been previously termed "irreversibly" bound [15]. The above analysis effectively confirms this description for a typical bivalently bound antibody, because the antibody will be endocytically consumed prior to dissociation.

Bivalent binding to fixed antigens on a solid surface When considering antibody binding in the previous section, we assumed that a second receptor antigen could always diffuse within reach of the unbound site of a singly bound antibody, so that we could use mass-action kinetics to describe the association reaction $C1 + R \rightarrow C2$. In certain situations, however, the antigen is scattered at random but fixed locations, on the surface of a solid phase.

To describe such a situation, we must divide the immobilized antigens into two classes: those close enough to be bound by both arms of the antibody, and those too

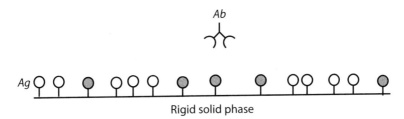

Figure 3.22 Antibody binding to immobilized antigen that is not free to diffuse laterally. Some of the antigen molecules (unshaded) are close enough to another antigen to allow bivalent antibody binding. Others (shaded) are too far from their nearest neighbor to allow bivalent binding and can only be singly bound.

distant from their nearest neighbor to allow bivalent binding. These two categories are represented in figure 3.22. The random distribution of sites throughout a surface is well described by Poisson statistics, which will be considered in chapter 9. Let's assume that P_{mono} is the probability that a particular antigen molecule has no neighbor near enough for bivalent binding. One can use the analysis from the previous section for equilibrium binding to all of the remaining antigen, replacing $[R]_o$ with $(1 - P_{mono})[R]_o$ wherever it occurs. Binding to the monovalently bindable antigen $P_{mono}[R]_o$ can be described straightforwardly by the expression

$$\frac{d[C1_{mono}]}{dt} = 2k_{on,1}[P][R_{mono}] - k_{off,1}[C1_{mono}] \qquad (3.98)$$

where $[R_{mono}]_o = P_{mono}[R]_o$.

Muller et al. [16] have used numerical evaluation of this model to describe experimental data for bivalent antibody binding to antigen immobilized on a surface plasmon resonance (SPR) biosensor device.

3.4 Fast or Complex Reaction Measurements

Very fast reactions present a particular measurement challenge. If the reaction is completed in the mixing time of the experiment, it is not possible to measure the rate by simply introducing reactants, mixing, and measuring. Instead, a general methodology has been developed that involves allowing the reaction to reach equilibrium, perturbing it slightly away from equilibrium, and watching it return to equilibrium (called relaxation). The importance of this class of methods was recognized in 1967 by the Nobel Prize in Chemistry awarded to Manfred Eigen, Ronald G. W. Norrish, and George Porter "for their studies of extremely fast chemical reactions, effected by disturbing the equilibrium by means of very short pulses of energy."

3.4.1 Relaxation Kinetics

The association of protein and ligand to form a 1:1 complex was introduced at the beginning of this chapter. The reaction rate, as monitored by the change in concentration of complex with respect to time, is given by

$$\frac{d[C]}{dt} = k_{\text{on}}[L][P] - k_{\text{off}}[C] \tag{3.99}$$

which is zero at equilibrium. If the temperature is now changed by a small amount in a very short time span, the concentrations will not have an opportunity to change. High-powered lasers allow temperature changes on the order of 10°C in less than 10 nanoseconds. If the position of equilibrium is different between the new and old temperatures, then the system is effectively instantaneously out of equilibrium (without the need for mixing). The change in position of equilibrium can be described by the van't Hoff equation (chapter 2),

$$\left(\frac{\partial \ln K_d}{\partial T}\right)_P = \frac{\Delta H^\circ}{RT^2} \tag{3.100}$$

which indicates that any reaction with nonzero ΔH° will have a temperature-dependent equilibrium position, where the ΔH° is for the dissociation reaction (matching the direction for K_d). If ΔH° is approximately constant over the small temperature span considered, then one can accurately calculate the new equilibrium position. To follow the relaxation back to the new equilibrium at the new temperature, let $\bar{C}$, $\bar{P}$, and $\bar{L}$ represent concentrations at the new equilibrium, and let ΔC, ΔP, and ΔL represent the differences between the current concentration and its equilibrium value (e.g., $\Delta C = [C] - \bar{C}$). The stoichiometry of the chemical reaction dictates that

$$\Delta C = -\Delta P = -\Delta L \tag{3.101}$$

to conserve matter. We can rewrite equation 3.99 as

$$\frac{d(\bar{C} + \Delta C)}{dt} = k_{\text{on}}(\bar{L} + \Delta L)(\bar{P} + \Delta P) - k_{\text{off}}(\bar{C} + \Delta C) \tag{3.102}$$

Because $\bar{C}$ is a constant, it has no time dependence. This can be used to simplify the left-hand side of the equation. The right-hand side is multiplied out to give

$$\frac{d(\Delta C)}{dt} = k_{\text{on}}(\bar{L}\bar{P} + \bar{L}\Delta P + \bar{P}\Delta L + \Delta L \Delta P) - k_{\text{off}}(\bar{C} + \Delta C) \tag{3.103}$$

which can be simplified by recognizing that at equilibrium, the rate of change of complex is zero,

$$k_{on}\bar{L}\bar{P} = k_{off}\bar{C} \qquad (3.104)$$

giving

$$\frac{d(\Delta C)}{dt} = k_{on}\left[\bar{L}\Delta P + \bar{P}\Delta L + \mathcal{O}(\Delta^2)\right] - k_{off}\Delta C \qquad (3.105)$$

The term $\mathcal{O}(\Delta^2)$ indicates that the $\Delta P \Delta L$ term is second order in the perturbation: For small enough perturbations, this term will be negligible with respect to the rest of the equation and can be dropped. Making use of equation 3.101 to replace ΔP and ΔL gives

$$\frac{d(\Delta C)}{dt} = -\Delta C\left[k_{on}(\bar{L} + \bar{P}) + k_{off}\right] \qquad (3.106)$$

which can be easily solved to give the relaxation of C to the new equilibrium:

$$\Delta C(t) = \Delta C_o e^{-[k_{on}(\bar{L}+\bar{P})+k_{off}]t} \qquad (3.107)$$

The relaxation follows a single exponential transient with $k_{obs} = k_{on}(\bar{L} + \bar{P}) + k_{off}$. Values for k_{on} and k_{off} can be straightforwardly extracted by carrying out a series of perturbations resulting in a variety of values for $\bar{L} + \bar{P}$ and then plotting k_{obs} versus $(\bar{L} + \bar{P})$; the slope of the resulting line is k_{on}, and the intercept is k_{off}. Note the similarity of this relaxation solution to the pseudo-first-order approximation for association kinetics (equation 3.16).

In this way, the measurement of very fast reaction rates can be achieved by applying small perturbations and observing the relaxation to equilibrium. The illustration here was based on a temperature jump acting as the perturbation away from equilibrium, which is among the most common. In principle, changes of pressure, electric field, or other conditions could also serve as the perturbation.

3.4.2 Double-Jump Experiments to Detect Hidden Rate Processes

It should be remembered that experimental measurements often do not distinguish between kinetically distinct species. For example, consider the formation of an intermediate "encounter complex" C^* en route to a more stable complex C, as shown below:

$$R + L \underset{k_{-1}}{\overset{k_1}{\rightleftharpoons}} C^* \underset{k_{-2}}{\overset{k_2}{\rightleftharpoons}} C \qquad (3.108)$$

Some measurement methods do not distinguish between the signal from C^* and the signal from C. For example, a biosensor reports the additional mass of captured ligand and consequently gives identical mass signals for both the encounter complex and the final complex, reporting a total signal for $[C^*] + [C]$. In such a case, one can examine the underlying interconversion of C^* and C by perturbing the system twice rather than once. The operational definition of zero time in all kinetic experiments corresponds to some perturbation—for example, the mixing of protein and ligand solutions. Adding a second perturbation (e.g., removing free ligand) at a series of different times in different experiments sheds light on the "invisible" interconversion of C^* and C.

3.5 Theory and Practice of Biomolecular Measurements

When you can measure what you are speaking about, and express it in numbers, you know something about it; but when you cannot measure it, when you cannot express it in numbers, your knowledge is of a meagre and unsatisfactory kind; it may be the beginning of knowledge, but you have scarcely in your thoughts advanced to the state of Science, whatever the matter may be.
—Lord Kelvin

Measure what is measurable, and make measurable what is not so.
—Galileo Galilei

A mechanistic understanding of experimental methodology is essential for proper interpretation of thermodynamic and kinetic data. Often, dynamic aspects of the measurement process itself can obscure the underlying phenomena of interest. In the following sections, we discuss methodological approaches for measuring biomolecular interactions and numerically extracting parameter estimates (and uncertainties) from data. Building mechanistic models of biological phenomena, particularly at the level of cells and tissues, requires the ability to measure the amounts, rates, and activities of a variety of species in a range of contexts. Additionally, it is important to consider the location at which reactions occur, the identity and disposition of binding partners, and potential rearrangements of molecular and cellular structures. The types of problems that we are interested in require making measurements of kinetic rate constants and thermodynamic equilibrium constants. We describe a useful subset of measurement methodologies that permit coverage of a broad range of applications—considering theory, principles, commonly used technologies and reagents, and implementation information.

3.6 General Issues in Measuring Binding Affinity and Kinetics

To determine the parameters that characterize the equilibria and kinetics of a given biomolecular interaction (i.e., K_d, k_{off}, and k_{on}), one must measure the concentration of the complex [C] in the simultaneous presence of unbound protein and ligand. There are two basic approaches to this problem: (1) detecting complex formation by localization of the ligand to a distinct spatial compartment, generally the surface of a solid phase; and (2) exploitation of differences in the intrinsic properties of the complex relative to its individual components. Certain general issues must always be considered when making such measurements:

Choice of an appropriate binding model. A monovalent binding isotherm is generally first attempted as the simplest mathematical description of protein-ligand binding equilibrium. However, this model may not be correct due to incomplete approach to equilibrium, depletion of bulk ligand, nonspecific adsorption, diffusion limitations, multivalent interactions, cellular processes (such as receptor internalization), or the presence of unidentified binding partners. Any of these complicating factors can make a monovalent isotherm model an inadequate description of the system. In such cases, it may be possible to alter the experimental design to eliminate these complications; if not, the mathematical description must incorporate these additional processes. In section 3.9, we will consider quantitative metrics for a model's *goodness of fit* to data.

Insufficient equilibration time. What degree of error might be expected from not allowing an experimental binding titration to come to equilibrium at all of its concentration points? As an example, let's consider a protein-ligand interaction with the binding parameters $K_d = 1$ nM, $k_{on} = 10^5$ M^{-1}s^{-1}, and $k_{off} = 10^{-4}$ s^{-1}. Twenty-two samples are evenly log-spaced in $[L]_o$ from 0.01 to 100 nM; assume that pseudo-first-order conditions apply throughout this range (i.e., $[P]_o \ll ([L]_o + K_d)$). The time required to allow samples to reach near-equilibrium (i.e., $t = 5/k_{obs}$) increases as ligand concentration decreases, because $k_{obs} = k_{on}[L]_o + k_{off}$ (figure 3.23). From this relationship, it appears that the sample at $[L]_o = 100$ nM effectively reaches equilibrium at 500 seconds, whereas those samples at $[L]_o < 1$ nM do not reach equilibrium until $t > 20,000$ seconds. The errors due to insufficient equilibration are not evenly distributed across the concentration range; much greater inaccuracy is present at the lowest concentrations.

The apparent isotherms (obtained before equilibrium is actually reached at all ligand concentrations) were calculated using $y = y_{eq}(1 - e^{-k_{obs}t})$ at the times indicated by each of the five labels A–E in figure 3.23. These isotherms are plotted in figure 3.24. Only the set corresponding to the leftmost point appears fully

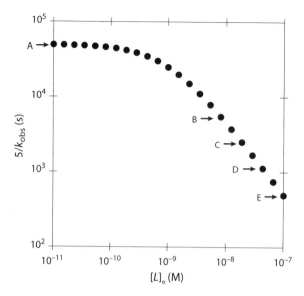

Figure 3.23 Time required to effectively reach equilibrium for each of the ligand concentrations in a titration curve. Equilibrium time was calculated from the pseudo-first-order relationships $y = y_{eq}(1 - e^{-k_{obs}t})$ and $k_{obs} = k_{on}[L]_o + k_{off}$, with $5/k_{obs}$ arbitrarily chosen as the time required to effectively reach equilibrium (~99% of the approach). The five labels (A–E) indicate the equilibration times plotted in figure 3.24.

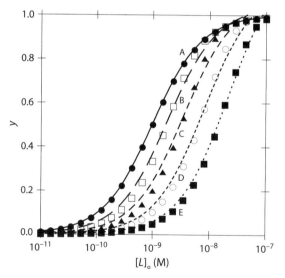

Figure 3.24 Apparent binding isotherms for nonequilibrated samples at the five time points corresponding to the labels A–E in figure 3.23. The lines represent the nonlinear least squares curve fit to a pseudo-first-order equilibrium isotherm.

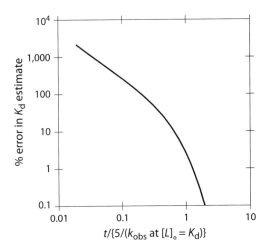

Figure 3.25 Percentage error in the estimate of K_d for the example under consideration, when insufficient time is allowed to reach equilibrium. If the equilibration time required for the ligand concentration at the midpoint (i.e., $\approx K_d$) is used, error $< 10\%$ is obtained. However, equilibration for lesser times can lead to fairly dramatic errors.

equilibrated. All others could be viewed as insufficiently equilibrated for ligand concentrations that have lower values of k_{obs}. The general quality of each curve fit appears good by eye, yet they differ in their estimate of K_d; all but the leftmost curve are incorrect, some by a substantial amount. (The careful observer may note that the curve fits to the nonequilibrated data sets systematically over-fit at low ligand concentrations and under-fit at high ligand concentrations, resulting in ordered residuals. However, given the error associated with real biological measurements, this resolution of poor curve fitting may not be evident in practice.) The magnitude of error in the estimate of K_d as a function of equilibration time is shown in figure 3.25.

So long as the equilibration time is at least as great as that calculated for $[L]_o = K_d$, the error is less than 10%. However, for shorter equilibration times, dramatic errors can result, with as much as an order of magnitude overestimate of K_d.

Example 3-7 A hypothetical biotechnology company, with a vested interest in claiming high affinities for its antibody, performs SPR measurements of binding kinetics by flowing bivalent IgG antibody across a sensor chip that is densely derivatized with antigen. Assume that the single-site interaction with soluble antigen is characterized by the parameters $K_d = 0.1$ nM, $k_{on} = 10^5$ M^{-1}s^{-1}, and $k_{off} = 10^{-5}$ s^{-1}. If the effective concentration of antigen immobilized on the chip perceived by bound bivalent IgG is $[A]_{eff} = 1$ mM, what values of k_{on} and k_{off} will be measured in the SPR experiment, and what estimate for K_d will be obtained by their ratio? If this same antibody is used to label cells with 10^5 receptors on their surface, what will the apparent k_{off} be, and at what antibody concentration will the cells be half-maximally labeled?

Solution Because the chip is densely derivatized, let's assume that all bound antibodies are within reach of a second antigen molecule to bind. We therefore need to choose a binding model that accounts for the bivalency of the antibody. Appropriate estimates of the kinetic parameters can be obtained using equations 3.67 and 3.68:

$$k_{on,app} = 2 \times k_{on} \tag{3.109}$$

$$= 2 \times 10^5 \text{ M}^{-1}\text{s}^{-1} \tag{3.110}$$

$$k_{off,app} = 2 \times k_{off} \frac{K_d}{[A]_{eff}} \tag{3.111}$$

$$= 2 \times 10^{-5} \text{ s}^{-1} \times \frac{10^{-10} \text{ M}}{10^{-3} \text{ M}} \tag{3.112}$$

$$= 2 \times 10^{-12} \text{ s}^{-1} \tag{3.113}$$

The ratio of these apparent kinetic constants gives the phenomenal apparent $K_d = 10^{-17}$ M; an *attomolar* affinity. Of course, dissociation constants smaller than 10^{-6} s^{-1} cannot be measured directly with SPR biosensors, because after 29 hours, only 10% of the complex has dissociated for $k_{off} = 10^{-6}$ s^{-1}. Let's now examine the properties of this antibody's interaction with cells. Using equation 3.90,

$$k_{off,app} = 7 \times 10^9 \frac{K_{d,1}(\text{mol/L})}{[R](\#/\text{cell})} k_{off,1} \tag{3.114}$$

$$= 7 \times 10^9 \frac{10^{-10} \text{ M}}{10^5 \text{ \#/cell}} 10^{-5} \text{ s}^{-1} \tag{3.115}$$

$$= 7 \times 10^{-11} \text{ s}^{-1} \tag{3.116}$$

and so the dissociation kinetics on the cell surface for bivalently bound antibody will indeed be very slow (effectively irreversible, as metabolic turnover of the antigen is likely to occur before dissociation). However, recall that the coexistence of singly and doubly bound antibody on a cell surface leads to a binding isotherm that is not consistent with the simple monovalent relationship (figure 3.17). The half-maximal labeling concentration for the antibody will be at $[Ab] = K_d/2 = 5 \times 10^{-11}$ M.

Nonspecific protein adsorption. It is an unfortunate reality of experimental biochemistry that proteins stick to essentially any surface. In general, a protein will eventually coat any solid-phase material it is in contact with; as a rule of thumb, 0.5–1.0 µg of protein will coat 1 cm^2 of a solid surface [17]. Thus, some fraction of any soluble protein of interest is lost through nonspecific adsorption. *Passivating* a surface by attaching a hydrophilic polymer, such as polyethylene glycol, can decrease the rate of protein adsorption somewhat. Zwitterionic polymers have been invented that provide the greatest measured passivation activity for surfaces [18]. One can also add *carrier proteins* (such as albumin) that compete with the protein of interest for surface adsorption sites, or nonionic detergents (e.g., Tween or Triton X-100) that accomplish the same goal. Protein stickiness can be problematic at both high concentrations (by excess sticking to cells) and low concentrations (by depletion from solution). It is often necessary to account for nonspecific sticking in cell-surface binding measurements at protein concentrations in the µM range, where weak nonspecific protein adsorption can contribute significantly to cell-surface labeling. At low solution-phase protein concentrations, surface adsorption must also be considered, because losing 0.5 µg per 1 cm^2 of surface can constitute a significant fraction of the protein in the liquid sample.

Example 3-8 Estimate potential losses from 100 µL of a 50 pM solution of a 50 kDa protein by adsorption to a micropipette tip.

Solution The total mass of protein in the sample is

$$100 \text{ µL} \times \frac{1 \text{ L}}{10^6 \text{ µL}} \times 50 \frac{\text{pmol}}{\text{L}} \times 50{,}000 \frac{\text{pg}}{\text{pmol}} = 250 \text{ pg} \quad (3.117)$$

Although many micropipette tips are in fact conical, let's approximate the liquid column in the pipette tip as a cylinder of diameter D and length L:

$$\text{Volume} = \frac{\pi D^2 L}{4} \quad (3.118)$$

$$\text{Area} = \pi D L \quad (3.119)$$

> If we assume that the aspect ratio (length:diameter) of the liquid column is 10, then for a 100 μL volume, $D = 2.3$ mm and $L = 23$ mm, resulting in an area $= 1.7$ cm^2. A monolayer of protein bound to such a surface would constitute approximately 1 μg of protein, more than three orders of magnitude greater than the total amount of protein in the pipetted volume. Fortunately, materials of manufacture for pipette tips have been chosen to decrease the rate of such protein adsorption, such that expeditious handling and the judicious use of "carrier" proteins can minimize what are clearly potentially very large errors in protein concentration due to adsorptive losses. A carrier protein is a protein inactive in the assay of interest that can be added in excess—albumin is often used for this purpose.

Labeling artifacts. Many methods for detecting complex formation rely on the covalent attachment of a fluorophore or affinity ligand, such as biotin. A common point of attachment to proteins is through primary amines present on lysine residues and at the amino terminus of a polypeptide chain. Most proteins possess several surface lysines, some of which may lie directly in the binding interface of interest. Clearly, attaching a probe molecule to such a site is likely to hinder binding, but more subtle effects may also result from attachment adjacent to the binding site, through electrostatic interactions or partial steric hindrance of the protein-ligand interaction. If the average number of conjugated lysines per molecule is kept low, the fraction of protein inactivated by probe attachment can be minimized. Alternatively, a single site for attachment can often be introduced into a protein through addition of a cysteine residue at the N- or C-terminus. In extracellular proteins, the great majority of cysteine side chains are either oxidized to disulfide bonds or buried in the hydrophobic core, so an artificially appended terminal cysteine often presents a single exposed thiol for conjugation to probe molecules. Site-specific attachment provides a better characterized system for analysis compared to a mixture of molecules derivatized at random subsets of surface lysines. Nevertheless, the latter approach is most often satisfactory and is generally simpler to implement.

Sections 3.7 and 3.8 describe several alternative methods for measuring the kinetics and equilibria of macromolecular interactions. A minimum expectation is that the parameters measured by one method will be directly comparable to those measured by another; of course, this may not be so, due to artifacts and peculiarities of each technique. In a thorough validation case study, Myszka and coworkers [19] have compared three of these methods (SPR biosensors, isothermal titration calorimetry (ITC), and stopped flow fluorescence (SFF) spectroscopy) for a

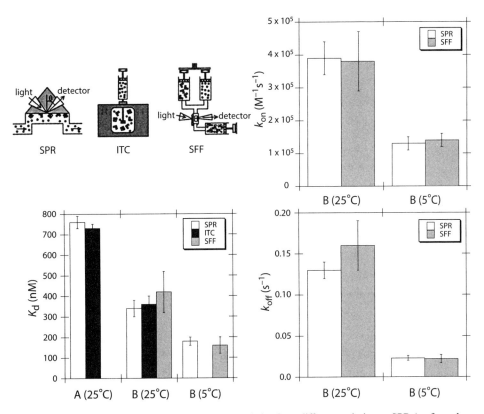

Figure 3.26 Comparison of binding measurements made by three different techniques: SPR (surface plasmon resonance biosensor; section 3.7.2); ITC (isothermal titration calorimetry; section 3.8.2); and SFF (stopped flow fluorescence; section 3.8.1) for the interaction between carbonic anhydrase II and two compounds (A = 4-carboxybenzenesulfonamide; B = dansylamide), at 5 and 25°C. Each measurement was performed in two to eight independent replicates, and the error bars represent standard deviations from those replicates. When the methods are applied and interpreted correctly, they give equivalent results [19].

protein-ligand interaction. When performed properly, these three methods produce statistically indistinguishable parameter estimates (figure 3.26).

3.7 Methods That Detect Altered Localization on Binding

If a protein is tethered to a solid surface or is in a semipermeable membrane, then complex formation may be inferred by localization of its ligand to the same site. The solid phase in such an assay may be a biosensor device, a microplate well, or a cell surface.

3.7.1 Cell-Surface Titrations and Competition

A common method for measuring binding affinity and dissociation kinetics for surface receptors is to titrate the protein-ligand interaction in situ on the cell plasma membrane, with detection of ligand binding by flow cytometry, fluorescence microscopy, or radioactive scintillation counting. Advantages of this approach include direct analysis in the cellular milieu (most likely of greatest relevance) and obviating the need for separate expression and purification of the receptor ectodomain (which in any case may be extensively challenging for multiple-pass membrane proteins). There are several potential disadvantages. There may be unknown participants in the binding reaction on the cell surface, which would make simple interpretation of the binding equilibrium difficult. There is the potential for nonspecific binding of the ligand to sticky sites on the cell surface. The receptor may be involved in linked oligomerization equilibria that would complicate analysis. If the ligand is multivalent, then avidity phenomena must be considered as described earlier in this chapter. Finally, if receptor endocytic trafficking occurs during the analysis, then the process cannot reach equilibrium—for this reason, many such titrations must be performed on ice and the results extrapolated to physiological temperature.

Ligand depletion One simple artifact that must be avoided is the stoichiometric depletion of ligand by binding to the cells; the simplest experimental design is to use, when possible, excess soluble ligand such that pseudo-first-order equilibria and kinetics apply. When this is impractical, ligand depletion can be explicitly accounted for by the following relationship:

$$y = \frac{(1-\eta y)[L]_o}{K_d + (1-\eta y)[L]_o} \qquad (3.120)$$

where

$$\eta \equiv \frac{n[R]_o}{N_A[L]_o} \qquad (3.121)$$

and $n \equiv$ number of cells/volume, $[R]_o \equiv$ number of receptors per cell, and $N_A \equiv$ Avogadro's number, 6.02×10^{23} molecules/mol. Thus, η is simply the ratio of the receptor molarity ($n[R]_o/N_A$) to the ligand molarity ($[L]_o$). Note that as the receptor molarity increases (either via cell concentration n or receptor expression level $[R]_o$), ligand depletion increases. This correction need not be applied if the assay can be performed with $n \ll \frac{N_A[L]_o}{[R]_o}$. (Note that equation 3.120 is actually a rearranged form of equations 3.6 and 3.7.)

Example 3-9 Yeast cells expressing an antibody fragment on their surface were labeled with biotinylated protein antigen, followed by a streptavidin-phycoerythrin conjugate, and the average single-cell fluorescence was measured by flow cytometry. The fluorescence versus antigen concentration data shown below were obtained:

[Antigen] (nM)	Mean fluorescence units
0.020	25.4
0.050	47.8
0.200	139
0.600	255
1.20	332
3.00	518
10.0	600
27.0	684
74.0	699
200	700

Estimate K_d for the antibody-antigen interaction using these data.

Solution Plotting the data semilogarithmically and using nonlinear least squares curve fitting gives the curve shown in figure 3.27 and the following fitted equation:

$$\text{Fluorescence} = \text{Background} + \text{Constant} \times \frac{[Ag]}{K_d + [Ag]}$$

$$= 26.7 + 677 \times \frac{[Ag]}{1.27 + [Ag]}$$

So the estimated K_d is 1.27 nM.

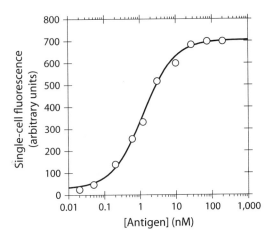

Figure 3.27 Single-cell fluorescence as a function of antigen concentration. The circles represent the data, and the solid curve is the semilogarithmic curve fit to the equation, given the values shown.

Binding competition How can one determine whether the measured binding curve is perturbed by nonspecific sticking of the ligand? If an identical cell line lacking only the receptor of interest is available, a negative control titration can be performed. Alternatively, competition with unlabeled ligand is considered definitive evidence for binding specificity. Depending on the detection modality, such competitors could be: the ligand unlabeled by fluorophores, radioactive isotopes, or a peptide epitope tag. Let's consider how the presence of a competitor alters the binding isotherm, adding an asterisk to the variables for unlabeled ligand (L^*) and the undetected complex with unlabeled ligand (C^*). Formally considering the possibility that the binding constant of the unlabeled ligand may differ (K_d^*), it can be shown that the detectable binding saturation under pseudo-first-order conditions for both L and L^* will be

$$y = \frac{[L]_\circ}{K_d + [L]_\circ + \frac{K_d}{K_d^*}[L^*]_\circ} \qquad (3.122)$$

Example 3-10 If a cell is labeled to 95% saturation with a fluorescently labeled ligand, how much unlabeled competitor must be added to reduce labeling to 50% saturation? Assume that the labeled and unlabeled ligand have identical equilibrium binding constants ($K_d = K_d^* = 10$ nM) and that the unlabeled competitor stock solution is highly concentrated so that its addition negligibly changes the volume.

Solution If $y = 0.95$ when $[L^*] = 0$, then $[L]_\circ = yK_d/(1-y) = 190$ nM. From equation 3.122,

$$0.5 = \frac{190}{10 + 190 + [L^*]_\circ} \qquad (3.123)$$

So $[L^*] = 180$ nM to reduce binding to half saturation.

Measurement of dissociation kinetics is often performed by adding excess unlabeled ligand and measuring the decrease in cell labeling with time, which is $y = y_{eq}e^{-k_{off}t}$ (equation 3.19; where y_{eq} is the equilibrium value of y). Such an experiment is often called "competing off" the ligand, which unfortunately connotes an active role for the competitor in prying the ligand from the receptor. In fact, the competitor simply replaces labeled ligand that dissociates at the same rate regardless of whether the competitor is added.

Nonspecific binding Residual binding that occurs in the absence of the pertinent receptor (or presence of excess competitor) is generally due to nonspecific binding that is characterized by a large number of cell-surface binding sites with very weak affinity. Particular ligands and cell types display variable proclivities for such "sticky" binding. The nonspecific component of a labeling isotherm can often pragmatically be described by a linear term as follows:

$$[B] = K_N[L]_o \tag{3.124}$$

where $[B] \equiv$ nonspecifically bound labeled ligand (number per cell) and $K_N \equiv$ nonspecific association binding constant (number per cell/M). This linear approximation to the nonspecific isotherm is valid in most cases when $[L]_o \ll [R]_{o,ns}/K_N$, where $[R]_{o,ns}$ is the total number of sites for nonspecific binding. The overall detected signal is given by

$$[C] + [B] = \frac{[L]_o}{K_d + [L]_o}[R]_o + K_N[L]_o \tag{3.125}$$

This relationship (equation 3.125) is often used to curve-fit a slow upward drift at higher concentrations in a binding isotherm (figure 3.28).

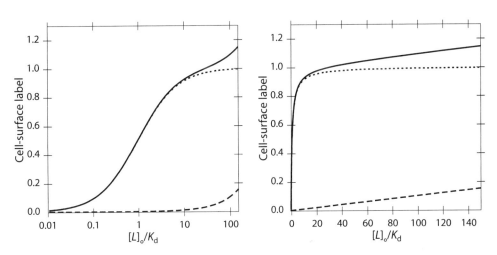

Figure 3.28 Simulation of a cell-surface labeled-ligand titration in the presence of nonspecific binding. The left panel is plotted as a semilog curve, the right on linear coordinates. The solid curve is the observable total cell-associated ligand label, the dotted curve is that portion due to specific binding, and the dashed curve is nonspecific ligand binding with $K_N K_d/[R]_o = 0.001$. The observed slow upward drift at high ligand concentration is very often observed in actual experiments, and consequently, equation 3.125 is used to curve-fit such data to estimate K_d.

A dimensionless nonspecific binding capacity v can be defined as follows:

$$v \equiv \frac{nK_N}{N_A} \qquad (3.126)$$

Experimental values for v can vary widely, but are generally in the range 0.001–0.1.

Case Study 3-3   E. T. Boder and K. D. Wittrup. Optimal screening of surface-displayed polypeptide libraries. *Biotechnol. Prog.* **14: 55–62 (1998) [20].**

Directed evolution is a method for improving the properties of a protein by screening large, genetically diverse populations of mutant proteins (often called libraries) for those few individuals with the desired characteristics. Libraries of antibodies or other binding molecules can be displayed on the surface of cells and then screened by flow cytometry. In this paper, the authors consider what labeling conditions will best discriminate improved mutants from the excess background of wild-type (wt) clones.

Two methods are in common use for screening libraries of binding proteins: equilibrium screening or "off-rate" kinetic screening, represented in the top-left and top-right panels of figure 3.29, respectively. The design variable in equilibrium screening is the ligand concentration, whereas the design variable in kinetic screening is the time of competition with unlabeled ligand. In both cases, it is clear that at one extreme, all cells are fully labeled (high ligand concentration in equilibrium screening, zero competition time in kinetic screening), and at the other extreme (low ligand concentration or long competition times), the signal of all cells has fallen to a common background level. At an intermediate value for each of these design variables, the discrimination between the desired mutant and the wild-type background will be maximized. Although there is an inappropriate tendency to use the word "optimize" as synonymous with "improve," this is one instance where an optimal labeling approach actually exists. Note that the maximum discrimination obtainable by kinetic screening ($\approx 2.8\times$) is superior to that for equilibrium screening ($\approx 1.55\times$) in the following example experimental conditions.

The single-cell fluorescence for clone i in equilibrium screening is given by the expression

$$\text{Fluorescence} = F_{bg} + (F_{max} - F_{bg})\frac{[L]}{[L] + K_{d,i}} \qquad (3.127)$$

where F_{bg} is the background fluorescence value, and F_{max} is the saturation signal. The extreme value (maximum or minimum) for the ratio of fluorescence for two different clones is found by setting the derivative of the ratio with respect to $[L]$ to zero. In this case, the following appealingly simple relationship is obtained for the optimal ligand labeling concentration:

$$[L]_{opt} = \sqrt{\frac{K_{d,mut} K_{d,wt}}{F_{max}/F_{bg}}} \qquad (3.128)$$

Binding Equilibria and Kinetics

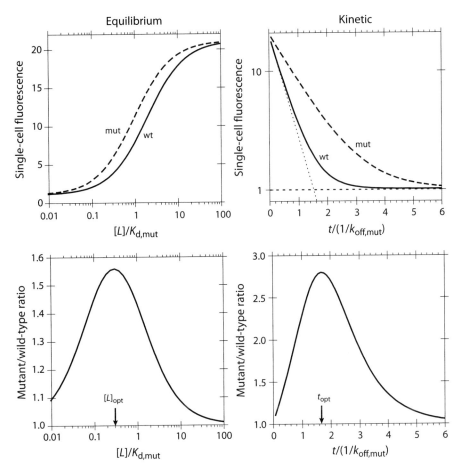

Figure 3.29 Optimal labeling for equilibrium or kinetic combinatorial library screening, in which the mutant ligand has a twofold stronger affinity resulting from a twofold slower dissociation rate constant.

In kinetic screening, the single-cell fluorescence for clone i is given by the expression

$$\text{Fluorescence} = F_{bg} + (F_{max} - F_{bg})e^{-k_{off,i}t} \qquad (3.129)$$

where t is the competition time. Although no closed-form solution for the optimum competition time is available, the following approximation (obtained by extrapolating the tangent, as shown in the top-right panel of figure 3.29) is useful:

$$t_{opt} = \frac{\ln \frac{F_{max}}{F_{bg}}}{k_{off,wt}} \qquad (3.130)$$

In qualitative terms, for both equilibrium and kinetic screening the optimal mutant/wild-type discrimination is obtained when the signal from wild-type cells is effectively at the background level. Further decreases in ligand concentration (or increases in competition time) serve only to decrease the mutant signal without any further decrease in the wild-type signal. The general concept of optimal equilibrium ligand concentration or competition time is applicable to other screening methodologies, such as phage display. These optimal labeling formulas have been validated experimentally [21].

3.7.2 Biosensors

A useful type of label-free biosensor detects macromolecular complex formation near a solid surface by detecting alterations in the local optical properties. The two main approaches use surface plasmon resonance (SPR) or biolayer interferometry (BLI) as their physical detection principles. These two techniques share many common features; we describe the SPR approach in more detail below.

SPR-based biosensors operate by detecting macromolecular complex formation on the surface of a thin (35–200 nm) gold film (figure 3.30) [22]. The binding partner is immobilized on one side of the gold film, often in a dextran hydrogel 100–400 nm thick. A white light source is shined on the opposite face of the gold film at an angle sufficiently acute to be completely reflected, and an evanescent wave or *surface plasmon* forms in the solution adjacent to the gold surface on the binding-partner side. The energy of this surface plasmon drops exponentially with distance from the gold film, extending only a few hundred nanometers and therefore interacting primarily with the aqueous volume in which the binding partner is immobilized. At a particular resonant wavelength, most of the light is absorbed rather than reflected. This wavelength is a strong function of the index of refraction of the aqueous medium above the gold film, which increases with increasing macromolecular mass in the volume. When ligand is flowed past the surface and complexes are formed, the index of refraction near the surface increases and changes the resonant wavelength, which is then detected with a diode array. The shifting resonant wavelength is reported in *resonance units* (RUs), with 1 RU equivalent to approximately 1 pg protein/mm^2. The value in RUs is proportional to macromolecular mass up to surface densities of approximately 50 ng/mm^2.

Advantages of SPR and BLI biosensors include label-free detection and simultaneous estimation of both equilibrium and kinetic parameters. Potential problems include diffusion limitations, immobilization artifacts, and avidity effects.

Analysis of biosensor data most often applies pseudo-first-order kinetic and equilibrium relationships, because the continuously flowing bulk ligand solution

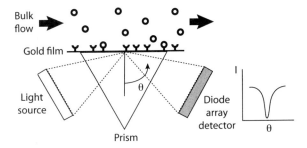

Figure 3.30 Schematic diagram of an SPR biosensor.

maintains $[L] \approx [L]_\circ$ unless binding at the surface depletes ligand more rapidly than replenishment by diffusion. The relevant relationships are

Association phase

$$RU = RU_\circ + (RU_{eq} - RU_\circ)\left(1 - e^{-(k_{on}[L]_\circ + k_{off})t}\right) \quad (3.131)$$

Equilibrium phase

$$RU_{eq} = RU_\circ + (RU_{max} - RU_\circ)\frac{[L]_\circ}{K_d + [L]_\circ} \quad (3.132)$$

Dissociation phase

$$RU = RU_\circ + (RU_{eq} - RU_\circ)e^{-k_{off} \cdot t} \quad (3.133)$$

Notice that the three phases of the experiment contain redundant information. For example, k_{on}, k_{off}, and K_d (as k_{off}/k_{on}) can theoretically be estimated from the association phase alone. K_d, k_{off}, and k_{on} (as k_{off}/K_d) can be estimated from the equilibrium and dissociation phases together. This redundancy enables one to perform consistency checks for parameter estimates [23]. The two estimates of k_{off} must agree:

$$k_{off}|_{\text{association phase fit}} = k_{off}|_{\text{dissociation phase fit}} \quad (3.134)$$

And the equilibrium constant K_d must agree with the kinetic parameters:

$$K_d|_{\text{equilibrium phase fit}} = \frac{k_{off}}{k_{on}}\bigg|_{\text{association and dissociation phase fit}} \quad (3.135)$$

This agreement is often enforced by fitting one consistent set of kinetic parameters simultaneously to all three phases of the experiment (figure 3.31), across a series of varied $[L]_\circ$, by nonlinear global least squares curve fitting.

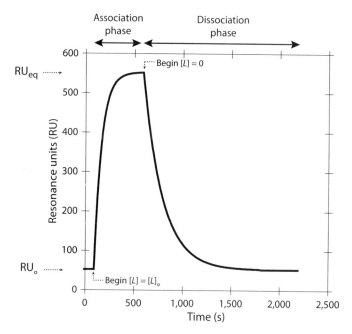

Figure 3.31 Typical kinetic experiment with an SPR biosensor.

Example 3-11 By inspection of figure 3.31, estimate k_{on} and k_{off} for the interaction being measured. Assume that $[L]_o = 10$ nM during the association phase.

Solution Recall that the half-life for an exponential decay is $t_{1/2} = (\ln 2)/k$. During the association phase in the figure, RU increases from 50 to 550, passing through its halfway point of 300 at approximately $t_{1/2} = 150$ s; remembering to first subtract the initial time of 100 s gives $k_{obs} = (\ln 2)/(50 \text{ s}) = 0.014 \text{ s}^{-1}$. Similarly, during the dissociation phase, the half-point of 300 is reached at approximately $t = 750$ s, giving a rough approximation $k_{off} \approx (\ln 2)/(750 - 600) \text{ s} = 0.005 \text{ s}^{-1}$. Because $k_{obs} = k_{on}[L]_o + k_{off}$, we can estimate that $k_{on}[L]_o \approx (0.014 - 0.005) \text{ s}^{-1} = 0.009 \text{ s}^{-1}$. Therefore, $k_{on} \approx 9 \times 10^5 \text{ M}^{-1}\text{s}^{-1}$.

3.7.3 Microplate Immunoassays

Microplates are a convenient format for performing many separate assays in parallel. They are flat plastic dishes, often with 96 wells of ≈ 300 μL capacity arrayed in an 8×12 grid (although the inexorable march of miniaturization for analytical

Binding Equilibria and Kinetics

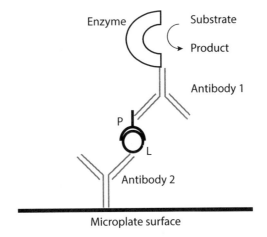

Figure 3.32 Example of a microplate sandwich immunoassay to detect a complex of protein (P) and ligand (L). Two common enzymes used in such assays are horseradish peroxidase and alkaline phosphatase, and the enzymes may be covalently linked to antibodies. These enzymes catalyze the conversion of colorigenic, fluorogenic, and chemiluminescent substrates to produce color, fluorescence, or light, respectively. Antibody 1 recognizes a surface ("epitope") on the protein that does not interfere or compete with ligand binding. Antibody 2 recognizes such a noncompeting epitope on the ligand. These epitopes on the protein and ligand can be short peptides artificially expressed as fusions to the N or C terminus. Antibodies against such "epitope tags" as c-Myc, FLAG, HA, $(His)_6$, V5, and others are commercially available. Antibody 2 may be attached to the microplate through simple adsorption or captured by another antibody or antibody-binding protein.

biochemistry has made 384- and even 1,536-well plates increasingly common). Many different surface chemistries are available for specifically or nonspecifically localizing proteins to the microplate surface. A panoply of supporting devices and reagents is commercially available for loading, washing, labeling, and reading out signals from such microplates. The most common assay performed with microplates is an enzyme-linked immunosorbent assay (ELISA); hence, they are often known colloquially as ELISA plates. The same format is often used for radioimmunoassays and fluorescence immunoassays.

The most sensitive and specific assays are sandwich immunoassays, shown schematically in figure 3.32. Because a signal requires two specific antibody-antigen interactions, nonspecific signals attributable to simple sticky adsorption are reduced. Although the individual steps in microplate assays are ostensibly describable by mathematical models of the kinetics and equilibria, this is impractical due to the presence of nonlinearities throughout the process: nonspecific protein adsorption, substrate depletion in enzyme-based detection schemes, detector saturation for strong signals, and binding-site saturation on the plate surface. Microplate assay quantification is therefore generally restricted to interpolation on log-log plots of signal versus standard sample concentrations. Nevertheless, in their calibrated ranges, microplate assays can be used as the readout for solution-phase binding equilibria. For example, the protein-ligand mixture in figure 3.32 could be equilibrated in solution, then captured on microplates with immobilized Antibody 2 and detected with Antibody 1.

3.7.4 Microfluidic Assays

Devices for biological assays can be miniaturized such that the size of individual features are in the micrometer–millimeter length scale. Such "microfluidic" devices present various potential advantages over more macroscale approaches [24–26].

1. *Reduced sample volume.* Smaller sample volumes generally enable each of the advantages enumerated below (i.e., speed, parallelization, integration, and disposability). To a patient providing bodily fluids for diagnosis, the desirability of small sample volume requirements is self-evident. In nonclinical applications, the use of expensive, rare, and difficult-to-produce reagents can be minimized by decreasing sample volumes. However, several hard physical constraints prevent limitless shrinking of volumes—for example, at least one molecule of analyte must be in the sample. In figure 3.33, the volume of sample containing a single molecule on average is shown for some analytes of practical interest. For example, any device designed to diagnose the presence of HIV virions from a sample of 1 µL of blood cannot succeed, because on average, there is less than one virus particle per sample. This theoretical limit is relevant, because devices with picoliter sample volumes can be constructed [27].

2. *Fast analysis time.* Mixing can be greatly accelerated by miniaturization. As will be considered in more depth in chapter 8, the time scale for diffusion is proportional to length squared. Also, the time for transport from one compartment to another in a microfluidic device is clearly reduced with decreasing distance.

3. *Highly parallel analyses and integration of multiple components.* Miniaturization can enable parallelization of analyses of multiple samples and inclusion of a variety of different functions on one device (e.g., mixing, reaction, and spectroscopy).

4. *Single-use disposability.* The manufacture of small microfluidic devices can be scaled up and automated so as to enable cheap, single-use applications. In addition to the consequent economic benefits, cross-contamination of biological samples can be avoided.

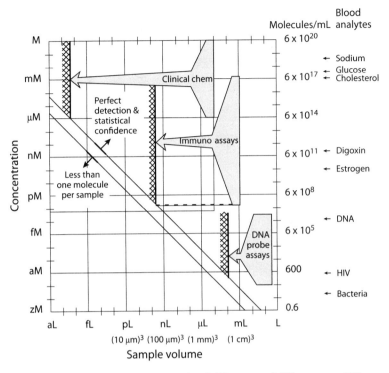

Figure 3.33 Feasible regions and limitations for microfluidic assays of different types [26].

3.7.5 Dialysis

If the molecular weight of a complex is significantly larger than that of either the protein or ligand, then they can be separated on the basis of their ability to diffuse across a semipermeable dialysis membrane. This method has been applied extensively to radiolabeled small-molecule ligands.

Case Study 3-4 T. J. Silhavy, S. Szmelcman, W. Boos, and M. Schwartz. On the significance of the retention of ligand by protein. *Proc. Natl. Acad. Sci. U.S.A.* 72: 2120–2124 (1975) [28].

In this work, the authors examined how the presence of a binding protein in dialysis tubing slows the apparent diffusion rate of a ligand out of the tubing. The system is shown schematically in figure 3.34. A radiolabeled small ligand (*L*) is free to diffuse across the dialysis membrane, while the binding protein (*P*) is retained in the tubing.

Typical data obtained with such an experimental setup are shown in figure 3.35 with $[L_{out}] \approx 0$ (by dilution in a large volume), and with maltose binding protein and ^{14}C-labeled

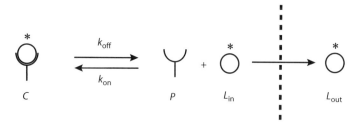

Figure 3.34 Schematic diagram of dialysis analysis for small ligand binding to proteins.

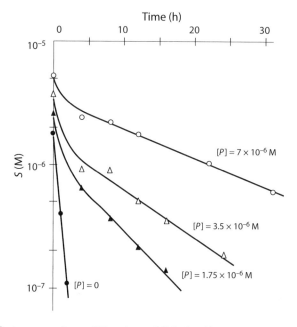

Figure 3.35 Data for transport of a small ligand out of dialysis tubing versus time.

maltotriose as the ligand. The rate of transport of ligand out of the tubing is clearly a function of the concentration of maltose binding protein, $[P]$.

The signal S versus time consists of the sum of both bound and unbound ligand:

$$S = [C] + [L_{in}] \qquad (3.136)$$

Transport across dialysis membranes is generally proportional to the membrane surface area and the concentration gradient across the membrane. When $[L_{out}] \ll [L_{in}]$,

$$\frac{dS}{dt} = -\alpha [L_{in}] \qquad (3.137)$$

Binding Equilibria and Kinetics

where α is the product of a mass-transfer coefficient and the surface area/volume ratio for the interior dialysis compartment, with units of time^{-1}. To solve equation 3.137, we need to determine the relationship between S and $[L_{in}]$. One approximate solution is to assume that the binding protein and ligand reach equilibrium rapidly relative to the rate of transport across the membrane:

$$\alpha \ll (k_{on}[P] + k_{off}) \qquad (3.138)$$

such that at all times the protein-ligand interaction is at equilibrium:

$$K_d = \frac{[L_{in}][P]}{[C]} \qquad (3.139)$$

Substituting this relationship into the definition of the detected signal (equation 3.136), we have

$$S = [P]_o \frac{[L_{in}]}{K_d + [L_{in}]} + [L_{in}] \qquad (3.140)$$

Equation 3.140 and the relationship $\frac{d[L_{in}]}{dt} = \frac{dS}{dt} / \frac{dS}{d[L_{in}]}$ yield

$$\frac{d[L_{in}]}{dt} = \frac{-\alpha [L_{in}]}{1 + \frac{K_d[P]_o}{(K_d + [L_{in}])^2}} \qquad (3.141)$$

This rather complicated form is not analytically solvable, necessitating numerical solution. Unfortunately, such numerical solutions are often not easy to generalize or apply; they also provide little in the way of insights to the rate-limiting processes. Consequently, the authors derived a simpler form that is approached at long times, when $[L_{in}] \ll [P]$, such that $[P] \approx [P]_o$. Hence,

$$[C] \approx \frac{[L_{in}][P]_o}{K_d} \qquad (3.142)$$

which, placed into equation 3.136, yields

$$S \approx [L_{in}] \left(1 + \frac{[P]_o}{K_d}\right) \qquad (3.143)$$

or rearranging,

$$[L_{in}] = \frac{S}{1 + [P]_o / K_d} \qquad (3.144)$$

which can be inserted into equation 3.137 to allow the following analytical solution for $S(t)$, shown in figure 3.36:

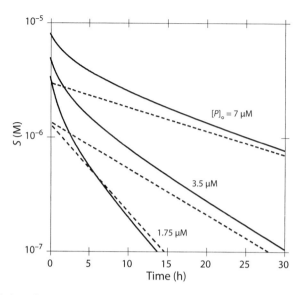

Figure 3.36 Simulation of transport of a small ligand out of dialysis tubing vs. time. Both equation 3.145 (dashed lines) and the numerical solution of equation 3.141 (solid lines) show good qualitative and quantitative agreement with the experimental data shown in figure 3.35 for maltose binding protein and maltotriose.

$$S = S_o e^{-\left(\frac{\alpha}{1+\frac{[P]_o}{K_d}}\right)t}$$
(3.145)

However, equation 3.145 permits immediate visual assessment of how changing protein concentration $[P]_o$ will affect the rate of ligand leakage from the tubing, as well as a mechanistic interpretation that ligand transport is delayed due to equilibrium sequestration by binding protein.

(For those who wish to delve further into this paper, the first expression in the following nomenclature relationships is that used by Silhavy et al. [28], and the second expression is the one used above: $k_d \equiv k_{off}$; $k_a \equiv k_{on}$; $[L_f] \equiv [L_{in}]$; $[L] \equiv S$; and $[P] \equiv [P]_o$.)

3.8 Methods to Detect Changes in Intrinsic Properties on Binding

Changes in the intrinsic optical or thermodynamic properties of the complex relative to its constituents can be used to signal binding, instead of the altered localization approaches outlined in section 3.7.

3.8.1 Optical Spectroscopy

The intensity of fluorescence emission from a tryptophan side chain is a function of its local environment: polarity (e.g., exposure to solvent), rotational flexibility, and proximity to quenching moieties all change tryptophan fluorescence emission characteristics. Consequently, this property generally changes on formation of a protein-ligand complex. Tryptophan fluorescence also changes on transition from the folded to unfolded conformation of a protein and is used as a probe for protein-folding kinetics.

Let's consider how to use a change in tryptophan fluorescence to infer complex formation. Let $\hat{F}_i$ be the fluorescence intensity emission per mole per liter of component i, such that the total detected fluorescence intensity F, summed over all the fluorescent species, is given by

$$F = \hat{F}_P[P] + \hat{F}_L[L] + \hat{F}_C[C] \qquad (3.146)$$

In terms of the fractional saturation parameter y:

$$F = \hat{F}_P(1-y)[P]_o + \hat{F}_L([L]_o - y[P]_o) + \hat{F}_C y[P]_o \qquad (3.147)$$

Solving for y yields

$$y = \frac{F - \hat{F}_P[P]_o - \hat{F}_L[L]_o}{(\hat{F}_C - \hat{F}_P - \hat{F}_L)[P]_o} \qquad (3.148)$$

Clearly, y is indeterminate in equation 3.148 if the intrinsic fluorescence intensity does not change on complex formation (i.e., if $\hat{F}_C = \hat{F}_P + \hat{F}_L$). Equating equation 3.148 and equation 3.7, we obtain the following rather complex relationship:

$$F = \hat{F}_P[P]_o + \hat{F}_L[L]_o + (\hat{F}_C - \hat{F}_P - \hat{F}_L) \times$$
$$\frac{(K_d + [L]_o + [P]_o) - \sqrt{(K_d + [L]_o + [P]_o)^2 - 4[P]_o[L]_o}}{2} \qquad (3.149)$$

In a typical experiment, $[P]_o$ is constant, increments $\delta[L]_o$ are added and allowed to equilibrate, and the experimental output is F versus $[L]_o$. An estimate for K_d can be obtained by nonlinear least squares regression, with F as the dependent variable, $[P]_o$ and $[L]_o$ as the independent variables, and K_d, $\hat{F}_C$, $\hat{F}_P$, and $\hat{F}_L$ as the fitted parameters. A rough initial estimate for K_d can be obtained from the value of $[L]_o$ at 50% of the maximum fluorescence, and independent estimates of $\hat{F}_P$ and $\hat{F}_L$ are obtainable from measuring the fluorescence intensity of P and L separately.

3.8.2 Isothermal Titration Calorimetry

When a reaction occurs with a change in enthalpy, heat is released or absorbed by the system. In chapter 2, we considered the enthalpy change that occurs during reversible, noncovalent, macromolecular interactions. The quantity of heat released or taken up at feasible protein-ligand concentrations is far below what one could sense by touch. However, instrumentation has been developed that precisely measures this flow of heat. Calorimetric measurement of binding heat can be used to determine K_d as well as the enthalpy of the interaction [29, 30]. A schematic diagram of such an instrument, called an "isothermal titration calorimeter," is shown in figure 3.37.

The advantages of isothermal titration calorimetry (ITC) include:

1. *Complete thermodynamic information.* For interactions with affinity in the general range 50 nM $< K_d <$ 1 mM, an ITC measurement can allow determination of both $\Delta H°$ (from the integration of the power peaks versus the moles of complex formed) and $\Delta G°$ for the reaction (by fitting K_d to the ΔH versus $[L]$ isotherm). Determination of $\Delta H°$ and $\Delta G°$ also allows direct calculation of $\Delta S°$. The approximate criterion for obtaining an isotherm useful for estimating K_d is

$$10 < \frac{[P]}{K_d} < 100 \qquad (3.150)$$

The parameter $c \equiv [P]/K_d$ is varied by changing the concentration of protein in the sample cell. When $c < 10$, the binding isotherm is a shallow line without curvature to fit a value for the binding constant. When $c > 100$, each incremental pulse of ligand added to the sample binds to completion, and so a series of equal-height peaks is observed, with an intermediate final peak before complete complexing of all protein in the sample. It should be noted that for spectroscopic methods, by contrast, a value of $c < 1$ is more generally desirable.

2. *Label-free, no immobilization.* ITC allows determination of solution-phase interactions without any alteration of the binding partners. (As we have discussed in this chapter, some artifacts potentially can arise from conjugation to labels or surfaces.)

3. *Generality.* ITC can be applied to proteins, peptides, nucleic acids, carbohydrates, lipids, metals, drugs, and other organic compounds—because neither bioconjugation chemistry nor particular spectroscopic signature changes on binding are required. Ionic strength, pH, and temperature can be varied across wide physiological ranges.

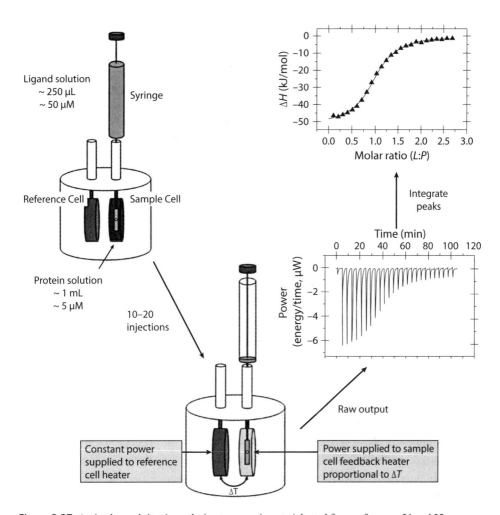

Figure 3.37 An isothermal titration calorimetry experiment. Adapted from references 31 and 32.

Disadvantages of ITC include:

1. *Not generally suitable for high-affinity interactions.* The concentrations of analyte necessary to satisfy the c criterion listed above result in too small a heat signature for accurate detection when $K_d < 10$ nM. However, indirect measures of binding, by linked equilibrium with protonation or displacer binding, can be used in such cases [29, 30, 33].

2. *Large quantities of material necessary.* A single ITC experiment can consume hundreds of micrograms of protein, because micromolar concentrations are typical;

often these reagents are expensive to purchase or require considerable effort to prepare. Furthermore, many proteins aggregate at micromolar concentrations, which can lead to large transients or drifting baselines.

3. *Linked ionization equilibria.* The heat signature reports on the sum total of all events in the sample cell, which can include the net addition or removal of protons from the analytes of interest and the buffer salts. Such ionization equilibria generally exhibit considerable enthalpies, so titrations must be performed in a series of different buffers with different heats of ionization to subtract the buffer's heat signature.

3.8.3 Altered Mobility

A class of measurements relies on changes in the transport properties of a complex relative to its substituent parts. Changes in diffusivity, permeation into pores, or charge/mass ratio can provide a signal of complex formation [33].

1. *Analytical ultracentrifugation.* In analytical ultracentrifugation, macromolecules in a solution are subjected to a sedimenting force (proportional to the mass and acceleration of the particle) and opposing buoyancy and frictional forces. The net force results in movement of the macromolecules toward one end of the centrifuged sample (either bottom or top, depending on the relative densities of the macromolecules and the solvent). At equilibrium, this force-driven accumulation is countered by molecular diffusion, setting up a time-invariant concentration gradient. For a sample containing a protein and a ligand that interact, the quantitative shape of this gradient contains information about protein, ligand, and complex concentrations as a function of radial distance in the sample and can be used to extract the equilibrium binding constant.

2. *Chromatography.* Size-exclusion chromatography separates macromolecules on the basis of their partitioning into pores in the column solid phase. Larger molecules are carried through the column more rapidly by convection, because they are not delayed by diffusion into side channels. Binding curves can be extracted from the shape of chromatographic elution profiles.

3. *Electrophoresis.* Electrophoresis separates macromolecules on the basis of their charge/mass ratio. For proteins, sodium dodecyl sulfate-polyacrylamide gel electrophoresis (SDS-PAGE) exploits the fairly constant binding capacity of proteins for the denaturing ionic detergent SDS. Nondenaturing SDS-free electrophoresis exploits the protein's native charges, which avoids perturbing binding equilibria when separating complexes from binding partners to obtain binding isotherms. Capillary electrophoresis allows more rapid and precise separations.

3.9 Fitting Models to Data

The full behavior of a mechanistic model can only be specified by the addition of a set of parameters that specifies the rate constant values and other quantitative factors, such as total concentrations. For any given model structure (sometimes called model topology), the parameters can be adjusted for the best fit to available experimental data. In any such exercise, one generally tries to answer the following questions:

1. *Is the model consistent with the data?* In other words, does a set of physically realistic parameter values exist, such that the disagreement between the model predictions and the actual data is not systematically greater than the noise in the data? Of course, one must always recall that consistency is not proof of correctness, because an infinite number of different models display equivalent consistency with a given set of data. Consistency is necessary but not sufficient for the correctness of a model.

2. *What is the most probable value of the model parameters?* What particular set of parameters has the highest likelihood of reflecting the true values? Obtaining numerical parameter estimates is often the explicit goal of a model-fitting exercise, in order to characterize the system under study. For example, estimating K_d of a protein-ligand interaction is often the goal of fitting a model to a binding isotherm, with the correctness of the underlying model implicitly assumed.

3. *How valid is the best estimate for the parameter values?* A useful way to rephrase this question, so as to obtain a quantitative answer, is: Can we define a range of parameter values, in which the correct value will lie to a certain percentage likelihood? Such a range is called the confidence limits of the estimate.

3.9.1 Data Collection

Before diving into the methods used for model fitting, it is worth noting that even a correct model cannot salvage a poorly collected dataset. As an illustrative example, let us consider the simple equilibrium binding model between protein (P) and ligand (L), neglecting ligand depletion: $[C] = \frac{[P]_o[L]_o}{K_d+[L]_o}$. In our experimental setup, let's say that the ligand concentrations are known, and equilibrium complex concentrations are readily measurable, but the initial protein concentration (which is the same across all samples) is initially unknown (e.g., the protein is just one component in a complex mixture).

If we expect a K_d value in the nanomolar range, it is essential that we include ligand concentrations that are significantly above and below the nanomolar range to accurately estimate the two unknown parameters in the model, $[P]_o$ and K_d. If we use ligand concentrations only up to the picomolar range, the resulting *binding curve* ($[C]$ versus $[L]_o$) will actually be linear, because all ligand concentrations used in the experiment are much less than K_d, so $[C] \approx \frac{[P]_o}{K_d}[L]_o$. This is problematic, because an infinite number of combinations of $[P]_o$ and K_d can match the slope of the experimental data, so neither of our unknown parameters can be accurately estimated from this dataset. By contrast, if we only use ligand concentrations that are micromolar or greater, the resulting *binding curve* will essentially be a horizontal line, because all ligand concentrations are much greater than K_d, so $[C] \approx [P]_o$. In this case, $[P]_o$ could be estimated, but K_d could not.

Now, if we were to combine these two experiments such that we had a full range of ligand concentrations from picomolar to micromolar, we would be able to accurately estimate both $[P]_o$ and K_d. The high ligand concentrations would uniquely constrain $[P]_o$, because the binding curve would asymptote to this value. Then the slope of the binding curve at low ligand concentrations, $[P]_o/K_d$, would allow unique estimation of K_d, because $[P]_o$ is now independently constrained by the high-ligand data in the binding curve. As is evident from this analysis, an accurate estimate of $[P]_o$ is necessary to obtain an accurate estimate of K_d (but not vice versa). In theory, only high and low ligand concentrations are needed to estimate $[P]_o$ and K_d in this binding model. However, given the experimental realities of measurement error (and uncertainties about model appropriateness), intermediate ligand concentrations are also very useful, because they provide additional constraints that improve parameter estimation (or guide model selection).

More generally, analyzing a model of interest prior to experimentation can help identify regimes in which certain parameters (or combinations of parameters) dominate the response. This can be used to guide the ranges of independent variables (e.g., ligand concentration, time) that are experimentally tested to ensure the most accurate estimates of the desired parameters when fitting the model to the collected experimental data.

3.9.2 Nonlinear Least Squares

Generally, the disagreement between model-predicted data and observed data is represented through the quantity χ^2 (*chi squared*), which is a dimensionless quantity that weights each data point by the inverse of its uncertainty, represented by the variance of the data measurement:

$$\chi^2 = \sum_{i=1}^{n_{\text{data}}} \left(\frac{(data)_i - (model)_i}{\sigma_i} \right)^2 \qquad (3.151)$$

where n_{data} is the number of data points, $(data)_i$ and $(model)_i$ are the experimental data and model data value for the *i*th point, respectively, and σ_i is the standard deviation of the experimental data point i. Notice that this weighting emphasizes those data points known most certainly (i.e., with the smallest σ_i). In many cases, the experimental uncertainty is unknown for each individual point and must therefore be considered a constant $1/\sigma^2$ that weights all of the data points equally or sometimes, when appropriate, the experimental uncertainty (variance) is treated as a fixed percentage of the measured value squared.

Methods to minimize χ^2 with respect to the adjustable parameters can be used to produce realizations for a given model topology with good agreement with experimental data. If such realizations are in poor agreement with experimental data, it may be evidence that the model topology is incorrect, incomplete, or inadequate in some other way to capture the experimental phenomenon. The parameter values that minimize χ^2 for linear models can be solved analytically. For nonlinear models, one relies on local optimization, starting from an initial guess set of parameters to numerically search for the parameter values that minimize χ^2. This approach is strengthened by using a large number of local minimizations, starting from different initial guesses, a technique called multistart.

It can be shown that minimizing χ^2 will provide parameter estimates with the highest probability of being correct [34] if the system is consistent with the following assumptions:

1. *The model is correct.* This condition is of course necessary, because the concept of a most probable parameter value in an incorrect model is not meaningful.

2. *There is no noise in the independent variable(s).* In other words, the values for variables set by the investigator (e.g., time, concentration) are known to far greater precision than the measured output or dependent variables.

3. *Experimental noise in the dependent variable(s) obeys a Gaussian, or normal, distribution.* Normally distributed noise is bell shaped and symmetric around zero. Many biochemical measurements generate uncertainties that are consistent with Gaussian noise.

4. *The noise for each data point is independent of the noise for all other data points.* In other words, there is no correlation between noise in sequential sets of data.

Evaluation of model-fitting quality Two widely applied methods are used for evaluating whether a particular combination of parameters in a model is consistent with a data set. A quantitative approach is to determine the probability that the numerically obtained minimized value of χ^2 would have been observed by chance if the model were correct. A more qualitative, but quite useful, method is to visually examine the difference between model and data for visible trends.

1. *Acceptable values for χ^2 of a least squares fit.* A reduced value of χ^2 is usually used to assess goodness of fit as follows:

$$\chi_R^2 = \frac{\chi^2}{n-m} \qquad (3.152)$$

where $n =$ number of data points and $m =$ number of fitted parameters, so $n - m =$ number of degrees of freedom. χ_R^2 is a ratio of the model error to the experimental uncertainty, and for a good fit, $\chi_R^2 \to 1$. Values of $\chi_R^2 < 1$ are

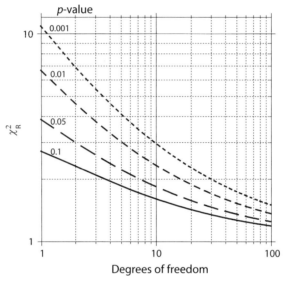

Figure 3.38 Curves of constant p-value for χ_R^2 versus the degrees of freedom [35].

Table 3.1 Critical values of χ_R^2 for $p = 0.05$ [35].

Degrees of freedom	χ_R^2 for $p = 0.05$	Degrees of freedom	χ_R^2 for $p = 0.05$
1	3.8410	20	1.5705
2	2.9955	30	1.4591
3	2.6050	40	1.3939
4	2.3720	50	1.3501
5	2.2140	60	1.3180
6	2.0987	70	1.2911
7	2.0096	80	1.2735
8	1.9384	90	1.2572
9	1.8799	100	1.2434
10	1.8307		

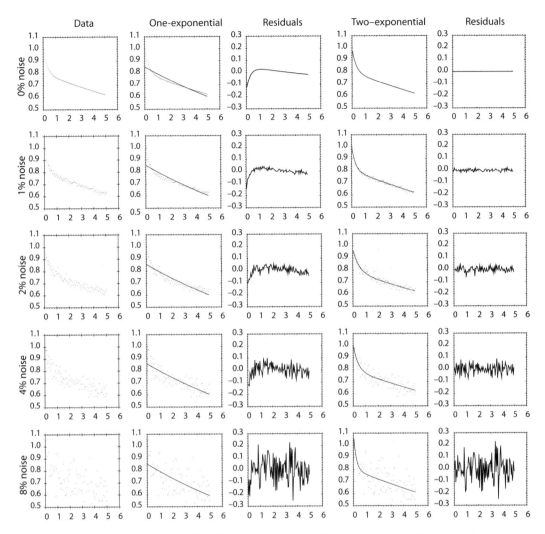

Figure 3.39 Examination of residuals for patterns can be used to test for systematic variation between experimental data and the model being curve fit. A data set was generated with the functional dependence $y = 0.2e^{-3t} + 0.8e^{-0.05t}$. Gaussian-distributed noise with standard deviation 0.01, 0.02, 0.04, or 0.08 was added to this function (sequential rows). Each data set was then fit by nonlinear least squares to a one-exponential function $(y = c_1 e^{-c_2 t})$ or a two-exponential function $(y = c_1 e^{-c_2 t} + c_3 e^{-c_4 t})$. For the one-exponential fit, there is clear nonrandom patterning in the residuals for the data sets with 0, 1, 2, and 4% noise. However, when the noise level is similar to the systematic variation (e.g., the one-exponential fit to the data set with 8% noise), no strong pattern is apparent in the residuals. For the two-exponential fit, there is no asymmetric variation in the residuals around zero for any of the noise levels.

indicative of *overfitting* the data, or forcing the model curve to follow random noise trends in the data. The larger the value of χ_R^2, the worse the model fit is. One approach to evaluating the goodness of fit of the parameters obtained by nonlinear least squares is to look up the value of χ_R^2 in a table to determine a *p*-value (figure 3.38; table 3.1). This *p*-value is the probability that at least the observed level of model disagreement with the data would have occurred by chance if all the observed disagreement were the result solely of Gaussian distributed noise with the specified standard deviations σ_i. The smaller this *p*-value is, the lower the probability will be that the model error is due to random noise; instead it must be due to the application of an incorrect underlying model topology, Gaussian noise with a larger standard deviation than expected, or the presence of non-Gaussian noise.

2. *Visual examination of residuals.* If the noise in successive data points is independent, then one should not be able to observe trends or patterns in the *residuals* (differences between model and data). Instead, if the noise is in fact Gaussian, the residuals should be distributed in a bell curve around zero, with no clear patterns. An example of residuals analysis for a synthetic data set is shown in figure 3.39. Although subjective, the human eye is a valuable tool for identifying such trends as those shown in the figure.

Example 3-12 What are the values for χ_R^2 for each of the eight nonlinear least squares model fits in the presence of noise in figure 3.39? What is the probability, in each case, that this value of χ_R^2 or higher would be observed due simply to experimental noise?

Solution χ_R^2 was calculated from equations 3.151 and 3.152. σ_i is a constant value σ for each of the simulated data sets, as the noise was simulated with a Gaussian random number generator. *One-exponential* refers to a fit of the equation $y = c_1 e^{-c_2 x}$, whereas *two-exponential* refers to a fit of the equation $y = c_1 e^{-c_2 x} + c_3 e^{-c_4 x}$. Because we have 100 data points, there are 98 degrees of freedom for the one-exponential fits and 96 for the two-exponential fits.

% noise of generator	σ from realization	One-exponential χ_R^2	Two-exponential χ_R^2
1	0.0093	8.13	0.97
2	0.0202	2.47	0.98
4	0.0385	1.53	1.01
8	0.0948	1.12	1.02

Comparing these values of χ_R^2 to the p-values in figure 3.38, the one-exponential model with 1%, 2%, and 4% noise can be rejected. For the 1% and 2% data sets, $p \ll 0.001$, so it is highly unlikely that the observed data set would be observed randomly when Gaussian noise is added to the predicted model form. For the 4% data set, the one-exponential fit exhibits $\chi_R^2 = 1.53$, which is very close to the $p = 0.001$ contour at 98 degrees of freedom in figure 3.38. Because there is only a one-in-a-thousand probability that the observed disagreement between model and data would occur for Gaussian noise, we can reject this fit as well. However, $\chi_R^2 = 1.12$ for the one-exponential fit to the 8% noise data set, corresponding to $p > 0.1$. Because there is a better than 10% chance that the observed model error is simply Gaussian noise, we cannot reject the one-exponential fit for the data set with 8% noise.

Both the χ_R^2 criterion and the visual examination of residuals convergently discriminate between the one- and two-exponential fits for data sets with 0, 1, 2, or 4% Gaussian noise. In each of these cases, both criteria clearly favor the correctness of the two-exponential fit. On the contrary, neither method can unambiguously discriminate between the one- and two-exponential fits when the noise is at the 8% level. Does the apparent equivalence of the χ_R^2 values and residuals plots for 8% noise mean that both models are correct? No—it simply means that both the one- and two-exponential fits are equally consistent with the provided data, because the higher levels of noise mask the systematic variation between the model and the underlying trend. In such a situation, unless one has independent mechanistic support for the more complicated two-exponential fit, the simpler one-exponential fit would be preferred by Occam's razor.

Example 3-13 How is model discrimination and parameter estimation affected by adding an adjustable baseline to the curve fit in the example shown in figure 3.39?

Solution The equation $c_1 e^{-c_2 t}$ was fit in the second column of figure 3.39. Let's see what happens when we instead fit the equation $c_1 e^{-c_2 t} + c_3$. As shown in figure 3.40, the single-exponential model with variable baseline now provides an adequate fit to the data at noise levels of 2% and 4%—even though we know this model is not correct, because the synthetic data set was generated

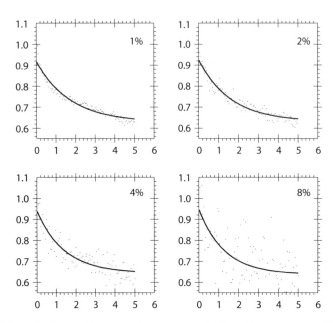

Figure 3.40 In each of the nonlinear curve fits shown as solid lines, the equation $c_1 e^{-c_2 t} + c_3$ was fit, introducing a variable baseline parameter c_3. Comparing these fits to those in figure 3.39, it is clear that allowing the baseline to vary produces a misleadingly good fit of the data to a single exponential model.

from a two-exponential function. This example illustrates the importance of not carelessly curve-fitting parameters that can be independently measured experimentally: Baselines can generally be independently determined rather than freely curve-fit.

3.9.3 Confidence Intervals on Fitted Parameters

How certain can we be that the parameter estimates obtained by a nonlinear least squares curve-fitting exercise are correct? This question can be posed and answered in a quantitative fashion as follows. Can we define a range of values for each parameter that we are 95% certain contains the true parameter value? (Assume independent Gaussian noise, model correctness, and a noiseless independent variable.) Actually, two statistical confidence levels are commonly used: 68% and 95%. The perhaps surprising 68% value is actually the implicit confidence level when data are presented with error bars (e.g., 4.5 ± 0.2), because such error bars are typically given as one standard deviation σ, and the range $\pm \sigma$ encompasses 68% of a Gaussian normal distribution. Confidence intervals at the 95% level will of course be larger, because we need to consider an expanded range of possible parameter values if we wish to be more confident that the correct value is within it.

We outline here a procedure that provides not only a confidence interval but also intuitive insight into the dependence of the overall model fit on each parameter [36]. Most nonlinear least squares computational programs will provide estimated confidence intervals for the fitted parameters. However, the method used to calculate these often assumes that the parameter effects on χ_R^2 are uncorrelated, which is seldom true. Such package-provided estimates are almost always smaller than those obtained by the procedure described here. Therefore, the present procedure is more conservative.

Procedure for determining parameter confidence limits:

1. Fix the parameter of interest at a particular numerical value.
2. Minimize χ_R^2 by varying all other parameters.
3. Record the least squares value of χ_R^2 at the fixed value for the parameter of interest.
4. Repeat steps 1–3, mapping out a curve for the ratio $\chi_R^2/\chi_{R,min}^2$ versus the value for the parameter of interest (note that this ratio will be 1 at the least squares estimate for the parameter).
5. Determine the critical value for the F distribution statistic $F_{0.05}(d.f., n)$ (for 95% confidence interval) or $F_{0.32}(d.f., n)$ (for 68% confidence interval) from figure 3.41, where n is the number of other fitted parameters (i.e., excluding the parameter of interest) and degrees of freedom $(d.f.)$ is number of data points $- n$.
6. Determine the upper and lower confidence limits, which will be where

$$\frac{\chi_R^2}{\chi_{R,min}^2} = 1 + \frac{n}{d.f.} F_{0.05 \text{ or } 0.32}(d.f., n) \qquad (3.153)$$

Example 3-14 Determine the 95% and 68% confidence intervals for the rate constant c_2 for each of the noise levels in figure 3.39.

Solution As described in the text, the parameter c_2 is fixed at a variety of values, and χ_R^2 is minimized by varying the remaining parameters (i.e., c_1, c_3, c_4). There are 100 data points and three parameters being fit, so $d.f. = 97$ and $n = 3$. Either estimating from figure 3.41 or using the F.INV function in Microsoft Excel, $F_{0.05}(97, 3) = 2.7$, and $F_{0.32}(97, 3) = 1.2$. By equation 3.153, the critical values of $\chi_R^2/\chi_{R,min}^2$ are

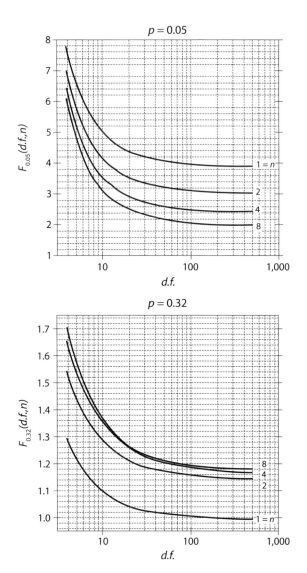

Figure 3.41 Critical values for the F distribution, to be used for calculating confidence intervals for parameters estimated by nonlinear least squares. Degrees of freedom ($d.f.$); $n =$ number of parameters in the model. (F distribution critical values were calculated using the F.INV function in Microsoft Excel.)

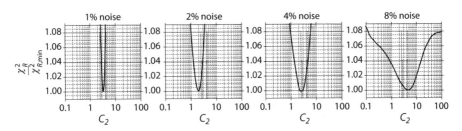

Figure 3.42 Confidence interval calculation for the parameter estimation cases of figure 3.39. The estimated 68% confidence intervals for the parameter c_2 are as follows (respectively for the 1%, 2%, 4%, and 8% noise data sets): (3.0, 4.0), (1.4, 3.0), (1.5, 4.8), and (0.9, 16). The estimated 95% confidence intervals for the parameter c_2 are as follows (respectively, for the 1%, 2%, and 4% noise data sets): (2.8, 4.3), (1.1, 3.5), and (1.0, 6.2). No upper limit to the 95% confidence interval can be calculated for the 8% noise data set, indicating that one cannot bracket the value of this parameter with 95% confidence.

$$\left(\frac{\chi_R^2}{\chi_{R,min}^2}\right)_{0.05} = 1 + \left(\frac{3}{97}\right) 2.7 \tag{3.154}$$

$$= 1.08 \tag{3.155}$$

and

$$\left(\frac{\chi_R^2}{\chi_{R,min}^2}\right)_{0.32} = 1 + \left(\frac{3}{97}\right) 1.2 \tag{3.156}$$

$$= 1.04 \tag{3.157}$$

Plots of $\chi_R^2/\chi_{R,min}^2$ versus c_2 at each noise level are shown in figure 3.42, with the critical values $\chi_R^2/\chi_{R,min}^2 = 1.08$ or 1.04 marked as horizontal dashed lines on each plot. The intersection points between the solid curve and the horizontal dashed line represent the lower and upper confidence limits for c_2 at the specified p-value. Note that the confidence limits are not symmetric around the most probable estimate for c_2 in any of the cases. For the 1% noise data set, there is a 68% probability that $3.0 < c_2 < 4.0$ and a 95% probability that $2.8 < c_2 < 4.3$. For the 2% noise data set, there is a 68% probability that $1.4 < c_2 < 3.0$ and a 95% probability that $1.1 < c_2 < 3.5$. For the 4% noise data set, there is a 68% probability that $1.5 < c_2 < 4.8$ and a 95% probability that $1.0 < c_2 < 6.2$. For the 8% noise data set, there is a 68% probability

> that $0.9 < c_2 < 16$, but no confidence interval can be calculated at the 95% probability level, because no value of the chi-squared ratio reaches the critical F value 1.08 on the high side.

3.9.4 Global Nonlinear Least Squares Fitting

When multiple independent experiments are performed on one system, usually a unifying mechanistic thread connects the interpretation. For example, if a binding isotherm is measured three separate times, the same value for K_d is expected to apply to each of the three experiments, even if the instrumental baseline or background noise varies from experiment to experiment. One approach to estimating a common parameter present in the model describing each of n data sets is to independently estimate that parameter by curve-fitting the model to each data set separately, obtaining n independent estimates of the parameter, and then to average the individual estimates. However, a potentially more powerful approach is to globally fit all the data sets simultaneously to a single value for the parameter. This method is called global nonlinear least squares curve fitting.

It has been found that aggregating all available data and constraining the overall fit to a comprehensive model by global fitting considerably tightens up the confidence limit on the estimated parameter [37]. This result is intuitively reasonable, because there is bound to be a narrower range of values for any parameter that will simultaneously agree with many data sets compared to agreement with a single data set.

Implementation of global curve fitting is straightforward. Two types of parameters are fit: some that are present in the model for all data sets (e.g., K_d from the earlier example) and some that are specific to individual data sets (e.g., baseline and signal-to-noise ratio). The overall χ^2 parameter to be minimized is defined by summing over all data sets, each with its own particular model that contains both common and specific parameters.

Suggestions for Further Reading

C. R. Cantor and P. R. Schimmel. *Biophysical Chemistry, Part III: The Behavior of Biological Macromolecules*. W. H. Freeman & Company, 1980.

Energetics of biological macromolecules. Series in *Methods in Enzymology*, parts A–E, volumes 259 (1995), 295 (1998), 323 (2000), 379 (2004), and 380 (2004).

A. R. Fersht, *Structure and Mechanism in Protein Science: A Guide to Enzyme Catalysis and Protein Folding*. Worth Publishing, 1998.

S. E. Harding and B. Chowdhry (eds.). *Protein–Ligand Interactions: Hydrodynamics and Calorimetry.* Practical Approach Series. Oxford University Press, 2000.

I. M. Klotz, *Ligand-Receptor Energetics*, John Wiley & Sons, 1997.

D. A. Lauffenburger and J. J. Linderman. *Receptors: Models for Binding, Trafficking, and Signaling.* Oxford University Press, 1996.

A. S. Perelson, C. DeLisi, F. W. Wiegel (eds.). *Cell Surface Dynamics: Concepts and Models.* Marcel Dekker, 1984.

Problems

3-1 You have purchased an antibody against a cell-surface protein and perform a cell-surface titration to estimate that $K_d = 3$ nM. After labeling your cells at 500 nM antibody, you wash and resuspend them in antibody-free buffer at room temperature and then leave for lunch. Estimate how many minutes will pass before half of the bound antibody has dissociated from the cells.

3-2 A ligand binds to four sites on a macromolecule. The apparent macroscopic dissociation constants $K_{d,2}$ and $K_{d,3}$ are identical (within experimental error.) Is there an interaction energy between sites 2 and 3? If not, why not? If so, calculate the interaction energy.

3-3 Derive an analytical expression for the percentage error incurred by neglecting ligand (A) depletion in a calculation of the equilibrium fraction of bound receptor protein (B). State your answer in terms of the dimensionless concentrations:

$$\alpha = \frac{[A]_\circ}{K_d} \qquad (3.158)$$

$$\beta = \frac{[B]_\circ}{K_d} \qquad (3.159)$$

For $\beta = 0.1$, 1, and 10, plot the percentage error as a function of α.

3-4 Show that, when $[P]_\circ \ll K_d$, ligand depletion is negligible for any value of $[L]_\circ$ (i.e., even if $[L]_\circ$ is not much greater than $[P]_\circ$). (This observation is useful for experimentally generating full binding isotherms that do not exhibit ligand depletion in any of the samples.)

3-5 You incubate 10 nM ligand (L) in a tube with immobilized receptor protein (P). You do not know the receptor protein concentration precisely, but you know that it is present at a substantially lower concentration than ligand. These molecules reversibly associate to form C complex. After 500 seconds, the liquid is removed and replaced with a very large volume of buffer without ligand. The fraction of receptor that is present as ligand-receptor complex is shown in the graph below.

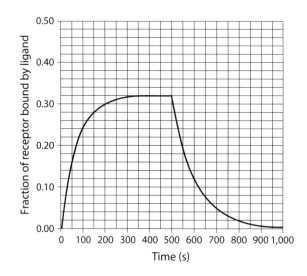

a. What is the dissociation rate constant of the C complex? Derive any equation needed to answer this question.

b. What is the equilibrium dissociation constant? Derive any equation needed to answer this question.

c. Consider a situation in which you have engineered the receptor protein to have a dissociation rate constant that is lower than the original value. Draw the expected fraction versus t curve if you repeated the experiment with this receptor. Qualitatively, identify key similarities and/or differences from the results with the original receptor.

3-6 You are studying a protein ligand that you suspect forms multimers, but you are unsure of the valency of these multimers. You hypothesize that each monomeric unit binds to its target irrespective of whether the other ligands are bound or unbound. You perform a series of titrations and find that the equilibrium dissociation constants for binding the first and second targets are 2.0 nM and 4.8 nM, respectively. Unfortunately, you could not complete the experiments to determine whether any additional targets could be bound.

a. What is the valency of the multimers?

b. What is the intrinsic affinity of each individual ligand for its target?

c. You mutate an amino acid on a ligand's surface that improves hydrogen bonding at the ligand-receptor interface. This results in a 1.2 kcal/mol improvement in the standard state Gibbs free energy change on binding. What is the affinity of the new ligand-protein interaction?

3-7 Some patients with prostate cancer have elevated levels of a protein called prostate-specific antigen (PSA) in the blood. A concentration above 2 ng/mL often leads to further evaluation, although many factors, including PSA dynamics, must be considered. You aim to develop a diagnostic test to quantify PSA concentration from blood samples. You have a ligand that binds to PSA with 0.5 nM affinity. The association rate constant of this ligand for PSA is 3×10^5 M^{-1} s^{-1}.

You are characterizing a pilot version of your blood test using a patient's blood sample. You adsorb 10 femtomoles of ligand inside a test tube. 500 μL of blood is added to the tube, and a long wait allows association and dissociation of the PSA-ligand complex to near equilibrium. You rinse the tube to remove unbound PSA, while PSA-ligand complex remains adsorbed. An enzymatic detection scheme enables you to quantify that 500 million PSA molecules are present in the tube.

a. Is this patient's PSA level above the 2 ng/mL limit?

b. What fraction of ligand binding sites would be complexed if the blood concentration were 2 ng/mL?

c. You needed to make a quick estimate of the bound ligand fraction at the normal limit, so you assumed that the PSA was present in large excess relative to ligand. How much does your estimate of the bound fraction deviate from the true value?

d. What should be done to reduce the error to 1%?

e. You would like to speed up the assay. If the blood sample is incubated in the test tube for only 1 hour prior to washing, how much do you underestimate the PSA concentration in 2 mL of a 2 ng/mL sample? How long would you have to wait to reduce this error to 10%?

3-8 A protein is a tetramer of four equivalent domains. Each domain has 10 nM affinity for a small-molecule binding partner. Binding of a molecule does not impact the other domains.

a. If 1 pM of protein is placed in 2 nM of small molecule, what are the concentrations of proteins with 0, 1, 2, 3, and 4 molecules bound?

b. What concentration of small molecule would maximize singly bound protein?

3-9 You wish to inhibit a particular IgM antibody, which is present in the blood at a concentration of 10 pM. You engineer a protein to interact—with 3 nM affinity—with the IgM antibody's natural binding region, thereby preventing the natural IgM-antigen interaction. You cannot block all the natural binding sites without infinite engineered protein. Because much of the IgM functionality is predicated on its high multivalency, you aim to block enough natural binding sites such that the average IgM molecule has only one unblocked natural binding site. What concentration of your engineered protein is needed to achieve this goal?

3-10 You aim to make an antidote to a toxin, which is present in the bloodstream at 5 nM concentration. You engineer a small protein to bind the toxin with 2.5 nM affinity.

a. What concentration of antidote protein must be initially present in the bloodstream to bind 90% of the toxin at equilibrium?

b. After the system has equilibrated, unbound antidote protein is removed from the bloodstream, while toxin-protein complex remains. Within 12 minutes, 15% of the toxin-protein complex has dissociated. Neglecting any rebinding of dissociated protein (because it is efficiently diluted and removed in the bloodstream), what is the association rate constant for the toxin-protein interaction? Is the association rate constant relatively fast, slow, or normal?

c. In an effort to create large molecular complexes to speed clearance of bound toxin, you link three antidote proteins together into a trivalent molecule. The three independent domains do not interact, even on toxin binding. What is the correct equation relating the equilibrium dissociation constant for the ith binding event ($K_{d,i}$) to the intrinsic affinity (K_d^μ)?

d. If you generate a steady concentration of 10 nM trivalent antidote, what is the ratio—at equilibrium—of antidote binding exactly two toxins relative to the antidote binding fewer than two toxins?

e. Now you create an alternative construct in which the three antidote proteins are again linked in a trivalent construct, but now binding one toxin molecule enhances binding of the second and third toxins. In the limit of perfect cooperativity, if you generate a steady concentration of 10 nM trivalent antidote, what is the ratio—at equilibrium—of antidote binding at least two toxins relative to the antidote binding fewer than two toxins?

3-11 When cardiac tissue is damaged, troponin I is released into the blood. You are developing a diagnostic blood test to detect and quantify troponin I. In a test tube, you immobilize 1 picomole of an antibody fragment that binds troponin with 5 nM affinity. You add 100 μL of a blood sample containing 100 nM of troponin and incubate extensively to nearly reach equilibrium. You then remove the blood and wash away the unbound molecules, leaving the tube sitting in 1 mL of buffered saline solution, which is sufficient to neglect rebinding of dissociated troponin.

a. How much time passes before 10% of the bound troponin has dissociated from the adsorbed antibody?

b. How would your answer differ if you used an antibody with 0.5 nM affinity?

c. If the wash were performed in 100 μL of buffered saline solution, how much error would be incurred by neglecting rebinding?

3-12 A trivalent ligand has three uniform, independent binding domains. These binding domains can each interact with identical but separate receptors on the cell surface. Simultaneous binding of three receptors by a single trivalent ligand localizes the receptors, leading to a signaling cascade. Thus, you are interested in the dynamics of formation of the complex of triply bound ligand with three receptors. Write out the differential equations and material balances needed to describe the rates of formation of singly bound, doubly bound, and triply bound ligand. Define all terms (rate constants and concentrations), and provide units. You do not need to solve the system of equations.

3-13 For a protein with n identical and energetically equivalent ligand binding sites, the aggregate parameter v is defined as follows:

$$v \equiv \frac{\text{mol ligand bound}}{\text{mol protein}} \tag{3.160}$$

$$= \frac{\sum_{i=1}^{n} i[P_i]}{\sum_{i=0}^{n} [P_i]} \tag{3.161}$$

$$= n \cdot y \tag{3.162}$$

for $0 \leq v \leq n$.

Derive the following simplified relationship:

$$v = \frac{n[L]}{K_d^{\mu} + [L]} \tag{3.163}$$

3-14 You have engineered an IgG antibody with two equivalent sites that reversibly bind a monomeric soluble cytokine. Binding at one site does not impact the energetics of binding at the other site. What concentration of the cytokine maximizes the amount of singly bound antibody? Derive the answer in terms of total antibody concentration and the monovalent affinity K_d.

3-15 The monovalent Fab fragment of an antibody is determined to bind to a cell-surface antigen with affinity $K_d = 3$ nM. What is the observed affinity for the original bivalent antibody?

3-16 Consider a monomeric protein that can bind to two different ligands. The binding sites for the two ligands overlap, such that the second ligand cannot bind to a protein molecule already complexed with the first ligand, and vice versa. Derive an expression for the equilibrium fraction of protein complexed with the first ligand when both ligands are present. You need to measure the affinity of the protein for the first ligand. You have an experimental setup in which the protein-ligand$_1$ complex generates a signal, but the protein-ligand$_2$ complex does not. Because your computer's battery is dead, you decide to estimate the K_d by finding the ligand concentration at which the signal is half of the value approached using extremely high ligand concentrations. Unfortunately, you do not know that your vial of the first ligand is contaminated with the second ligand. How does this affect your estimate? What if your protein solution were contaminated instead?

3-17 A binding isotherm is obtained for a protein-ligand interaction under conditions of ligand excess and is shown in the following table. Do these binding data indicate the presence of cooperativity in the interaction? (Answer both qualitatively and quantitatively.) If possible, estimate the apparent valency of the interaction.

$[L]_o$ (nM)	y
0.0065	4.22×10^{-9}
0.0390	9.13×10^{-7}
0.0975	1.43×10^{-5}
0.260	0.000270
0.650	0.00421
1.625	0.0619
3.90	0.477
9.75	0.934
26.0	0.996
65.0	1.00

3-18 A protein P reversibly binds a ligand L to form a complex C. The table below lists complex concentration (in pM) measured as a function of time t with varying initial ligand concentrations ($[L]_o = 1$, 5, or 15 µM). The initial protein concentration $[P]_o$ was always 1 nM. Estimate k_{on}, k_{off}, and K_d of the reaction. (Contributed by P. Bransford.)

t (s)	$[L]_o = 1\ \mu M$	$[L]_o = 5\ \mu M$	$[L]_o = 15\ \mu M$
0.0	0	0	0
0.1	95	392	774
0.2	180	627	945
0.3	256	768	982
0.4	324	953	991
0.5	385	904	993

3-19 Derive dimensionless equation 3.26 from equation 3.22. (Recall that $K_d = k_{off}/k_{on}$.)

3-20 Consider a bivalent antibody binding at equilibrium (monovalent $K_{d1} = 1$ nM) to a cell-surface antigen (10^5 antigen molecules per cell; cell diameter, 10 microns). Plot the number of antibody molecules bound to the cell by only one binding site as a function of bulk antibody concentration in the range 0.001–1,000 nM. Plot the number of antibody molecules bound by two binding sites on the same plot. If the bivalently bound antibody crosslinks the antigen and sends an intracellular signal, at what bulk concentration of the antibody will this signal be maximal?

3-21 A small-molecule ligand was synthesized as a dimer linked by polypeptides of varying chain length. If a cell with diameter 10 microns has 2.5×10^7 ligand receptors on its surface, what linker length will provide the greatest avidity enhancement to the apparent binding affinity?

3-22 You are interested in experimentally determining the ligand binding affinity of a cell-surface receptor with $K_d = 10$ pM and association rate constant $k_{on} = 10^5\ M^{-1}s^{-1}$. You incubate nine separate cell suspensions with the following concentrations of fluorescently labeled ligand: 0.1, 0.3, 1.0, 3.0, 10.0, 30.0, 100.0, 300.0, and 1,000.0 pM. Simulate the binding isotherm (binding curve) you would obtain if you measured cell-bound fluorescence at the following times: 15 minutes, 1 hour, 4 hours, 16 hours, and 64 hours. You may assume pseudo-first-order kinetics. Curve-fit each of these five isotherms to obtain an estimate of K_d. Is it necessary for every sample to reach equilibrium in order to obtain an accurate K_d estimate?

3-23 A macromolecule has six sites, identical by symmetry, for the ligand L to bind. After careful experimentation, three of the macroscopic dissociation constants for ligand binding, K_1, K_4, and K_5 were measured. The results indicate $K_5 = 15K_1$, and $K_4 = 8K_1$. Do these results indicate cooperativity, anticooperativity, or independence of the binding sites? Why? Is it possible to predict the values of the other macroscopic dissociation constants in terms of K_1? If so, make the predictions. If not, explain why none can be made.

3-24 A macromolecule M has two nonidentical binding sites, one for binding $L1$ (one ligand) and the other for $L2$ (another ligand). The concentrations of $L1$ and $L2$ are such that at equilibrium, half the $L1$ sites and half the $L2$ sites are filled. Demonstrate that at equilibrium under these conditions, $[M] = [M \cdot L1 \cdot L2]$, and that $[M \cdot L1] = [M \cdot L2]$.

3-25 Consider a monomer that is self-complementary, such that it can either add another monomer or cyclize, as shown in the figure below. Let K_d be the dissociation constant of an open dimer going to two open monomers. Let $[L]_{\text{eff},1}$ be the effective concentration of the two ends for each other in the monomer. Let $[L]_{\text{eff},2}$ be the effective concentration of the two ends for each other in the open dimer.

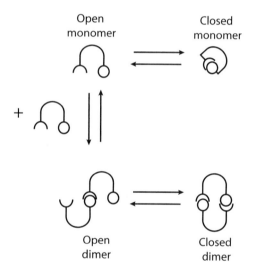

a. Assuming that the only four species that can exist are those in the figure above, solve for the ratio of the equilibrium concentration of closed dimer to closed monomer as a function of the effective concentrations, the dissociation constant, and the concentration of open monomer.

b. Using your result from part a, state three limits for which [*closed dimer*] $\gg$ [*closed monomer*]. How could each be achieved in practice?

c. For this part, assume that the diagram continues down the page, with open dimers reacting to form open trimers, which can then cyclize or go on to form open tetramers, and so forth. Let the same K_d describe the addition of a single open monomer to an open chain, independent of its length. Let $[L]_{\text{eff},j}$ be a function of the chain length j that describes the effective concentration of the ends of an open chain for each other. Give an example, and justify your choice of a function $[L]_{\text{eff},j}$ that will lead to a distribution of cyclized polymers that is peaked at its center.

3-26 The rate constants describing reversible binding between a protein and a ligand to form a complex are determined using relaxation analysis. Suppose that a series of experiments is performed in which the pressure of a system is slightly perturbed, and the system is then allowed to settle to a new equilibrium state. The time required for the concentration to reach the midpoint between its undisturbed value and the new equilibrium it reaches is recorded as $t_{1/2}$, and the final protein and ligand concentrations in the new system are somehow measured. The following table displays the results of these experiments:

$[P]_f$ (nM)	$[L]_f$ (nM)	$t_{1/2}$ (s)
0.1	4.2	25
0.2	9.0	12
1.0	16	6.3
3.0	23	4.2
5.0	29	3.2
6.0	40	2.4
8.0	43	2.2
11	47	1.9
14	53	1.7
19	64	1.3

Determine the association and dissociation rate constants for this protein-ligand complex. (Contributed by J. Spangler.)

3-27 Consider a cell line with 10^5 particular receptors per cell. You have radiolabeled a monovalent ligand, which has an affinity $K_d = 1$ nM for the receptor. Radiolabeled ligand is incubated with 10^4, 10^5, or 10^6 cells in a 100 microliter microplate well, and unbound ligand is washed away.

a. Plot the number of radiolabeled ligand molecules bound per cell as a function of initial ligand concentration (across the range 0.01–100 nM ligand) for each of the cell densities.

b. Curve-fit a simple K_d (without accounting for depletion effects) to each of the three data sets. What is the apparent binding constant at each of the three cell densities?

3-28 One method for estimating a binding constant is to compete a labeled ligand with an unlabeled ligand. Consider the system of the previous problem, with radiolabeled ligand fixed at an initial concentration of 1 nM.

a. Plot the number of radiolabeled ligand molecules bound per cell as a function of initial unlabeled competitor ligand concentration, in the range of initial concentration 0.01–1,000 nM. You may assume that radiolabeling the ligand does not appreciably alter its affinity for the receptor. Perform this analysis for each of 10^4, 10^5, or 10^6 cells in 100 microliter microplate wells.

b. Curve-fit a simple K_d (without accounting for depletion effects) using equation 3.122 for binding competition. What is the apparent binding constant at each of the three cell densities?

3-29 Derive equation 3.7 starting from equation 3.120. Which form is most useful for fitting data of y versus $[L]_o$?

3-30 A Biacore SPR biosensor experiment is performed with 10 nM flowing ligand and a receptor ectodomain immobilized on the chip, to give the result shown in the graph below. If this same ligand were used at 10 pM to label cells expressing the ligand on their surface (under conditions in which the receptor would not be internalized), how long would it take to reach 95% of the equilibrium fraction of bound ligand?

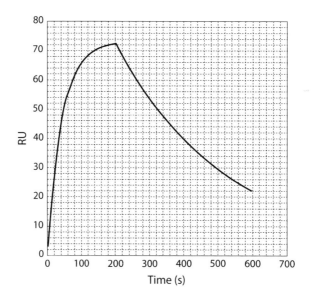

3-31 For a particular protein–small-molecule interaction, $\Delta G° = -8.18$ kcal/mol. In a typical isothermal titration calorimetry experiment, the volume of the cell is 1.5 mL. Concentration of the protein in the cell is 70 µM, and the concentration of the small molecule in the syringe is 1.4 mM. Assume 10 µL injections 400 seconds apart. The experiment takes place at 25°C. Calculate $\Delta H°$ and $\Delta S°$, given that the fifth injection generates 19.86 µcal. At what time will the signal reach baseline (reaction > 99% complete)?

3-32 A given protein-ligand interaction has a $\Delta G°$ of –8.18 kcal/mol and a $\Delta S°$ of 10 cal/(mol·K). Calculate $\Delta H°$ (at 25°C), and use this value to calculate the heat released by the second injection of ligand (10 µL of a 300-µM solution) into a 1.5 mL sample cell containing protein at a concentration of 20 µM in an ITC experiment.

3-33 You have recently generated a small-molecule (ATP-analog) kinase inhibitor for a given target protein in the lab. Your initial indications give a binding affinity (dissociation constant) of 20 nM. To completely characterize the inhibitor and provide some indication of how it may be improved in second- and third-stage iterations, you want to determine all thermodynamic parameters describing this protein-ligand interaction. For an ITC experiment (sample cell volume, 1.5 mL), how much protein (assume 50 kDa for molecular weight) will be needed for a successful experiment? How much of your small-molecule ligand (molecular weight = 342.3 g/mol) is necessary, and what concentration should you use to generate a final mole ratio of 2:1? This same small-molecule inhibitor is toxic to all cells (not the desired effect!) at concentrations above 1 micromolar. What is the maximum inhibition possible, given a kinase-ATP dissociation constant of 75 mM and $[ATP] = 3$ mM?

3-34 The following ITC data were collected for a certain protein-ligand binding reaction (table 3.2 and figure 3.43). The first column in table 3.2 is the molar ratio and is unitless; the second column is the heat term (in units of kcal/mol), which is the integration of the area under each peak in the upper plot in figure 3.43.

Estimate $\Delta H°$, K_d, $\Delta G°$, and $\Delta S°$ for this binding reaction. This ITC experiment was performed at a temperature of 300 K. Assume that the concentration of protein stays constant at 0.019 mM in the sample cell throughout the experiment. (*Note:* This is not exactly true. The amount of protein in the sample cell does not change throughout the experiment. But a small amount of buffer is added with each ligand injection, so the concentration

Table 3.2 ITC data for problem 3-34.

Molar ratio ($[L]/[P]$)	Heat (kcal/mol)
0.081	−7.95
0.16	−7.40
0.24	−6.88
0.32	−6.33
0.41	−5.99
0.49	−5.21
0.57	−4.66
0.65	−4.13
0.74	−3.70
0.82	−3.07
0.90	−2.75
0.99	−2.55
1.07	−2.13
1.15	−1.81
1.24	−1.67
1.32	−1.52
1.41	−1.17
1.49	−1.13
1.58	−1.06
1.67	−1.01
1.75	−0.81
1.84	−0.73
1.93	−0.78
2.01	−0.61
2.10	−0.68

of protein actually decreases during the experiment from 0.020 mM to 0.018 mM.)

3-35 For the data in table 3.3, use nonlinear least squares to fit a value for K_d for the wild-type protein-ligand interaction and for the mutant protein-ligand interaction. Assume that $[P]_o \ll [L]_o$. Plot the residuals and a comparison of the fits to the data. Remember to consider the relation between measured signal intensity and actual binding fraction. Discuss your model and the results of the fitting.

3-36 The purpose of this exercise is to implement a global nonlinear least squares routine to simulate kinetic data. The association phase data for three SPR-type experiments are provided in table 3.4, under pseudo-first-order conditions at ligand concentrations of 10 nM, 20 nM, and 30 nM. Analyze these data with

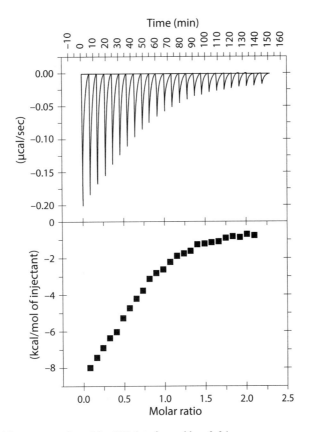

Figure 3.43 Graphic representation of the ITC data for problem 3-34.

two separate methods: (1) Fit a single exponential to each of the three experiments to obtain k_{obs}, then from a plot of k_{obs} versus $[L]_o$, estimate k_{on} and k_{off}. (2) Globally fit all three data sets to pseudo-first-order kinetics by fitting the parameters k_{on} and k_{off} simultaneously to all three data sets. How well do the two methods agree? Demonstrate the goodness of fit for both approaches by plotting the residuals, as well as the model and data curves overlaid. Obtain 95% confidence limits on the estimates for the individually fitted parameters k_{obs} from method (1), and on the k_{on} and k_{off} estimates from method (2). Contrast the confidence interval from MATLAB function nlparci with that obtained by the F statistic and the χ^2 sensitivity method. (*Hint*: These data incorporate such inconvenient experimental realities as a baseline that varies from experiment to experiment and a recorded time that is offset from zero.)

Table 3.3 Cell-surface binding isotherm data for problem 3-35.

Wild type		Mutant	
$[L]_o$ (nM)	Signal	$[L]_o$ (nM)	Signal
700	1.14	350	0.99
350	1.01	87.0	1.04
87.5	0.93	35.0	1.00
35.0	0.89	8.75	0.95
8.75	0.86	3.50	0.81
5.00	0.75	2.33	0.69
3.50	0.75	1.75	0.60
2.30	0.70	0.88	0.52
1.75	0.54	0.58	0.45
0.88	0.40	0.44	0.37
0.58	0.32	0.35	0.31
0.44	0.32	0.27	0.31
0.35	0.21	0.22	0.34
0.18	0.15	0.18	0.20
0.088	0.11	0.088	0.095

Table 3.4 SPR kinetic data for problem 3-36.

Time (s)	RU ($[L]_o = 30$ nM)	RU ($[L]_o = 20$ nM)	RU ($[L]_o = 10$ nM)
14.8	7.28	13.1	11.4
29.6	5.07	14.1	11.6
44.4	3.86	15.4	9.74
59.2	7.37	13.9	9.12
74.0	5.02	15.5	12.3
88.8	32.5	35.9	19.5
104	47.7	48.9	27.9
118	56.3	52.4	29.0
133	66.2	58.0	37.2
148	71.6	68.7	46.6
163	77.7	69.1	47.5
178	80.0	75.0	52.5
192	85.3	76.8	49.8
207	75.5	75.2	55.5
222	79.2	78.9	56.0
237	83.3	88.7	55.6
252	87.2	84.6	60.7
266	83.8	83.3	61.7
281	88.2	83.2	65.0
296	82.5	87.7	62.6
311	85.3	86.6	62.4
326	86.7	88.0	64.0

(*continued*)

Table 3.4 (*continued*)

Time (s)	RU ($[L]_0 = 30$ nM)	RU ($[L]_0 = 20$ nM)	RU ($[L]_0 = 10$ nM)
340	83.3	85.8	60.7
355	88.9	91.4	67.1
370	82.0	87.1	66.7
385	83.8	89.2	70.9
400	84.5	83.4	69.4
414	87.0	85.1	69.3
429	84.1	90.9	71.0
444	90.3	87.1	63.1
459	86.2	83.6	65.1
474	84.2	82.7	69.2
488	83.0	87.4	66.6
503	90.4	85.3	76.8
518	83.9	85.4	74.8
533	86.0	86.6	72.2
548	87.9	88.0	70.8
562	88.1	88.3	73.1
577	86.6	85.6	72.8
592	82.5	85.6	70.2
607	85.7	87.3	73.7
622	86.9	81.5	68.7

Table 3.5 Tryptophan fluorescence quench isotherm data.

[*Ligand*] (nM)	Signal with $0.5 \times [P]$	Signal with $1 \times [P]$	Signal with $2 \times [P]$
0.0300	0.347	0.696	1.39
0.100	0.341	0.688	1.38
0.300	0.328	0.663	1.34
1.00	0.290	0.594	1.23
3.00	0.235	0.479	0.998
10.0	0.185	0.372	0.754
30.0	0.163	0.326	0.656
100	0.153	0.306	0.616

3-37 You have performed a tryptophan fluorescence quench titration to estimate the equilibrium binding constant for a protein-ligand interaction. Although you know the concentration of the ligand precisely and have performed the titrations by adding aliquots of that ligand, you have at best only a vague estimate of the protein concentration, believing it to be approximately 1 nM at the $1\times$ concentration in table 3.5. You perform three titrations, at $0.5\times$, $1\times$, and $2\times$ protein concentration. The data are represented in table 3.5. Assuming that the data are accurate to within 1% (no significant background or nonspecific signals), obtain estimates of the following:

a. The actual concentration of protein at $1\times$ concentration.

b. K_d for this interaction.

References

1. J. Foote and H. N. Eisen. Breaking the affinity ceiling for antibodies and T cell receptors. *Proc. Natl. Acad. Sci. U.S.A.* 97: 10679–10681 (2000).

2. G. Schreiber, G. Haran, and H.-X. Zhou. Fundamental aspects of protein-protein association kinetics. *Chem. Rev.* 109: 839–860 (2009).

3. C. Y. F. Huang and J. E. Ferrell. Ultrasensitivity in the mitogen-activated protein kinase cascade. *Proc. Natl. Acad. Sci. U.S.A.* 93: 10078–10083 (1996).

4. W. P. Jencks. On the attribution and additivity of binding energies. *Proc. Natl. Acad. Sci. U.S.A.* 78: 4046–4050 (1981).

5. P. Friedl and J. Storim. Diversity in immune-cell interactions: States and functions of the immunological synapse. *Trends Cell Biol.* 14: 557–567 (2004).

6. R. M. Stroud and J. A. Wells. Mechanistic diversity of cytokine receptor signaling across cell membranes. *Science STKE* 231: re7 (2004).

7. M. Mammen, S. K. Choi, and G. M. Whitesides. Polyvalent interactions in biological systems: Implications for design and use of multivalent ligands and inhibitors. *Angew. Chem. Int. Ed.* 37: 2755–2794 (1998).

8. H.-X. Zhou. Quantitative account of the enhanced affinity of two linked scFvs specific for different epitopes on the same antigen. *J. Mol. Biol.* 329: 1–8 (2003).

9. K. H. Pearce, Jr., B. C. Cunningham, G. Fuh, T. Teeri, and J. A. Wells. Growth hormone binding affinity for its receptor surpasses the requirements for cellular activity. *Biochemistry* 38: 81–89 (1999).

10. J. D. Stone, J. R. Cochran, and L. J. Stern. T-cell activation by soluble MHC oligomers can be described by a two-parameter binding model. *Biophys. J.* 81: 2547–2557 (2001).

11. D. M. Crothers and H. Metzger. The influence of polyvalency on the binding properties of antibodies. *Immunochemistry* 9: 341–357 (1972).

12. E. N. Kaufman and R. K. Jain. Effect of bivalent interaction upon apparent antibody affinity: Experimental confirmation of theory using fluorescence photobleaching and implications for antibody binding assays. *Cancer Res.* 52: 4157–4167 (1992).

13. T. R. Sosnick, D. C. Benjamin, J. Novotny, P. A. Seeger, and J. Trewhella. Distances between the antigen-binding sites of three murine antibody subclasses measured using neutron and X-ray scattering. *Biochemistry* 31: 1779–1786 (1992).

14. Y. Mazor, A. Hansen, C. Yang, P. S. Chowdhury, J.Wang, G. Stephens, H.Wu, and W. F. Dall'Acqua. Insights into the molecular basis of a bispecific antibody's target selectivity. *mAbs* 7: 461–469 (2015).

15. M. J. Mattes. Binding parameters of antibodies: Pseudo-affinity and other misconceptions. *Cancer Immunol. Immunother.* 54: 513–516 (2005).

16. K. M. Muller, K. M. Arndt, and A. Plückthun. Model and simulation of multivalent binding to fixed ligands. *Anal. Biochem.* 261: 149–158 (1998).

17. C. J. Burke, B. L. Steadman, D. B. Volkin, B. K. Tsai, M. W. Bruner, and C. R. Middaugh. The adsorption of proteins to pharmaceutical container surfaces. *Int. J. Pharmacol.* 86: 89–93 (1992).

18. S. Jiang and Z. Cao. Ultralow-fouling, functionalizable, and hydrolyzable zwitterionic materials and their derivatives for biological applications. *Adv. Mater.* 22: 920–932 (2010).

19. Y. S. Day, C. L. Baird, R. L. Rich, and D. G. Myszka. Direct comparison of binding equilibrium, thermodynamic, and rate constants determined by surface- and solution-based biophysical methods. *Protein Sci.* 11: 1017–1025 (2002).

20. E. T. Boder and K. D. Wittrup. Optimal screening of surface-displayed polypeptide libraries. *Biotechnol. Prog.* 14: 55–62 (1998).

21. J. J. VanAntwerp and K. D. Wittrup. Fine affinity discrimination by yeast surface display and flow cytometry. *Biotechnol. Prog.* 16: 31–37 (2000).

22. P. Schuck. Use of surface plasmon resonance to probe the equilibrium and dynamic aspects of interactions between biological macromolecules. *Annu. Rev. Biophys. Biomol. Struct.* 26: 541–566 (1997).

23. P. Schuck and A. P. Minton. Kinetic analysis of biosensor data: Elementary tests for self-consistency. *Trends Biochem. Sci.* 21: 458–460 (1996).

24. N. Lion, F. Reymond, H. H. Girault, and J. S. Rossier. Why the move to microfluidics for protein analysis? *Curr. Opin. Biotechnol.* 15: 31–37 (2004).

25. T. P. Burg and S. R. Manalis. Suspended microchannel resonators for biomolecule detection. *Appl. Phys. Lett.* 83: 2698–2700 (2003).

26. K. E. Petersen, W. A. McMillan, G. T. A. Kovacs, M. A. Northrup, L. A. Christel, and F. Pourahmadi. Toward next generation clinical diagnostic instruments: Scaling and new processing paradigms. *Biomed. Microdevices* 1: 71–79 (1998).

27. T. Thorsen, S. J. Maerkl, and S. R. Quake. Microfluidic large-scale integration. *Science* 298: 580–584 (2002).

28. T. J. Silhavy, S. Szmelcman, W. Boos, and M. Schwartz. On the significance of the retention of ligand by protein. *Proc. Natl. Acad. Sci. U.S.A.* 72: 2120–2124 (1975).

29. I. Jelesarov and H. R. Bosshard. Isothermal titration calorimetry and differential scanning calorimetry as complementary tools to investigate the energetics of biomolecular recognition. *J. Mol. Recognit.* 12: 3–18 (1999).

30. L. Indyk and H. F. Fisher. Theoretical aspects of isothermal titration calorimetry. *Methods Enzymol.* 295: 350–364 (1998).

31. J. E. Ladbury. Application of isothermal titration calorimetry in the biological sciences: Things are heating up! *Biotechniques* 37: 885–887 (2004).

32. A. Cooper, C. M. Johnson, J. H. Lakey, and M. Nöllmann. Heat does not come in different colours: Entropy-enthalpy compensation, free energy windows, quantum confinement, pressure perturbation calorimetry, solvation and the multiple causes of heat capacity effects in biomolecular interactions. *Biophys. Chem.* 93: 215–230 (2001).

33. S. E. Harding and B. Chowdhry (eds.). *Protein-Ligand Interactions: Hydrodynamics and Calorimetry. A Practical Approach.* Oxford University Press (2000).

34. M. L. Johnson. Why, when, and how biochemists should use least squares. *Anal. Biochem.* 206: 215–225 (1992).

35. NIST/SEMATECH eHandbook of Statistical Methods. http://www.itl.nist.gov/div898/handbook/.

36. J. R. Lakowicz. *Principles of Fluorescence Spectroscopy*, third edition. Springer (2006).

37. J. M. Beechem. Global analysis of biochemical and biophysical data. *Methods Enzymol.* 210: 37–54 (1992).

4
Enzyme Kinetics

The problem of the chemical nature of the substances which control the reactions occurring in living cells has been a subject of research, and also of controversy, for nearly two hundred years.
—John Northrop, Nobel Lecture, 1946

We still share genes around, and the resemblance of the enzymes of grasses to those of whales is a family resemblance.
—Lewis Thomas, *The Lives of a Cell: Notes of a Biology Watcher*

Enzymes are biological catalysts that ensure that required chemical reactions occur at the proper time and place to convert metabolic energy, build biopolymers and cellular structures, transmit signals culminating in altered gene expression, and carry out most other biological events involving covalent transformations. In this chapter, we construct a mathematical framework for understanding the rates of enzyme-catalyzed reactions at varying concentrations of reactants and inhibitors. We then extend this framework to integrate the role of enzymatic reactions in signal transduction and metabolism.

4.1 Enzymes Are Catalysts

Enzymes are protein or nucleic acid catalysts that accelerate chemical reactions through participation in the reaction without being progressively consumed. A chemical catalyst is defined by the following properties:

1. Accelerates a reaction by lowering its activation energy,
2. Is not progressively consumed or created by the reaction, and
3. Does not alter the overall thermodynamics of the reaction (i.e., the equilibrium of the overall reaction).

These properties are graphically reflected in the reaction coordinate diagram in figure 4.1.

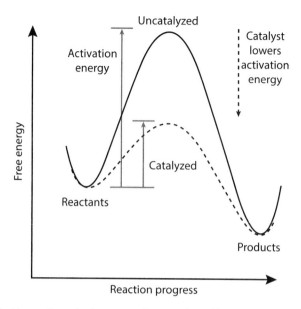

Figure 4.1 A catalyst lowers the activation energy for a reaction without changing the energy states of the reactants or products; hence, the equilibrium is not altered, but the rate of interconversion between the states is accelerated.

The word *enzyme* comes from the Greek term for *leavened*, because it was first discovered that extracts of yeast could catalyze fermentation in the absence of living cells. The great majority of enzymes are proteins; however, RNA enzymes have also been shown to perform critical functions in gene expression, and artificial DNA enzymes have been created in the laboratory. The tremendous diversity of enzymatic reactions in biology can be classified in six broad categories (a surprisingly small number, based on the chemical transformation catalyzed, as shown in table 4.1 [1–6]).

The common name of an enzyme generally describes its function, ending with the suffix *-ase*. For example, DNA polymerase catalyzes the polymerization of nucleotides to form DNA; DNA ligase catalyzes the ligation of two DNA segments; a protease hydrolyzes peptide bonds.

Enzymes possess an active site (consisting of a patch, pit, or groove) that binds reactants (generally called substrates), thereby localizing the necessary chemical groups to accelerate the given reaction and then releasing the products when the reaction is complete. For example, the structure of chymotrypsin (a protease, which catalyzes the hydrolysis or cleavage of peptide bonds) in complex with and without its peptide substrate bound is shown in figure 4.2.

Table 4.1 IUPAC-IUBMB enzyme classifications according to Enzyme Commission (EC) number.

EC 1	*Oxidoreductases*	*EC 3*	*Hydrolases*
EC 1.1	CH–OH donors	EC 3.1	ester bonds
EC 1.2	aldehyde or oxo donors	EC 3.2	glycosylases
EC 1.3	CH–CH donors	EC 3.3	ether bonds
EC 1.4	CH–NH_2 donors	EC 3.4	peptide bonds (peptidases)
EC 1.5	CH–NH donors	EC 3.5	carbon-nitrogen bonds, other than peptide bonds
EC 1.6	NADH or NADPH		
EC 1.7	other nitrogenous compounds as donors	EC 3.6	acid anhydrides
		EC 3.7	carbon-carbon bonds
EC 1.8	sulfur group donors	EC 3.8	halide bonds
EC 1.9	heme group donors	EC 3.9	phosphorus-nitrogen bonds
EC 1.10	diphenols and related substances as donors	EC 3.10	sulfur-nitrogen bonds
		EC 3.11	carbon-phosphorus bonds
EC 1.11	peroxide as acceptor	EC 3.12	sulfur-sulfur bonds
EC 1.12	hydrogen as donor	EC 3.13	carbon-sulfur bonds
EC 1.13	single donors with incorporation of molecular oxygen (oxygenases)		
		EC 4	*Lyases*
EC 1.14	paired donors with incorporation or reduction of molecular oxygen	EC 4.1	carbon-carbon lyases
		EC 4.2	carbon-oxygen lyases
EC 1.15	superoxide radicals as acceptors	EC 4.3	carbon-nitrogen lyases
EC 1.16	oxidizing metal ions	EC 4.4	carbon-sulfur lyases
EC 1.17	CH or CH_2 group	EC 4.5	carbon-halide lyases
EC 1.18	iron-sulfur proteins as donors	EC 4.6	phosphorus-oxygen lyases
EC 1.19	reduced flavodoxin as donor	EC 4.7	carbon-phosphorous lyases
EC 1.20	phosphorus or arsenic in donors	EC 4.99	other lyases
EC 1.21	X–H and Y–H form an X–Y bond		
EC 1.22	halogen in donors	*EC 5*	*Isomerases*
EC 1.23	reducing C–O–C group as acceptor	EC 5.1	racemases and epimerases
EC 1.97	other oxidoreductases	EC 5.2	*cis-trans* isomerases
		EC 5.3	intramolecular isomerases
		EC 5.4	intramolecular transferases (mutases)
EC 2	*Transferases*		
EC 2.1	one-carbon group	EC 5.5	intramolecular lyases
EC 2.2	aldehyde or ketonic group	EC 5.99	other isomerases
EC 2.3	acyltransferases		
EC 2.4	glycosyltransferases	*EC 6*	*Ligases*
EC 2.5	alkyl or aryl group, other than methyl groups	EC 6.1	carbon-oxygen bonds
		EC 6.2	carbon-sulfur bonds
EC 2.6	nitrogenous groups	EC 6.3	carbon-nitrogen bonds
EC 2.7	phosphorus-containing groups	EC 6.4	carbon-carbon bonds
EC 2.8	sulfur-containing groups	EC 6.5	phosphoric ester bonds
EC 2.9	selenium-containing	EC 6.6	nitrogen-metal bonds
EC 2.10	molybdenum- or tungsten-containing groups		

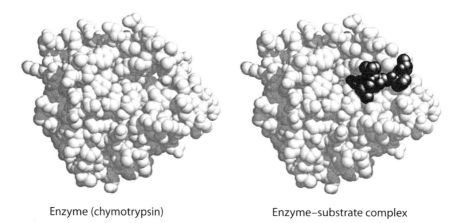

Enzyme (chymotrypsin)   Enzyme–substrate complex

Figure 4.2 The structure of an enzyme, looking into its active site (left), and its complex with substrate bound in the active site (right). The tertiary structure of the enzyme presents amino acid side chains in a shape that is complementary to the substrate and organizes reactive groups to interact with the substrate when bound. (Left) Bovine γ-chymotrypsin, which is a serine protease (an enzyme that cleaves peptide and protein substrates by a mechanism in which a serine residue initially attacks the peptide bond). (Right) Bovine γ-chymotrypsin in complex with a pentapeptide substrate (TPGVY), rendered in a darker shade. Formation of an enzyme-substrate complex is an intermediate step that is universal to all enzymatic catalysis. PDB ID 1AB9; EC number 2.4.21.1 [7].

4.2 Enzymatic Rate Laws

In chapter 3, we derived mass-action expressions that describe how concentration affects binding rates and equilibrium levels of complexation. We now consider how the concentrations of enzyme and reactants determine the rate of enzymatically catalyzed reactions.

4.2.1 Michaelis-Menten Rate Law

How does the rate of an enzyme-catalyzed reaction depend on the concentrations of enzyme and of substrate? The first part of this question has a simple answer: Enzyme-catalyzed reaction rates are essentially always proportional to the concentration of added enzyme. This result can be rationalized by recognizing that (1) formation of an enzyme-substrate complex is a necessary first step for enzymatic catalysis and (2) mass-action kinetics predicts that the rate of complex formation will be proportional to enzyme concentration. However, the second part of the question turns out to require more analysis: In general, the rates of enzyme-catalyzed reactions are found to be proportional to substrate concentration at low concentrations (i.e., first order in $[S]$) but are concentration independent at high concentrations (i.e., zeroth order in $[S]$).

Enzyme Kinetics

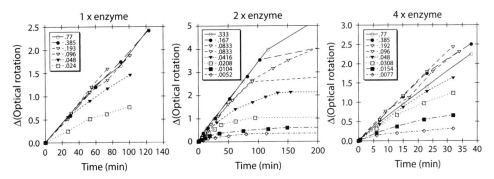

Figure 4.3 Empirically observed product accumulation curves at varying initial substrate concentrations (given in M in the inset legend in each panel). The enzyme is invertase, the substrate sucrose, and the products are glucose and fructose. The data for the middle panel used an enzyme concentration twice that of the first panel; the third panel used an enzyme concentration four times that of the first panel. Reaction progress was followed by the change in circular polarization of light by the sugar solution, because the products (glucose and fructose) interact with polarized light differently than does the reactant (sucrose) [9].

Why might a reaction be zeroth order? How can the rate of conversion of substrate actually be independent of its concentration? The answer is that each enzyme molecule has a maximum rate at which it can do chemistry. Therefore, once each enzyme molecule in a volume is bound in complex with substrate, additional substrate molecules must wait their turn for an enzyme either to release substrate or to catalyze a reaction and release products.

Let's consider a simple rate form that captures this mechanism quantitatively, first demonstrated by Victor Henri in 1902 [8] and refined by Leonor Michaelis and Maud Menten in 1913 [9]. Michaelis and Menten examined the reaction of the enzyme invertase with table sugar (sucrose), using circular dichroism to measure the lumped change in concentration of reactants and products. This data set is presented in figure 4.3. As one of the last sets of enzyme kinetic data to be taken in the absence of a mechanistically explanatory rate law, it's an interesting exercise to examine the features of the data that led to derivation of the Michaelis-Menten rate form.

A proportional increase in the reaction rate with increasing enzyme concentration is apparent from comparisons among individual points in the three panels of figure 4.3. For substrate concentrations above 0.1 M, 1× enzyme catalyzes an increase in Δ(optical rotation) of 2 by 100 minutes; for 2× enzyme, this level is reached at 50 minutes; for 4× enzyme, this level is reached at 25 minutes. Clearly, the rate form for this reaction should be proportional to enzyme concentration, as we mentioned at the outset.

However, note the saturation behavior with respect to dependence on substrate. At each of the enzyme concentrations tested, the reaction rate increases with substrate

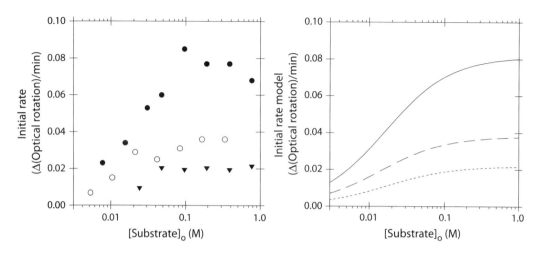

Figure 4.4 (Left panel) Initial rates of product formation as a function of initial substrate concentration, from the data in figure 4.3. Note that the substrate concentration has been varied in a geometric series, so that the data points are evenly spaced on a logarithmic scale. (Right panel) Nonlinear least squares curve fit of equation 4.12. Parameters: $v_{max} = 0.081, 0.038, 0.022$ M s^{-1}; $K_m = 0.017, 0.014, 0.017$ M for filled circles, solid line; open circles, dashed line; and triangles, short-dashed line, respectively.

concentration when it is below 0.1 M; at higher substrate levels, the reaction proceeds at a maximum rate proportional to the enzyme level. Such behavior shows zeroth-order concentration dependence with respect to substrate.

One feature of the data from batch reactions such as these is that the reactants are eventually consumed (e.g., note the plateau in the curve for 2× enzyme and 0.0416 M substrate). Consequently, the relationship between concentrations and rates is easiest to interpret at the earliest times, when the concentrations are effectively unaltered from the original mixture. Also, many enzyme-catalyzed reactions are reversible, and by considering only initial rate data before substantial product is accumulated, one can parse out the properties of the forward rate in the absence of substantial reverse reaction.

Initial rate data can be extracted by measuring the slope of the curves at the earliest times, attempting to determine the slope of the curve as time asymptotically approaches zero. These estimated initial rates for the Michaelis-Menten invertase-sucrose data in figure 4.3 are plotted as a function of substrate concentration in figure 4.4.

The plot has a logarithmic scale for substrate concentration to spread the data usefully, as we have also seen to be preferable for protein-ligand binding isotherms. Note the plateaus in initial reaction rates at substrate concentrations above 0.1 M, indicating zeroth-order kinetics. Before the advent of digital computers, it was common to plot transformed variables to linearize the data, but this serves little current purpose. For completeness, we'll nevertheless consider these methods below.

Michaelis and Menten derived a simple mathematical description of the rate form shown in figure 4.4 by hypothesizing the following mechanism for enzyme-catalyzed reactions:

$$E + S \underset{k_{-1}}{\overset{k_1}{\rightleftharpoons}} ES \overset{k_{\text{cat}}}{\longrightarrow} E + P \tag{4.1}$$

where E is the enzyme, S is the substrate, P is the product, and ES is the enzyme-substrate complex. Note the assumptions that: (1) reaction and release are simultaneous and (2) the reverse reaction does not occur. The first assumption, as will be shown shortly, is not necessary to obtain the particular mathematical form of the Michaelis-Menten rate law, but it changes the mechanistic interpretation of the constants. As mentioned above, the second assumption is approximately valid at the very earliest time points, before substantial product has accumulated.

The following differential equations can be written for this system by constructing the dynamic material balance for each of the species:

$$\frac{d[S]}{dt} = -k_1[E][S] + k_{-1}[ES] \tag{4.2}$$

$$\frac{d[ES]}{dt} = k_1[E][S] - (k_{-1} + k_{\text{cat}})[ES] \tag{4.3}$$

$$\frac{d[P]}{dt} = k_{\text{cat}}[ES] \tag{4.4}$$

Michaelis and Menten obtained an approximate solution to these equations by assuming that the complex ES is in rapid equilibrium with free substrate and enzyme. Briggs and Haldane [10] subsequently demonstrated that the same rate form is valid under a broader set of conditions—using the quasi-steady-state approximation (QSSA) for the concentration of intermediate [ES]. Some call the resulting rate law the Michaelis-Menten-Briggs-Haldane form, or even the Henri-Michaelis-Menten-Briggs-Haldane, to reflect Henri's precedence. However, we will follow the most common convention of terming this the *Michaelis-Menten rate law*.

The QSSA corresponds to setting

$$\frac{d[ES]}{dt} = 0 \tag{4.5}$$

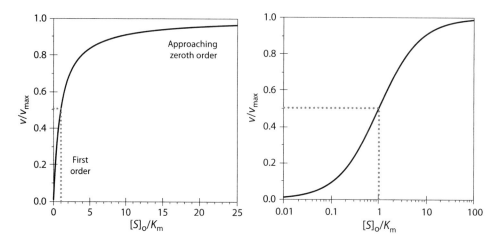

Figure 4.5 Initial rates of product formation as a function of initial substrate concentration on (left panel) linear and (right panel) semilog scales. Note that the half-maximal rate is achieved at $[S]_o = K_m$. Also, note that the rate is first order with respect to substrate concentration at low concentrations and zeroth order at high concentrations.

so

$$0 = k_1[E][S] - (k_{-1} + k_{cat})[ES]_{QSSA} \qquad (4.6)$$

or

$$[ES]_{QSSA} = \frac{k_1[E][S]}{k_{-1} + k_{cat}} \qquad (4.7)$$

If enzyme is not in great excess over substrate, then $[E] \neq [E]_o$, and depletion of free enzyme concentration to the complexed form must be accounted for:

$$[E] = [E]_o - [ES] \qquad (4.8)$$

Substituting this result into equation 4.7 and solving for $[ES]_{QSSA}$, we have

$$[ES]_{QSSA} = \frac{[E]_o[S]}{\frac{k_{-1}+k_{cat}}{k_1} + [S]} \qquad (4.9)$$

As we are deriving a rate law for initial reaction rates, we assume that $[S] = [S]_o$, which will be approximately true only before the reaction has consumed much substrate. Initial rates such as this are presented in figures 4.4 and 4.5. Substituting equation 4.9 into equation 4.4, the reaction rate is obtained (with the nomenclature v for reaction velocity):

$$v \equiv \frac{d[P]}{dt} \qquad (4.10)$$

$$= \frac{k_{\text{cat}}[E]_\circ[S]}{\frac{k_{-1}+k_{\text{cat}}}{k_1} + [S]} \quad (4.11)$$

$$= \frac{v_{\max}[S]}{K_{\text{m}} + [S]} \quad (4.12)$$

where $v_{\max} \equiv k_{\text{cat}}[E]_\circ$ is the maximum reaction velocity, and $K_{\text{m}} \equiv (k_{-1} + k_{\text{cat}}/k_1)$ is the Michaelis-Menten constant. The substrate concentration at which the reaction velocity is half-maximal is K_{m}. Note that this rate form is consistent with the experimental observation of first-order kinetics at low substrate concentration ($v \approx \frac{v_{\max}}{K_{\text{m}}}[S]$ when $[S] \ll K_{\text{m}}$) and zeroth-order kinetics at high substrate concentration ($v \approx v_{\max}$ when $[S] \gg K_{\text{m}}$).

Examining equation 4.12, one can see that plotting $1/v$ versus $1/[S]$ would yield a linear plot with slope $K_{\text{m}}/v_{\max}$ and intercept $1/v_{\max}$. Such a plot is called a Lineweaver-Burk plot and historically, it was often used to fit constants to data and determine patterns of inhibition. However, the nonlinear transformation skews the data and invalidates the assumptions that lead to nonlinear least squares procedures producing the most probable parameter estimates. Monte Carlo calculations have directly demonstrated greater inaccuracy in parameter estimates from curve-fitting Lineweaver-Burk plots than from direct fitting of v versus $[S]$ data [11]. Pattern matching to v versus $[S]$ semilog plots is more intuitive, and we will adopt this convention here.

Many alternative mechanisms lead to the Michaelis-Menten functional form
Michaelis and Menten assumed rapid equilibration of $E + S \rightleftharpoons ES$ rather than QSSA and obtained the same mathematical form $v = \frac{v_{\max}[S]}{K_{\text{m}}+[S]}$, except with $K_{\text{m}} \equiv k_{-1}/k_1 = K_{\text{d}}$, the equilibrium dissociation constant for the enzyme-substrate interaction. Both the Michaelis-Menten assumption of rapid equilibrium and the Briggs-Haldane assumption of the quasi-steady state lead to the identical rate form $v([S], [E]_\circ) = \frac{k_{\text{cat}}[E]_\circ[S]}{K_{\text{m}}+[S]}$, another illustration of the principle that a mathematical model's consistency with data cannot prove the correctness of the underlying mechanism from which the model was derived.

Example 4-1 As it happens, another mechanism with an additional QSSA intermediate also yields the Michaelis-Menten rate form. Consider the following mechanism, in which reaction and release are not simultaneous (i.e., relaxing the first assumption that follows equation 4.1 above):

$$E + S \underset{k_{-1}}{\overset{k_1}{\rightleftharpoons}} ES \underset{k_{-2}}{\overset{k_2}{\rightleftharpoons}} EP \xrightarrow{k_{\text{cat}}} E + P \quad (4.13)$$

Show that this mechanism leads to a rate law mathematically identical to the Michaelis-Menten form.

Solution Applying the QSSA to both intermediate species, we have

$$\frac{d[EP]}{dt} = k_2[ES] - (k_{-2} + k_{cat})[EP] = 0 \quad (4.14)$$

$$\implies [ES] = \frac{k_{-2} + k_{cat}}{k_2}[EP] \quad (4.15)$$

and

$$\frac{d[ES]}{dt} = k_1[E][S] + k_{-2}[EP] - (k_{-1} + k_2)[ES] = 0 \quad (4.16)$$

$$\implies k_1[S]\,([E]_\circ - [ES] - [EP]) + k_{-2}[EP] - (k_{-1} + k_2)[ES] = 0 \quad (4.17)$$

Substituting equation 4.15 into equation 4.17 and solving for $[EP]$, the reaction velocity ($v = \frac{d[P]}{dt} = k_{cat}[EP]$) does indeed adopt the familiar rate form:

$$v = \frac{v_{max,app}[S]}{K_{m,app} + [S]} \quad (4.18)$$

where

$$v_{max,app} \equiv \frac{k_2 k_{cat}}{k_2 + k_{-2} + k_{cat}}[E]_\circ \quad (4.19)$$

and

$$K_{m,app} \equiv \frac{k_{-1}k_{-2} + k_{-1}k_{cat} + k_2 k_{cat}}{k_1(k_2 + k_{-2} + k_{cat})} \quad (4.20)$$

When we consider reversible inhibitors (section 4.3), we will find that the hyperbolic functional form of the Michaelis-Menten rate law still holds, but the parameters are affected by inhibitor concentration in ways that depend on their mechanism of action.

We'll also see that the most common multisubstrate enzyme reactions give rate forms mathematically equivalent to Michaelis-Menten, when all but one substrate concentration is held constant. It bears repeating here that there are an infinite number of models consistent with any data set; one can *disprove* a model's hypothesized mechanism by lack of consistency, but one cannot *prove* a model's correctness by consistency. Consistency is necessary, but not sufficient, for a model to be correct.

Typical kinetic parameter values Let's now consider typical ranges for the parameters K_m and k_{cat}.

- *Michaelis-Menten parameter, K_m*

    K_m values are very often greater than the physiological concentrations of the substrate of interest (figure 4.6). This can be rationalized in evolutionary terms by observing that when mutants with $K_m < [S]_o$ arise, the catalyzed reaction approaches zeroth-order kinetics, diminishing evolutionary pressure for further reduction in K_m [12]. A major exception to this trend is with kinases that bind ATP, which very often exhibit $K_{m,ATP} \ll [ATP]$ (right panel of figure 4.6). This observation is indicative of evolutionary pressure for kinase activities to be independent of cellular ATP concentrations.

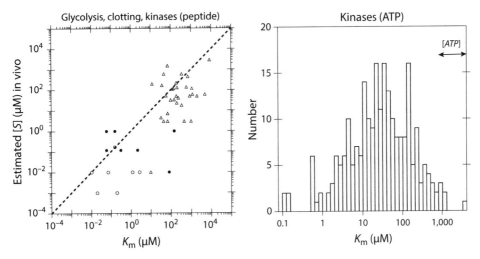

Figure 4.6 Substrate concentrations in vivo are generally below the K_m values of enzymes. Examples are shown for glycolytic enzymes (triangles [12]); MAP kinase signaling pathway enzymes versus their peptide substrates (solid circles [14]); and blood clotting cascade proteases (open circles [15]). A strong exception to this trend is the K_m of cellular kinases for ATP, a co-substrate in the great majority of kinase reactions. In this case, almost all values of K_m lie below the intracellular concentration of ATP [16], which is usually in the 1–5 mM range [17].

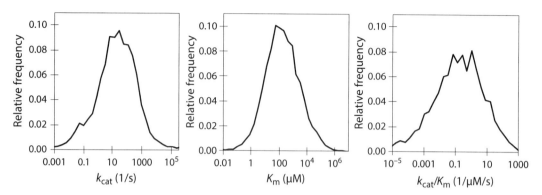

Figure 4.7 The distributions of enzyme parameter values are presented for natural enzymes in the BRENDA database [18].

- *Turnover number, k_{cat}*

  The parameter k_{cat} is also called the turnover number, because it represents the maximum number of substrate reaction events per unit time per enzyme active site. At high substrate concentrations (i.e., $[S]_\circ \gg K_m$), the reaction velocity in equation 4.12 reduces to $k_{cat}[E]_\circ$, so the turnover number is also the apparent first-order rate constant under these conditions. Turnover number varies widely, with typical values from $0.1\ \text{s}^{-1}$ to $1{,}000\ \text{s}^{-1}$ (figure 4.7), as well as numerous exceptions beyond this range. Turnover numbers for solid inorganic catalysts of organic reactions can be similar to those for enzymes, but they generally require temperatures in the hundreds of degrees Celsius to attain these rates and exhibit far less stereospecificity [13].

  Enzymes with the function of detoxifying reactive or harmful compounds often have very high turnover numbers. For example, catalase is an enzyme that degrades hydrogen peroxide to water, and its turnover number approaches $10^7\ \text{s}^{-1}$. The turnover number cannot exceed the collision rate between enzyme and substrate molecules, of course; hence, it is limited by the diffusive collision rate as considered in chapter 8 [19]. Electrostatic steering can, however, accelerate the collision rate beyond that obtainable with uncharged or suboptimally charged molecules [20].

- *Apparent second-order rate constant k_{cat}/K_m*

  At low substrate concentrations (i.e., $[S]_\circ \ll K_m$), the initial rate law becomes first order in substrate concentration and enzyme concentration, respectively, yielding the following limiting form:

$$v_\circ = \frac{k_{cat}}{K_m}[E]_\circ[S]_\circ \qquad (4.21)$$

Table 4.2 Characteristic enzyme turnover numbers k_{cat} and k_{cat}/K_m.

Enzyme	k_{cat} (s^{-1})	k_{cat}/K_m (M^{-1}s^{-1})
Catalase	$1 \times 10^7 - 4 \times 10^7$	$4 \times 10^7 - 4 \times 10^8$
Superoxide dismutase	1×10^6	$3 \times 10^9 - 7 \times 10^9$
Cytochrome p450 [22]	$0.01 - 10$	$1 \times 10^4 - 2 \times 10^6$
Fumarase	$8 \times 10^2 - 9 \times 10^2$	$4 \times 10^7 - 2 \times 10^8$
Triose phosphate isomerase	4×10^3	2×10^8
Chymotrypsin	$5 \times 10^{-2} - 2 \times 10^2$	$1 \times 10^4 - 3 \times 10^5$
Coagulation factor proteases [23]	$0.3 - 8$	$1 \times 10^6 - 1 \times 10^8$
p38 MAP kinase [24]	0.08	2×10^5
MAPK phosphatase [25]	$0.01 - 0.07$	$0.7 - 1,500$
Ribozymes in vivo [26]	$0.002 - 0.02$	$2 \times 10^2 - 3 \times 10^4$

Note: Sources where not specifically cited: Voet et al. [21] and Fersht [12].

This limiting case is approximately valid in many physiological situations, as can be seen by inspection of the left panel of figure 4.6. In this case, the expression k_{cat}/K_m can be considered to be an effective second-order rate constant for the reaction. Example parameter values are shown in table 4.2, and the distributions of values across thousands of enzymes are shown in figure 4.7.

> **Example 4-2** At time zero, a solution of 1 μM catalase is mixed with 1 mM H$_2$O$_2$. At what time will the hydrogen peroxide concentration have been reduced by six orders of magnitude?
>
> **Solution** From table 4.2, the value of K_m can be calculated by dividing the value in the second column by the value in the third column, indicating a very high $K_m \approx 1$ M. Because $[S] \ll K_m$, the reaction can be viewed as approximately first order in $[S]$, with a rate constant $\frac{k_{cat}}{K_m}[E]_o \approx (1 \times 10^8$ M^{-1}s$^{-1})(10^{-6}$ M$) = 10^2$ s^{-1}. For a first-order reaction, $[S]/[S]_o = e^{-kt}$, or, rearranging, $t = \frac{\ln[S]_o - \ln[S]}{k} = 0.14$ seconds! Clearly, catalase's role in detoxifying damaging intracellular hydrogen peroxide has led natural selection to drive the evolution of an extraordinarily fast enzyme.

Conditions for validity of the Michaelis-Menten rate form Under what condition is the QSSA valid, such that the Michaelis-Menten rate form can be applied? Let's compare an exact numerical solution to the approximate form (figure 4.8). At early times, the QSSA does not hold, because the concentration of enzyme-substrate complex has not yet reached its quasi-steady value. Also note that the complex concentration does in fact change with time once the QSSA has been reached,

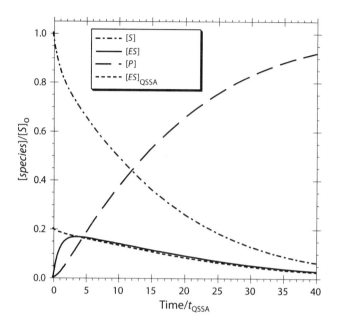

Figure 4.8 Numerical solutions of the dynamic material balance for each species in the reaction for parameter values where QSSA is approximately valid. *ES* rapidly approaches a value well described by the QSSA. The characteristic values $[ES]_{QSSA}$ and t_{QSSA} are given by equations 4.26 and 4.29, respectively.

even though we assumed that $d[ES]/dt = 0$ in the derivation. This seeming paradox results because the QSSA requires balancing adjustments in the complex concentration that are *much faster* than changes in substrate concentration—not that $[ES]$ literally remains constant. A useful analogy is a runner on a treadmill: if one resets the speed of the conveyor belt, the runner rapidly adjusts her foot speed to match it, while her position remains constant or only slowly changes. This becomes clear by examination of equation 4.9; $[ES]_{QSSA}$ cannot remain constant, because $[S]$ is a function of time.

Following an analysis by Segel [27], a simple criterion for the validity of the Michaelis-Menten rate form can be derived by first clearly framing the QSSA assumptions:

1. The time for $[ES]$ to reach its QSSA value is much shorter than the overall time for substrate to be depleted.

2. Negligible substrate is consumed during the approach of $[ES]$ to $[ES]_{QSSA}$. (Note that this assumption is standard practice when initial reaction rates are plotted versus $[S]_o$, as we have done in figure 4.4.)

How can we estimate the time period t_{QSSA} for $[ES]$ to approach $[ES]_{QSSA}$? If the second assumption is satisfied (which we will enforce for QSSA validity in any case), so that $[S] \approx [S]_o$ for time $t < t_{QSSA}$, a linear differential equation for $[ES]$ results:

$$\frac{d[ES]}{dt} = k_1[E][S]_o - (k_{-1} + k_{cat})[ES] \qquad (4.22)$$

$$= k_1([E]_o - [ES])[S]_o - (k_{-1} + k_{cat})[ES] \qquad (4.23)$$

$$= k_1[E]_o[S]_o - k_1(K_m + [S]_o)[ES] \qquad (4.24)$$

with K_m as already defined. The solution to this linear ODE is

$$[ES] = [ES]_{QSSA}(1 - e^{-\lambda t}) \qquad (4.25)$$

where

$$[ES]_{QSSA} \equiv \frac{[E]_o[S]_o}{K_m + [S]_o} \qquad (4.26)$$

and

$$\lambda \equiv k_1(K_m + [S]_o) \qquad (4.27)$$

Because equation 4.25 shows an exponential approach to $[ES]_{QSSA}$, let's define the time for the initial fast transient as

$$t_{QSSA} = \frac{1}{\lambda} \qquad (4.28)$$

$$= \frac{1}{k_1(K_m + [S]_o)} \qquad (4.29)$$

Now, let's obtain an estimate of the overall time for substrate depletion, $t_{[S]}$, to compare with t_{QSSA} in order to determine whether assumption 1 is satisfied, as follows:

$$t_{[S]} \approx \frac{\text{Total change in } [S]}{\text{Maximum rate of depletion}} \qquad (4.30)$$

Because the total change in $[S]$ from the initial amount to zero is $[S]_o$, and the maximum rate of depletion is the reaction velocity when $[ES]$ initially reaches quasi-steady state $v = k_{cat}[ES]_{QSSA}$, we have

$$t_{[S]} = \frac{[S]_\circ}{k_{\text{cat}}[ES]_{\text{QSSA}}} \tag{4.31}$$

$$= \frac{[S]_\circ}{k_{\text{cat}} \frac{[E]_\circ [S]_\circ}{K_{\text{m}}+[S]_\circ}} \tag{4.32}$$

$$= \frac{K_{\text{m}} + [S]_\circ}{k_{\text{cat}}[E]_\circ} \tag{4.33}$$

Assumption 1 can now be mathematically stated as

$$t_{\text{QSSA}} \ll t_{[S]} \tag{4.34}$$

or

$$\frac{1}{k_1(K_{\text{m}}+[S]_\circ)} \ll \frac{K_{\text{m}}+[S]_\circ}{k_{\text{cat}}[E]_\circ} \tag{4.35}$$

which, for reasons that will become clear momentarily, can be rearranged as follows, to provide the first criterion for the validity of QSSA:

$$\frac{[E]_\circ}{K_{\text{m}}+[S]_\circ} \ll \left(1 + \frac{k_{-1}}{k_{\text{cat}}}\right)\left(1 + \frac{[S]_\circ}{K_{\text{m}}}\right) \tag{4.36}$$

Let's return now to our second assumption, that substrate is not significantly depleted during the fast transient approach to QSSA. Defining $\Delta[S]$ as the drop in $[S]$ for $t < t_{\text{QSSA}}$, the fractional drop in substrate during the fast transient can be approximated as

$$\frac{\Delta[S]}{[S]_\circ} \approx \frac{\left(\frac{-d[S]}{dt}\right)_{\max} \cdot t_{\text{QSSA}}}{[S]_\circ} \tag{4.37}$$

$$= \frac{k_1[E]_\circ [S]_\circ \cdot \frac{1}{k_1(K_{\text{m}}+[S]_\circ)}}{[S]_\circ} \tag{4.38}$$

$$= \frac{[E]_\circ}{K_{\text{m}}+[S]_\circ} \tag{4.39}$$

So the condition that $\Delta[S]/[S]_\circ \ll 1$ can be restated as

$$\frac{[E]_\circ}{K_{\text{m}}+[S]_\circ} \ll 1 \tag{4.40}$$

Because it would be impossible to satisfy equation 4.40 without simultaneously satisfying equation 4.36, equation 4.40 is a sufficient criterion for validity of the Michaelis-Menten approximate rate form. (It should be noted that the often-invoked condition that $[E]_\circ/[S]_\circ \ll 1$ is a sufficient, but not necessary, criterion that is a limiting case of equation 4.40 when $[S]_\circ \gg K_m$.)

To numerically probe the validity of this criterion, let's first nondimensionalize the equations so that the concentration variables and time are dimensionless and less than one. The following equations are obtained:

$$\frac{ds}{d\tau} = \varepsilon \left(\frac{\sigma}{\sigma+1} c \cdot s + \frac{\kappa}{\kappa+1} \frac{1}{\sigma+1} c - s \right) \tag{4.41}$$

$$\frac{dc}{d\tau} = s - \frac{\sigma}{\sigma+1} c \cdot s - \frac{1}{\sigma+1} c \tag{4.42}$$

$$\frac{dp}{d\tau} = \varepsilon \frac{1}{\sigma+1} \frac{1}{\kappa+1} c \tag{4.43}$$

where the dimensionless variables are defined as follows:

$$s \equiv \frac{[S]}{[S]_\circ}, \quad c \equiv \frac{[ES]}{[ES]_{QSSA,\circ}}, \quad p \equiv \frac{[P]}{[S]_\circ}, \quad \tau \equiv \frac{t}{t_{QSSA}} \tag{4.44}$$

and the following three dimensionless parameters emerge from the nondimensionalization:

$$\varepsilon \equiv \frac{[E]_\circ}{K_m + [S]_\circ}, \quad \sigma \equiv \frac{[S]_\circ}{K_m}, \quad \kappa \equiv \frac{k_{-1}}{k_{cat}} \tag{4.45}$$

The dimensionless form of the validity criterion equation 4.40 is now $\varepsilon \ll 1$. The dimensionless form of the Michaelis-Menten rate form is $v/v_{max} = \frac{\sigma}{\sigma+1}$. To estimate the degree of error in the QSSA, the dimensionless ODEs were solved for varying values of the parameters ε and σ (the parameter κ will be considered shortly). The results are shown in figure 4.9. The Michaelis-Menten prediction of the enzymatic reaction rate (denoted by QSSA) is always greater than the actual numerical value. This is because $[ES]$ does not instantaneously reach its QSSA value, and so the observed reaction rate is slower than predicted by the QSSA. As predicted by equation 4.40, the QSSA provides a good approximation for $\varepsilon = 0.1$. Given how widespread the use of this rate form is, let's examine its error dependence on parameters more closely.

The QSSA error as a function of the three dimensionless parameters ε (the key parameter for QSSA validity, according to equation 4.40), σ (the dimensionless substrate concentration), and κ (the ratio of catalytic and dissociation rate constants, which speaks to the efficiency of the reaction) is plotted in figure 4.10. As expected,

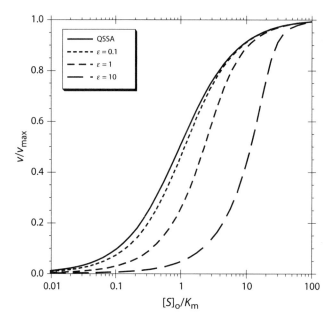

Figure 4.9 Numerical solution of equation 4.41 and comparison to the QSSA prediction (i.e., the Michaelis-Menten rate form). To mimic experimental rate versus substrate concentration curves, the value for the parameter ε was varied as a function of $[S]_o$ by the relationships in equation 4.45 so as to maintain constant $[E]_o$. The ε value in the legend is that at $[S]_o/K_m = 1$. The initial reaction rate, v, was determined by the maximum value of $dp/d\tau$ at its inflection point, where c attains its maximum value.

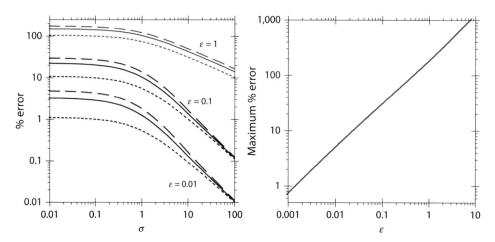

Figure 4.10 Errors in QSSA as a function of the parameters ε, σ, and κ. In the left panel, the percentage error in the QSSA is plotted versus σ, with ε and κ as parameters. For each of three values of ε, three curves are plotted for different values of κ (long dashes, $\kappa = 0.01$; solid line, $\kappa = 1$; short dashes, $\kappa = 100$). The right panel indicates the maximum percentage error as a function of ε, in the limit of very low κ and very low σ.

ε exerts the greatest impact on QSSA error. By contrast, κ has little effect on the error of the approximation. The maximum error for a given value of ε is found at low values of κ and σ, and these values are plotted in the right panel of figure 4.10. A useful take-home message is that for $\varepsilon < 0.1$, the maximum error due to the quasi-steady-state approximation is less than 30%, which is likely to be comparable to errors in parameter estimates and experimental rate determinations.

4.2.2 Bisubstrate Kinetics

Although we have only considered enzyme reactions with a single substrate up to this point, most enzymatic reactions involve at least two substrates. With the exception of the isomerase enzyme class, most enzymes catalyze the covalent transfer of a group from one molecule to another. For example, hydrolases transfer backbone bonds of biopolymers to water; kinases transfer phosphate groups from nucleosides to proteins; and oxidoreductases transfer electrons between substrates, one of which is often a cofactor, such as NADH or coenzyme A. There are a few canonical bisubstrate enzyme mechanisms commonly observed, distinguished by the order of binding and reaction events.

A general nomenclature proposed by Cleland [28] has been broadly adopted for multiple-substrate enzymatic mechanisms. The terms *Uni* (one), *Bi* (two), *Ter* (three), and *Quad* (four) are used to denote the number of substrates and products. For example, a Uni Bi reaction would have one substrate and two products. Detailed treatments of a bewildering variety of alternative multisubstrate mechanisms (e.g., the Bi Uni Uni Bi Ping Pong Ter Ter mechanism) are available elsewhere [29, 30] but are beyond the scope of this textbook.

Sequential mechanisms

- *Ordered Bi Bi*

  In this mechanism, the two substrates (A and B) bind to the enzyme in only one order (i.e., A before B), followed by reaction, then ordered dissociation of the two products (i.e., P, then Q). This mechanism is shown in figure 4.11, using Cleland's graphic nomenclature for representing such reactions.

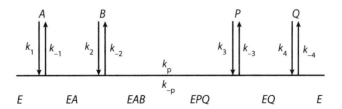

Figure 4.11 Cleland notation for the Ordered Bi Bi enzyme mechanism. Forward binding, reaction, and release steps are represented from left to right relative to a horizontal line representing the enzyme and complexes with it.

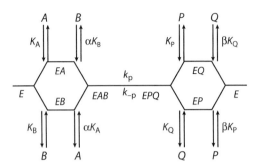

Figure 4.12 Random Bi Bi enzyme mechanism.

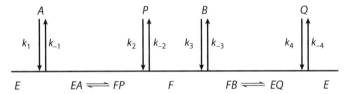

Figure 4.13 Ping Pong Bi Bi reaction mechanism.

- *Random Bi Bi*

    In this mechanism (figure 4.12), the two substrates (A and B) bind to the enzyme in random order. That is, one molecule of A and one molecule of B each bind before reaction can occur, but they can bind in either order. Reaction then occurs, and the pair of products P and Q dissociate in either order.

- *Ping Pong Bi Bi (double displacement reaction)*

    In this mechanism (figure 4.13), two reactions occur in strict sequential order, with a modified enzyme intermediate existing between the two. That is, the first substrate binds, reaction occurs, and the first product dissociates (leaving modified enzyme); then the second substrate binds, reacts, and the second product dissociates (regenerating the enzyme in its original form). Note that in this example, the original form of the enzyme is alternatively consumed and created, but at the end of a complete enzyme cycle, there is no net change in the enzyme.

General bisubstrate rate forms Dalziel [31] first showed that the initial forward-rate laws for many bisubstrate mechanisms were of the following form:

$$\frac{v_{\max}}{v} = \Phi_\circ + \frac{\Phi_A}{[A]} + \frac{\Phi_B}{[B]} + \frac{\Phi_{AB}}{[A][B]} \qquad (4.46)$$

Enzyme Kinetics

Table 4.3 Constants in equation 4.46 for different bisubstrate mechanisms.

Mechanism	$\Phi_\circ$	Φ_A	Φ_B	Φ_{AB}
Ordered Bi Bi	1	$\dfrac{k_3 k_4 k_p}{k_1(k_3 k_4 + k_3 k_p + k_4 k_p + k_4 k_{-p})}$	$\dfrac{k_4(k_{-2} k_3 + k_{-2} k_{-p} + k_3 k_p)}{k_2(k_3 k_4 + k_3 k_p + k_4 k_p + k_4 k_{-p})}$	$\dfrac{k_{-1}}{k_1}\Phi_B$
Random Bi Bi	1	αK_A	αK_B	$\alpha K_A K_B$
Ping Pong Bi Bi	1	$\dfrac{k_{-4}(k_{-1}+k_{-2})}{k_1(k_{-2}+k_{-4})}$	$\dfrac{k_{-2}(k_{-3}+k_{-4})}{k_3(k_{-2}+k_{-4})}$	0

where the individual constants Φ_i vary for the different mechanisms as shown in table 4.3. These equations were derived for initial rates under conditions with zero product initially.

In many bisubstrate reactions, one of the substrates is present in excess, for example, water (55 M), ATP (1–5 mM in vivo), NADH (≈ 0.1 mM), or coenzyme A (≈ 0.1 mM). In such cases, if B is considered to be the substrate present in excess, then equation 4.46 obligingly rearranges to the familiar Michaelis-Menten rate form:

$$v = \frac{v_{\max,\text{app}}[A]}{K_{m,\text{app}} + [A]} \quad (4.47)$$

where

$$v_{\max,\text{app}} \equiv \frac{k_{\text{cat}}[E]_\circ[B]}{\Phi_\circ[B] + \Phi_B} \quad (4.48)$$

and

$$K_{m,\text{app}} = \frac{\Phi_A[B] + \Phi_{AB}}{\Phi_\circ[B] + \Phi_B} \quad (4.49)$$

One common application of this form is to model intracellular protein signaling cascades, in which kinases transfer a phosphate group from ATP to a protein substrate. Formally, these are bisubstrate reactions with a pair of substrates (a protein and ATP) and a pair of products (a phosphorylated protein and ADP). Many models successfully simplify the kinetics to a simple Michaelis-Menten reaction, involving only the protein substrate and phosphorylated product, through the use of the apparent rate constant formalism described here. Moreover, as noted several times previously, various very different underlying molecular mechanisms are all capable of producing a given rate form; this goes far toward explaining why the hyperbolic Michaelis-Menten form is so broadly applicable (and utilized).

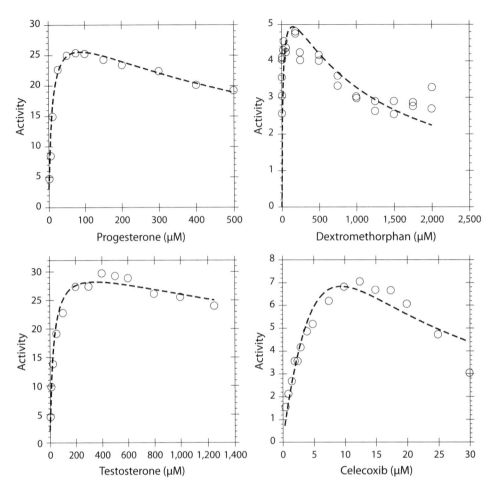

Figure 4.14 Examples of substrate inhibition. The liver metabolizes many drugs and endogenous organic molecules with cytochrome P450 enzymes. At high concentrations, many of these substances bind to the enzyme and lower its activity. Dashed lines are nonlinear least squares fits to equation 4.52 [22].

4.2.3 Substrate Inhibition

At high substrate concentrations, an additional substrate molecule can sometimes bind to the enzyme-substrate complex and thereby reduce the rate of progression to the product. This is called substrate inhibition, and examples are shown in figure 4.14.

A simplified model of substrate inhibition can be derived by assuming that the doubly bound enzyme is inactive, and that the second, inhibitory substrate molecule binds after the first:

$$E + S \underset{}{\overset{K_s}{\rightleftharpoons}} ES \xrightarrow{k_{cat}} E + P \qquad (4.50)$$

$$ES + S \underset{}{\overset{K_I}{\rightleftharpoons}} ES_2 \qquad (4.51)$$

If the substrate-enzyme binding reactions are assumed to rapidly reach equilibrium, the following dependence of the initial reaction rate on substrate concentration can be derived:

$$v = \frac{k_{cat}[E]_o[S]}{K_s + [S](1 + \frac{[S]}{K_I})} \qquad (4.52)$$

By setting the derivative $dv/d[S] = 0$, one can determine that the maximum rate is observed at $[S]_{max} = \sqrt{K_s K_I}$.

4.2.4 Cooperative Enzymes

Multisubunit enzymes often energetically communicate among their multiple active sites via allostery. When substrate binding at one site raises affinity at the other sites, this leads to positive cooperativity. Conversely, when binding at one site lowers the affinity at other sites, negatively cooperative behavior is observed. As discussed in chapter 3, a limiting case for perfect positive cooperativity can be described with a Hill binding isotherm for the enzyme-substrate complex:

$$[ES_n] = \frac{[E]_o[S]^n}{K_m^n + [S]^n} \qquad (4.53)$$

where, analogous to the analysis in chapter 3, the exponent n on K_m is included so that K_m has units of concentration and equals the substrate concentration at which $[ES_n]$ is half-maximal. Assuming steady state of substrate-enzyme binding followed by rate-limiting reaction leads to the following empirical rate law for a cooperative enzyme:

$$v = \frac{v_{max}[S]^n}{K_m^n + [S]^n} \qquad (4.54)$$

The hallmark of positive cooperativity is a steeper increase in activity versus substrate concentration than is predicted by the Michaelis-Menten rate law. However, as will be discussed in chapter 6, multiple mechanisms exist that can create switch-like responses to inputs. It should always be remembered that the exponent n in equation 4.54 is physically constrained to be less than or equal to the valency of a multimeric enzyme, and if treated as a completely free parameter to be curve-fit to switchlike responses, it may lead to values that are not physically meaningful.

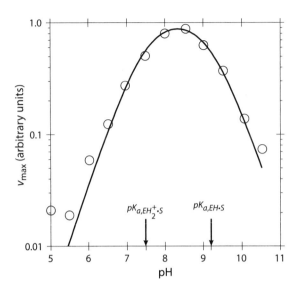

Figure 4.15 Most enzymes have an optimum activity with respect to pH. The data points shown here are for the enzyme fumarase [34]. The solid curve is a nonlinear least squares curve fit to equation 4.63, yielding $pK_{a,EH_2^+ \cdot S} = 7.5$ and $pK_{a,EH \cdot S} = 9.2$.

Two different structural theories of cooperativity have been developed: the Monod-Wyman-Changeux (MWC) model [32], and the Koshland-Némethy-Filmer (KNF) theory [33]. In MWC theory, the multimeric enzyme exists in two states, a *relaxed* and a *tensed* state that differ in their affinities for substrate. The tensed-relaxed equilibrium precedes substrate binding and is not directly affected by the substrate. By contrast, the KNF theory proposes that substrate binding directly favors formation of a higher-affinity state via induced fit. MWC theory requires fewer fitted parameters but cannot explain negative cooperativity.

4.2.5 pH Effects

Enzymes are often found to exhibit bell-shaped curves of activity versus pH, as shown in figure 4.15. The shape of such a curve indicates that the fully active enzyme must have a certain ionization state: As pH changes and particular side chains become protonated or deprotonated, the fraction of enzyme molecules in the fully active conformation goes through an optimum. A simplified model for this behavior can be constructed by assuming that there is one critical ionizable group with pK_a below the optimum pH, and one critical ionizable group with pK_a above the optimum pH. Such a kinetic scheme can be represented as follows [34]:

$$E^- + 2H^+ \xrightleftharpoons{pK_{a,EH}} EH + H^+ \xrightleftharpoons{pK_{a,EH_2^+}} EH_2^+ \qquad (4.55)$$

$$E^- \cdot S + 2H^+ \xrightleftharpoons{pK_{a,EH\cdot S}} EH \cdot S + H^+ \xrightleftharpoons{pK_{a,EH_2^+\cdot S}} EH_2^+ \cdot S \quad (4.56)$$

$$EH + S \underset{k_{-1}}{\overset{k_1}{\rightleftharpoons}} EHS \xrightarrow{k_{cat}} EH + P \quad (4.57)$$

where the ionization equilibria are as follows:

$$K_{a,EH} \equiv \frac{[H^+][E^-]}{[EH]} \quad (4.58)$$

$$K_{a,EH_2^+} \equiv \frac{[H^+][EH]}{[EH_2^+]} \quad (4.59)$$

$$K_{a,EH\cdot S} \equiv \frac{[H^+][E^- \cdot S]}{[EH \cdot S]} \quad (4.60)$$

$$K_{a,EH_2^+\cdot S} \equiv \frac{[H^+][EH \cdot S]}{[EH_2^+ \cdot S]} \quad (4.61)$$

and $pK \equiv -\log_{10} K$.

Assuming that the ionization reactions rapidly reach equilibrium and that the enzyme-substrate complex is at quasi-steady state, one can derive the following rate form for the initial rate:

$$v = \frac{v_{max}^*[S]}{K_m^* + [S]} \quad (4.62)$$

where

$$v_{max}^* = \frac{v_{max}}{1 + \frac{[H^+]}{K_{a,EH_2^+\cdot S}} + \frac{K_{a,EH\cdot S}}{[H^+]}} \quad (4.63)$$

and

$$K_m^* \equiv K_m \frac{1 + \frac{[H^+]}{K_{a,EH_2^+}} + \frac{K_{a,EH}}{[H^+]}}{1 + \frac{[H^+]}{K_{a,EH_2^+\cdot S}} + \frac{K_{a,EH\cdot S}}{[H^+]}} \quad (4.64)$$

with v_{max} and K_m defined as in the previous Michaelis-Menten derivation.

Table 4.4 pK_a for ionizable groups in proteins [35].

Group	pK_a
Amino terminus	6.8–8.0
Carboxyl terminus	3.5–4.3
Aspartate	3.9–4.0
Glutamate	4.3–4.5
Arginine	12
Lysine	10.4–11.1
Histidine	6.0–7.0
Cysteine	9.0–9.5
Tyrosine	10.0–10.3

One can straightforwardly demonstrate that equation 4.63 exhibits a maximum by setting the derivative $dv^*_{max}/d[H^+] = 0$ and solving for the $[H^+]_{opt}$:

$$pH_{opt} = \frac{1}{2}\left(pK_{a,EH_2^+ \cdot S} + pK_{a,EH \cdot S}\right) \quad (4.65)$$

Typical values for the ionization equilibrium constant pK_a for protein side chains are given in table 4.4. Of course, pK_a values well outside these ranges can be observed, due to local electrostatic effects in particular proteins.

4.2.6 Temperature Effects

Enzyme activity is dependent on temperature. Within a narrow range, increased temperature results in more rapid enzyme activity in accordance with the Arrhenius equation:

$$k_{cat} = Ae^{-\frac{E_a}{RT}} \quad (4.66)$$

where A is a pre-exponential factor, and E_a is the activation energy. Yet at slightly higher temperatures, often only $\approx 50°C$, enzyme activity is reduced because of enzyme denaturation (figure 4.16).

Considering an equilibrium between the active and inactive states, the maximal reaction rate can be written as

$$v_{max} = \frac{\beta T e^{-\frac{E_a}{RT}}}{1 + e^{\frac{\Delta S_d}{R}} e^{\frac{-\Delta H_d}{RT}}} \quad (4.67)$$

where β is a proportionality constant and ΔS_d and ΔH_d are the entropy and enthalpy of denaturation, respectively.

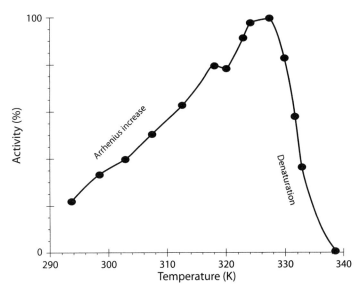

Figure 4.16 Rate of hydrolysis of o-nitrophenyl β-D-galactopyranoside by β-galactosidase as a function of temperature. Data from reference 36.

4.3 Reversible Enzyme Inhibition

Enzymes are often regulated in vivo by inhibitors that bind to them and either block substrate binding (competitive mechanism) or decrease the reactivity of the enzyme-substrate complex (noncompetitive and uncompetitive mechanisms). Many drugs are being developed as rational competitive inhibitors against viral proteases and signaling kinases in tumor cells, by designing organic molecules that take the place of natural substrates in active sites. High-throughput screening of natural and synthetic compound pools also can identify noncompetitive inhibitors that influence reactivity in unpredicted ways.

4.3.1 Reversible Inhibitors

A common class of enzyme inhibitors binds reversibly to the enzyme with significantly lower affinity than the substrate-enzyme interaction. Such inhibitors can be classified according to whether they inhibit the substrate binding affinity or slow the progression of the reaction of the enzyme-substrate complex to form product.

Competitive inhibition The mechanism of a competitive inhibitor is shown schematically in figure 4.17. An example of a designed competitive inhibitor is shown in figure 4.18, where a molecule has been designed to block access of the HIV protease to its natural substrates. The mechanism of figure 4.17 translates to

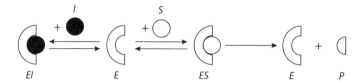

Figure 4.17 Schematic mechanism of a competitive inhibitor, which binds to an enyzme's active site and sterically blocks substrate access.

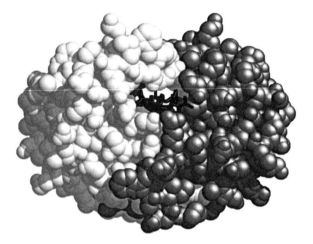

Figure 4.18 Crystal structure of the HIV protease dimer with competitive inhibitor ritonavir (dark stick figure), which binds in the active site groove and thereby sterically blocks access to the protease's polypeptide substrates (PDB ID 2B60).

the following equations:

$$E + S \underset{k_{-1}}{\overset{k_1}{\rightleftharpoons}} ES \xrightarrow{k_{cat}} E + P$$
$$+$$
$$I$$
$$\updownarrow K_I$$
$$EI$$

We assume rapid equilibrium for inhibitor binding:

$$K_I = \frac{[E][I]}{[EI]} \quad (4.68)$$

For the non-inhibitor-bound species, we have the same rate equations as before (equations 4.2–4.4):

$$\frac{d[S]}{dt} = -k_1[E][S] + k_{-1}[ES]$$

$$\frac{d[ES]}{dt} = k_1[E][S] - (k_{-1} + k_{cat})[ES]$$

$$\frac{d[P]}{dt} = k_{cat}[ES]$$

Using the QSSA for the concentration of enzyme-substrate complex $[ES]$ and the new material balance for total enzyme concentration ($[E]_\circ = [E] + [ES] + [EI]$), the following relationship is derived:

$$v = \frac{k_{cat}[E]_\circ [S]}{K_m \left(1 + \frac{[I]}{K_I}\right) + [S]} \qquad (4.69)$$

where K_m is as defined previously, and K_I is the equilibrium dissociation constant for the enzyme-inhibitor complex. It should be noted that a competitive inhibitor increases the apparent value of the Michaelis constant $K_{m,app} = K_m(1 + [I]/K_I) > K_m$, without altering the apparent maximum reaction velocity $v_{max,app} = k_{cat}[E]_\circ = v_{max}$. This effect is shown in figure 4.19.

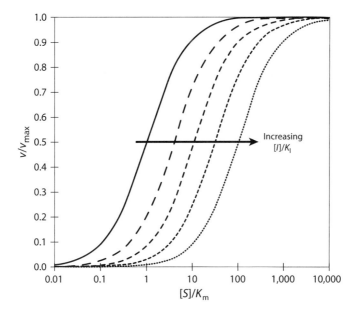

Figure 4.19 The effect of increasing concentration of a competitive inhibitor on the relationship between initial reaction rate v and substrate concentration $[S]$. From left to right in the plot, the $[I]/K_I$ values are 0, 3, 10, 30, and 100. The apparent value of v_{max} is unaltered, while the apparent value of K_m increases.

Noncompetitive inhibition A noncompetitive inhibitor binds to an enzyme at a site distinct from the active site but perturbs the enzyme's structure or dynamics in a way that prevents catalysis when substrate is bound. An ideal noncompetitive inhibitor binds with the same affinity whether substrate is bound or not. This mechanism is shown schematically in figure 4.20. An example of a pharmacological noncompetitive inhibitor is shown in figure 4.21, where an HIV reverse transcriptase inhibitor binds at a site distinct from the active site that inhibits catalysis.

The mechanism of figure 4.20 translates to the following equations:

$$E + S \underset{k_{-1}}{\overset{k_1}{\rightleftharpoons}} ES \overset{k_{cat}}{\rightarrow} E + P$$

$$+ \qquad\qquad +$$
$$I \qquad\qquad I$$
$$\updownarrow K_I \qquad\qquad \updownarrow K_I$$

$$EI + S \underset{k_{-1}}{\overset{k_1}{\rightleftharpoons}} EIS$$

Note that although we assumed only that the inhibitor binds to the enzyme with the same affinity whether substrate is bound or not (ideal noncompetitive inhibitor), the thermodynamic cycle above also requires that parallel affinity for

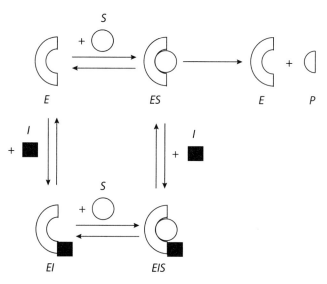

Figure 4.20 Schematic mechanism of a noncompetitive inhibitor, which binds to an enzyme at a site distinct from the active site and allosterically inhibits reaction of bound substrate.

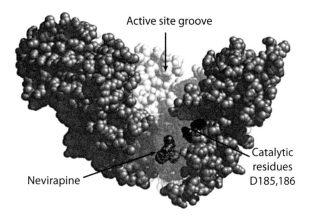

Figure 4.21 Crystal structure of the HIV-1 reverse transcriptase in complex with nevirapine, a noncompetitive inhibitor. The transcriptase residues surrounding the inhibitor were rendered as a dot surface to make the inhibitor visible. The inhibitor binds near the catalytically active residues but does not sterically block access of the substrate to them (PDB 1VRT) [37].

binding the substrate is the same whether the inhibitor is bound or not. Here we characterize parallel affinity by using the same forward and reverse rate constants for binding substrate in both cases, but in general only their ratio need be conserved.

We assume rapid equilibrium for inhibitor binding:

$$K_I = \frac{[E][I]}{[EI]} \tag{4.70}$$

$$K_I = \frac{[ES][I]}{[EIS]} \tag{4.71}$$

For the non-inhibitor-bound species, we apply a similar set of rate equations as before, but with the *ES* balance updated to include inhibitor interaction:

$$\frac{d[S]}{dt} = -k_1[E][S] + k_{-1}[ES]$$

$$\frac{d[ES]}{dt} = k_1[E][S] - (k_{-1} + k_{cat})[ES] - k_i[ES][I] + k_{-i}[EIS]$$

$$\frac{d[P]}{dt} = k_{cat}[ES]$$

Note that the binding and unbinding of inhibitor (the last two terms in the *ES* balance, which introduce the association and dissociation rate constants, k_i and k_{-i}, respectively) cancel each other, because they are treated as an infinitely fast equilibrium: $k_{-i}/k_i = K_I$, subject to the constraint represented by equation 4.71. For a noncompetitive inhibitor, the following rate expression is derived using QSSA for the enzyme-substrate complex [ES], and the updated material balance on enzyme $[E]_o = [E] + [ES] + [EI] + [EIS]$:

$$v = \frac{\left(\frac{k_{cat}}{1+\frac{[I]}{K_I}}\right)[E]_o[S]}{K_m + [S]} \quad (4.72)$$

In this case, the presence of a noncompetitive inhibitor reduces the apparent value of the maximum reaction velocity $v_{max,app} = \frac{k_{cat}}{1+[I]/K_I}[E]_o < v_{max}$, but it does not alter the Michaelis constant $K_{m,app} = K_m$. The effect of a noncompetitive inhibitor on the reaction rate v as a function of $[S]$ is shown in figure 4.22.

Uncompetitive inhibition An uncompetitive inhibitor binds the ES complex and blocks its reaction, as shown in figure 4.23.

The mechanism of figure 4.23 translates to the following equations:

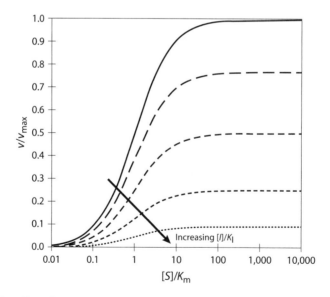

Figure 4.22 The effect of increasing concentration of a noncompetitive inhibitor on the relationship between initial reaction rate v and substrate concentration $[S]$. From top to bottom in the plot, the $[I]/K_I$ values are 0, 0.3, 1, 3, and 10. The apparent value of v_{max} is decreased, while the apparent value of K_m is unaltered.

Enzyme Kinetics

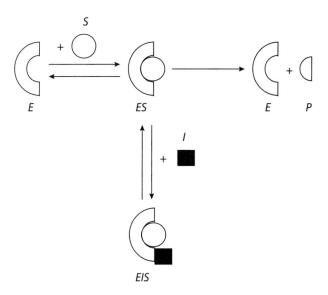

Figure 4.23 Schematic mechanism of an uncompetitive inhibitor, which binds to an enzyme at a site distant from the active site and allosterically inhibits reaction of bound substrate.

$$E + S \underset{k_{-1}}{\overset{k_1}{\rightleftharpoons}} ES \overset{k_{cat}}{\longrightarrow} E + P$$
$$+$$
$$I$$
$$\updownarrow K_I$$
$$EIS$$

We again assume rapid equilibrium for inhibitor binding:

$$K_I = \frac{[ES][I]}{[EIS]} \tag{4.73}$$

And we use the same rate equations as in noncompetitive inhibition for the non-inhibitor-bound species:

$$\frac{d[S]}{dt} = -k_1[E][S] + k_{-1}[ES]$$

$$\frac{d[ES]}{dt} = k_1[E][S] - (k_{-1} + k_{cat})[ES] - k_i[ES][I] + k_{-i}[EIS]$$

$$\frac{d[P]}{dt} = k_{cat}[ES]$$

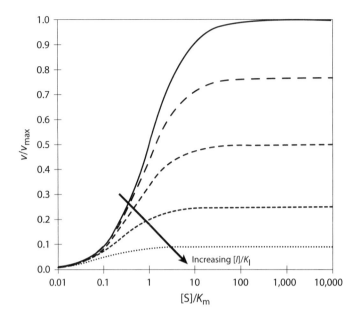

Figure 4.24 The effect of increasing concentration of an uncompetitive inhibitor on the relationship between initial reaction rate v and substrate concentration $[S]$. From top to bottom in the plot, the $[I]/K_I$ values are 0, 0.3, 1, 3, and 10. The apparent values of both v_{max} and K_m are decreased.

Using the QSSA for the concentration of enzyme-substrate complex $[ES]$, and the material balance for total enzyme concentration ($[E]_\circ = [E] + [ES] + [EIS]$), the following relationship is derived:

$$v = \frac{\left(\dfrac{k_{cat}}{1+\frac{[I]}{K_I}}\right)[E]_\circ [S]}{\left(\dfrac{K_m}{1+\frac{[I]}{K_I}}\right) + [S]} \quad (4.74)$$

where K_m is as defined previously, and K_I is the equilibrium dissociation constant for the enzyme-inhibitor complex. An uncompetitive inhibitor decreases the apparent values of both the Michaelis constant ($K_{m,app} = \frac{K_m}{1+[I]/K_I}$) and the maximum reaction velocity ($v_{max,app} = \frac{k_{cat}\,[E]_\circ}{1+[I]/K_I}$) by the same amount. This effect is shown in figure 4.24.

The distinction between noncompetitive and uncompetitive inhibitors is subtle when examining v versus $[S]$ plots, such as in figures 4.22 and 4.24. This makes sense, given the subtle difference between their mechanisms (figures 4.20 and 4.23);

the sole distinction is whether the inhibitor binds the enzyme in the absence of substrate. At high substrate concentrations, this difference obviously becomes moot (because few enzyme molecules are not bound to substrate).

> **Example 4-3** An enzyme inhibitor alters the apparent K_m without affecting v_{max}. Can a single enzyme molecule simultaneously bind both the inhibitor and the substrate?
>
> **Solution** A competitive inhibitor has the described effect on the measured parameters, and its mechanism is to bind at the enzyme active site and thereby block access to the substrate. Therefore, the inhibitor and substrate are not expected to simultaneously bind the enzyme in this case.

4.3.2 Pharmacological Inhibition

Drugs act by binding to a particular molecular target and altering its biological function. If the drug directly stimulates a biological response from its receptor, it is called an agonist. Drugs that act by inhibiting their receptors, however, are called antagonists. Many drug targets are enzymes; significant examples include cell-surface and intracellular kinases, viral processing proteases, and viral reverse transcriptases. Such antagonist drugs are enzyme inhibitors, and the kinetics of their action can be described by the steady-state relationships derived earlier in this chapter.

The *potency* of a drug is expressed as the concentration required to achieve a certain fraction of its maximum effect. For agonists, EC_{50} is the effective concentration that elicits 50% of the maximal response, and ED_{50} is the effective dose that achieves the same thing. (ED_{50} is used when the dose—i.e., mass of drug administered per mass of subject organism—is known but not the precise concentration at the site of action in the whole organism.) For antagonists, the IC_{50} is defined as the drug concentration at which an agonist's effect is reduced to 50% of its value in the absence of the antagonist.

If an antagonist is a competitive enzyme inhibitor, then its effect on receptor activity can be described by equation 4.69. Rewritten with $[A]$ denoting the agonist, $[I]$ denoting the antagonist, and p denoting the proportion of maximal receptor effect (v/v_{max}), as well as substituting EC_{50} for the Michaelis-Menten constant for the uninhibited agonist, one obtains

$$p = \frac{[A]}{EC_{50}\left(1 + \frac{[I]}{K_I}\right) + [A]} \qquad (4.75)$$

In the pharmacology literature, equation 4.75 is known as the Gaddum equation.

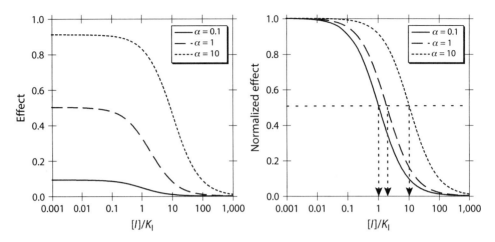

Figure 4.25 Illustration that $IC_{50} \neq K_I$ in general. In the left panel are plotted antagonism curves at three different agonist concentrations, $\alpha = [A]/EC_{50} = 0.1$, 1, and 10. In the right panel, the effect is normalized to that obtained in the absence of inhibitor. By inspection, the IC_{50} values (marked with arrows) shift to higher values with increasing agonist concentration, including 11 times K_I when $[A]/EC_{50} = 10$.

The action of a particular antagonist is described by the inhibition constant K_I. However, experiments are often performed by varying the antagonist concentration $[I]$ at a fixed concentration of agonist $[A]$ and identifying the antagonist concentration for half-maximal activity, IC_{50}.

IC_{50} and K_I are not in fact equal, however. This is because the half-maximal effect is a function of both the agonist and antagonist concentrations, and so at high agonist concentration, it outcompetes the antagonist, requiring higher antagonist concentrations to drop the effect to 50%. This phenomenon is illustrated schematically in figure 4.25.

The relationship between IC_{50} and K_I can be derived as follows. In the absence of antagonist, the effect p is given by

$$p = \frac{[A]}{EC_{50} + [A]} \tag{4.76}$$

When antagonist $[I] = IC_{50}$, the observed activity is halved. Setting half of equation 4.76 equal to equation 4.75 at $[I] = IC_{50}$, we have

$$\frac{1}{2} \cdot \frac{[A]}{[A] + EC_{50}} = \frac{[A]}{EC_{50}\left(1 + \frac{IC_{50}}{K_I}\right) + [A]} \tag{4.77}$$

which can be rearranged to give

$$IC_{50} = K_I \left(1 + \frac{[A]}{EC_{50}}\right) \qquad (4.78)$$

Equation 4.78 is called the Cheng-Prusoff equation. It should be remembered that this relationship is only valid in the case of a competitive antagonist; for an uncompetitive inhibitor antagonist, the following relationship can be derived:

$$IC_{50} = K_I \left(1 + \frac{EC_{50}}{[A]}\right) \qquad (4.79)$$

And for a noncompetitive inhibitor, it turns out that $IC_{50} = K_I$.

It is clear that reporting a value for IC_{50} without also specifying the agonist concentration $[A]$ gives an incomplete picture of the system. By contrast, the inhibition constant K_I is independent of agonist concentration.

> **Example 4-4** A biotechnology company enters into a contract to find an antibody that binds a cell-surface receptor and blocks its interaction with a soluble ligand. A term in the contract requires the company to be paid if the antibody meets the milestone that IC_{50} for receptor antagonism is below 100 pM. Why is this an ambiguous criterion?
>
> **Solution** As per equation 4.78, the observed IC_{50} is highly assay dependent, unlike the equilibrium binding constant K_I. For example, if the soluble ligand has potency $EC_{50} = 100$ pM and the ligand is present at 1 nM in the assay, then the observed $IC_{50} = K_I \left(1 + \frac{1.0}{0.1}\right)$, or $IC_{50} = 11 K_I$! One can imagine the awkward discussion between the discoverer of an antibody with a binding constant $K_I = 50$ pM and the person performing the bioassay, who informs the discoverer that the antibody only has an $IC_{50} = 550$ pM. In fact, no competitive inhibitor can have a single well-defined IC_{50}, because this value depends strongly on the potency and concentration of the agonist it is competing with.

A single inhibition curve of effect versus $[I]$ cannot identify the antagonist mechanism (competitive versus noncompetitive versus uncompetitive). To make this determination, one must measure data for multiple agonist response curves at varying antagonist concentration, as shown in the left panel of figure 4.26. Visual examination of the curves makes it clear that the maximum effect is the same for each antagonist concentration, consistent with a competitive inhibitor but inconsistent with noncompetitive or uncompetitive inhibition.

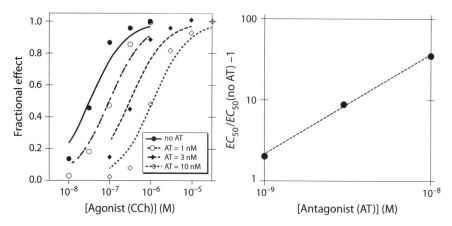

Figure 4.26 Schild analysis of a competitive antagonist. The experiments consisted of bathing strips of Guinea pig stomach tissue in the muscarinic agonist CCh and measuring the tension on the tissue with a microdynamometer (the fractional effect plotted in the left panel). The experiment was repeated at three concentrations of the competitve antagonist atropine (AT), and the EC_{50} for CCh was fit by nonlinear least squares to each of the four experiments [39].

The most rigorous procedure to test consistency of a data set such as this with equation 4.75 is to perform a global nonlinear least squares fit and examine the statistical goodness of fit by the procedures outlined in chapter 3. However, before the advent of facile access to nonlinear least squares fits, a procedure termed *Schild analysis* was developed to test for consistency with competitive antagonism [38]. Although one might argue that this approach has outlived most of its usefulness in the same fashion that Lineweaver-Burk plots have, the concept permeates the pharmacology literature sufficiently to merit mention here. In brief, equation 4.75 can be transformed to the following:

$$\log_{10}(DR - 1) = -\log_{10}(K_\mathrm{I}) + \log_{10}([I]) \qquad (4.80)$$

where

$$DR = \frac{EC_{50} \text{ with antagonist}}{EC_{50} \text{ without antagonist}} \qquad (4.81)$$

For a competitive antagonist, a plot of $\log_{10}(DR-1)$ versus $\log_{10}([I])$ will therefore be a straight line with a slope of one and an intercept of $-\log_{10}(K_\mathrm{I})$. The term pA_2, which is defined as $-\log_{10}(K_\mathrm{I})$, is often substituted in this equation. The Schild plot in the right panel of figure 4.26 exhibits a slope close to one, consistent with the antagonist atropine being a competitive antagonist.

4.3.3 Tight-Binding Inhibitors

Inhibitors that can be described by the relationships in section 4.3.1 share the property of binding sufficiently weakly that they must be present in great excess over the concentration of enzyme to exert a significant effect (i.e., $K_I \gg [E]_o$). Consequently, the inhibitor is not depleted by complex formation, and $[I] \approx [I]_o$. However, when an inhibitor binds tightly enough that $K_I \approx [E]_o$, inhibitor depletion by complexation must be taken into account, and so different rate laws are appropriate [40].

A simple and direct way to identify the presence of tight-binding inhibition is to obtain inhibition curves at varying initial enzyme concentration and constant substrate concentration. For weakly binding inhibitors, the apparent IC_{50} is not a function of enzyme concentration; by contrast, for a tight-binding inhibitor, the following relationship holds:

$$IC_{50} = \frac{1}{2}[E]_o + K_I^{app} \qquad (4.82)$$

where the apparent inhibition constant K_I^{app} varies with the mode of inhibition (competitive, noncompetitive, or uncompetitive). For competitive inhibitors, we have

$$K_I^{app} = K_I \left(1 + \frac{[S]}{K_m}\right) \qquad (4.83)$$

for noncompetitive inhibitors,

$$K_I^{app} = K_I \qquad (4.84)$$

and for uncompetitive inhibitors,

$$K_I^{app} = K_I \left(1 + \frac{K_m}{[S]}\right) \qquad (4.85)$$

At the IC_{50} of a high-affinity inhibitor, half of all enzyme molecules are occupied by inhibitor, and these enzyme-bound inhibitor molecules must be explicitly accounted for when K_I^{app} is not much greater than $[E]_o$, thus explaining the $[E]_o/2$ term in equation 4.82.

Example 4-5 A noncompetitive enzyme inhibitor is determined to have a binding constant $K_I = 10$ pM for the enzyme it inhibits. If an enzyme assay is performed with $[E]_o = 2$ nM and the inhibitor concentration is varied, what inhibitor concentration will be required to inhibit half the observed enzyme activity?

> **Solution** A reflexive estimate might be that an inhibitor concentration $\approx K_\mathrm{I}$, or 10 pM, would decrease the observed activity by half. However, this neglects the stoichiometry of the reaction: How could 10 pM inhibitor bind to half of the enzyme, which corresponds to a 1 nM concentration of targets? Referring to equation 4.82, $IC_{50} = \frac{1}{2}(2\ \mathrm{nM}) + 0.01\ \mathrm{nM} = 1.01\ \mathrm{nM}$, or approximately half the original enzyme concentration $[E]_\circ$.

The enzyme rate law for tight-binding inhibitors is often called the Morrison equation and is derived in similar fashion to the binding isotherm in the presence of ligand depletion as described in chapter 3, to give the fractional inhibition:

$$\frac{v(\text{inhibited})}{v(\text{no inhibitor})} = 1 - \frac{[E]_\circ + [I]_\circ + K_\mathrm{I}^{\mathrm{app}} - \sqrt{([E]_\circ + [I]_\circ + K_\mathrm{I}^{\mathrm{app}})^2 - 4[E]_\circ[I]_\circ}}{2[E]_\circ} \tag{4.86}$$

Case Study 4-1 R. A. Spence, W. M. Kati, K. S. Anderson, and K. A. Johnson. Mechanism of inhibition of HIV-1 reverse transcriptase by nonnucleoside inhibitors. *Science* 267: 988–993 (1995) [41].

HIV produces long-lived infections in humans that can eventually lead to diminished immune function and death, often from other, opportunistic infections that take advantage of reduced immunity. When HIV was initially discovered in human populations in the early 1980s, there was no effective therapy. Over time, mechanistic studies of the viral life cycle led to the identification of essential proteins that were targeted in drug discovery efforts.

One such target is the HIV reverse transcriptase (RT), which converts the single-stranded virus RNA into double-stranded DNA for integration into the host DNA complement. This target is unique to the virus, in that human cells have no enzyme with this activity, and it is essential: Without it, the virus can not integrate and proceed along the infectious pathway. The viral RNA template generally combines with a host tRNA molecule that serves as a primer, and this template-primer binds to the enzyme and initiates the polymerization reaction to synthesize the first DNA strand. In the case of HIV RT, a particular Lys-tRNA serves as primer. The reverse transcriptase also contains an RNase H activity (both an endonuclease and a $3' \rightarrow 5'$ exonuclease); the enzyme digests the original RNA strand to allow synthesis of the second DNA strand and also removes the tRNA primer. The second DNA strand is synthesized using the first as template, and is self-primed by means of a hairpin that forms at the 3' end of the template. Because of the nature of the deoxynucleotide substrate for the DNA polymerization activity, various nucleosides or nucleoside analogs are found to be excellent competitive inhibitors

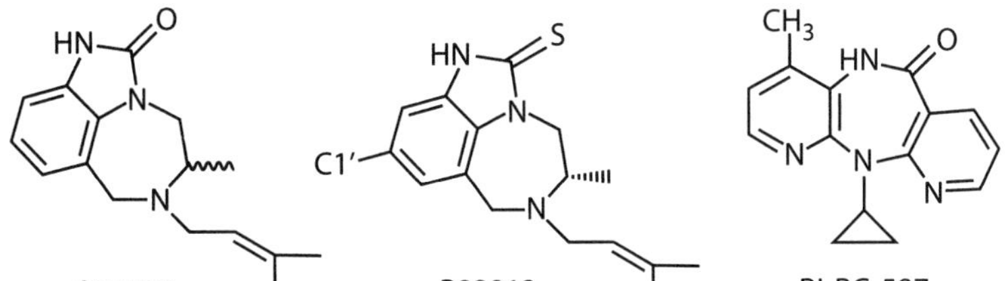

Figure 4.27 Chemical structures of the three nonnucleoside inhibitors studied by Spence et al. [41].

that occupy this substrate binding site. Unfortunately, some human enzymes also take nucleotide and nucleoside substrates (e.g., DNA polymerase and RNA polymerase); cross-reactivity of nucleoside-based HIV RT inhibitors with human enzymes produces side effects and toxicity.

Research has identified entirely different inhibitors that bind specifically to HIV RT. The paper by Spence et al. contributed to understanding how this important class of inhibitors functions [41]. Three specific inhibitors were studied. Two were chemically related to tetrahydro-benzodiazepine (TIBO), called O-TIBO and Cl-TIBO; the third was a dipyridodiaepinone called nevirapine (figure 4.27).

The authors applied pre-steady-state analysis. Whereas steady-state kinetic analysis leads to measurements of k_{cat}, K_M, and K_I, pre-steady-state analysis measures the rates of individual reaction steps, often through the examination of enzyme reactions under single-turnover conditions. Here, an excess of DNA substrate (consisting of a 45-base template annealed to a 25-base primer strand) was pre-equilibrated with enzyme. Reaction was initiated by adding and rapidly mixing an excess of the next deoxynucleoside triphosphate (dNTP) required to extend the primer strand by one base. The reaction was prevented from extending a further base, because a different dNTP that was required for the subsequent step was not present in the reaction mixture. The reaction was quenched, and products were quantified by running them out on a sequencing gel, which could distinguish the 25-base reactant primer from the 26-base product. This fundamental experiment was run in the absence of inhibitor and then was repeated with various concentrations of inhibitor pre-equilibrated with the enzyme and DNA substrate.

The results are shown in figure 4.28A. Each reaction started very rapidly, over the first roughly 0.2 s, and then continued at a slower rate. The early behavior is referred to as a burst phase; the subsequent, slower phase appears linear. The burst phase represents the rapid processing of a single substrate to a single product once by essentially every enzyme in the reaction vessel with its pre-equilibrated substrate. The linear phase represents a slow step required before each enzyme is capable of processing a second substrate. Here, the slow step has been associated with release of the product from enzyme, by separate measurement of that quantity [42]. The size of the burst phase corresponds to the amount of active enzyme–DNA complex present, the rate of the burst phase corresponds to the rate of the kinetics of binding dNTP and turning over substrate, and the rate of the linear phase corresponds to the steady-state rate. Fitting the kinetic trajectories to the standard

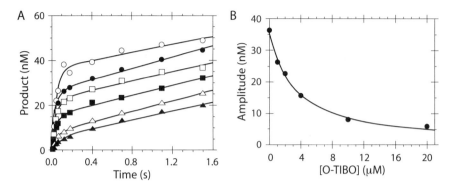

Figure 4.28 Burst kinetics taken from Spence et al. [41]. Enzyme was pre-equilibrated with DNA substrate and the nonnucleoside inhibitor O-TIBO; the reaction was initiated by mixing with dNTP, the next deoxynucleoside triphosphate to be added to the substrate. (A) Concentration of product measured as a function of time. Open circles correspond to 0 μM inhibitor; closed circles, 1.0 μM inhibitor; open squares, 2.0 μM inhibitor; closed squares, 4.0 μM inhibitor; open triangles, 10.0 μM inhibitor; and closed triangles, 20.0 μM inhibitor. (B) The sizes of the burst phases from panel A are plotted as a function of inhibitor concentration and fit to a binding isotherm to yield a value of $K_d = 3.0 \pm 0.2$ μM for the O-TIBO inhibitor binding the complex between enzyme and DNA substrate.

burst kinetics equation, $[P] = A[1 - e^{-kt} + mt]$, provides a turnover rate of 22 ± 3 s^{-1} from k and a steady-state rate of 0.26 s^{-1} from m that roughly matched the separately measured steady-state k_{cat} [42].

Increasing inhibitor concentration decreased the amplitude of the burst phase but left the rates of both the burst and linear phases unchanged. This behavior is consistent with a model in which binding inhibitor reduces the amount of active enzyme–DNA complex available for rapid reaction, and in which dissociation of inhibitor from the enzyme-DNA-inhibitor complex is slow relative to the rate of turnover. Associating burst size with amount of active enzyme complex, a plot of active enzyme complex as a function of inhibitor concentration (figure 4.28B) was fit to the equation $y = [I]/(K_d + [I])$ (or more properly, $1 - y = K_d/(K_d + [I])$), to give $K_d = 3.0 \pm 0.2$ μM for inhibitor binding to enzyme.

To measure the rate of inhibitor binding and release to the enzyme-DNA complex, an experiment was constructed in which enzyme-DNA complex was rapidly mixed with inhibitor. The binding reaction was continued for a controlled period of time, after which the next dNTP was added, rapidly mixed, and quenched after a single-turnover reaction time (0.2 s). The amount of product formed represented the amount of uninhibited enzyme–DNA complex, which was measured as a function of the time of the inhibitor binding reaction. The results were fit to a single exponential, for which $k_{obs} = k_{on}[I] + k_{off} = k_{on}([I] + K_d)$, to obtain $k_{on} = 1.2 \times 10^5$ M^{-1} s^{-1} and $k_{off} = 0.36$ s^{-1}.

Reactions carried out in saturating concentrations of inhibitor, which effectively reduced the size of the burst phase to zero, showed the linear phase continued with only a modest decrease in rate, from 0.26 s^{-1} in the absence of inhibitor to 0.11 s^{-1} in the presence of 20 μM of the inhibitor O-TIBO. Further analysis of results of these studies and their

Enzyme Kinetics

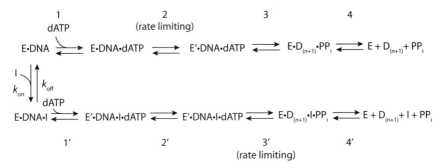

Figure 4.29 The experiments described in the text, together with further detailed measurements probing additional reaction steps, result in the scheme shown here [41].

dependence on inhibitor concentration confirmed that inhibitor-bound enzyme complex continued to catalyze the polymerization reaction, albeit more slowly, rather than reaction occurring during periods of inhibitor dissociation. The overall reaction scheme found to be consistent with these analyses is shown in figure 4.29.

In the absence of inhibitor, the rate-limiting step is a conformational change (step 2) that activates the enzyme-DNA-dNTP complex. In the presence of nonnucleoside inhibitors, the chemical polymerization step is sufficiently slowed that it becomes rate limiting (step 3′).

4.4 Signaling Pathways

4.4.1 Tyrosine Kinases

Kinases transduce many intracellular signals by adding phosphates to hydroxyl-containing side chains (Tyr, Ser, Thr) of other proteins and are themselves the substrates for other kinases (or themselves in some cases). Predictions from sequence homology [43] indicate that 119 of 6,241 *Saccharomyces cerevisiae* genes, 388 of 18,424 *Caenorhabditis elegans* genes, and 249 of 13,601 *Drosophila melanogaster* genes possess at least one eukaryotic protein kinase domain; so in yeast, worms, and flies, about 2% of proteins have kinase activity. In addition to modulation of their activity by binding to inhibitors or activators, many kinases often become activated following phosphorylation of key tyrosines, threonines, or serines. The structural theme for phosphorylation-mediated activation is repeated in most of these proteins. A very common theme in signaling is that an inactive kinase enzyme becomes activated following phosphorylation [44]. Shown in figure 4.30 is one mechanism by which this can occur.

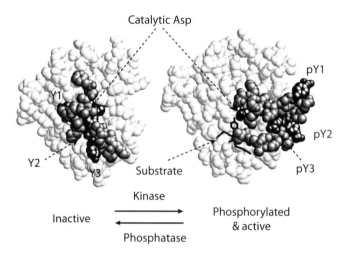

Figure 4.30 A particular aspartate residue (D1132) of the insulin receptor intracellular tyrosine kinase domain is directly involved in the addition of inorganic phosphate to a tyrosine residue [45]. In the inactive conformation (left), a loop of the kinase covers the peptide substrate binding site and blocks access to the active site. Tyrosine 1162 (Y2 in the figure) is in nearly the same location that the substrate tyrosine must be for phosphorylation to occur. Following autophosphorylation of three tyrosines in the loop (Y1, Y2, and Y3 in the figure), the loop undergoes a dramatic movement away from the active site (right), allowing access of a substrate peptide.

Case Study 4-2   P. A. Schwartz, P. Kuzmic, J. Solowiej, S. Bergqvist, B. Bolanos, C. Almaden, A. Nagata, K. Ryan, J. Feng, D. Dalvie, J. C. Kath, M. Xu, R. Wani, and B. W. Murray. Covalent EGFR inhibitor analysis reveals importance of reversible interactions to potency and mechanisms of drug resistance. *Proc. Natl. Acad. Sci. U.S.A.* 111: 173–178 (2014) [46].

Mutations in the epidermal growth factor receptor (EGFR) (e.g., L858R) result in oncogenic activation of this signaling pathway in a significant percentage (10–30%) of individuals with non-small cell lung cancer. Patients treated with drugs that act as ATP competitive inhibitors, such as erlotinib or gefitinib, often respond to this treatment, but half of the patients exhibit drug resistance due in part to an additional mutation in the active site (T790M). Given that EGFR presents a cysteine nucleophile in the ATP binding cleft of EGFR, an alternative way to ablate kinase activity is to use an inhibitor with an electrophile that chemically reacts with this EGFR cysteine to produce a covalent adduct. Such covalent inhibitors appear to be attractive drug candidates, but their performance in clinical trials has been mixed, in part due to our limited understanding of their mechanisms of action.

In this study, the authors combined quantitative experiments and mathematical modeling to elucidate the roles of reversible binding affinity and chemical reactivity on overall potency for a large panel of covalent EGFR inhibitors, including afatinib, neratinib, CI-1033, and WZ4002 (figure 4.31).

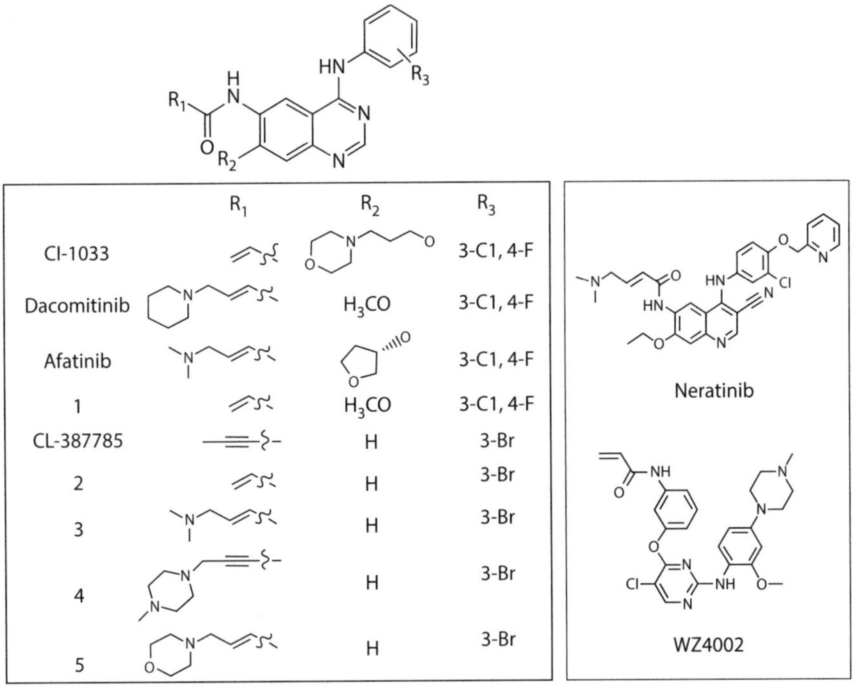

Enzyme Kinetics

Figure 4.31 Structures of covalent EGFR inhibitors examined.

The mechanism of action of these covalent inhibitors can be described by the following reaction scheme:

$$E + I \underset{k_{off}}{\overset{k_{on}}{\rightleftharpoons}} E \cdot I \overset{k_{inact}}{\longrightarrow} E - J$$
$$+$$
$$S$$
$$\downarrow k_{sub}$$
$$E + P$$

The authors quantified enzymatic activity using a fluorescent peptide substrate and found that the reaction velocity at early times ($t < 7$ min) was clearly dependent on the inhibitor concentration. That dependence indicates that the inhibition proceeds in two steps: First, the enzyme is rapidly and reversibly bound by inhibitor in a concentration-dependent manner; and second, these complexes slowly and irreversibly convert to the covalent adduct. With the initial velocities from these experiments, the Morrison equation (equation 4.86) was used to calculate K_I^{app} values. Because the mode of inhibition is competitive (and reversible at short times), the true K_I values were obtained using equation 4.83.

Separately, an ODE model based on the reaction scheme above was constructed to globally fit full time-series data and extract underlying rate constants:

$$\frac{d[E]}{dt} = -k_{on}[E][I] + k_{off}[E \cdot I] \tag{4.87}$$

$$\frac{d[S]}{dt} = -k_{sub}[E][S] \tag{4.88}$$

$$\frac{d[P]}{dt} = k_{sub}[E][S] \tag{4.89}$$

$$\frac{d[I]}{dt} = -k_{on}[E][I] + k_{off}[E \cdot I] \tag{4.90}$$

$$\frac{d[E \cdot I]}{dt} = k_{on}[E][I] - k_{off}[E \cdot I] - k_{inact}[E \cdot I] \tag{4.91}$$

$$\frac{d[E - J]}{dt} = k_{inact}[E \cdot I] \tag{4.92}$$

From a global fit of data, k_{inact} and k_{sub} were well reproduced across independent replicates. Given the rapid equilibrium binding of inhibitor to enzyme noted earlier, it was not

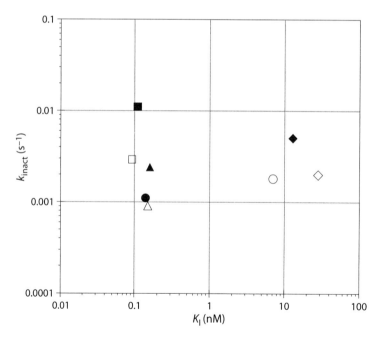

Figure 4.32 Biochemical parameters K_I and k_{inact} for covalent EGFR inhibitors afatinib (triangles), neratinib (circles), CI-1033 (squares), and WZ4002 (diamonds) against both wild-type EGFR (open symbols) and the L858R/T790M EGFR variant (closed symbols).

possible to extract independent k_{on} and k_{off} values, but their ratio, K_I, was again highly consistent across replicates. Thus, the K_I and k_{inact} values could be separately extracted (figure 4.32).

Interestingly, it was found that covalent inhibitor potency is only modestly correlated with the irreversible step (i.e., the value of k_{inact}) but highly correlated with the *reversible* binding affinity (i.e., the value of K_I). For example, WZ4002 and CI-1033 have the same reactive substituent, which has an approximately fivefold higher intrinsic chemical reactivity in WZ4002 than in CI-1033, based on reactivity to glutathione. However, WZ4002 is considerably less potent than CI-1033 in receptor phosphorylation inhibition assays: In A549 tumor cells expressing wild-type EGFR, $IC_{50,WZ4002} = 1,400$ nM, and $IC_{50,CI-1033} = 4.9$ nM; in H1975 tumor cells expressing L858R/T790M EGFR, $IC_{50,WZ4002} = 75$ nM, and $IC_{50,CI-1033} = 2.3$ nM. This observation cannot be explained by k_{inact}, which is essentially the same for both compounds, but appears to correlate with K_I (figure 4.32).

To further test this hypothesis, the authors examined 154 covalent inhibitors from six structural classes and plotted their IC_{50} values against their K_I values, corroborating the importance of reversible binding affinity in determining the potency of covalent inhibitors (figure 4.33).

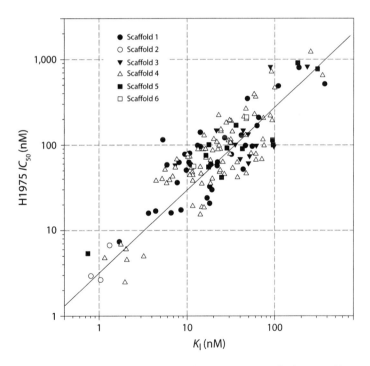

Figure 4.33 Inhibition of EGFR autophosphorylation in H1975 tumor cells for 154 cell-permeable, covalent inhibitors from six different structural classes. The IC_{50} values correlate strongly with the reversible binding affinity K_I [46].

Reversible binding is important in the design of covalent inhibitors for multiple reasons: (1) It serves to orient the reactive moieties on the two molecules, (2) it increases the effective concentration of the reacting species, (3) it determines the number of binding events that are needed before a productive reaction occurs, and (4) it can result in enzyme inhibition even if the subsequent chemical reaction does not occur (e.g., if the EGFR cysteine is oxidized).

4.4.2 Trimeric G Proteins and G Protein-Coupled Receptors

A significant number of signals are transduced from the extracellular to the intracellular space via a class of enzymes known as trimeric GTPases, or G proteins. The broad spectrum of ligands that initiate particular G protein signaling pathways includes small organic molecules, lipids, amino acids, peptides, and proteins.

As shown schematically in figure 4.34, binding of an extracellular ligand triggers a conformational change in a G protein–coupled receptor (GPCR), causing exchange of GDP for GTP on an associated G protein and producing the active form of the α subunit of the G protein. The activated α subunit then binds to a target effector protein, causing it to become activated. This period of activation is clocked by the presence of GTP in the α subunit—once it becomes hydrolyzed to GDP, the G protein subunit dissociates from the target protein and reforms an inactive trimer with the β and γ subunits. The processes of GTP/GDP exchange

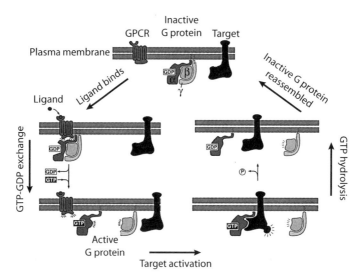

Figure 4.34 Steps in signal transduction by G protein-coupled receptors (GPCRs). The enzymatic GTPase activity of the G protein α subunit provides a critical timing component, determining the duration of the active signaling period [47].

and GTP hydrolysis are in turn catalyzed by auxiliary proteins. It is estimated that half of all current drugs mediate their effects via GPCRs.

4.5 Metabolism

Cellular energy and biosynthetic metabolism are carried out enzymatically, and attempts to understand or engineer these processes can be facilitated by mathematical descriptions. The presence of extensive allosteric feedback, both positive and negative, severely limits the utility of simple Michaelis–Menten-type rate laws. Metabolic control analysis (MCA) is a framework for describing metabolic pathways, and incorporating stoichiometric limitations into constraint-based models can also provide useful information.

4.5.1 Flux-Control Coefficients

Metabolic control analysis is mathematically equivalent to local sensitivity analysis, as described in chapter 1. MCA was developed independently by Heinrich and Rapaport [48] and Kacser, Burns, and coworkers [49, 50].

If one defines the steady-state rate of substrate undergoing a particular reaction as the flux J, then the effect of changing the activity of a particular enzyme e_i on this flux is given by the flux-control coefficient C_i^J, which is defined as follows:

$$C_i^J \equiv \frac{\partial \ln J}{\partial \ln e_i} \tag{4.93}$$

Thus, C_i^J is a normalized local objective sensitivity, with the objective function being the reaction rate of interest. Experimentally determined flux-control coefficients can be helpful in engineering pathways to maximize productivity of a metabolite of interest. A central result of MCA is the *flux-control summation theorem*:

$$\sum_{i=1}^{N} C_i^J = 1 \tag{4.94}$$

The effect of reactant concentrations on fluxes is defined by elasticity coefficients:

$$\varepsilon_{ji} \equiv \frac{\partial \ln r_i}{\partial \ln s_j} \tag{4.95}$$

where r_i is the net reaction rate for the reaction catalyzed by enzyme i, and s_j is a particular reactant j. The flux-control connectivity theorem is as follows:

$$\sum_{i=1}^{N} C_i^J \varepsilon_{ji} = 0; j = 1, \ldots, N-1 \qquad (4.96)$$

Detailed treatments of MCA may be found elsewhere [51, 52]. MCA has been applied in such divergent areas as yeast glycolysis [53], heart muscle energy metabolism [54], penicillin production [55], and trypanosomal parasite therapy [56].

4.5.2 Constraint-Based Models

A more general framework that subsumes many of the concepts and techniques of flux-balance analysis and related methods into a larger conceptualization is known as constraint-based modeling. The essential idea, as with flux balance, is to extend the reach of network topological descriptions and experimental data through application and propagation of a variety of constraints. If the underlying topology is sufficiently descriptive and accurate and the constraints powerful enough, this approach can provide a basis for placing limits on the range of achievable biological function in a system. Appropriately tight limits correspond to behavior predictions, and with appropriate techniques, the models may be used to design desired changes in biological function. Reviews outlining the approach and some of the results achieved can be found in the literature [57–59].

Constraint-based computational models have mainly been used for genome-level analysis of microbial organisms, largely to study metabolism and metabolism-related regulation and thus to provide a more detailed description of the underlying tensions leading to cellular phenotype. The constraints applied come from four areas—environmental, physicochemical, evolutionary, and regulatory—which, taken together, represent both fixed and time-varying constraints, due to both internal and external sources and imposed by both outside and self. We briefly describe a few such constraints and how their application can limit the overall range of model behavior.

Environmental constraints represent a combination of external and internal (i.e., extracellular and intracellular) sources that can change with time. Externally applied constraints include available nutrients, physical conditions (e.g., temperature, pressure, pH, ionic strength, incident light, availability of water), and neighboring influences (e.g., crowding, signaling molecules, waste products from other individuals in the same environment). Internal constraints include such limitations as the cell volume and defined physical properties (e.g., molecular crowding and osmotic pressure). An attractive feature of this viewpoint on cellular network function is that it naturally accommodates such notions as that the minimal gene complement for an organism may depend strongly on nutritional constraints—a smaller set of genes may be essential in rich medium relative to minimal medium.

Physicochemical constraints are due to laws of nature and fundamental principles; like other universal limits, such as the speed of light, they are invariant and do not change with time. Constraints that fit in this category include conservation of mass, energy, and momentum; thermodynamic laws; kinetic principles; and kinetically achievable rates. These are some of the most powerful constraints; conservation of mass is sufficiently limiting on its own to be the sole constraint applied in the family of techniques related to flux-balance analysis.

Whereas environmental constraints and physicochemical ones are not generally viewed as coming from the function of the cell itself (although one might argue that the internal environment is largely a result of cellular function), evolutionary and regulatory constraints are viewed as a result of cellular function. In fact, in some sense, they lie on a continuum from short-term to long-term adaptations of living systems to cope with conditions and challenges. Stated another way, in this view, regulatory and evolutionary constraints are self-imposed as part of the solution developed by living systems to carry out their functions.

Evolutionary constraints include the replicative error rate per cell division, which partially mediates the tension between fidelity of replication (corresponding to genome stability) and adaptability (corresponding to evolutionary change in moderation and genome instability in the extreme). The error rate may itself be under variable control and may change in response to stress and adaptive need. We tend to think of evolution as undirected, and the results of evolutionary processes as random accidents that are unlikely to occur if the experiment were repeated again. With this in mind, it is useful to carry out a back-of-the-envelope calculation of the likelihood of a particular single-base mutation in *Escherichia coli*. Covert et al. [57] show that the probability of acquiring a particular mutation P_m in n replication events is given by

$$P_m = 1 - (1 - P_{err})^n \tag{4.97}$$

where P_{err} is the mutation rate per base pair per cell division in *E. coli* (roughly 10^{-10} when unstressed). The form of this equation is that the probability of finding the mutation is one minus the probability of always avoiding the mutation (a double negative is a positive in this case). To achieve a 99.9% chance of finding the mutation requires roughly 7×10^{10} cell divisions. Although this sounds like a big number, it is only 36 generations, or roughly 12 hours of evolution in favorable growth conditions. The interpretation of this simple calculation is that our mutation is extremely likely to be found in a very short time; in fact, because this mutation is arbitrary, the descendants of a single *E. coli* cell sample every possible single mutation on a daily basis.

Regulatory constraints include a variety of effects. One important concept involves regulatory states that control which parts of the genome are expressed (and at which levels) and which portions of the genome are turned off. In principle, every gene could be expressed independently—at any level from off to highly expressed. In practice, biological control mechanisms include parallel control (where the same transcription factor turns on some genes simultaneously), cooperative control (where turning on the expression of some genes leads to turning off expression of others), and a variety of other effects that lead to only a limited set of these states being possible. In the language of constraint-based modeling, the existing evolutionary control mechanisms place significant constraints on the sets of levels of gene products that can be achieved in a cell.

The constraints described tend to be represented as either a bound or a balance. Bounds represent physical limits (such as the upper bound set by diffusion limits or the lower bound set by the nonnegativity of concentrations); balances result from continuity equations (such as flux balance equations). A variety of optimization tools exists to operate on constraints of this sort, particularly if the objective function and constraints are both linear in form, and these tools provide a useful mechanism for probing constraint-based models.

4.6 Hydrolytic Regulatory Enzymes

Some significant irreversible biological responses are initiated by the action of regulatory hydrolytic enzymes that specifically cleave at particular sites in their substrates, which are often also hydrolases activated by this cleavage. The end result is an abrupt, catastrophic transition in the system's state. Examples of this type of transformation include apoptosis and blood clotting.

4.6.1 Caspases

Apoptosis, the orderly and programmed self-destruction of a cell, must be: (1) tightly regulated to avoid random cellular self-destruction throughout the organism and (2) definitively progressive such that the commitment to apoptose is irreversible (otherwise, partially self-destroyed cells could trigger inflammation and other dysfunctions). Caspases are regulatory proteases that play an essential role in apoptosis.

In contrast to lysosomal and digestive proteases that metabolically break down proteins somewhat indiscriminately, caspases cleave at specific recognition sites, with cleavage always C-terminal to an aspartic acid (from which derives the name "*casp*ase"). Caspases are themselves caspase substrates, initially synthesized as inactive zymogens or pro-enzymes. Following cleavage at several particular sites in their own sequence, caspases assemble into a homodimer of heterodimers, which is catalytically active.

Caspase self-activation occurs autocatalytically, following an increase in local concentration by binding to a signaling complex. There are two general classes

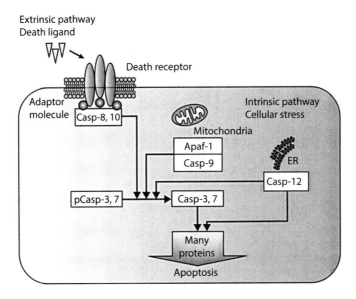

Figure 4.35 Pathways for caspase activation leading to apoptosis. Death receptors (such as TNFR, Fas/CD95, DR3, DR4, DR5, or DR6) bind death ligands (such as TNFα, FasL, or TRAIL), assembling a death-inducing signaling complex (DISC) that recruits initiator caspase zymogens and raises their local concentration sufficiently to trigger dimerization, cleavage, and activation. The intrinsic mitochondrial pathway initiates with release of cytochrome C from the mitochondria, which then binds to apoptotic protease activating factor 1 (Apaf-1) to assemble the apoptosome. This raises the local concentration of initiator caspases sufficiently to result in proteolytic activation [60].

of caspase activation: (1) extrinsic, resulting from exposure to an extracellular death ligand; and (2) intrinsic, resulting from cellular stress and mediated through the mitochondria or endoplasmic reticulum (figure 4.35). Following the proximal initiating event, these pathways overlap and crosstalk substantially, converging on a common set of effector caspases that specifically cleave major cytoskeletal and nuclear proteins, leading to cell disassembly.

The kinetics of death-ligand binding, leading to effector caspase activation, have been studied by mathematical modeling [61–63].

4.6.2 Blood Coagulation Cascades

The rapid and irreversible (on a short time scale) formation of a blood clot is an essential response to injury. However, it must also be tightly controlled to avoid inappropriate interrruption of blood supply to healthy tissues. This control is effected by a cascade of proteolytic enzymatic reactions that activate other proteases.

The extrinsic pathway of coagulation is activated on exposure of tissue factor (TF) to the bloodstream by mechanical injury (figure 4.36; [64]). TF is a surface

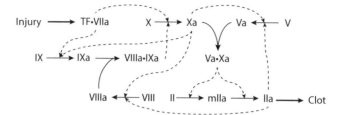

Figure 4.36 Core elements of the coagulation cascade. Dashed lines represent enzymatic catalysis, and solid lines represent conversion from a zymogen to an activated protease by proteolysis. Proteases exist in an inactive zymogen form and an activated form (denoted by appending a small "a"). Following injury, TF·VIIa activates factors X and IX; Xa activates more IX, as well as factors V and VIII. VIIIa and IXa combine to form a complex VIIIa·IXa, known as tenase (i.e., "X"-ase), which activates more factor X, providing positive feedback. Xa and Va combine to form prothrombinase, which activates prothrombin (factor II) to an intermediate form mIIa (meizothrombin) and the final thrombin form (IIa). Thrombin provides further positive feedback by activating factors V and VIII. The presence of nested positive feedback loops gives the impression of an unstable system poised on the brink of clot formation on the slightest fluctuation. However, anticoagulation molecules throughout the pathway are not represented in this simple schematic and add stability to the unclotted state.

protein expressed on cells surrounding the vasculature, which binds to circulating protease factor VII and its active form, factor VIIa. The TF·VIIa complex catalyzes a series of further proteolytic reactions that culminates in the activation of thrombin (factor IIa), which in turn cleaves fibrinogen to form fibrin, which rapidly polymerizes to form the framework for a clot.

Mathematical modeling of the regulation of clot formation has contributed significantly to an understanding of the system's control and dynamics [65]. In the following case study, kinetic modeling is used to show how the appropriate combination of enzyme-catalyzed pro-enzyme activation reactions that had been individually characterized in vitro combines to give overall thrombin activation kinetics consistent with the threshold-like kinetics of the combined in vitro model system.

Case Study 4-3 K. C. Jones and K. G. Mann. A model for the tissue factor pathway to thrombin. II. A mathematical simulation. *J. Biol. Chem.* 269: 23,367–23,373 (1994) [23].

This paper is one of a series of kinetic models of the coagulation cascade from the group of K. G. Mann, including the first, published in 1984 [66], a notable paper in 2002 [67], and a 2012 perspective [68]. The purpose of their 1994 model was to combine the rates of individually measured reactions in the cascade [69] and demonstrate consistency with the observed overall rate of thrombin activation [23]. The authors clearly laid out the goal of the modeling exercise as follows:

> The results of this paper do *not* suggest new knowledge about previously unknown reactions or kinetic constants. Rather, the prediction data comply reasonably in form and magnitude

Enzyme Kinetics

$$IX + TF \cdot VIIa \underset{k_{16}}{\overset{k_6}{\rightleftharpoons}} IX \cdot TF \cdot VIIa \overset{k_{11}}{\rightarrow} TF \cdot VIIa + IXa$$

$$X + TF \cdot VIIa \underset{k_{17}}{\overset{k_6}{\rightleftharpoons}} X \cdot TF \cdot VIIa \overset{k_{12}}{\rightarrow} TF \cdot VIIa + Xa$$

$$X + VIIIa \cdot IXa \underset{k_{18}}{\overset{k_6}{\rightleftharpoons}} X \cdot VIIa \cdot IXa \overset{k_{13}}{\rightarrow} VIIIa \cdot IXa + Xa$$

$$IX + Xa \overset{k_{15}}{\rightarrow} Xa + IXa$$

$$V + Xa \overset{k_1}{\rightarrow} Xa + Va$$

$$VIII + Xa \overset{k_3}{\rightarrow} Xa + VIIIa$$

$$V + IIa \overset{k_2}{\rightarrow} IIa + Va$$

$$VIII + IIa \overset{k_4}{\rightarrow} IIa + VIIIa$$

$$II + Va \cdot Xa \underset{k_{19}}{\overset{k_6}{\rightleftharpoons}} II \cdot Va \cdot Xa \overset{k_{14}}{\rightarrow} Va \cdot Xa + mIIa$$

$$mIIa + Va \cdot Xa \overset{k_5}{\rightarrow} Va \cdot Xa + IIa$$

$$VIIIa + IXa \underset{k_9}{\overset{k_7}{\rightleftharpoons}} VIIIa \cdot IXa$$

$$Va + Xa \underset{k_{10}}{\overset{k_8}{\rightleftharpoons}} Va \cdot Xa$$

Figure 4.37 Reactions used to describe the coagulation cascade [23].

to the empirical progress curves for the reaction.... The rate constants used in the model for this paper do show that no additional kinetic events must be invoked to explain the empirical data for α-thrombin generation.

The kinetic scheme consists of the reactions depicted in figure 4.36, represented by the pertinent parameters in figure 4.37. Enzymatic kinetics are treated either by explicitly describing the formation of a reversible complex followed by first-order conversion to product, or as a second-order reaction if formation of the complex is the rate-limiting step. Translation of these reactions into the derivative vector for the numerical analysis package MATLAB is shown in figure 4.38. Such a dizzying array of subscripts is usual for models with this number of variables and raises the specter of typographical errors (one of which was made by Jones and Mann in the initial implementation—the rate constant k_6 was mistakenly applied in the balance equations for Va and Xa, where k_8 is the appropriate constant).

```
xdot = [k(11)*x(12)-k(6)*x(1)*x(2)+k(16)*x(12)+k(12)*x(13)-k(6)*x(1)*x(3)+k(17)*x(13);...
    k(16)*x(12)-k(6)*x(1)*x(2)-k(15)*x(2)*x(16)-k(15)*x(2)*x(8);...
    k(17)*x(13)-k(6)*x(1)*x(3)-k(6)*x(7)*x(3)+k(18)*x(14);...
    -k(1)*x(4)*x(16)-k(2)*x(4)*x(9)-k(2)*x(4)*x(11);...
    -k(3)*x(5)*x(16)-k(4)*x(5)*x(9)-k(4)*x(5)*x(11);...
    k(19)*x(10)-k(6)*x(8)*x(6);...
    k(7)*x(18)*x(15)-k(9)*x(7)-k(6)*x(7)*x(3)+k(18)*x(14)+k(13)*x(14)-k(20)*x(7);...
    k(8)*x(16)*x(17)-k(10)*x(8)+k(19)*x(10)-k(6)*x(8)*x(6)+k(14)*x(10);...
    k(5)*x(8)*x(11);...
    k(6)*x(8)*x(6)-k(19)*x(10)-k(14)*x(10);...
    k(14)*x(10)-k(5)*x(8)*x(11);...
    k(6)*x(1)*x(2)-k(16)*x(12)-k(11)*x(12);...
    k(6)*x(1)*x(3)-k(17)*x(13)-k(12)*x(13);...
    k(6)*x(7)*x(3)-k(18)*x(14)-k(13)*x(14);...
    k(9)*x(7)-k(7)*x(18)*x(15)+k(11)*x(12)+k(15)*x(2)*x(16)+k(15)*x(2)*x(8);...
    k(10)*x(8)-k(8)*x(16)*x(17)+k(12)*x(13)+k(13)*x(14);...
    k(10)*x(8)-k(8)*x(16)*x(17)+k(1)*x(4)*x(16)+k(2)*x(4)*x(9)+k(2)*x(4)*x(11);...
    k(9)*x(7)-k(7)*x(18)*x(15)+k(3)*x(5)*x(16)+k(4)*x(5)*x(9)+k(4)*x(5)*x(11)];
```

Figure 4.38 MATLAB code for the derivatives of an ODE model of the reactions in figure 4.37.

It is good practice in kinetic modeling to insert error-checking flags to notify the user when nonphysical results are obtained, either due to typographical errors or numerical instability. For the present implementation, the following sums should be satisfied within tolerance:

$$[II]_\circ = [II] + [IIa] + [mIIa];$$

$$[TF \cdot VIIa]_\circ = [TF \cdot VIIa] + [TF \cdot VIIA \cdot IX] + [TF \cdot VIIA \cdot X];$$

$$[X]_\circ = [X] + [Va \cdot Xa] + [Va \cdot Xa \cdot II] + [TF \cdot VIIa \cdot X] + [VIIIa \cdot IXa \cdot X] + [Xa];$$

and

$$[V]_\circ = [V] + [Va \cdot Xa] + [Va \cdot Xa \cdot II] + [Va].$$

Satisfaction of these equalities is *necessary* for the correctness of a numerical solution, and their violation should be prominently flagged to the user in good code. ODE solvers, such as those in MATLAB, also often include a user-settable option for variables that cannot physically assume negative values. Of course, satisfaction of material balances is not *sufficient* for correctness of the numerical solution.

The variables, initial values, parameters, and their estimates are shown in table 4.5. The initial concentrations are intended to reflect those found in blood in vivo; note that the initial concentrations span six orders of magnitude. The rate constants are mostly measured in the companion experimental paper or were reported earlier in the literature. The authors utilized a nonphysical mechanism to account for VIIIa·IXa degradation that had been observed experimentally; a simple first-order term with the same rate constant (k_{20}) can be substituted.

Simulations of thrombin activation kinetics as a function of input [TF·VIIa]$_\circ$ are shown in figure 4.39. They show qualitative agreement with the experimental results.

Peering inside the simulation results at the concentrations of species not experimentally reported (figure 4.40), one can see that factor V is fully activated by ≈50 seconds, and

Table 4.5 Variables and parameters of coagulation cascade model.

Variable #	Species	Initial value	Parameter	Description	Estimate
1	TF·VIIa	5×10^{-12} M	k_1	Xa activation of V	2×10^7 M^{-1} s^{-1}
2	IX	9×10^{-8} M	k_2	IIa or mIIa activation of V	2×10^7 M^{-1} s^{-1}
3	X	1.7×10^{-7} M	k_3	Xa activation of VIII	1×10^7 M^{-1} s^{-1}
4	V	2×10^{-8} M	k_4	IIa or mIIa activation of VIII	2×10^7 M^{-1} s^{-1}
5	VIII	7×10^{-10} M	k_5	mIIa to IIa by Va·Xa	1×10^7 M^{-1} s^{-1}
6	II	1.4×10^{-6} M	k_6	Fast k_{on} for several reactions	1×10^8 M^{-1} s^{-1}
7	VIIIa·IXa	0	k_7	k_{on} for VIIIa·IXa	1×10^7 M^{-1} s^{-1}
8	Va·Xa	0	k_8	k_{on} for Va·Xa	4×10^8 M^{-1} s^{-1}
9	IIa	0	k_9	k_{off} for VIIIa·IXa	0.005 s^{-1}
10	Va·Xa·II	0	k_{10}	k_{off} for Va·Xa	0.4 s^{-1}
11	mIIa	0	k_{11}	k_{cat} for IX activation by TF·VIIa	0.3 s^{-1}
12	TF·VIIA·IX	0	k_{12}	k_{cat} for X activation by TF·VIIa	1.15 s^{-1}
13	TF·VIIa·X	0	k_{13}	k_{cat} for X activation by VIIIa·IXa	8.2 s^{-1}
14	VIIIa·IXa·X	0	k_{14}	k_{cat} for mIIa activation by Va·Xa	32 s^{-1}
15	IXa	0	k_{15}	Xa activation of Ix	1×10^5 M^{-1} s^{-1}
16	Xa	0	k_{16}	k_{off} for IX·TF·VIIa	24 s^{-1}
17	Va	0	k_{17}	k_{off} for X·TF·VIIa	44 s^{-1}
18	VIIIa	0	k_{18}	k_{off} for X·VIIIa·IXa	0.001 s^{-1}
			k_{19}	k_{off} for II·Va·Xa	70 s^{-1}
			k_{20}	First-order VIIIa·IXa degradation	0.02 s^{-1}

factor X is fully activated by ≈200 seconds, approximately coincident with thrombin (factor II) activation. These results are consistent with experiments [69].

When constructing a model with 18 variables and 20 parameters, an essential question is whether the model is as simple as possible while still capturing the known mechanistic components of the system of interest. In the present case, the kinetics of many of the individual reactions had been characterized in vitro previously, and the qualitative existence of each reaction included in the model had been independently confirmed. Given the varying uncertainties among the 20 parameters, it is nevertheless of interest to determine how much impact each parameter's value has on the overall thrombin activation kinetics shown in figure 4.39. The way to do this is by using sensitivity analysis, a subject discussed in chapter 1. Sensitivity of the thrombin activation time to each parameter is presented in figure 4.41. A local normalized objective sensitivity for each parameter, defined as

$$S(t_{max}; k_i) \equiv \frac{\partial \ln(t_{max})}{\partial \ln(k_i)} \tag{4.98}$$

was determined numerically by finite differences.

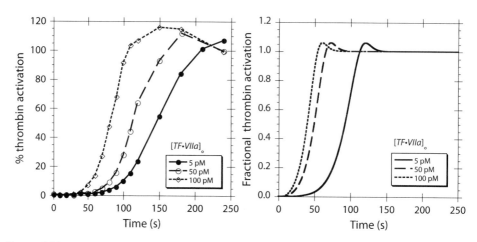

Figure 4.39 Experimental (left panel) and simulation (right panel) results for thrombin activation as a function of input signal $[TF \cdot VIIa]_o$.

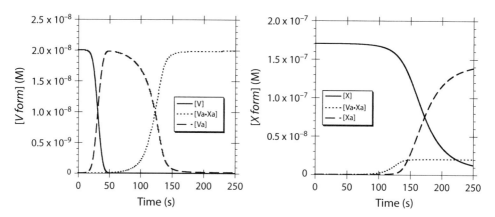

Figure 4.40 Simulation results for factors V and X.

Examination of the parametric sensitivities indicates that little detail would be lost by considering all interactions to be irreversible; none of the six k_{off} parameters affects t_{max} to a great extent. By contrast, the forward catalysis rate constants exert a more substantial effect on the coagulation delay time (with the exception of IIa or Xa activations of VIII or V). This is perhaps not surprising for such a forward-biased reaction scheme, which contains multiple positive feedback loops and no restraining negative feedback inhibition.

Sensitivities that are very low should raise the question: Need this parameter be included in the model at all? Although one cannot zero out such seemingly insensitive parameters

Enzyme Kinetics

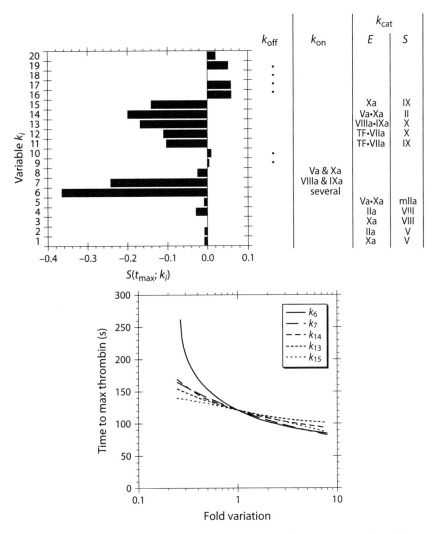

Figure 4.41 Local normalized objective sensitivities $S(t_{max}; k_i)$ where t_{max} is the time for achievement of peak thrombin activity. (There is a peak in thrombin activity, because the mIIa intermediate form has 20% higher specific activity than the final form factor IIa.)

in every case, often much can be learned in the attempt. For the present case, a reduced model can be constructed with $k_2 = k_3 = k_9 = k_{10} = k_{18} = k_{20} = 0$. The overall thrombin activation kinetics are essentially unaffected, as is the activation of factor X (figure 4.42). Factor V exhibits significantly lower activation before 100 seconds, however, at which time, sufficient Xa is present to activate V directly.

It is not advisable in general, of course, to discard validated reactions from an overall scheme for the sole purpose of obtaining a simpler model. However, it can be useful

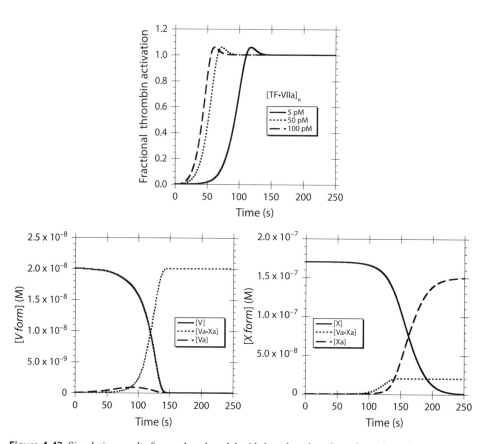

Figure 4.42 Simulation results for a reduced model with $k_2 = k_3 = k_9 = k_{10} = k_{18} = k_{20} = 0$.

to learn which of the reactions have meaningful impact on the particular objective function of interest. Note that the sensitivities may be very different for a different objective function—so one should be very cautious when eliminating reactions that have been independently verified, even if their effect on one objective function is negligible.

One take-home message from the sensitivity analysis is that the parameter most affecting thrombin activation time is the association rate constant k_6 for four of the reactions in the kinetic scheme (figure 4.37). (This parameter was also used implicitly to calculate k_{16}, k_{17}, k_{18}, and k_{19} from reported Michaelis-Menten parameters.) This result is rather unfortunate, because k_6 was pulled somewhat out of thin air to represent fast interactions. It would be a high priority to directly measure these association rate constants, say, by stopped-flow spectrometry (chapter 2).

4.6.3 Extracellular Metalloproteases

Matrix metalloproteases (MMPs) have long been known to be important in tissue remodeling in growth, development, wound healing, and cancer, but their main function was thought to be to "clear the brush" to make room for cell migration or growth. In fact, MMPs play a major role in sending specific receptor-mediated signals via release of ligands that are inactive prior to cleavage (figure 4.43). Growth factors and bioactive fragments—such as endostatin, vascular endothelial growth factor (VEGF), and transforming growth factor β (TGFβ)—are liberated from the extracellular matrix or the cell surface after cleavage by MMPs [70].

The ligands for the ErbB extracellular growth factor receptor family are largely expressed in an initially membrane-anchored form on the cell surface. Cleavage by a class of cell-surface metalloproteases known as ADAMs (the acroynm for a disintegrin and metalloprotease) is a critical step in liberation of these pregrowth factors into a form capable of binding their receptors and signaling [71]. The receptor ectodomain may also be cleaved and shed, influencing capturing and buffering free ligand outside the cell (figure 4.44). Altered expression of extracellular protease activity can provide a positive feedback loop by upregulating protease activity in response to a first wave of receptor signaling [72] or can provide cross-talk between different signaling pathways, such as GPCR and ErbB signaling [73].

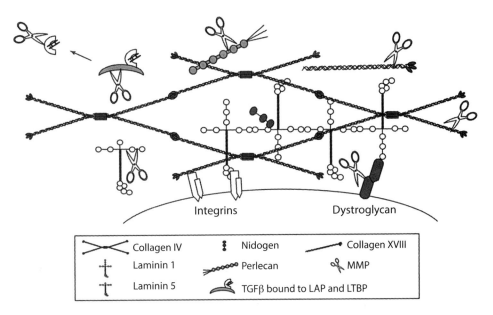

Figure 4.43 Known and hypothesized regulatory actions catalyzed by MMPs [70].

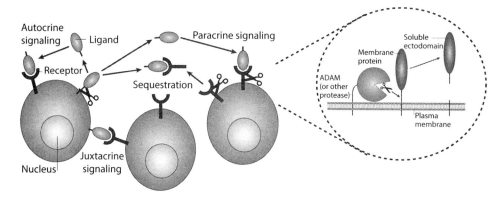

Figure 4.44 Release of growth factors and their receptor binding domains from the cell surface is catalyzed by a family of cell surface proteases known as ADAMs [71].

Suggestions for Further Reading

H. W. Blanch and D. S. Clark. *Biochemical Engineering*. Marcel Dekker, 1996.

P. F. Cook and W. W. Cleland. *Enzyme Kinetics and Mechanism*. Garland Science, 2007.

R. A. Copeland. *Enzymes—A Practical Introduction to Structure, Mechanism, and Data Analysis*, second edition. John Wiley & Sons, 2000.

I. H. Segel. *Enzyme Kinetics—Behavior and Analysis of Rapid Equilibrium and Steady-State Enzyme Systems*. John Wiley & Sons, 1975.

Problems

4-1 The enzyme urease catalyzes the reaction $NH_2CONH_2 + H_2O \rightarrow 2NH_3 + CO_2$ with the kinetic parameters: $k_{cat} = 30,800$ s^{-1}, $k_1 = 1 \times 10^7$ M^{-1} s^{-1}, and $k_{-1} = 10,000$ s^{-1}.

What is the minimum concentration of urease in a batch reactor that will catalyze the destruction of 99% of the urea in a 10 mM solution within 10 seconds?

4-2 For the data in the previous problem, let's examine the range of validity of the QSSA. Plot the concentration of urea versus time, calculated two ways: (a) from the implicit solution to the Michaelis-Menten rate form and (b) from numerical solution of the ODEs describing the system, without the QSSA. What is the maximum discrepancy in the concentration of urea calculated by the two methods? Repeat this analysis for urease concentrations both tenfold higher than and one-tenth the base-case concentration.

Enzyme Kinetics

4-3 Chase et al. [74] measured the hydrolysis kinetics of sucrose by invertase and obtained the results in table 4.6:

Table 4.6 Sucrose hydrolysis rate data.

Initial sucrose (M)	Initial rate (M/s)
0.0292	0.182
0.0584	0.265
0.0876	0.311
0.117	0.330
0.146	0.349
0.175	0.372
0.205	0.347
0.234	0.371

Estimate v_{max} and K_m from this data.

4-4 A particularly unstable enzyme exhibits the following specific activity versus time profiles as a function of temperature shown in table 4.7:

Table 4.7 Enzyme denaturation curves.

Time (min)	5°C	15°C	25°C	35°C
0	1	1	1	1
1	0.88	0.77	0.60	0.36
2	0.77	0.60	0.36	0.13
3	0.68	0.47	0.22	0.05
4	0.60	0.36	0.13	0.02
5	0.53	0.28	0.08	0.01
6	0.47	0.22	0.05	0.00
7	0.41	0.17	0.03	0.00
8	0.36	0.13	0.02	0.00
9	0.32	0.10	0.01	0.00
10	0.28	0.08	0.01	0.00

What is the half-time for loss of activity at 52°C?

4-5 Firefly luciferase is often used as a reporter enzyme in cell and molecular biology. It catalyzes the oxidation of the substrate luciferin, with the simultaneous emission of a photon. The kinetic parameters for luciferase are $k_{cat} = 0.04$ mol oxyluciferin/mol luciferase/s and $K_m = 100$ μM [75]. Consider an assay wherein 10^5 cells are suspended in a 100 μL microplate well, each cell is expressing 10^5 molecules of luciferase, and 1 mM luciferin is added to the well at time zero. At what time will the light-generation intensity drop to 90% of its maximum value? At what time will it drop to 50% of its maximum value?

4-6 Kinases function by binding ATP and transferring a phosphate to a substrate. ATP analogs function as kinase inhibitors by blocking the ATP-binding pocket. For a particular tyrosine kinase (Met), the ATP dissociation constant ($K_{d,ATP}$) is 5.7×10^{-6} M. Staurosporine is a competitive kinase inhibitor versus ATP with a K_I for this kinase of 190×10^{-9} M. The ATP concentration for an in vitro kinase reaction (total volume = 50 µL) is 100×10^{-6} M. How much staurosporine (MW = 466.5 Da) should be added to achieve 90% inhibition?

4-7 An enzyme at initial concentration $[E]_o = 0.1$ µM is found to exhibit Michaelis-Menten kinetics with the parameters $v_{max} = 0.6$ µM s^{-1} and $K_m = 80$ µM. By stopped-flow analyses, the substrate-enzyme association rate constant is determined to be 2.0×10^6 M^{-1} s^{-1}. Estimate the values of k_{cat} and k_{-1}. On average, what fraction of enzyme-substrate binding events results in product formation? (Adapted from a problem by J. Haugh.)

4-8 In the presence of a particular concentration of inhibitor, the observed v_{max} and K_M are both found to be reduced by a factor of two. What class of inhibition is this consistent with, and what molecular species are likely to be bound by the inhibitor? (Adapted from a problem by J. Haugh.)

4-9 Derive the expression for $v = d[P]/dt$ for the mechanism shown here (note that ES and ES' are simply two different 1:1 complexes of enzyme and the same substrate):

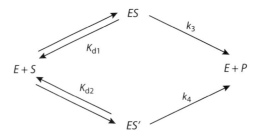

Treat the two reversible reactions each as a fast equilibrium with dissociation constants as indicated (rather than using the QSSA). Express the result in terms of $[E]_o$, $[S]$, and the four constants shown above. Does your expression collapse to the Michaelis-Menten rate law form?

4-10 HIV protease inhibitors target the catalytic site of the viral aspartyl protease, which contains two carboxylate moieties (RCOO$^-$) in close proximity to each other. Here we examine the expected pH dependence of inhibition given certain simplifying assumptions.

The thermodynamic cycle below focuses on the titration state of a single functional group on the enzyme (E^- and EH represent the unprotonated and protonated forms, respectively) that can bind substrate (S) to form a Michaelis complex and then proceed to products and can also bind a competitive inhibitor (I). For simplicity, the substrate is assumed to bind only to the protonated form of the enzyme:

$$IE^- + H^+ + S \rightleftharpoons E^- + H^+ + S + I$$

$$K_a^I \updownarrow \qquad\qquad K_a \updownarrow$$

$$IEH + S \;\overset{K_2}{\rightleftharpoons}\; EH + S + I \;\overset{K_1}{\rightleftharpoons}\; SEH + I \;\overset{k_{cat}}{\longrightarrow}\; \text{products} + I$$

a. Derive a Michaelis-Menten-type expression for the rate of product formation. Recall that this is done by first writing a first-order expression for the rate of product formation from the last step on the bottom line of the above thermodynamic cycle, then substituting in an expression for the concentration of enzyme-containing species in terms of the total enzyme concentration (that is, the concentration of enzyme in all forms, $[E]_o$), and finally rearranging to have the form of the Michaelis-Menten equation ($v = \frac{k_{cat}[E]_o[S]}{K_m + [S]}$). You should assume that all the above equilibria are rapid. Your final result should contain equilibrium and rate constants as well as the concentrations $[E]_o$, $[S]$, $[I]$, and $[H^+]$.

b. Check your result in part a by showing that it collapses to the Michaelis-Menten equation in the appropriate limit ($K_a = K_a^I \to \infty$ (fully protonated), $K_2 \to \infty$ (no inhibitor binding), in which case K_1 is associated with K_m).

c. For the general case (not the limit in part b), is the apparent k_{cat} dependent on pH? Does the apparent K_m for the reaction depend on pH?

d. From your expression for the general reaction rate, does it matter whether the inhibitor I binds more tightly to the unprotonated or protonated form of the enzyme? Explain.

4-11 Assuming that the kinetics follow the Michaelis-Menten expression, derive an expression for the IC_{50} for a competitive inhibitor to the ATP binding site of a protein kinase in terms of the background concentration of ATP substrate, the binding constant for the inhibitor K_I, and the effective Michaelis-Menten constant for the uninhibited reaction K_m.

4-12 Lansoprazole (marketed as Prevacid) is a therapeutic option for gastroesophageal reflux disease. Aside from its dominant biological activity, it also uncompetitively inhibits an alkaline phosphatase enzyme. Note that alkaline phosphatase binds its natural substrate with 5 µM affinity.

a. Derive an expression for the relationship between the equilibrium dissociation constant (for enzyme/inhibitor) and the IC_{50} at a given substrate concentration $[S]$.

b. If the substrate is present in the cell at 20 µM, what concentration of lansoprazole must be delivered to reduce natural product formation by 10% from its uninhibited value? Assume the enzyme-inhibitor equilibrium dissociation constant is 1.5 mM.

4-13 You performed an experiment to characterize an enzyme of interest, which contains 917 amino acids. You added 0.25 g/L of purified enzyme to a test tube along with the indicated amounts of glucose substrate. You measured the initial rates of phosphorylation, which are also indicated in table 4.8.

Table 4.8 Enzyme initial rate data.

Glucose (mg/L)	Initial rate (µM/s)
0.36	0.013
3.6	0.136
36	0.719
360	2.129
3,600	2.244
36,000	2.550

a. What initial phosphorylation rate do you expect to observe if you use 10 mg/L glucose?

b. How will this change if you double the amount of enzyme to 0.5 g/L?

c. Alternatively, how will the initial phosphorylation rate change if you double the amount of glucose (i.e., 0.25 g/L enzyme and 20 mg/L glucose)?

d. What is the value of the catalytic rate constant k_{cat}?

e. As an alternative to doubling the amount of enzyme or substrate, you decide to engineer an improved enzyme. Glucose dissociates from the improved enzyme at half the rate at which it dissociates from the original enzyme. What is the initial phosphorylation rate with 0.25 g/L of new enzyme and 10 mg/L glucose? You may assume that the association rate constant remains 3.7×10^6 M^{-1} s^{-1}.

f. How much, if at all, does the answer to part e differ if you assume rapid equilibrium or quasi-steady-state complex?

4-14 Consider an enzymatic mechanism in which product formation is reversible (i.e., enzyme can catalyze conversion of product to substrate):

$$E + S \underset{k_{-1}}{\overset{k_1}{\rightleftharpoons}} ES \underset{k_{-2}}{\overset{k_2}{\rightleftharpoons}} E + P \qquad (4.99)$$

a. Derive an expression for the net rate of product formation.

b. Compare this rate to the rate observed if product formation is irreversible.

c. Evaluate the rate (with reversible product formation) in the limit of low product concentration.

d. Evaluate the rate (with reversible product formation) in the limit of a low rate constant for product conversion to enzyme-substrate complex.

4-15 You are studying an enzyme that catalyzes substrate dimerization. The enzyme has two identical binding sites, each with an intrinsic affinity K_d for the substrate (i.e., the equilibrium dissociation constant for a single binding site and a single substrate, in the absence of the other binding site, equals K_d). Binding of substrate at one site does not change the free energy of binding at the other site. When two substrates are bound, the enzyme catalyzes dimerization with a first-order rate constant k_{cat}. The dimerized product immediately dissociates from the enzyme and does not rebind to the enzyme.

Derive an expression for the rate of product formation as a function of the total enzyme concentration ($[E]_o$), unbound substrate concentration ($[S]$), and the constants K_d and k_{cat}. You may assume rapid equilibrium for the binding events. At what concentration of unbound substrate is the rate of product formation half of its maximal value?

4-16 Enzyme and substrate reversibly bind to form a complex. Product is irreversibly formed from the complex. Yet if either the enzyme or the substrate is protonated, they do not associate to form a complex. The complex itself cannot become protonated. Show that the expression for the rate of product formation is

$$\frac{d[P]}{dt} = \frac{v_{\text{max,app}}[S]_u}{K_{\text{m,app}} + [S]_u} \qquad (4.100)$$

in which $[S]_u$ is the concentration of substrate that is not in complex with enzyme; this includes the deprotonated and protonated forms. Write a mechanism that fits this description.

4-17 You have engineered a protease to catalytically break the backbone amide bond of a peptide (histidine-alanine-glutamine) substrate. The substrate binds to the enzyme active site only if the substrate's histidine is positively charged. Conversely, an aspartic acid residue in the enzyme's active site must be negatively charged to enable binding.

  a. Draw a schematic of the reaction mechanisms.
  b. Derive an expression for the rate of product formation as a function of rate constants and concentrations of total protease, unbound peptide (in either protonation state), and H^+.
  c. Plot the initial reaction rate as a function of pH for the following parameter values:
     protease-substrate association rate constant = $10^6 \text{ M}^{-1} \text{ s}^{-1}$
     protease-substrate dissociation rate constant = 2 s^{-1}
     catalytic turnover number = 10 s^{-1}
     total enzyme concentration = 1 nM
     unbound peptide concentration = 2 μM
  d. What pH yields the maximum reaction rate?

4-18 An enzyme noncovalently binds substrate and irreversibly forms product. This catalytic behavior is inhibited by noncovalent binding of a small molecule, which can bind *cooperatively* in multiple locations on the enzyme. Small-molecule binding prevents substrate binding.

  a. Derive an expression for the rate of reaction as a function of constants, substrate concentration, and inhibitor concentration.
  b. You performed a series of inhibitor titrations in which you measured initial reaction rate at various inhibitor concentrations. Each titration was performed in the presence of a different amount of substrate. Determine the effective valency of inhibitor binding to enzyme based on the data in the table below.

$[S]_o$ (μM)	$[I]_o$ (μM)	v (μM/s)
0.2	0.1	0.175
0.2	0.2	0.173
0.2	0.4	0.143
0.2	0.8	0.085
0.2	1.6	0.02
0.2	3.2	0.003
0.2	6.4	0.0
0.2	12.8	0.0
1	0.1	0.524
1	0.2	0.471
1	0.4	0.493
1	0.8	0.314
1	1.6	0.103
1	3.2	0.016
1	6.4	0.002
1	12.8	0.0
5	0.1	0.84
5	0.2	0.873
5	0.4	0.857
5	0.8	0.676
5	1.6	0.335
5	3.2	0.073
5	6.4	0.009
5	12.8	0.001

References

1. Nomenclature Committee of the International Union of Biochemistry and Molecular Biology (NC-IUBMB) (ed.). *Enzyme Nomenclature*. Academic Press (1992).

2. Nomenclature Committee of the International Union of Biochemistry and Molecular Biology (NC-IUBMB). Enzyme nomenclature. Recommendations 1992. Supplement: Corrections and additions. *Eur. J. Biochem.* 223: 1–5 (1994).

3. Nomenclature Committee of the International Union of Biochemistry and Molecular Biology (NC-IUBMB). Enzyme nomenclature. Recommendations 1992. Supplement 2: Corrections and additions (1994). *Eur. J. Biochem.* 232: 1–6 (1995).

4. Nomenclature Committee of the International Union of Biochemistry and Molecular Biology (NC-IUBMB). Enzyme nomenclature. Recommendations 1992. Supplement 3: Corrections and additions (1995). Eur. J. Biochem. 237: 1–5 (1996).

5. Nomenclature Committee of the International Union of Biochemistry and Molecular Biology (NC-IUBMB). Enzyme nomenclature. Recommendations 1992. Supplement 4: Corrections and additions (1997). *Eur. J. Biochem.* 250: 1–6 (1997).

6. Nomenclature Committee of the International Union of Biochemistry and Molecular Biology (NC-IUBMB). Enzyme supplement 5 (1999). *Eur. J. Biochem.* 264: 610–650 (1999).

7. N. H. Yennawar, H. P. Yennawar, and G. K. Farber. X-ray crystal structure of γ-chymotrypsin in hexane. *Biochemistry* 33: 7326–7336 (1994).

8. V. Henri. Théorie générale de l'action de quelques diastases. *C. R. Acad. Sci. Paris* 135: 916–919 (1902).

9. L. Michaelis and M. L. Menten. Die Kinetik der Invertinwirkung. *Biochem. Z.* 49: 333–369 (1913).

10. G. E. Briggs and J. B. S. Haldane. A note on the kinetics of enzyme action. *Biochem. J.* 19: 338–339 (1925).

11. R. J. Ritchie and T. Prvan. A simulation study on designing experiments to measure the K_m of Michaelis-Menten kinetics curves. *J. Theor. Biol.* 178: 239–254 (1996).

12. A. R. Fersht. *Enzyme Structure and Mechanism*, second edition. W. H. Freeman (1985).

13. R. W. Maatman. Enzyme and solid catalyst efficiencies and solid catalyst site densities. *Catal. Rev.* 8: 1–28 (1974).

14. M. Hatakeyama, S. Kimura, T. Naka, T. Kawasaki, N. Yumoto, M. Ichikawa, J. H. Kim, K. Saito, M. Saeki, M. Shirouzu, S. Yokoyama, and A. Konagaya. A computational model on the modulation of mitogen-activated protein kinase (MAPK) and Akt pathways in heregulin-induced ErbB signalling. *Biochem. J.* 373: 451–463 (2003).

15. A. L. Kuharsky and A. L. Fogelson. Surface-mediated control of blood coagulation: The role of binding site densities and platelet deposition. *Biophys. J.* 80: 1050–1074 (2001).

16. Z. A. Knight and K. M. Shokat. Features of selective kinase inhibitors. *Chem. Biol.* 12: 621–637 (2005).

17. T. W. Traut. Physiological concentrations of purines and pyrimidines. *Mol. Cell. Biochem.* 94: 1–22 (1994).

18. A. Bar-Even, E. Noor, Y. Savir, W. Liebermeister, D. David, D. Tawfik, and R. Milo. The moderately efficient enzyme: Evolutionary and physicochemical trends shaping enzyme parameters. *Biochemistry* 50: 4402–4410 (2011).

19. M. V. Smoluchowski. Versuch einer mathematischen Theorie der Koagulationskinetik kolloider Lösungen. *Z. Phys. Chem.* 92: 129–168 (1917).

20. C. J. Camacho, Z. Weng, S. Vajda, and C. DeLisi. Free energy landscapes of encounter complexes in protein-protein association. *Biophys. J.* 76: 1166–1178 (1999).

21. D. Voet, J. G. Voet, and C. W. Pratt. *Fundamentals of Biochemistry: Life at the Molecular Level*, fourth edition. Wiley (2015).

22. Y. Lin, P. Lu, C. Tang, Q. Mei, G. Sandig, A. D. Rodrigues, T. H. Rushmore, and M. Shou. Substrate inhibition kinetics for cytochrome P450-catalyzed reactions. *Drug Metab. Dispos.* 29: 368–374 (2001).

23. K. C. Jones and K. G. Mann. A model for the tissue factor pathway to thrombin. II. A mathematical simulation. *J. Biol. Chem.* 269: 23,367–23,373 (1994).

24. P. V. LoGrasso, B. Frantz, A. M. Rolando, S. J. O'Keefe, J. D. Hermes, and E. A. O'Neill. Kinetic mechanism for p38 MAP kinase. *Biochemistry* 36: 10,422–10,427 (1997).

25. B. Zhou and Z.-Y. Zhang. Mechanism of mitogen-activated protein kinase phosphatase-3 activation by ERK2. *J. Biol. Chem.* 274: 35,526–35,534 (1999).

26. R. S. Yadava, E. M. Mahen, and M. J. Fedor. Kinetic analysis of ribozyme-substrate complex formation in yeast. *RNA* 10: 863–879 (2004).

27. L. A. Segel. On the validity of the steady state assumption of enzyme kinetics. *Bull. Math. Biol.* 50: 579–593 (1988).

28. W. W. Cleland. The kinetics of enzyme-catalyzed reactions with two or more substrates or products. I. Nomenclature and rate equations. *Biochim. Biophys. Acta.* 67: 104–137 (1963).

29. D. Purich. *Enzyme Kinetics: Catalysis and Control*. Elsevier (2010).

30. I. Segel. *Enzyme Kinetics: Behavior and Analysis of Rapid Equilibrium and Steady-State Enzyme Systems*. Wiley (1993).

31. K. Dalziel. Initial steady state velocities in the evaluation of enzyme-coenzyme-substrate reaction mechanisms. *Acta Chem. Scand.* 11: 1706–1723 (1957).

32. J. Monod, J.Wyman, and J.-P. Changeux. On the nature of allosteric transitions: A plausible model. *J. Mol. Biol.* 12: 88–118 (1965).

33. D. E. Koshland, Jr., G. Némethy, and D. Filmer. Comparison of experimental binding data and theoretical models in proteins containing subunits. *Biochemistry* 5: 365–385 (1966).

34. R. A. Alberty. Kinetic effects of the ionization of groups in the enzyme molecule. *J. Cell. Comp. Physiol.* 47: 245–281 (1956).

35. T. E. Creighton. *Proteins: Structures and Molecular Properties*, second edition. W. H. Freeman (1992).

36. D. J. Hei and D. S. Clark. Estimation of melting curves from enzymatic activity-temperature profiles. *Biotechnol. Bioeng.* 42: 1245–1251 (1993).

37. J. Ren, R. Esnouf, E. Garman, D. Somers, C. Ross, I. Kirby, J. Keeling, G. Darby, Y. Jones, D. Stuart, and D. Stammers. High resolution structures of HIV-1 RT from four RT-inhibitor complexes. *Nat. Struct. Biol.* 2: 293–302 (1995).

38. O. Arunlakshana and H. O. Schild. Some quantitative uses of drug antagonists. *Br. J. Pharmacol. Chemother.* 14: 48–58 (1959).

39. V. Calderone, B. Baragatti, M. Breschi, and E. Martinotti. Experimental and theoretical comparisons between the classical Schild analysis and a new alternative method to evaluate the pA2 of competitive antagonists. *Naunyn-Schmiedeberg's Arch. Pharmacol.* 360: 477–487 (1999).

40. J. W. Williams and J. F. Morrison. The kinetics of reversible tight-binding inhibition. *Methods Enzymol.* 63: 437–467 (1979).

41. R. A. Spence, W. M. Kati, K. S. Anderson, and K. A. Johnson. Mechanism of inhibition of HIV-1 reverse transcriptase by nonnucleoside inhibitors. *Science* 267: 988–993 (1995).

42. W. M. Kati, K. A. Johnson, L. F. Jerva, and K. S. Anderson. Mechanism and fidelity of HIV reverse transcriptase. *J. Biol. Chem.* 267: 25,988–25,997 (1992).

43. G. M. Rubin, M. D. Yandell, J. R. Wortman, G. L. G. Miklos, C. R. Nelson, I. K. Hariharan, M. E. Fortini, P. W. Li, R. Apweiler, W. Fleischmann, J. M. Cherry, S. Henikoff, M. P. Skupski, S. Misra, M. Ashburner, E. Birney, M. S. Boguski, T. Brody, P. Brokstein, S. E. Celniker, S. A. Chervitz, D. Coates, A. Cravchik, A. Gabrielian, R. F. Galle, W. M. Gelbart, R. A. George, L. S. B. Goldstein, F. C. Gong, P. Guan, N. L. Harris, B. A. Hay, R. A. Hoskins, J. Y. Li, Z. Y. Li, R. O. Hynes, S. J. M. Jones, P. M. Kuehl, B. Lemaitre, J. T. Littleton, D. K. Morrison, C. Mungall, P. H. O'Farrell, O. K. Pickeral, C. Shue, L. B. Vosshall, J. Zhang, Q. Zhao, X. Q. H. Zheng, F. Zhong, W. Y. Zhong, R. Gibbs, J. C. Venter, M. D. Adams, and S. Lewis. Comparative genomics of the eukaryotes. *Science* 287: 2204–2215 (2000).

44. M. Huse and J. Kuriyan. The conformational plasticity of protein kinases. *Cell* 109: 275–282 (2002).

45. S. R. Hubbard. Crystal structures of the activated insulin receptor tyrosine kinase in complex with peptide substrate and ATP analog. *EMBO J.* 16: 5572–5581 (1997).

46. P. A. Schwartz, P. Kuzmic, J. Solowiej, S. Bergqvist, B. Bolanos, C. Almaden, A. Nagata, K. Ryan, J. Feng, D. Dalvie, J. C. Kath, M. Xu, R.Wani, and B.W. Murray. Covalent EGFR inhibitor analysis reveals importance of reversible interactions to potency and mechanisms of drug resistance. *Proc. Natl. Acad. Sci. U.S.A.* 111: 173–178 (2014).

47. B. Alberts, A. D. Johnson, J. Lewis, D. Morgan, M. Raff, K. Roberts, and P. Walter. *Molecular Biology of the Cell*, sixth edition. W. W. Norton & Company (2014).

48. R. Heinrich and T. A. Rapaport. A linear steady-state treatment of enzymatic chains. *Eur. J. Biochem.* 42: 89–95 (1973).

49. H. Kacser and J. A. Burns. The control of flux. *Symp. Soc. Exp. Biol.* 27: 65–104 (1973).

50. H. Kacser and J. A. Burns. The control of flux. *Biochem. Soc. Trans.* 23: 341–366 (1973).

51. D. A. Fell. Metabolic control analysis: A survey of its theoretical and experimental development. *Biochem. J.* 286: 313–330 (1992).

52. G. N. Stephanopoulos, A. A. Aristidou, and J. Nielsen. *Metabolic Engineering: Principles and Methodologies.* Academic Press (1998).

53. J. L. Galazzo and J. E. Bailey. Fermentation pathway kinetics and metabolic flux control in suspended and immobilized *Saccharomyces cerevisiae*. *Enzyme Microb. Technol.* 12: 162–172 (1990).

54. Y. Kashiwaya, K. Sato, N. Tsuchiya, S. Thomas, D. A. Fell, R. L. Veech, and J. V. Passonneau. Control of glucose utilization in working perfused rat heart. *J. Biol. Chem.* 269: 25,502–25,514 (1994).

55. P. de Noronha Pissara, J. Nielsen, and M. J. Bazin. Pathway kinetics and metabolic control analysis of a high-yielding strain of *Penicillium chrysogenum* during fed batch cultivations. *Biotechnol. Bioeng.* 51: 168–176 (1996).

56. B. M. Bakker, P. A. M. Michels, F. R. Opperdoes, and H. V. Westerhoff. What controls glycolysis in bloodstream form *Trypanosoma brucei*? *J. Biol. Chem.* 274: 14,551–14,559 (1999).

57. M. W. Covert, I. Famili, and B. Ø. Palsson. Identifying constraints that govern cell behavior: A key to converting conceptual to computational models in biology? *Biotechnol. Bioeng.* 84: 763–772 (2003).

58. J. Schellenberger, R. Que, R. M. T. Fleming, I. Thiele, J. D. Orth, A. M. Feist, D. C. Zielinski, A. Bordbar, N. E. Lewis, S. Rahmanian, J. Kang, D. R. Hyduke, and B. Ø. Palsson. Quantitative prediction of cellular metabolism with constraint-based models: The COBRA Toolbox v2.0. *Nat. Protoc.* 6: 1290–1307 (2011).

59. N. E. Lewis, H. Nagarajan, and B. Ø. Palsson. Constraining the metabolic genotype-phenotype relationship using a phylogeny of in silico methods. *Nat. Rev. Microbiol.* 10: 291–305 (2012).

60. M. G. Grutter. Caspases: Key players in programmed cell death. *Curr. Opin. Struct. Biol.* 10: 649–655 (2000).

61. M. Fussenegger, J. E. Bailey, and J. Varner. A mathematical model of caspase function in apoptosis. *Nat. Biotechnol.* 18: 768–774 (2000).

62. M. Bentele, I. Lavrik, M. Ulrich, S. Stöber, D. W. Heermann, H. Kalthoff, P. H. Krammer, and R. Eils. Mathematical modeling reveals threshold mechanism in CD95-induced apoptosis. *J. Cell Biol.* 166: 839–851 (2004).

63. F. Hua, M. G. Cornejo, M. H. Cardone, C. L. Stokes, and D. A. Lauffenburger. Effects of Bcl-2 levels on Fas signaling-induced caspase-3 activation: Molecular genetic tests of computational model predictions. *J. Immunol.* 175: 985–995 (2005).

64. B. Dahlbäck. Blood coagulation. *Lancet* 355: 1627–1632 (2000).

65. F. I. Ataullakhanov and M. A. Panteleev. Mathematical modeling and computer simulation in blood coagulation. *Pathophysiol. Haemost. Thromb.* 34: 60–70 (2005).

66. M. E. Nesheim, R. P. Tracy, and K. G. Mann. "Clotspeed," a mathematical simulation of the functional properties of prothrombinase. *J. Biol. Chem.* 259: 1447–1453 (1984).

67. M. F. Hockin, K. C. Jones, S. J. Everse, and K. G. Mann. A model for the stoichiometric regulation of blood coagulation. *J. Biol. Chem.* 277: 18,322–18,333 (2002).

68. K. G. Mann. Is there value in kinetic modeling of thrombin generation? Yes. *J. Thromb. Haemost.* 10: 1463–1469 (2012).

69. J. H. Lawson, M. Kalafatis, S. Stram, and K. G. Mann. A model for the tissue factor pathway to thrombin. I. An empirical study. *J. Biol. Chem.* 269: 23,357–23,366 (1994).

70. J. D. Mott and Z. Werb. Regulation of matrix biology by matrix metalloproteinases. *Curr. Opin. Cell Biol.* 16: 558–564 (2004).

71. C. P. Blobel. ADAMs: Key components in EGFR signalling and development. *Nat. Rev. Mol. Cell Biol.* 6: 32–43 (2005).

72. S. Y. Shvartsman, M. P. Hagan, A. Yacoub, P. Dent, H. S. Wiley, and D. A. Lauffenburger. Autocrine loops with positive feedback enable context-dependent cell signaling. *Am. J. Physiol. Cell Physiol.* 282: C545–C559 (2002).

73. H. Ohtsu, P. J. Dempsey, and S. Eguchi. ADAMs as mediators of EGF receptor transactivation by G protein-coupled receptors. *Am. J. Physiol. Cell Physiol.* 291: C1–C10 (2006).

74. A. M. Chase, H. C. V. Meier, and V. J. Menna. The non-competitive inhibition and irreversible inactivation of yeast invertase by urea. *J. Cell. Phys.* 59: 1–13 (1962).

75. J. M. Ignowski and D. V. Schaffer. Kinetic analysis and modeling of firefly luciferase as a quantitative reporter gene in live mammalian cells. *Biotechnol. Bioeng.* 86: 827–834 (2004).

5
Gene Expression and Protein Trafficking

The strict application of the balance sheet-quantitative analysis method permitted to trace their respective distribution among the various cellular compartments and thus, determine the specific role they performed in the life of the Cell.... We have entered the cell, the Mansion of our birth, and started the inventory of our acquired wealth.
—Albert Claude, Nobel Lecture, 1974

Now that we know the nature of such mechanisms, we have the possibility of learning to master them, with all the consequences which that will surely entail.
—Sven Gard, Presentation of 1965 Nobel Prize to Francois Jacob, Andre Lwoff, and Jacques Monod

So far, we've considered the rate processes involved in protein binding events and enzyme-catalyzed reactions. Now, let's turn our attention to rates of gene expression and protein trafficking throughout the cell. Although these are complex processes built up from numerous separate reactions, they can often be usefully described with simplified empirical mass-action rate forms.

5.1 Synthesis, Degradation, and Growth

To a first approximation, the events of mRNA transcription, polypeptide translation, protein folding, protein degradation, and protein trafficking in membrane vesicles can be described by mass-action kinetics. However, doing so requires defining the relevant concentrations in terms of their true intracellular values rather than on a culture volume basis. Because the cellular volume in a culture continually increases in a growing cell population, intracellular concentrations can change even in the absence of fresh synthesis or degradation. Let's write the transient material balance for a particular mRNA species to quantitatively consider this effect. We will see that it is useful to track the number of mRNA molecules rather than their concentration. We define $[mRNA]$ as the number of molecules (or moles) of a particular mRNA transcript per unit of intracellular volume, and V_i as the intracellular volume in the culture (for example, the average intracellular volume per cell × the number

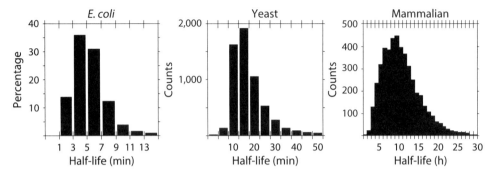

Figure 5.1 mRNA half-lives ($\ln(2)/\gamma_R$) vary markedly across species (*Escherichia coli* [1], yeast [2], and human [3]) and also exhibit significant variability within a species. Notably, there appears to be a scaling law between mRNA half-life and cell doubling time, with average mRNA half-life approximately one-fifth of the cell doubling time [4].

of cells in the culture). It should be noted that this intracellular volume is quite a bit smaller than the liquid culture volume, except for very dense cell suspensions or tissues. The material balance in terms of numbers of mRNA molecules is then as follows:

$$Accumulation = Synthesis - Degradation \qquad (5.1)$$

$$\frac{d\,([mRNA]V_i)}{dt} = k_R V_i - \gamma_R [mRNA] V_i \qquad (5.2)$$

Note in particular that the rate constant k_R will be a function of gene copy number, the specific transcription factors bound to the gene's promoter, the histone binding status of the promoter in a eukaryotic cell, the availability of RNA polymerase components, and a host of other factors that influence a gene's transcription. For the purposes of a model this simple, however, we will assume that k_R is constant during the time period of interest. We also assume that γ_R, the first-order rate constant for transcript degradation, is constant, but its specific value is strongly transcript and species dependent (figure 5.1).

Expanding the derivative in equation 5.2 by the chain rule,

$$[mRNA]\frac{dV_i}{dt} + V_i\frac{d[mRNA]}{dt} = k_R V_i - \gamma_R [mRNA] V_i \qquad (5.3)$$

and rearranging, we have

$$\frac{d[mRNA]}{dt} = k_R - \gamma_R [mRNA] - [mRNA]\frac{1}{V_i}\frac{dV_i}{dt} \qquad (5.4)$$

However, as we shall see in chapter 7, the rate of increase in cell mass, volume, or number is often described by a specific growth rate μ:

$$\frac{dV_i}{dt} = \mu V_i \tag{5.5}$$

When substituted into equation 5.4, the result is

$$\frac{d[mRNA]}{dt} = k_R - (\gamma_R + \mu)[mRNA] \tag{5.6}$$

Had we neglected the change in intracellular volume over time (i.e., assumed constant V_i), equation 5.2 would have become

$$\frac{d[mRNA]}{dt} = k_R - \gamma_R[mRNA] \tag{5.7}$$

A comparison of equations 5.6 and 5.7 reveals that an additional term, $-\mu[mRNA]$, arises when accounting for the change in intracellular volume. This term represents dilution of the mRNA by cell growth; the practical necessity of accounting for this process depends on the relative magnitudes of the terms μ and γ_R. To a first approximation, figure 5.1 suggests that $\mu \approx \frac{1}{5}\gamma_R$ for an average transcript, so the dilution term can be reasonably neglected in many cases. There is a large distribution of γ_R values, though, so dilution may be significant for a specific transcript of interest.

Continuing with the general case (i.e., not neglecting dilution), the steady-state mRNA concentration, $[mRNA]_{ss}$, can be obtained by setting the time derivative in equation 5.6 to zero:

$$[mRNA]_{ss} = \frac{k_R}{\gamma_R + \mu} \tag{5.8}$$

We can also obtain $[mRNA]$ as a function of time. With the initial condition that $[mRNA] = 0$ at $t = 0$, equation 5.6 can be solved analytically to yield:

$$[mRNA] = \frac{k_R}{\gamma_R + \mu}\left(1 - e^{-(\gamma_R + \mu)t}\right) \tag{5.9}$$

It is noteworthy that the time required to reach steady state is independent of the magnitude of the constant rate of transcription (i.e., k_R) and depends only on the rates of degradation (γ_R) and dilution by growth (μ). Experimental validation of equation 5.9 has been obtained by examination of the induction time versus turnover time for 25 mammalian mRNA species (figure 5.2) [5].

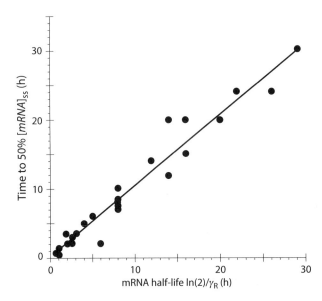

Figure 5.2 Correlation ($R^2 = 0.95$) between half-times for mRNA induction and degradation for 25 mammalian transcripts. Adapted from reference 5.

Example 5-1 Mouse L cells were grown in culture (doubling time 15 h), and ^{3}H-uridine was added to the culture at time zero [6]. At subsequent times, polyA-mRNA was extracted from cell aliquots, and the following degree of radiolabeling was measured as a function of time:

Time (h)	Fractional radiolabeling
0.5	0.026
1.0	0.052
2.0	0.19
3.0	0.23
5.0	0.45
10	0.69
20	0.89
25	0.81
30	0.99

Estimate the average half-life of polyA-mRNA in mouse L cells.

Solution We expect the data to approach steady state exponentially, with an observed rate constant $(\gamma_R + \mu)$. However, in the actual experiment, there are also delays in the uptake of ^{3}H-uridine and equilibration with the nucleotide precursor pool before any synthesis can be detected. We can account for these delays by modifying equation 5.9 to introduce a time delay constant, τ:

$$\frac{[mRNA]}{[mRNA]_{ss}} = 1 - e^{-(\gamma_R + \mu)(t-\tau)} \tag{5.10}$$

Plotting the data and performing a nonlinear least squares fit with this equation yields parameter estimates $(\gamma_R + \mu) = 0.11$ h^{-1} and $\tau = 0.29$ h, giving the fit shown in figure 5.3. Because $\mu = (\ln 2)/(15 \text{ h}) = 0.046$ h^{-1}, $\gamma_R = 0.088$ h^{-1}, and the half-life for degradation of polyA-mRNA is $(\ln 2)/0.088$ h$^{-1} = 8$ h.

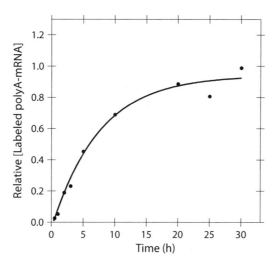

Figure 5.3 Data (circles) and nonlinear least squares fit (line) of equation 5.10 for incorporation of radiolabel into newly synthesized polyA-mRNA [6].

Let's now turn to the material balance for the protein $[P]$ encoded by this particular mRNA of interest. We assume that the rate of new protein synthesis is proportional to mRNA concentration and that other required components for translation, such as ribosomes, are in excess. By a process analogous to that for deriving the mRNA balance, we obtain

$$\frac{d[P]}{dt} = k_P[mRNA] - (\gamma_P + \mu)[P] \tag{5.11}$$

Note that in contrast to mRNA- and protein-specific rate constants for synthesis and degradation, dilution due to cell growth globally affects [mRNA] and [P] equivalently, so µ is the same. Substituting equation 5.9 into equation 5.11 yields

$$\frac{d[P]}{dt} = \frac{k_P k_R}{\gamma_R + \mu} \left(1 - e^{-(\gamma_R + \mu)t}\right) - (\gamma_P + \mu)[P] \qquad (5.12)$$

We can find the steady-state protein level $[P]_{ss}$ by setting $\frac{d[P]}{dt} = 0$ and letting $t \to \infty$:

$$[P]_{ss} = \frac{k_P k_R}{(\gamma_P + \mu)(\gamma_R + \mu)} \qquad (5.13)$$

$$\frac{[P]_{ss}}{[mRNA]_{ss}} = \frac{k_P}{\gamma_P + \mu} \qquad (5.14)$$

Examination of equation 5.14 suggests that for a given protein, steady-state protein levels should be proportional to steady-state mRNA levels. Figure 5.4 shows the experimentally determined levels of protein and mRNA in exponentially growing *Saccharomyces cerevisiae*. There is a slightly greater than proportional increase in protein level with increasing mRNA. However, because different mRNAs are translated with varying efficiency (i.e., varying k_P), and different proteins have different

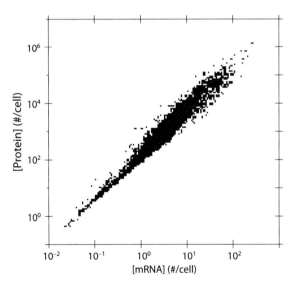

Figure 5.4 Protein and mRNA levels in *S. cerevisiae*, from 24 different studies that were rigorously analyzed to minimize the impact of noisy measurements [8]. mRNA levels are found to account for approximately 85% of the variation in protein levels.

metabolic half-lives (i.e., varying γ_P), a single proportionality constant does not apply for all protein:mRNA ratios. These proportionality constants have been measured at steady state across 20 different human cell lines and tissues, and they were found to range from ≈ 200 to $\approx 200{,}000$ protein molecules per mRNA molecule [7]. However, for a given mRNA and its corresponding protein, this proportionality constant was intriguingly found to be largely independent of its cellular origin, suggesting that the ratio given by equation 5.14 is an intrinsic gene-product property and is not strongly cell-type dependent.

Case Study 5-1 H. C. Towle, C. N. Mariash, H. L. Schwartz, and J. H. Oppenheimer. Quantitation of rat liver messenger ribonucleic acid for malic enzyme during induction by thyroid hormone. *Biochemistry* 20: 3486–3492 (1981) [9].

Levels of mRNA and protein for malic enzyme in rat liver were found to respond to treatment with a thyroid hormone with the kinetics shown in figure 5.5. The apparent delay in accumulation of enzyme activity had previously been attributed to a refractory period.

Deriving a form of equation 5.15, the authors tested the hypothesis that the observed enzyme activity accumulation rate is consistent with simple zeroth-order transcription and translation, and with first-order degradation of mRNA and enzyme. No dilution due to growth was included, because the liver cells are not expected to grow significantly during the course of the experiment. The model predictions are shown in figure 5.6.

Given the reasonable agreement between the experimental data and the model (with no adjustable parameters) based on equation 5.15, the apparent delay can be explained by

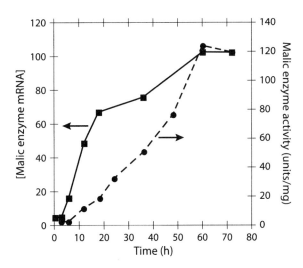

Figure 5.5 Enzyme activity rises more slowly than its mRNA level.

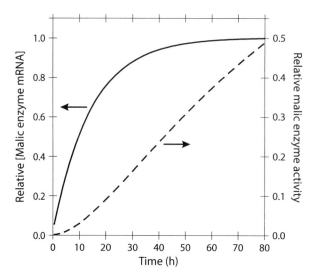

Figure 5.6 Equation 5.15 is plotted using the experimentally determined values of the turnover rate constants $\gamma_R = 0.0693\ h^{-1}$ and $\gamma_P = 0.0103\ h^{-1}$.

the requirement to accumulate a sufficient concentration of mRNA before protein synthesis is significant, and no additional biochemical processes need be invoked. It should also be noted that the model predicts that the enzyme has reached only approximately half of its eventual steady-state level, which could be tested with additional experiments.

The transient solution when $[P] = 0$ at $t = 0$ is straightforwardly solved to yield the following general expression:

$$[P] = [P]_{ss}\left(1 + \frac{(\gamma_R + \mu)e^{-(\gamma_P + \mu)t} - (\gamma_P + \mu)e^{-(\gamma_R + \mu)t}}{\gamma_P - \gamma_R}\right) \quad (5.15)$$

Equation 5.15 can be simplified for the common case of proteins that persist far longer than their corresponding mRNA transcripts. Protein turnover rates can vary widely; regulated proteolysis is one means of controlling some protein levels, and proteins damaged by heat shock or other environmental insults are generally degraded rapidly. However, it is not unusual for bacterial and mammalian proteins to exhibit turnover half-lives from hours to days [10–12]. Let's consider the particular case where the mRNA is much less stable than the protein (i.e., $\gamma_R \gg \gamma_P$),

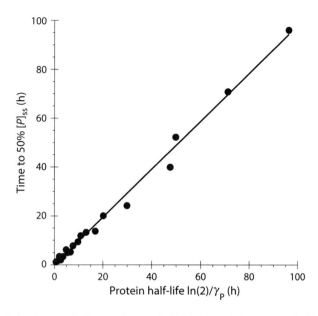

Figure 5.7 Correlation between half-times for protein induction and degradation for 20 mammalian proteins (adapted from reference 5). Fewer proteins were analyzed than mRNAs in figure 5.2, because several proteins were secretory and therefore did not accumulate intracellularly.

and the mRNA is degraded more rapidly than it is diluted by growth (i.e., $\gamma_R \gg \mu$). Under these conditions, equation 5.15 reduces to

$$[P] = \frac{k_P k_R}{\gamma_R (\gamma_P + \mu)} \left(1 - e^{-(\gamma_P + \mu)t}\right) \tag{5.16}$$

As for mRNA, the time scale for approach of $[P]$ to a new steady-state value depends on the degradation constant γ_P but is independent of the magnitude of the transcription (k_R) and translation (k_P) rates. This relationship has been validated experimentally (figure 5.7).

So far, we have used a continuum approximation when describing the events of transcription and translation, but transcription in particular is often stochastic in nature, with *bursts* of expression. For example, note how many of the mRNAs in figure 5.4 are present at an average of ten molecules per cell or less. One might expect considerable fluctuations in this number from cell to cell and over time, and in chapter 9, we introduce the underlying theory for such probabilistic events and present relevant stochastic models of gene transcription. For now, let us continue with some examples using the continuum models derived in this chapter.

Example 5-2 Derive the relative changes in steady-state protein concentration and half-time for approach to steady state for the following variations in a cellular system: halved growth rate, doubled protein stability, doubled transcription, and doubled mRNA half-life.

Solution From equation 5.13, the ratio of steady-state protein concentrations for two states is

$$\frac{[P]_{ss,2}}{[P]_{ss,1}} = \frac{k_{P,2} k_{R,2} (\gamma_{R,1} + \mu_1)(\gamma_{P,1} + \mu_1)}{k_{P,1} k_{R,1} (\gamma_{R,2} + \mu_2)(\gamma_{P,2} + \mu_2)} \quad (5.17)$$

Thus, for a halved growth rate ($\mu_2 = 0.5\mu_1$):

$$\frac{[P]_{ss,2}}{[P]_{ss,1}} = \frac{(\gamma_R + \mu_1)(\gamma_P + \mu_1)}{(\gamma_R + 0.5\mu_1)(\gamma_P + 0.5\mu_1)} = \frac{(\frac{\gamma_R}{\mu_1} + 1)(\frac{\gamma_P}{\mu_1} + 1)}{(\frac{\gamma_R}{\mu_1} + 0.5)(\frac{\gamma_P}{\mu_1} + 0.5)} \quad (5.18)$$

For a case with $\gamma_P = 0.01$ h^{-1}, $\gamma_R = 0.08$ h^{-1}, and $\mu_1 = 0.04$ h^{-1}, halving the growth rate would double the steady-state protein concentration.

Instead of a modified growth rate, consider doubling protein stability:

$$\frac{[P]_{ss,2}}{[P]_{ss,1}} = \frac{\gamma_{P,1} + \mu}{0.5\gamma_{P,1} + \mu} = \frac{\frac{\gamma_{P,1}}{\mu} + 1}{0.5\frac{\gamma_{P,1}}{\mu} + 1} \quad (5.19)$$

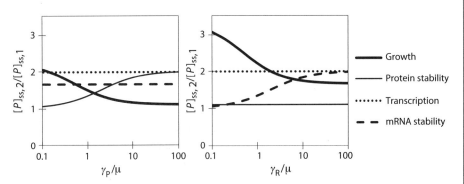

Figure 5.8 The ratios of steady-state protein concentrations for two different systems—halved growth rate, doubled protein stability, doubled transcription, and doubled mRNA half-life—are presented. The left panel varies protein stability for a fixed γ_R/μ value of 4. The right panel varies mRNA stability for a fixed γ_P/μ value of 0.25.

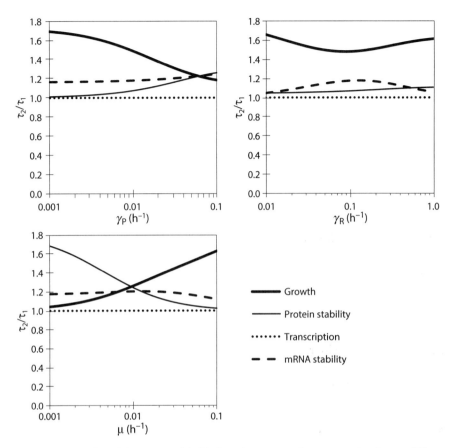

Figure 5.9 The ratios of times to reach half of steady-state protein concentration for two different systems—halved growth rate, doubled protein stability, doubled transcription, and doubled mRNA half-life—are presented. The upper-left panel varies protein stability for a fixed $\gamma_R = 0.08$ h^{-1} and $\mu = 0.04$ h^{-1}. The upper-right panel varies mRNA stability for a fixed $\gamma_P = 0.01$ h^{-1} and $\mu = 0.04$ h^{-1}. The lower panel varies growth rate for a fixed $\gamma_R = 0.08$ h^{-1} and $\gamma_P = 0.01$ h^{-1}.

In this case, for the aforementioned initial parameter values, the steady-state protein concentration increases by only 11%.

For a doubled rate of transcription, the steady-state protein concentration doubles, regardless of other parameter values.

For doubled mRNA half-life, we have

$$\frac{[P]_{ss,2}}{[P]_{ss,1}} = \frac{\gamma_{R,1} + \mu}{0.5\gamma_{R,1} + \mu} = \frac{\frac{\gamma_{R,1}}{\mu} + 1}{0.5\frac{\gamma_{R,1}}{\mu} + 1} \qquad (5.20)$$

For the example parameters, the steady-state protein concentration would increase by 50%.

Clearly, the impact on relative steady-state protein concentration is dependent on the relationships between the degradation rates and the growth rate. The impact of each system variation for a range of degradation rates is presented in figure 5.8.

The time (τ) to reach half of the steady-state protein concentration is readily calculated from equation 5.15:

$$\frac{1}{2}[P]_{ss} = [P]_{ss}\left(1 + \frac{(\gamma_R + \mu)e^{-(\gamma_P + \mu)\tau} - (\gamma_P + \mu)e^{-(\gamma_R + \mu)\tau}}{\gamma_P - \gamma_R}\right) \quad (5.21)$$

The impacts of each system variation—halved growth rate, doubled protein stability, doubled transcription, and doubled mRNA half-life—are presented for the case of $\gamma_P = 0.01\ h^{-1}$, $\gamma_R = 0.08\ h^{-1}$, and $\mu_1 = 0.04\ h^{-1}$, as well as single variations of each parameter (figure 5.9).

Example 5-3 Consider a gene expression experiment in which mRNA and protein are both changed by some perturbation. Demonstrate that the ratio of $[P]$ to $[mRNA]$ may appear to be altered if the new steady state is not reached.

Solution Integrating equation 5.6 with an initial mRNA concentration of $[mRNA]_o$ yields

$$[mRNA] = \frac{k_R}{\gamma_R + \mu} + \left([mRNA]_o - \frac{k_R}{\gamma_R + \mu}\right)e^{-(\gamma_R + \mu)t} \quad (5.22)$$

This expression can be reframed in terms of the steady-state mRNA concentration, $[mRNA]_{ss}$ (equation 5.8), and the perturbation ($\Delta[mRNA] = [mRNA]_o - [mRNA]_{ss}$):

$$[mRNA] = [mRNA]_{ss} + \Delta[mRNA]e^{-(\gamma_R + \mu)t} \quad (5.23)$$

Equation 5.11 can be integrated with the newly solved mRNA balance, and a protein perturbation of $\Delta[P] = [P]_o - [P]_{ss}$ yields

$$[P] = [P]_{ss} + \Delta[P]e^{-(\gamma_P+\mu)t}$$
$$+ \frac{[P]_{ss}\Delta[mRNA]}{[mRNA]_{ss}} \frac{\gamma_P+\mu}{\gamma_P-\gamma_R} \left(e^{-(\gamma_R+\mu)t} - e^{-(\gamma_P+\mu)t}\right) \quad (5.24)$$

Thus, the protein:mRNA ratio can deviate from the steady-state ratio:

$$\frac{[P]}{[mRNA]} =$$
$$\frac{[P]_{ss} + \Delta[P]e^{-(\gamma_P+\mu)t} + \frac{[P]_{ss}\Delta[mRNA]}{[mRNA]_{ss}} \frac{\gamma_P+\mu}{\gamma_P-\gamma_R} \left(e^{-(\gamma_R+\mu)t} - e^{-(\gamma_P+\mu)t}\right)}{[mRNA]_{ss} + \Delta[mRNA]e^{-(\gamma_R+\mu)t}}$$
(5.25)

Figure 5.10 demonstrates the deviation from steady state for the protein:mRNA ratio, following a 10% reduction in mRNA and varying reductions in protein.

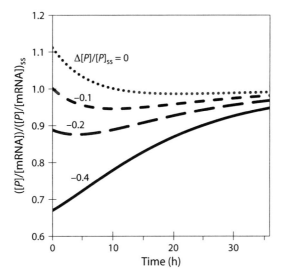

Figure 5.10 The dynamic protein:mRNA ratio, normalized to the steady-state ratio, is shown for a 10% reduction to [$mRNA$] and varying reductions to [P] for typical rate constant values of a mammalian cell: $\mu = 0.04$ h^{-1}; $\gamma_R = 0.1$ h^{-1}; $\gamma_P = 0.02$ h^{-1}.

Example 5-4 Pulse-chase experiments are often used to assess transcript and protein dynamics using radiolabeled ribonucleotides or amino acids. In these experiments, the amount of radiolabeled mRNA or protein is per volume of total culture, rather than per cellular volume. Derive equations for $[mRNA]'_V$ and $[P]'_V$, the radiolabeled concentrations per culture volume.

Solution As noted in equation 5.2, mRNA accumulation is a balance of synthesis and degradation:

$$\frac{d[mRNA]'_V}{dt} = k_R V_i - \gamma_R [mRNA]'_V \tag{5.26}$$

Substituting the exponential growth of cell volume (from an initial value of $V_{i,o}$ at the time of pulse labeling) and integrating, and assuming no radiolabeled material at $t = 0$ results in

$$[mRNA]'_V = \frac{k_R V_{i,o}}{\gamma_R + \mu}(e^{\mu t} - e^{-\gamma_R t}) \tag{5.27}$$

Likewise, protein accumulation is a balance of synthesis (proportional to mRNA concentration) and degradation:

$$\frac{d[P]'_V}{dt} = k_P [mRNA]'_V - \gamma_P [P]'_V \tag{5.28}$$

Considering initiation of transcription coincident with the radiolabeled pulse (i.e., assuming no initial transcripts or protein), we obtain

$$[P]'_V = \frac{k_P k_R V_{i,o}}{\gamma_R + \mu} \left(\frac{e^{\mu t} - e^{-\gamma_P t}}{\gamma_P + \mu} + \frac{e^{-\gamma_P t} - e^{-\gamma_R t}}{\gamma_P - \gamma_R} \right) \tag{5.29}$$

If transcription is initiated prior to the radiolabeled pulse (with time delay t_i), then

$$[P]'_V = \frac{k_P k_R V_{i,o}}{\gamma_R + \mu} \left(\frac{e^{\mu t_i}(e^{\mu t} - e^{-\gamma_P t})}{\gamma_P + \mu} + \frac{e^{-\gamma_R t_i}(e^{-\gamma_P t} - e^{-\gamma_R t})}{\gamma_P - \gamma_R} \right) \tag{5.30}$$

We have thus far assumed that the rate constant for protein synthesis is indeed constant, but the rate constant actually lumps together the processes of translation initiation, elongation, and termination. Bergmann and Lodish [13] developed a kinetic model to understand how the parameters describing these underlying processes impact the overall rate of protein synthesis.

Case Study 5-2   S. B. Lee. and J. E. Bailey. Analysis of growth rate effects of productivity of recombinant *Escherichia coli* populations using molecular mechanism models. *Biotechnol. Bioeng.* 26: 66–73 (1984) [14].

An early and important capability resulting from biotechnology was the ability to clone genes into plasmids and express them in bacterial cells. An analysis of protein yield and how it depends on growth conditions requires consideration of a large number of factors, including the rate at which the plasmid replicates and the level of expression from each plasmid copy. Lee and Bailey developed a mechanistic model that takes into account these and other factors to study production as a function of growth rate from a λdv plasmid (related to phage λ through deletion) [14].

Plasmid λdv replication is a function of Cro repressor (C) and initiator (I) protein. Cro repressor affects transcription of both C and I, as well as activation of the origin of replication (θ). mRNA and protein expression are treated with terms representing synthesis, first-order degradation, and dilution due to cell growth. For an average cell, these features of the model are encoded in the following three sets of equations:

Cro repressor synthesis

$$\frac{d[mRNA_C]}{dt} = k_R \eta [G] - (\gamma_R + \mu)[mRNA_C] \tag{5.31}$$

$$\frac{d[C]}{dt} = k_P [mRNA_C] - (\gamma_C + \mu)[C] \tag{5.32}$$

Initiator synthesis

$$\frac{d[mRNA_I]}{dt} = k_R \eta (1-f)[G] - (\gamma_R + \mu)[mRNA_I] \tag{5.33}$$

$$\frac{d[I]}{dt} = k_P [mRNA_I] - (\gamma_I + \mu)[I] \tag{5.34}$$

Origin activation

$$\frac{d\theta}{dt} = k_R \eta (1-f)(1-\theta) - \mu \theta \tag{5.35}$$

The above model uses a common rate of degradation for all mRNA species but protein-specific degradation rates for Cro repressor and initiator proteins. Here η represents the state of transcriptional efficiency, which is a nonlinear function of [C], accounting for binding effects of repressor C to the $P_R O_R$ promoter-operator region; [G] is the concentration of plasmid; and f is the efficiency of a transcriptional terminator that lies after the *cro* gene.

The level of replication complex is treated as an equilibrium involving the activated origin region and initiator-host protein complex,

$$[REP] = \left(\frac{\theta}{VN_A}\right)\left(\frac{K_\theta [I]}{1 + K_\theta [I]}\right) \tag{5.36}$$

with plasmid replication beginning when [REP] reaches a critical threshold and progressing to completion instantaneously.

The cloned protein (Z) production is given by the expressions

$$\frac{d[mRNA_Z]}{dt} = k_R \eta [G] - (\gamma_R + \mu)[mRNA_Z] \quad (5.37)$$

$$\frac{d[Z]}{dt} = k_P [mRNA_Z] - (\gamma_Z + \mu)[Z] \quad (5.38)$$

Although the plasmid is treated as constitutively expressing the cloned gene in this reference, the authors note that it would be straightforward to simulate the behavior under transcriptional control by putting the appropriate model into equation 5.37.

Parameters were estimated from available experimental data, including mRNA and protein synthesis rates, and simulations were run from initial conditions representing a single plasmid introduced into a single cell in exponential growth. Simulations were run for a number of cell divisions until a steady state was reached, in which the trajectory for intracellular concentrations in one growth cycle matched that of the previous cycle. The concentrations and other behaviors in this cycle steady state were then analyzed.

Increasing the cell growth rate decreased the level of plasmid per cell, with the effect being largest for simulations with weaker interaction between the Cro repressor and the $P_R O_R$ promoter-operator region. Interestingly, this effect of reduced plasmid count with increasing growth rate was traced to increasing levels of Cro repressor with the increased growth rate, leading to less frequent activation of the origin region and thus less frequent plasmid replication.

If the expression of cloned protein were proportional to the plasmid number, these results would suggest that low growth rates would be best for maximal yield. But the simulations show that there is a maximum in the concentration of cloned protein per cell at intermediate growth rates, leading to an optimum growth rate corresponding to highest yield. Moreover, the simulations suggest that the position of the maximum, as well as its value, depends on the degradation rate of the cloned protein. Thus, optimum expression conditions depend somewhat on the stability of the cloned product and might be tuned for each protein cloned.

Finally, Lee and Bailey examined the effect of changing the growth rate. They observed that changing cells to higher growth-rate conditions produced a temporary overshoot in the concentration of protein produced per cell that lasts for a few cell divisions. The theoretical results obtained were supported by available experimental data at the time. Moreover, the results are a powerful and early testament to the power of mechanistic models to describe complex biological and bioengineering scenarios, and they provide guidance for obtaining maximal protein yields from experimental systems.

5.1.1 Pure Delays in Biosynthesis

If one initiates expression of a gene at time zero, some minimum time must pass before the first completed protein molecule emerges from the ribosome. This delay results from the finite speeds at which RNA polymerases move along DNA and ribosomes move along mRNA. RNA polymerases in bacteria typically synthesize mRNA at an elongation rate of 50–100 nucleotides per second, whereas yeast and higher eukaryotes synthesize mRNA at elongation rates of 20–30 nucleotides per second [15,16]. Consequently, a 1 kb mRNA requires a minimum of 10–20 seconds to complete transcription in a bacterium, and 33–50 seconds to complete transcription in a eukaryote. Similarly, ribosomes synthesize polypeptide chains with an elongation rate of 15–20 amino acids per second in bacteria [17], 20 amino acids per second in yeast [12], and 1–5 amino acids per second in mammalian cells [18, 19]. Therefore, a 400-amino-acid protein would be expected to be translated in $\approx$20 seconds in yeast or bacteria and $\approx$1–7 minutes in mammalian cells. The time required for a protein to fold to its functional conformation would also be added to this delay, which varies widely among proteins, and the rate of which is generally difficult to measure in vivo.

These delays have been observed experimentally, as shown in figure 5.11. The delay between addition of an inducer and commencement of accumulation of an enzymatic activity is often called an induction lag. Because bacterial mRNA can commence translation prior to completion of transcription, the delays from transcription and translation are not additive. By contrast, in eukaryotes, the completely synthesized mRNA is transported to the cytoplasm prior to commencement of translation, so delays are additive.

Notice that the delays in figure 5.11 are in general short by comparison to the time scales for approach to steady state, as examined in the previous section. However, delays have a particularly strong destabilizing influence on the stability of feedback loops and can determine the presence of oscillatory behavior.

5.2 Compartmental Models of Protein Sorting

Having considered the dynamics of protein synthesis, we now examine the shuttling of these biomolecules within eukaryotic cells, which, unlike prokaryotic cells, are partitioned into several distinct membrane-bound compartments, or organelles (figure 5.12, table 5.1).

The movement of proteins among these compartments is tightly regulated. Some proteins are functional only when they arrive in their final compartment. For example, lysosomal proteases become enzymatically active only when they reach the lysosome, where the acidic pH and regulatory proteases convert the proteases to their active forms. Other proteins shuttle from one compartment to another as part of their regulation: Transcription factors such as NF-κB are specifically activated via trafficking into the nucleus from the cytoplasm in response to particular stimuli. Receptors on the cell surface are downregulated by endocytic vesicular trafficking as part of a negative feedback response to extracellular ligands.

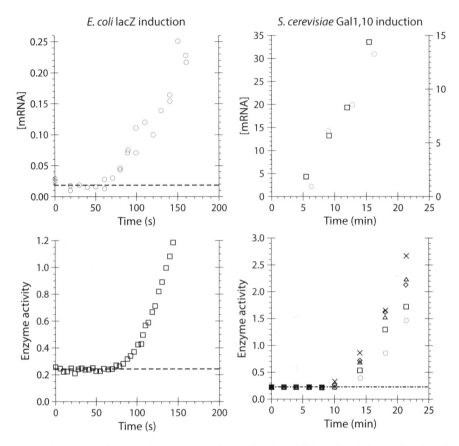

Figure 5.11 Biosynthetic delays in gene expression. Following addition of an inducer, there is a delay before appearance of completed mRNA molecules (top panels) and active enzyme (bottom panels). The left two panels represent induction of the lacZ operon by addition of IPTG in *E. coli*, and the right two panels the galactose-inducible GAL1 and GAL10 genes of *S. cerevisiae*. The upper-left panel is from reference 20; the lower-left panel is from reference 21. The upper-right panel [22] has two symbols representing GAL1 and GAL4 mRNA. The lower-right panel is data for galactokinase activity at five different galactose concentrations (different symbols) [23].

Regulated intercompartmental transport utilizes receptors for signal peptides on each protein. Table 5.2 lists typical compartmental sorting signal peptides.

The simplest mathematical treatment of each cellular compartment, such as those in figure 5.13, is as a well-mixed volume. In other words, proteins entering the compartment are considered to instantaneously distribute throughout the compartment to a uniform concentration. Because the doubling time for growth of eukaryotic cells is generally very slow relative to the sorting times under examination, compartmental volumes are considered to be constant. Let's examine the general properties

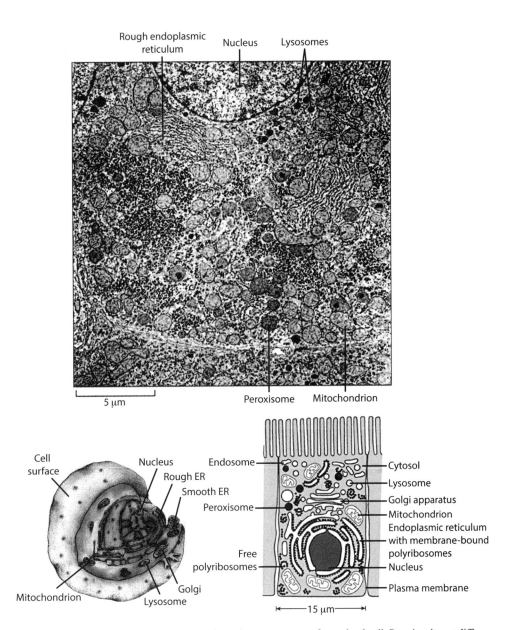

Figure 5.12 Three views of the membrane-bound compartments of an animal cell. Proteins do not diffuse across the membrane barriers defining each organelle; rather, protein traffic is specifically controlled through proteinaceous pores or vesicles that bud from one compartment and fuse with another. (Top panel) An electron micrograph of an animal cell [24]. (Bottom-left panel) Three-dimensional sketch of a eukaryotic cell. (Bottom-right panel) A schematic representation of the compartments represented in the top panel [24].

Table 5.1 Relative compartmental volumes and membrane areas in a liver cell (hepatocyte) (adapted from reference 24).

Compartment	Volume (%)	Membrane area (%)
Cytosol	54	N/A
Rough ER	9	35
Smooth ER & Golgi	6	23
Mitochondria	22	39
Nucleus	6	0.2
Lysosomes	1	0.4
Peroxisomes	1	0.4
Endosomes	1	0.4
Plasma membrane	N/A	2

Table 5.2 Sorting signal peptides (adapted from reference 24).

Location	Sequence
Import into ER	MMSFVSLLLVGILFWATEAEQLTKCEVFQ-
Retain in lumen of ER	-KDEL
Import into mitochondria	MLSLRQSIRFFKPATRTLCSSRYLL-
Import into nucleus	-PPKKKRKV-
Import into peroxisomes	-SKL-
Attach to membranes	GSSKSKPK-

of one such idealized compartment, as shown in figure 5.14. The simplest dynamic description of the signal-driven sorting events that result in a protein leaving a given compartment is via mass-action kinetics, with a first-order dependence on the protein's compartmental concentration $[P]_\alpha$ and an effective rate constant $k_{\text{out},\alpha}$. Entry of protein into the compartment is described by the constant term v_{in}. The material balance for protein P in compartment α is as follows:

$$\frac{d[P]_\alpha}{dt} = v_{\text{in}} - k_{\text{out},\alpha}[P]_\alpha \tag{5.39}$$

The steady-state value $[P]_{\alpha,\text{ss}}$ is found by setting the time derivative to zero:

$$[P]_{\alpha,\text{ss}} = \frac{v_{\text{in}}}{k_{\text{out},\alpha}} \tag{5.40}$$

and the transient solution for initial condition $[P]_\alpha = 0$ at $t = 0$ is

$$[P]_\alpha = [P]_{\alpha,\text{ss}}(1 - e^{-k_{\text{out},\alpha}t}) \tag{5.41}$$

Gene Expression and Protein Trafficking

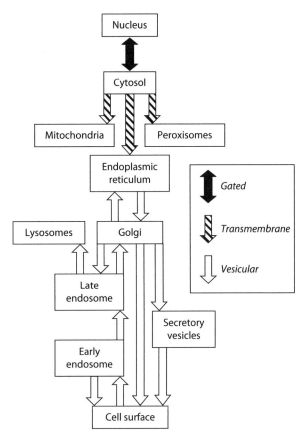

Figure 5.13 Schematic representation of protein transport between different subcellular membrane-bound compartments. Adapted from reference 24.

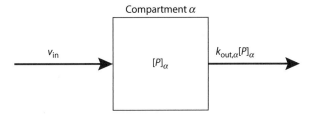

Figure 5.14 Schematic diagram of protein sorting into and out of a single compartment.

Note that, as for models of transcription (equation 5.9) and protein expression (equation 5.16), the time scale for approach to a new steady state is a function only of the rate processes that *decrease* concentration of the species of interest. This is because synthesis and protein entry terms are modeled as zeroth-order processes, whereas degradation, dilution, and protein exit are first order with respect to the dependent variable. As an example in which the time scale for approach depends on both increasing and decreasing rate processes, recall the rate of formation of complexes with constant ligand in chapter 3, for which the time constant is $1/(k_{on}[L]_o + k_{off})$.

Example 5-5 Examine the change in the steady-state protein concentration and the half-time to this steady state (i.e., time at which the protein concentration is half of its steady-state value) due to a doubling of the synthesis rate, v_{in}. Make the same examination for the case of a doubling of the compartment exit-rate constant, $k_{out,\alpha}$.

Solution From equation 5.40, we see that the steady-state concentration of protein in compartment α doubles when the synthesis rate doubles and halves when the exit-rate constant doubles. The half-time to steady state can be readily determined from equation 5.41: $t_{1/2} = (ln 2)/k_{out,\alpha}$, which is independent of the synthesis rate and halves when the exit rate doubles.

Let's now consider the case of two compartments with traffic from one to the other, as shown in figure 5.15. The material balance for both compartments, now explicitly tracking compartment volumes (again, assumed here to be constant), are

$$\frac{d}{dt}([P]_\alpha V_\alpha) = v_{in} V_\alpha - k_{out,\alpha}[P]_\alpha V_\alpha$$

$$\frac{d[P]_\alpha}{dt} = v_{in} - k_{out,\alpha}[P]_\alpha$$

(5.42)

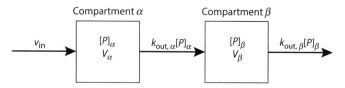

Figure 5.15 Schematic diagram of protein sorting from one compartment to another.

$$\frac{d}{dt}([P]_\beta V_\beta) = k_{\text{out},\alpha}[P]_\alpha V_\alpha - k_{\text{out},\beta}[P]_\beta V_\beta$$

$$\frac{d[P]_\beta}{dt} = k_{\text{out},\alpha}[P]_\alpha \frac{V_\alpha}{V_\beta} - k_{\text{out},\beta}[P]_\beta \tag{5.43}$$

The steady-state solution for $[P]_{\alpha,\text{ss}}$ is unaltered, because there is no transport from compartment β back to compartment α. The steady-state concentration of P in compartment β is

$$[P]_{\beta,\text{ss}} = \frac{v_{\text{in}}}{k_{\text{out},\beta}} \frac{V_\alpha}{V_\beta} \tag{5.44}$$

The transient solutions to equations 5.42 and 5.43 for initial conditions $[P]_\alpha = [P]_\beta = 0$ at $t = 0$ are

$$[P]_\alpha = [P]_{\alpha,\text{ss}} \left(1 - e^{-k_{\text{out},\alpha} t}\right) \tag{5.45}$$

$$[P]_\beta = [P]_{\beta,\text{ss}} \left(1 + \frac{k_{\text{out},\alpha} e^{-k_{\text{out},\beta} t} - k_{\text{out},\beta} e^{-k_{\text{out},\alpha} t}}{k_{\text{out},\beta} - k_{\text{out},\alpha}}\right) \tag{5.46}$$

Note that equation 5.46 has the same mathematical form as equation 5.15. This is because the underlying differential equations are the same, with $[mRNA] \equiv [P]_\alpha$ and $[P] \equiv [P]_\beta$.

As is apparent in figure 5.13, sorting of a protein may involve movement between multiple compartments. Furthermore, each compartment may itself consist of a series of distinct membrane-bound stacks or vesicles, as is the case for the Golgi apparatus shown in figure 5.16. Transport through such a complex structure might be described in two distinct ways: (1) entry into and random exit from a single well-mixed compartment or (2) movement at a characteristic velocity along a sequentially ordered path with a characteristic length. The first description is captured in figure 5.14 and equation 5.41. The second description represents a pure delay in transport across the compartment, equal to the distance traveled divided by the average velocity. The physical reality often lies somewhere between these two extreme representations.

One way to capture behavior intermediate between a single well-mixed compartment and a pure delay is to break the compartmental volume of interest into a series of smaller, equal-sized, well-mixed compartments (e.g., figure 5.17).

This approach was first developed in an attempt to describe complex flow patterns in chemical reactors and was dubbed the *tanks-in-series* model [26]. To maintain a constant flow through the overall system, the intercompartmental rate constants

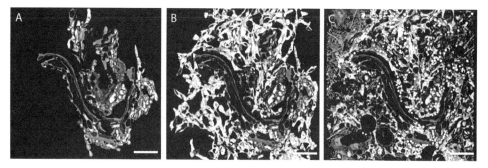

Figure 5.16 A three-dimensional reconstruction of the Golgi apparatus and its environment, from a series of electron micrographic images taken of 40–60 nm thick slices obtained by microtomy of insulinoma cells flash-frozen in 10–20 msec [25]. The scale bar is 0.5 μm. In panel A, several cisternae of the Golgi are shown. In panel B, images of the endoplasmic reticulum tubulovesicular network are shown surrounding and interpenetrating the Golgi apparatus. In panel C, additional electron-dense structures in the region are added to the picture: ER, membrane-bound ribosomes, free ribosomes, microtubules, dense core vesicles, clathrin-positive compartments and vesicles, clathrin-negative compartments and vesicles, and mitochondria. Given the extraordinary geometric complexity shown here, it is rather remarkable that a compartmental model describing the ER and the Golgi as individual homogenously well-mixed compartments can capture rather well some basic dynamic behaviors of cargo trafficking through them.

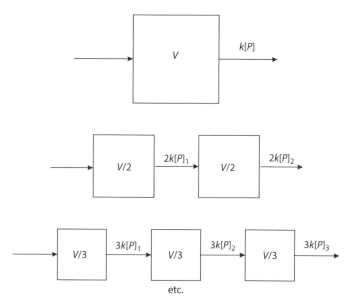

Figure 5.17 Behavior intermediate between a single compartment and a pure delay can be modeled by considering a fictitious subdivision of the volume of interest into a series of smaller well-mixed compartments.

Gene Expression and Protein Trafficking

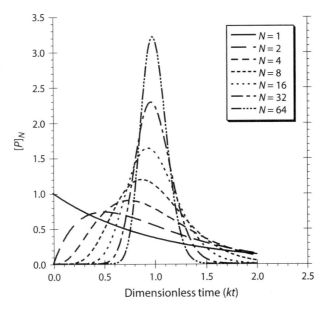

Figure 5.18 Tracer exit trajectories intermediate between a well-mixed compartment and a pure delay are reflected in the tanks-in-series model with increasing numbers of small subcompartments.

are scaled proportionally to the number of tanks (e.g., $2k$ for 2 compartments). It can be shown that if a given quantity of tracer is instantaneously placed in the first compartment at time zero, then the time trajectory for the concentration of that tracer exiting the final compartment as a function of time is given by

$$[P]_N(t) = [P]_0 N \frac{(Nkt)^{N-1}}{(N-1)!} e^{-Nkt} \qquad (5.47)$$

where N is the number of tanks in series. This relationship is graphed for increasing values of N in figure 5.18, illustrating the gradual shift from a well-mixed single compartment to a delayed system.

Thus, one approach to including an element of delay in the description of a compartmental system is to use the tanks-in-series formalism. This approach can also be extended to chemical reactions by simply considering each compartment to be an intermediate chemical species: A series of sequential reactions approximates the behavior of a time delay as the number of reactions increases. This behavior is apparent in biosynthesis of biopolymers, such as mRNA or proteins, which empirically exhibit a pure delay as a result of a large series of first-order polymerization reactions.

5.2.1 Receptor-Ligand Trafficking

An important class of localization processes is those involving cell-surface receptors and their ligands. This topic is covered comprehensively by Lauffenburger and Linderman [27]. Let us consider the system shown in figure 5.19. Writing down the dynamic material balances, we have

$$\frac{d[R]}{dt} = v_{s,R} - k_f[R][L] + k_r[C] - k_{e,R}[R] + k_{rec,R}[R_e] \tag{5.48}$$

$$\frac{d[C]}{dt} = k_f[R][L] - k_r[C] - k_{e,C}[C] + k_{rec,C}[C_e] \tag{5.49}$$

$$\frac{d[R_e]}{dt} = k_{e,R}[R] - k_{rec,R}[R_e] - k_{f,e}[R_e][L_e] + k_{r,e}[C_e] - k_{deg}[R_e] \tag{5.50}$$

$$\frac{d[C_e]}{dt} = k_{e,C}[C] - k_{rec,C}[C_e] + k_{f,e}[R_e][L_e] - k_{r,e}[C_e] - k_{deg}[C_e] \tag{5.51}$$

$$\frac{d[L_e]}{dt} = k_{r,e}[C_e] - k_{f,e}[R_e][L_e] - k_{rec,L}[L_e] - k_{deg}[L_e] \tag{5.52}$$

$$\frac{d[L]}{dt} = k_r[C] - k_f[R][L] + k_{rec,L}[L_e] \tag{5.53}$$

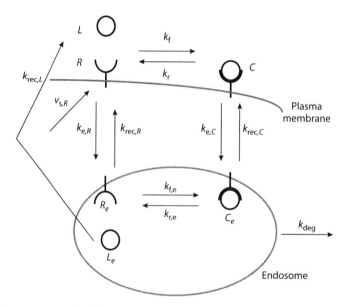

Figure 5.19 Rate processes in cell-surface receptor trafficking dynamics. Association and dissociation are denoted by k_f and k_r to account for local depletion effects should they occur, as examined in chapter 8. Endocytic trafficking (uptake and recycling) is described by mass-action kinetics rather than a detailed mechanical accounting for vesicle movement velocity and delays. Different rate constants for binding apply in the endosome due to a drop in pH.

Care must be taken to maintain proper units for evaluation of these equations. Receptor and complex concentrations are typically evaluated in molecules per cell. If ligand concentrations are also evaluated in molecules per cell, the equations can be implemented as written, though care should be taken to apply the association rate constants (k_f and $k_{f,e}$) in units of cells/molecule/time. Alternatively, if ligand concentration is evaluated in molar concentration, the external and endosomal ligand differential equations must be adjusted by multiplying k_r by molar cell density (cells per volume divided by Avogadro's number).

Steady-state receptor distribution Receptor synthesis, internalization, recycling, and degradation also proceed in the absence of ligand. It can be shown that the steady-state surface receptor concentration in the absence of ligand $([R]_{ss,o})$ is

$$[R]_{ss,o} = \frac{v_{s,R}}{k_{e,R}} \left(1 + \frac{k_{rec,R}}{k_{deg}}\right) \qquad (5.54)$$

The fraction of total receptors that reside on the cell surface at steady state in the absence of ligand is

$$\frac{[R]_{ss,o}}{[R]_{ss,o} + [R_e]_{ss,o}} = \frac{k_{rec,R} + k_{deg}}{k_{rec,R} + k_{deg} + k_{e,R}} \qquad (5.55)$$

Steady-state ligand trafficking Let's consider a situation where excess ligand is incubated with living cells, under conditions in which receptor and complex internalization occur. Over a sufficiently long period, the cells can be expected to consume the ligand completely. However, in practice, the pool of ligand may be maintained in sufficient excess to stay approximately constant over the time required to attain a steady-state level of receptor-ligand complex on the cell surface. Very approximately, the characteristic time to closely approach steady state should be no greater than the largest among the characteristic times for equilibration ($1/(k_f[L] + k_r)$), complex internalization ($1/k_{e,C}$), or receptor internalization ($1/k_{e,R}$).

Example 5-6 What is the ligand:receptor ratio in a six-well plate with 20 nM ligand and 10^5 receptors per cell?

Solution A 9 cm^2 well of a six-well plate will accommodate about one million adherent cells and has a typical liquid volume of 4 mL. Thus, the receptor concentration is

$$[R] = \frac{10^6 \text{ cells} \times 10^5 \frac{\text{receptors}}{\text{cell}}}{(6.02 \times 10^{23} \frac{\text{receptors}}{\text{mol}})(4 \times 10^{-3} \text{ L})} = 0.04 \text{ nM} \qquad (5.56)$$

> which results in a 500:1 initial ligand:receptor ratio. Substantial ligand binding and consumption must occur prior to an appreciable change in the free ligand concentration.

The steady-state fraction of receptor complexed with ligand will not be the same as the equilibrium fraction, due to the endocytic consumption of the receptor-ligand complex. Thermodynamic equilibrium is attainable only in a closed system, without sources or sinks for molecular species. Setting the derivative of $[C]$ in equation 5.49 equal to zero at steady state and assuming that recycling of the complex is negligible, we obtain

$$0 = k_f [R]_{ss}[L] - (k_r + k_{e,C})[C]_{ss} \quad (5.57)$$

which can be rearranged to the following expression for fractional saturation:

$$\frac{[C]_{ss}}{[C]_{ss} + [R]_{ss}} = \frac{[L]}{K_{d,app} + [L]} \quad (5.58)$$

where

$$K_{d,app} \equiv \frac{k_r + k_{e,C}}{k_f} \quad (5.59)$$

Note that $K_{d,app} > K_d$, so the steady-state fractional receptor occupancy will be lower than that predicted for equilibrium. This difference can be significant for very-high-affinity interactions, resulting in paradoxically low apparent affinity, as examined by J. Haugh in the next case study.

The phenomenon of receptor downregulation results if $k_{e,C} \neq k_{e,R}$ or $k_{rec,C} \neq k_{rec,R}$. The steady-state total number of receptors on the surface may be reduced in the presence of ligand due to: accelerated endocytosis of the complex relative to free receptor ($k_{e,C} > k_{e,R}$); decreased endosomal recycling of complex relative to free receptor ($k_{rec,C} < k_{rec,R}$); or decreased synthesis of receptor following downstream signaling events resulting from ligand binding. The effect of the first of these mechanisms can be predicted by assuming no change in receptor synthesis rate and neglecting recycling terms, as follows:

$$\frac{[C]_{ss} + [R]_{ss}}{[R]_\circ} = \frac{K_{d,app} + [L]}{K_{d,app} + \frac{k_{e,C}}{k_{e,R}}[L]} \quad (5.60)$$

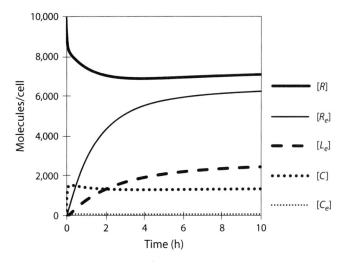

Figure 5.20 Dynamic simulation of equations 5.48–5.53 with the following parameters: 5 nM ligand concentration; $k_f = k_{f,e} = 8.3 \times 10^{-11}$ cell/molecule/s (2×10^5 M^{-1} s^{-1} with 250,000 cells/mL); $k_r = k_{r,e} = 5 \times 10^{-3}$ s^{-1}; $v_{s,R} = 0.65$ molecule/cell/s (to yield 10,000 surface receptors per cell at ligand-free steady state); 2 hour half-time for receptor internalization; 1 hour half-time for complex internalization; 4 hour half-time for receptor recycling; 8 hour half-time for complex recycling; no ligand recycling; and $k_{\text{deg}} = 1 \times 10^{-4}$ s^{-1}.

where $[R]_o$ is the steady-state surface receptor level in the absence of ligand. Considering equation 5.60 in the case of saturating ligand concentration ($[L] \gg K_{d,\text{app}}$), total surface receptor levels are reduced to a fraction $k_{e,R}/k_{e,C}$ of their original value.

Numerical integration of the system of equations (equations 5.48–5.53) lets us visualize typical system dynamics (figure 5.20). Given the number of parameters that are present in this daunting full system of equations, it is not practical to simultaneously and uniquely fit all the parameters in a given experiment. However, we can leverage the sequential nature of receptor-ligand trafficking events through intracellular compartments (e.g., from the extracellular space, to the endosome, to the lysosome) to parse individual rate constants. As an illustrative example, let's consider the measurement of the endocytotic rate constant, $k_{e,C}$.

By first incubating cells with radiolabeled ligand at 4°C, cell-surface complexes are created, and the lowered temperature prevents active trafficking from the plasma membrane. At $t = 0$, the temperature is raised to 37°C, at which point the number of surface-bound ligand molecules (i.e., $[C]$) and intracellular ligand molecules (i.e., $[C_e] + [L_e]$) can both be measured over time. This can be done by separating surface-associated and intracellular ligand molecules in a given sample, using acidic stripping conditions that efficiently dissociate surface complexes without lysing the cells. (Note that this approach does not explicitly discriminate between $[C_e]$ and

$[L_e]$, because they both contribute to the intracellular radioactive signal, but that is not a problem for measuring $k_{e,C}$, as will be imminently clear). For short experimental time courses (typically on the order of ≈10 min), there will be a significant amount of endocytosis but negligible recycling or degradation. Thus, under these experimental conditions, the relevant rate equation (sum of equations 5.51 and 5.52) reduces to

$$\frac{d([C_e]+[L_e])}{dt} = k_{e,C}[C] \tag{5.61}$$

Integrating and assuming $[C_e + L_e] = 0$ at $t = 0$, we obtain

$$[C_e + L_e](t) = k_{e,C} \int_0^t [C] d\tau \tag{5.62}$$

Thus, a plot of $[C_e + L_e]$ versus $\int_0^t [C] d\tau$ should yield a straight line with slope $k_{e,C}$. Notably, by visualizing the experimental data in this way, one can also rapidly identify the time period over which the model assumptions remain valid. At short times, the data should be linear, but at later times when downstream processes (such as recycling) become significant, the data will start to deviate from linearity. These latter data should not be included in the fit, because they reflect additional processes beyond what the model was intended to capture and will result in a less accurate estimate of $k_{e,C}$.

Other experimental setups can be used to quantitatively extract other trafficking parameters. These are beyond the scope of this textbook but are discussed in detail by Lauffenburger and Linderman [27].

Case Study 5-3 J. M. Haugh. Mathematical model of human growth hormone (hGH)-stimulated cell proliferation explains the efficacy of hGH variants as receptor agonists or antagonists. *Biotechnol. Prog.* 20: 1337–1344 (2004) [28].

Human growth hormone (hGH) is a polypeptide endocrine factor that stimulates a number of important processes related to growth. Binding of hGH to its cell-surface receptor (the human growth hormone receptor, hGHR) results in receptor homodimerization; the resulting juxtaposition of the intracellular Janus kinase 2 (Jak2) domains of the receptor leads to cross-phosphorylation and activation of signaling. The interaction between hormone and receptor has been extensively studied through structure determination, mutagenesis, affinity measurement, and directed evolution. Moreover, mutants that have been isolated and whose binding properties have been characterized serve as important reagents for studying receptor signaling behavior at the cellular level.

It has been established that the mechanism of binding involves a single hormone molecule associating with two receptor molecules. The hormone first binds to one monomeric receptor through the higher affinity (site 1) of two sites; a second monomeric receptor is recruited through site 2 to form the final ternary complex. It seems intuitively clear that mutants with enhanced binding affinity should lead to enhanced signaling through a greater propensity to dimerize. Because binding can saturate, a useful experimental measure for comparing different hormone mutants is the concentration leading to half-maximal proliferation (EC_{50}). Interestingly, experiments with enhanced affinity mutants of hGH do not show enhanced proliferation. This is true for affinity improvements at either the higher-affinity site 1 or the lower-affinity site 2. This type of seeming discrepancy between molecular and cellular behavior is not uncommon. Rather than representing an inconsistency, such observations frequently indicate a need to more fully describe biochemical and cell biological mechanisms and to account for their behavior in more detail. In the current study, Haugh built a model of receptor binding and activation that also includes receptor internalization and trafficking. This model was able to reconcile the binding and signaling behaviors of the mutants by accounting for increased internalization and receptor downregulation resulting from enhanced binding.

The model of receptor binding, trafficking, and activation is illustrated in figure 5.21. Assembly and disassembly of the active, dimeric receptor complex D occurs on the membrane surface. All receptor species are internalized through endocytosis, but the dimeric complex does so at a faster rate (k_e) and is immediately degraded, whereas other receptor species are internalized more slowly (k_t) and undergo a sorting decision with rates for recycling (k_{rec}) and degradation (k_{deg}). In figure 5.21, the tighter-binding site-1 face of the ligand is indicated by shading; the model treats disassembly through releasing either site, but initial release of the tighter site leads to immediate ligand dissociation from both sites.

This model can be described mathematically with ordinary differential equations by applying material balances at the membrane surface to free receptor P, bound-but-not-dimerized receptor complex C, and dimerized bound complex D, which is considered

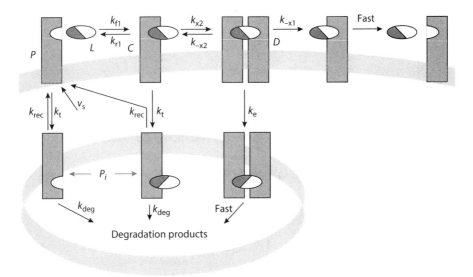

Figure 5.21 Trafficking rate processes and rate constants in a model for hGH receptor-ligand trafficking.

activated in the model); as well as to intracellular receptor P_i, which refers to the internalized nondimerized receptor, whether ligand is bound or not). Additionally, it is assumed that internalized, nondimerized ligand-receptor complex loses its ligand prior to recycling (i.e., the only species generated by recycling is P):

$$\frac{d[C]}{dt} = k_{f1}[L][P] + k_{-x2}[D] - (k_{r1} + k_{x2}[P] + k_t)[C] \tag{5.63}$$

$$\frac{d[D]}{dt} = k_{x2}[P][C] - (k_{-x2} + k_{-x1} + k_e)[D] \tag{5.64}$$

$$\frac{d[P]}{dt} = v_s + k_{r1}[C] + (k_{-x2} + 2k_{-x1})[D] - (k_{f1}[L] + k_{x2}[C] + k_t)[P] + k_{rec}[P_i] \tag{5.65}$$

$$\frac{d[P_i]}{dt} = k_t([P] + [C]) - (k_{deg} + k_{rec})[P_i] \tag{5.66}$$

In the absence of ligand, $[L] = [C] = [D] = 0$, and we can solve for the steady-state level of surface and internalized receptor by setting the time derivatives of $[P]$ and $[P_i]$ to zero:

$$\frac{d[P]}{dt} = 0 = v_s - k_t[P] + k_{rec}[P_i] \tag{5.67}$$

$$\frac{d[P_i]}{dt} = 0 = k_t[P] - (k_{deg} + k_{rec})[P_i] \tag{5.68}$$

which can be solved to give

$$[P]_o = \frac{v_s}{k_t}\left(1 + \frac{k_{rec}}{k_{deg}}\right) \tag{5.69}$$

$$[P_i]_o = \frac{v_s}{k_{deg}} \tag{5.70}$$

The steady-state behavior in the presence of ligand can also be solved analytically. It is convenient to combine equations 5.63–5.65 to produce an expression for $[P_i]$ to substitute into equation 5.66, where at steady state, all four rate equations are set to zero. Equation 5.63 plus two times equation 5.64 plus equation 5.65 gives

$$[P_i] = \frac{k_t([P] + [C]) + 2k_e[D] - v_s}{k_{rec}} \tag{5.71}$$

which on substitution into equation 5.66 in combination with Equation 5.69 yields

$$[P]_o = [P] + [C] + 2\frac{k_e}{k_t}\left(1 + \frac{k_{rec}}{k_{deg}}\right)[D] \tag{5.72}$$

From equation 5.64 at steady state, we obtain

$$[D] = \frac{k_{x2}[P][C]}{k_{-x2} + k_{-x1} + k_e} = K_X[P][C] \qquad (5.73)$$

where

$$K_X = \frac{k_{x2}}{k_{-x2} + k_{-x1} + k_e} \qquad (5.74)$$

From equations 5.63 and 5.73 at steady state, we have

$$[C] = \frac{k_{f1}[P][L]}{k_{r1} + k_t + (k_{x2} - k_{-x2}K_X)[P]} = \frac{k_{f1}[P][L]}{k_{r1} + k_t + (k_{-x1} + k_e)K_X[P]} \qquad (5.75)$$

Together, equations 5.72–5.75 can be combined to produce a quadratic equation for $[P]$ at steady state for any given $[L]$. This solution can then be used to obtain steady-state values for $[C]$, $[D]$, and $[P_i]$, which is a complete solution for the behavior of the system at steady state.

5.2.2 Vesicular Transport

In vesicular transport, spherical vesicles containing the soluble contents of a membrane-bound compartment (the *lumen*) bud off and are carried along cytoskeletal structures to another compartment. They then fuse with the new compartment, disgorging their contents (figure 5.22). Capture of the lumenal contents can be by passive volumetric sampling, or by binding to *cargo receptors* within the vesicle. For some transport processes (e.g., lysosomal or ER recovery), the cargo receptors recognize a specific signal (e.g., mannose-6-phosphate or C-terminal KDEL); however, the secretion of the majority of extracellular proteins appears to utilize cargo receptors that bind weakly and promiscuously to cargo proteins at sites that have not yet been well characterized.

Let's consider what simple kinetic rate forms might be appropriate to describe the complex process shown in figure 5.22. We define the following variables to describe the system:

$[P]$ = concentration of cargo protein in the lumenal space

$[R]_o$ = number of available cargo receptor molecules per vesicle

K_d = equilibrium dissociation constant for cargo/receptor interaction

$\dot{n}$ = number of vesicles exiting the compartment per unit time

V_c = compartment volume

V_v = volume of a single vesicle

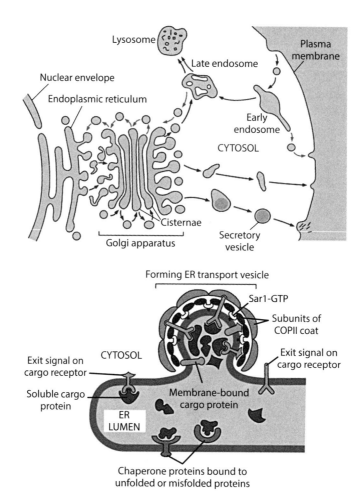

Figure 5.22 Schematic diagram of the mechanism of vesicular transport in eukaryotic cells. The soluble contents of one compartment are sampled into small spherical vesicles, which are transported to other compartments, where they fuse and disgorge their contents (top panel). Cargo is often loaded into the vesicles by receptors on the vesicle wall (bottom panel) [24].

Let's make the following assumptions: (1) the compartment and vesicle volumes are well mixed and at the same cargo protein concentration, (2) available cargo receptors are in excess over cargo ($[R]_\circ \gg V_v[P]$), and (3) the cargo/receptor interaction rapidly approaches equilibrium. With these assumptions, the following material balance on cargo protein can be written:

$$\frac{d(V_c[P])}{dt} = -\dot{n}V_v[P] - \dot{n}[R]_\circ \frac{[P]}{K_d + [P]} \quad (5.76)$$

If the compartmental volume is constant over the time of interest, this expression can be rearranged to yield

$$\frac{d[P]}{dt} = -\dot{n}\frac{V_v}{V_c}\left(1 + \frac{[R]_o}{V_v(K_d + [P])}\right)[P] \quad (5.77)$$

Let's consider the particular limiting case of secreted extracellular and cell-surface proteins, for which the cargo-receptor interaction appears to be very weak (i.e., $[P] \ll K_d$). Cargo protein exit is predicted to follow first-order kinetics:

$$\frac{d[P]}{dt} = -k_{sec}[P] \quad (5.78)$$

where $k_{sec} \equiv \dot{n}\frac{V_v}{V_c}\left(1 + \frac{[R]_o}{V_v K_d}\right)$. Such first-order kinetics have in fact been shown to describe experimental data for secretory cargo trafficking rather well, as shown in the following case study.

Case Study 5-4 K. Hirschberg, C. M. Miller, J. Ellenberg, J. F. Presley, E. D. Siggia, R. D. Phair, and J. Lippincott-Schwartz. Kinetic analysis of secretory protein traffic and characterization of Golgi to plasma membrane transport intermediates in living cells. *J. Cell Biol.* 143: 1485–1503 (1998) [29].

In this work, vesicular stomatitis virus G protein (VSVG) was used as a model membrane protein. In particular, a temperature-sensitive folding mutant (ts045) was used, which unfolds and is retained in the endoplasmic reticula at 40°C but folds and is exported at 32°C. To quantify cargo protein concentration by fluorescence microscopy, ts045 VSVG was fused at its C terminus to GFP. COS-7 cells were transfected with a vector for expression of the ts045 VSVG-GFP fusion construct, and the cells' endoplasmic reticula were loaded with unfolded cargo protein by expression for 20 hours at 40°C. At time zero, the cells were shifted to 32°C, and the cargo protein was quantified by fluorescence microscopy (figure 5.23).

The region delineated by the white curve in figure 5.23 (and most densely populated at the 40-minute time point) was attributed to the Golgi compartment. The total fluorescence in the darker outer boundary was quantified as total cellular cargo protein. No attempt was made to parse out the separate contributions from the endoplasmic reticulum compartment. Sample data for nine separate cells are shown in figure 5.24.

The compartmental model shown in figure 5.25 was hypothesized to represent the ts045 VSVG-GFP transport process during the experiment.

The first-order process with rate constant k_{PM} represents endosomal degradation of plasma membrane protein. Although the set of equations describing this model could be solved analytically, the authors solved them numerically and then used nonlinear least squares curve fitting to obtain the following parameter estimates: $k_{ER} = 0.028$ min^{-1}, $k_G = 0.030$ min^{-1}, and $k_{PM} = 0.0026$ min^{-1}.

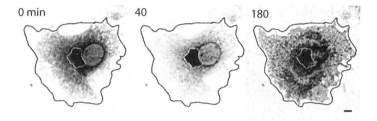

Figure 5.23 Fluorescence imaging of VSVG protein trafficking through the secretory pathway (scale bar = 5 μm) [29].

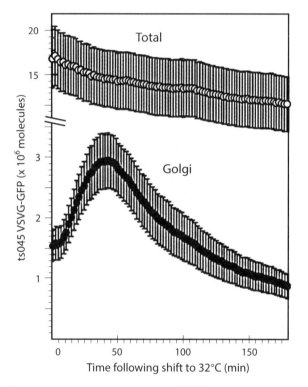

Figure 5.24 Quantification of the compartmental levels of VSVG versus time, from images such as those shown in figure 5.23.

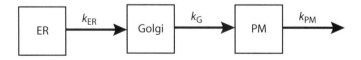

Figure 5.25 Multicompartmental model used to fit the data in figure 5.24.

The chief result of this paper is the demonstration that despite the apparent geometric complexity of the intracellular transport pathway, the rates of cargo movement through it are quantitatively well described by an extremely simple well-mixed compartmental model with first-order mass-action kinetics. The model does not attempt to capture the steps of vesicular movement between compartments, because such movement is much more rapid than the compartmental exit rates. Direct tracking of individual vesicles from the Golgi to the plasma membrane showed that this linear movement was completed in an average of 4 min, compared to the mean compartmental residence times of 39 (ER), 42 (Golgi), and 709 min (plasma membrane).

Suggestions for Further Reading

B. Alberts, A. D. Johnson, J. Lewis, D. Morgan, M. Raff, K. Roberts, and P. Walter. *Molecular Biology of the Cell*, sixth edition. W. W. Norton and Company, 2014.

D. A. Lauffenburger and J. J. Linderman. *Receptors: Models for Binding, Trafficking, and Signaling*. Oxford University Press, 1996.

R. Milo and R. Phillips. *Cell Biology by the Numbers*. Garland Science, 2015.

Problems

5-1 Derive the general time-dependent expression of a protein when $\gamma_R = \gamma_P$.

5-2 Derive an expression for the dynamic protein concentration when transcription is instantaneously and fully inhibited at $t = 0$. Assume that the system was at steady state prior to inhibition of expression of this particular gene.

5-3 a. Derive an expression for the number of cell surface receptors at steady state in the absence of ligand (equation 5.54).

b. Derive an expression for the fraction of total receptors that reside on the cell surface at steady state in the absence of ligand (equation 5.55).

5-4 If the rate constant for endocytosis of a receptor and for the receptor-ligand complex are identical, will the apparent equilibrium receptor-ligand binding constant be altered by endocytosis, relative to the value measured in a biosensor, for example?

5-5 Using quantum dot technology, Lidke and coworkers have studied receptor-mediated internalization of epidermal growth factor (EGF) by visualizing the subcellular location of EGF and quantifying the relative cell-surface and cell-interior amounts of EGF [30]. This problem concerns analysis of their data (table 5.3), which reports the fluorescence of EGF at the surface of the cell and the interior of the cell as a function of time. Based on the given data, estimate the value of endocytic rate constant ($k_{e,C}$). Compare your endocytic parameter to the ligand binding parameter, $k_f \approx 1.8 \times 10^8$ M^{-1} min^{-1} [31], for EGF binding to EGFR. The initial ligand concentration used in the experiment was 0.1–0.2 nM [30]. Does your analysis support the authors' conclusion: "Whereas surface binding followed an exponential course, internalization was linear. We conclude that the cellular endocytotic machinery and not ligand binding constitutes the rate-limiting process in the endocytosis of the receptor-ligand complex"? (Contributed by Judy Yeh.)

Table 5.3 Comparative kinetics of binding and internalization of EGF quantum dot.

Time (s)	Intensity (surface)	Intensity (interior)
0	0.09	0.09
50	0.28	0.15
100	0.45	0.20
150	0.61	0.25
200	0.72	0.31
250	0.79	0.37
300	0.84	0.43
400	0.90	0.57
500	0.95	0.70
600	0.98	0.81
700	0.99	0.92
800	1.00	1.05
900	1.00	1.19

5-6 A change in a microbe's environment results in a 20% increase in its transcription rate of a particular gene. Yet you observe a 20% *decrease* in the steady-state concentration of the corresponding protein. What other processes could have been impacted by the environmental change to account for this result? Be quantitative in your response.

5-7 A monovalent receptor binds a soluble ligand and sends a growth signal to the cell. Assume that the ligand is provided at a constant concentration in the growth medium and that the rate of receptor synthesis and the internalization rate constant are unaltered by ligand binding. Ignore receptor dilution by growth.

   a. At what concentration of ligand is half of the cell surface receptor bound by ligand at steady state? This concentration will be termed EC_{50}.

   b. If the ligand can be engineered by directed evolution to dissociate more slowly, what is the minimum achievable value for EC_{50}?

   c. If the ligand can be engineered by directed evolution to associate more rapidly, what is the minimum achievable value for EC_{50}?

5-8 You aim to achieve intracellular delivery of a drug ligand to a select cell type. You design a dosing mechanism to achieve a steady concentration of the ligand in the blood ($[L]_o = 4$ nM). Select cells accessible to the bloodstream have receptors on their surface ($[R]_o = 10^5$ receptors per cell). The ligand binds these receptors with 8 nM affinity. The association rate constant for this interaction is 3×10^5 M^{-1}s^{-1}. Ligand-receptor complex is internalized with a half-time of 8 min. After internalization, the ligand-receptor complex immediately dissociates, and receptor is immediately recycled—unbound—to the cell surface.

a. Derive an expression for the number of internalized ligands per cell as a function of time, $[R]_o$, $[L]_o$, and relevant rate constants. Note that these cells are relatively rare in the body, so that the ligand concentration in the blood is not depleted (i.e., it remains constant).

b. Calculate the number of ligands internalized in the first hour.

c. Calculate the steady-state fraction of receptors bound by ligand.

5-9 In this problem, we analyze autocrine signaling, which is a particular mode of signaling present in various biological networks. A protein ligand is synthesized by the ribosome and immediately secreted to the extracellular space. The ligand undergoes reversible binding with a receptor on the surface of the same cell. This binding and dissociation of the ligand and receptor are rapidly equilibrated relative to other processes. Ligand-receptor complex triggers transcription of numerous genes, including the ligand gene; the transcription rate is proportional to ligand-receptor complex concentration. The protein ligand can also be transported away from the vicinity of the cell. What fraction of receptors are bound by ligand at steady state?

5-10 You add a tracer molecule at concentration $[P]_0$ to a single tank of volume V. The concentration of tracer in the exit stream is

$$[P]_1 = [P]_0 e^{-kt} \qquad (5.79)$$

where k is the rate constant for efflux from the tank. Now consider the volume V to be distributed equally among N tanks in series. The rate constant for transport now scales with the number of tanks. Derive an expression for the concentration of tracer leaving the final tank ($[P]_N$) as a function of time (equation 5.47). Plot the dimensionless output concentration versus dimensionless time for $N = 1, 3, 10, 30,$ and 100 tanks.

5-11 Two identical monomeric receptors, which freely move throughout the cell membrane, reversibly bind to form a dimer. The dimer is capable of internalizing into the cell with first-order rate constant k_S, which simultaneously generates an activated intracellular signaling molecule (S). Immediately after internalization, the receptors return, as separated monomers, to the cell surface. Initially, the system has a receptor concentration of $[R]_o$; neither dimers nor activated signaling molecules are initially present.

   a. Solve for the concentration of dimerized receptor at a time long after initial startup.

   b. Separately, assume that dimerization does not substantially deplete the concentration of monomer present on the cell surface (i.e., dimers are rare relative to monomers). Derive an expression for the dynamic signal concentration in the cell.

   c. Extend the model's complexity to consider that the signaling molecule degrades at a rate described by first-order rate constant k_{deg}. Derive an expression for the steady-state concentration of signaling molecule. You may still assume nondepleting dimerization, as in part b.

   d. You are considering two different approaches to inhibit signaling. You have engineered one inhibitor that binds to monomeric receptor, with affinity K_I, and prevents dimerization. The inhibitor is incapable of binding to dimerized receptor. Separately, you have a different inhibitor that binds activated signaling molecule, also with affinity K_I, and prevents its activity. Which inhibitor yields a greater reduction in active signaling? Derive any equations needed to support your conclusion.

References

1. J. A. Bernstein, A. B. Khodursky, P. H. Lin, S. Lin-Chao, and S. N. Cohen. Global analysis of mRNA decay and abundance in *Escherichia coli* at single-gene resolution using two-color fluorescent DNA microarrays. *Proc. Natl. Acad. Sci. U.S.A.* 99: 9697–9702 (2002).

2. Y. Wang, C. L. Liu, J. D. Storey, R. J. Tibshirani, D. Herschlag, and P. O. Brown. Precision and functional specificity in mRNA decay. *Proc. Natl. Acad. Sci. U.S.A.* 99: 5860–5865 (2002).

3. B. Schwanhausser, D. Busse, N. Li, G. Dittmar, J. Schuchhardt, J. Wolf, W. Chen, and M. Selbach. Global quantification of mammalian gene expression control. *Nature* 473: 337–342 (2011).

4. R. Milo and R. Phillips. *Cell Biology by the Numbers*. Garland Science (2015).

5. J. L. Hargrove, M. G. Hulsey, and E. G. Beale. The kinetics of mammalian gene expression. *Bioessays* 13: 667–674 (1991).

6. J. R. Greenberg. High stability of messenger RNA in growing cultured cells. *Nature* 240: 102–104 (1972).

7. F. Edfors, F. Danielsson, B. M. Hallström, L. Käll, E. Lundberg, F. Pontén, B. Forsström, and M. Uhlén. Gene-specific correlation of RNA and protein levels in human cells and tissues. *Mol. Sys. Biol.* 12: 883 (2016).

8. G. Csárdi, A. Franks, D. S. Choi, E. M. Airoldi, and D. Drummond. Accounting for experimental noise reveals that mRNA levels, amplified by post-transcriptional processes, largely determine steady-state protein levels in yeast. *PLOS Genetics* 11: e1005206 (2014).

9. H. C. Towle, C. N. Mariash, H. L. Schwartz, and J. H. Oppenheimer. Quantitation of rat liver messenger ribonucleic acid for malic enzyme during induction by thyroid hormone. *Biochemistry* 20: 3486–3492 (1981).

10. A. L. Goldberg and J. F. Dice. Intracellular protein degradation in mammalian and bacterial cells. *Annu. Rev. Biochem.* 43: 835–869 (1974).

11. R. D. Mosteller, R. V. Goldstein, and K. R. Nishimoto. Metabolism of individual proteins in exponentially growing *Escherichia coli. J. Biol. Chem.* 255: 2524–2532 (1980).

12. B. Futcher, G. I. Latter, P. Monardo, C. S. McLaughlin, and J. I. Garrels. A sampling of the yeast proteome. *Mol. Cell. Biol.* 19: 7357–7368 (1999).

13. J. E. Bergmann and H. F. Lodish. A kinetic model of protein synthesis. Application to hemoglobin synthesis and translational control. *J. Biol. Chem.* 254: 11,927–11,937 (1979).

14. S. B. Lee and J. E. Bailey. Analysis of growth rate effects of productivity of recombinant *Escherichia coli* populations using molecular mechanism models. *Biotechnol. Bioeng.* 26: 66–73 (1984).

15. S. M. Uptain, C. M. Kane, and M. J. Chamberlin. Basic mechanisms of transcript elongation and its regulation. *Annu. Rev. Biochem.* 66: 117–172 (1997).

16. G. Orphanides and D. Reinberg. RNA polymerase II elongation through chromatin. *Nature* 407: 471–476 (2000).

17. F. Engbæk, N. O. Kjeldgaaed, and O. Maaløe. Chain growth rate of β-galactosidase during exponential growth and amino acid starvation. *J. Mol. Biol.* 75: 109–118 (1973).

18. J. W. Hershey. Translational control in mammalian cells. *Annu. Rev. Biochem.* 60: 717–755 (1991).

19. R. M. Winslow and V. M. Ingram. Peptide chain synthesis of human hemoglobins A and A2. *J. Biol. Chem.* 241: 1144–1150 (1966).

20. U. Vogel and K. F. Jensen. The RNA chain elongation rate in *Escherichia coli* depends on the growth rate. *J. Bacteriol.* 176: 2807–2813 (1994).

21. R. Schleif, W. Hess, S. Finkelstein, and D. Ellis. Induction kinetics of the L-arabinose operon of *Escherichia coli. J. Bacteriol.* 115: 9–14 (1973).

22. J. G. Yargera, H. O. Halvorson, and J. E. Hopper. Regulation of galactokinase (GAL1) enzyme accumulation in *Saccharomyces cerevisiae. Mol. Cell. Biochem.* 61: 173–182 (1984).

23. B. G. Adams. Induction of galactokinase in *Saccharomyces cerevisiae*: Kinetics of induction and glucose effects. *J. Bacteriol.* 111: 308–315 (1972).

24. B. Alberts, A. D. Johnson, J. Lewis, D. Morgan, M. Raff, K. Roberts, and P. Walter. *Molecular Biology of the Cell*, sixth edition. W. W. Norton & Company (2014).

25. B. J. Marsh, D. N. Mastronarde, K. F. Buttle, K. E. Howell, and J. R. McIntosh. Organellar relationships in the Golgi region of the pancreatic beta cell line, HIT-T15, visualized by high resolution electron tomography. *Proc. Natl. Acad. Sci. U.S.A.* 98: 2399–2406 (2001).

26. R. B. MacMullin and M. Weber. The theory of short-circuiting in continuous-flow mixing vessels in series and kinetics of chemical reactions in such systems. *Trans. Amer. Inst. Chem. Engr.* 31: 409–458 (1935).

27. D. A. Lauffenburger and J. J. Linderman. *Receptors: Models for Binding, Trafficking, and Signaling*. Oxford University Press (1996).

28. J. M. Haugh. Mathematical model of human growth hormone (hGH)-stimulated cell proliferation explains the efficacy of hGH variants as receptor agonists or antagonists. *Biotechnol. Prog.* 20: 1337–1344 (2004).

29. K. Hirschberg, C. M. Miller, J. Ellenberg, J. F. Presley, E. D. Siggia, R. D. Phair, and J. Lippincott-Schwartz. Kinetic analysis of secretory protein traffic and charaterization of Golgi to plasma membrane transport intermediates in living cells. *J. Cell Biol.* 143: 1485–1503 (1998).

30. D. S. Lidke, P. Nagy, R. Heintzmann, D. J. Arndt-Jovin, J. N. Post, H. E. Grecco, E. A. Jares-Erijman, and T. M. Jovin. Quantum dot ligands provide new insights into erbB/HER receptor-mediated signal transduction. *Nat. Biotechnol.* 22: 198–203 (2004).

31. C. M. Waters, K. C. Oberg, G. Carpenter, and K. A. Overholser. Rate constants for binding, dissociation, and internalization of EGF: Effect of receptor occupancy and ligand concentration. *Biochemistry* 29: 3563–3569 (1990).

6
Network Dynamics

Anybody that thought the genome was going to directly provide drugs was a fool. Biological networks are not simple, and making drugs to affect them won't be simple.
—Leroy Hood

Stir not until the signal.
—William Shakespeare, *Julius Caesar*

Our discussion of molecular interactions has primarily focused, thus far, on molecular pairs (e.g., ligand-receptor or enzyme-substrate). In these relatively simple cases, it is possible to develop reasonably comprehensive conceptual and mathematical descriptions of interaction dynamics. However, single interactions are generally not sufficient to explain more complex cellular dynamics, particularly those involving cellular sensing and responding to environmental cues. Simple two-species binding or enzymatic reactions described by mass-action kinetics do not produce the more complex dynamics that are often observed for biomolecular networks.

In this chapter, we cannot present an exhaustive treatment of even small biomolecular networks, given the large number of possibilities. For example, consider a two-component network in which each component can exert an activating, inhibitory, or null effect on the other component. Even in this simplest case (excluding self-interactions), there are six distinct motif types (figure 6.1). Expanding this network to three components enables 138 distinct motifs [1]. Because these interaction sets are generally not insulated from multiple outside regulatory interactions, one may question whether the choice to examine isolated motifs can capture the essential features of dynamic biochemical networks.

Instead of systematically evaluating different network architectures, here we focus our attention on a handful of commonly observed dynamic behaviors: switch-like responses, adaptation, and oscillations. These patterns are of particular relevance to cell signaling, and we highlight some illustrative examples of networks that can generate these dynamics. Before we dive into these cellular behaviors, we

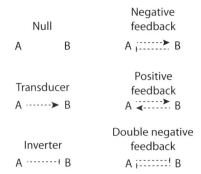

Figure 6.1 Shown are the six distinct motifs possible for interactions in a two-component network. The dashed lines signify activating (arrowhead) or inhibitory (bar) interactions between the two components. Self-interactions are not included in this accounting.

first present a mathematical framework that will enable us to quantitatively analyze the nonlinear dynamics of systems of ordinary differential equations.

6.1 Nonlinear Dynamics

Networked biological interactions generally lead to sets of nonlinear differential equations due to the bimolecularity ($2 \rightarrow 1$ nature) of binding reactions. Other common mathematical treatments that introduce nonlinear terms into the equations describing dynamics of biological systems include the Michaelis-Menten equation for enzyme kinetics and the Hill equation for cooperative binding. Although systems of nonlinear differential equations are generally more difficult to treat and solve than their linear counterparts, they also lead to a richer variety of behaviors. A useful theoretical framework describes the dynamical and steady-state behavior of nonlinear systems, and it provides an important set of tools for analyzing biological networks. Many excellent books provide an introduction as well as more advanced treatments of the material than will be included here. We refer the reader especially to the book by Strogatz [2], which provides a particularly accessible account.

We represent the concentrations of species in the set of N components $\{1, 2, 3, ..., N\}$ as $\vec{X}(t) = (x_1(t), x_2(t), x_3(t), ..., x_N(t))$. Ordinary differential equations describing the time-dependent behavior of the system would then be

$$\dot{x}_1(t) = f_1(x_1(t), x_2(t), x_3(t), ..., x_N(t)) \tag{6.1}$$

$$\dot{x}_2(t) = f_2(x_1(t), x_2(t), x_3(t), ..., x_N(t)) \tag{6.2}$$

$$\dot{x}_3(t) = f_3(x_1(t), x_2(t), x_3(t), ..., x_N(t)) \tag{6.3}$$

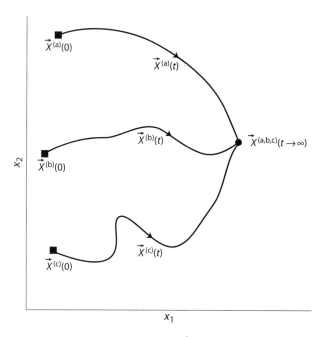

Figure 6.2 Three different trajectories (black curves) for $\vec{X}(t)$, starting from three different starting conditions at time $t=0$ (three black squares), representing $\vec{X}^{(a)}(0)$, $\vec{X}^{(b)}(0)$, and $\vec{X}^{(c)}(0)$. All three trajectories tend to the same point (black circle) at long time, which represents a steady state. Along each curve, the gradient represents the instantaneous change in concentrations at the corresponding time.

$$\vdots \qquad \vdots$$
$$\dot{x}_N(t) = f_N(x_1(t), x_2(t), x_3(t), \ldots, x_N(t)) \tag{6.4}$$

where $\dot{x}_i(t) \equiv dx_i(t)/dt$. The functions f_i represent the collection of kinetic terms that contribute to changes in the concentration x_i of species i, terms like $k_1 x_3 x_4$ and $-k_2 x_2$. The set of equations above can be represented more simply and compactly using vector notation as

$$\dot{\vec{X}}(t) = \vec{F}(\vec{X}(t)) \tag{6.5}$$

In this notation, a trajectory $\vec{X}(t)$, showing the concentrations of all species in the system as a function of time, can be visualized as a trace in N-dimensional space, with each dimension representing one of the $x_i(t)$. Figure 6.2 illustrates this idea for a system with two species, x_1 and x_2. The gradient $\dot{\vec{X}}(t) \equiv d\vec{X}(t)/dt$ is the velocity of the trajectory at time t, because $\dot{\vec{X}}(t) = (\dot{x}_1(t), \dot{x}_2(t), \ldots, \dot{x}_N(t))$. The gradient is tangent to the trajectory at any point.

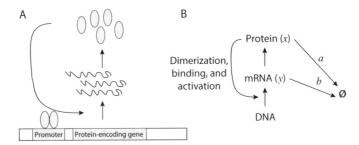

Figure 6.3 A model of gene autoexpression described by Griffith [3]. (A) Schematic diagram of a gene that is transcribed into mRNA and then translated into protein. The protein dimerizes and activates the expression of the orginal gene by binding to its promoter. (B) The protein concentration is represented as x and the mRNA as y. Degradation of protein and mRNA to nonfunctional states ("null" symbol) is included.

To help explore these ideas in more detail and to see their utility in a situation of practical interest, let us examine a model of gene autoregulation described by Griffith [3] and analyzed by Strogatz [2]. In this model, a gene is transcribed to its messenger RNA, which is subsequently translated to the corresponding protein; the protein folds and homodimerizes, and the protein dimer binds to the original gene's promoter, stimulating its transcription (figure 6.3A). This type of connectivity in a network is called a positive-feedback loop, because in at least some regimes, increasing the signal at some step produces changes in the network that tend to increase that signal further. For example, increasing the rate of gene transcription leads to an increased mRNA level, which leads to more protein, which results in more gene-activating protein dimer, which in turn enhances the rate of transcription. To complete the sketch, we note that Griffith also included degradation of both the protein and the mRNA (figure 6.3B). To turn this sketch into a model, mathematical forms and parameters must be assigned to all the rates indicated by arrows in figure 6.3B.

Letting x be proportional to the concentration of protein and y to the concentration of mRNA, Griffith's model proposes nondimensional equations

$$\dot{x} = y - ax \tag{6.6}$$

$$\dot{y} = \frac{x^2}{1+x^2} - by \tag{6.7}$$

(We will show in section 6.2.1 how homodimerization of the protein can, in principle, produce the Hill coefficient of 2, as suggested here.)

To understand how this system behaves, we construct a rough sketch of its *phase portrait*. A phase portrait is a representation of the dynamical and steady-state behavior of a particular model across all initial conditions. The construction of a phase portrait represents a general technique that can be applied broadly to dynamical systems. Although software packages exist that will produce a phase portrait easily, it can also be useful to construct one by hand. As a first step, we draw what are called the *nullclines* of the system. The nullclines are those points in phase space (that is, the space of concentrations), for which the velocity or flow (the instantaneous change in concentration with respect to time) is purely horizontal or purely vertical. Each nullcline is usually a continuous curve in the phase space (one for each of the concentrations) and represents points at which the instantaneous change in at least one concentration is zero.

For our model, one nullcline is defined by setting the instantaneous change in x with respect to time to zero. For example, setting equation 6.6 to zero gives the equation of a line $y = ax$ that passes through the origin with slope a, indicated by the dashed line in figure 6.4A. Along this line, the velocity is purely vertical because $\dot{x} = 0$ (i.e., it has no horizontal component). The second nullcline is found by setting the instantaneous change in y with respect to time, equation 6.7, to zero. This solution is given by $y = x^2/(b(1+x^2))$, which produces the sigmoidal curve sketched as the solid line in figure 6.4A. The velocity is purely horizontal along this curve, because it corresponds to points of $\dot{y} = 0$. Note that the graph is given only in the first quadrant, because x and y represent concentrations, which are nonnegative.

Points where the nullclines intersect are particularly significant, because they correspond to points in phase space where neither x nor y changes with time. That is, if the system reaches such a point, it will remain there indefinitely, in principle. Such points are called *stationary points* or *fixed points*.

Example 6-1 Find the values of x and y at the stationary points in the Griffith gene expression model.

Solution Returning to the model in equations 6.6 and 6.7 and setting both $\dot{x}$ and $\dot{y}$ to zero, we find $y = ax$ and $y = \frac{x^2}{b(1+x^2)}$. Setting these expressions for y equal to one another and rearranging, we find

$$x(abx^2 - x + ab) = 0 \tag{6.8}$$

whose roots are

$$x = 0, \; \frac{1 \pm \sqrt{1 - 4a^2b^2}}{2ab} \tag{6.9}$$

For each value of x, we plug into $y = ax$ to find y:

$$(x, y) = \left\{ (0, 0), \left(\frac{1 - \sqrt{1 - 4a^2b^2}}{2ab}, \frac{1 - \sqrt{1 - 4a^2b^2}}{2b} \right), \right.$$
$$\left. \left(\frac{1 + \sqrt{1 - 4a^2b^2}}{2ab}, \frac{1 + \sqrt{1 - 4a^2b^2}}{2b} \right) \right\} \quad (6.10)$$

Each of the three values in equation (6.10) needs to be checked to see whether it is feasible. Values could be infeasible because the quantity inside the square root is negative, leading to nonreal values for x and y (which does happen in this case for values of $2ab > 1$), or because they lead to negative real values for x or y (which does not happen in this case so long as a and b remain positive).

It is helpful to draw arrows on the nullclines to represent whether the horizontal spontaneous change (frequently referred to as a *flow*) is leftward or rightward, and whether the vertical flow is up or down. Looking at the $y = ax$ nullcline first, the vertical flow is given by $\dot{y} = x^2/(1 + x^2) - by$. In the limit of small, nonnegative values of x (i.e., $0 < x \ll 1$), the first term is essentially x^2 and is dominated by the second term ($-by$, which equals $-abx$ along this nullcline), because $O(x^2)$ terms are smaller in magnitude than $O(x)$ terms in this regime. This produces a negative $\dot{y}$, and the flow is downward along the $y = ax$ nullcline for small x. In the limit of large values of x (i.e., $x \gg 1$), the first term is essentially 1 and is again dominated in magnitude by the second term, $-abx$, in this regime. Thus, $\dot{y}$ is again negative, and the flow is downward along the $y = ax$ nullcline, but this time for large x. At intermediate values of x, the first term dominates the second, $\dot{y}$ is positive, and the flow along the nullcline is positive. The arrows in figure 6.4A indicate these three flows in the three regions separated by stationary points for the $y = ax$ nullcline, and similar logic leads to the three horizontal arrows shown on the other nullcline. Note that on a given segment of a nullcline that does not intersect the other nullcline, the directionality of all velocity arrows along that particular segment will be the same, because for the sign of the velocity to change, the velocity must pass through zero (but this would necessitate intersection with the other nullcline). Thus, the nullclines divide the phase space into regions defined by the sign of the flow (the signs of $\dot{x}$ and $\dot{y}$), which can be determined by examining the equations. Representative flow directions in the regions between the nullclines are also shown in the figure.

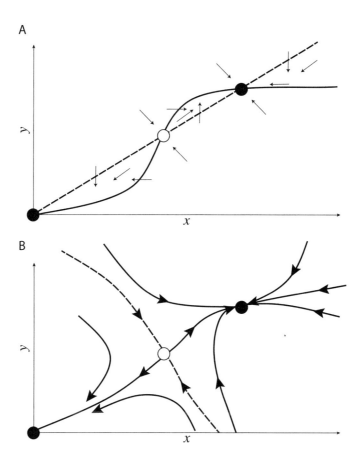

Figure 6.4 (A) The nullclines for the system given by equations 6.6 and 6.7. The dashed line is obtained by setting the former equation to zero, giving a purely vertical instantaneous change in state in the phase diagram. The solid line comes from setting the latter equation to zero, giving a purely horizontal flow. The three points of intersection are stationary states—if the system arrives or is placed there and remains undisturbed, it will not move away spontaneously. The two filled circles correspond to stable nodes and the open circle to an unstable one. The direction of flow (up or down; left or right) along each of the nullclines is indicated as well as in the regions between the nullclines. (B) The corresponding phase portrait, with representative flow lines and the direction of flow given. The dashed curve represents a *separatrix*: If the system is started from an initial condition to the left of this line, it will reach and remain at the lower steady state. If started from conditions to the right of this line, the system will reach and remain at the upper steady state.

Returning for a moment to the stationary points, they are important for many reasons. Key among these is this: If a stationary point is stable, then it corresponds to a steady state. To characterize a stationary state, one imagines an infinitesimal displacement in any direction and then tracks whether the local flow would return the system to the same stationary state or draw it away. A direction for which a displacement returns the system to the stationary state is referred to as *stable*. A direction for which a displacement carries the system away from the stationary state is *unstable*. If all directions from a stationary point return to it, then the point is said to be stable. If even one direction is unstable, then the stationary point is said to be unstable. Using the arrows in figure 6.4A as guides, one can see that the middle stationary point is unstable (along the dashed nullcline, a small displacement draws the system away from the point) and the other two are stable. The middle stationary point, although unstable along the direction of the dashed nullcline, is stable along the direction of the solid nullcline. A combination of stable and unstable directions is called a *saddle point*, because for a two-dimensional function, it corresponds to the contours of a saddle: From the seat, the saddle rises toward the head and tail of the horse and falls in the directions of the rider's feet.

The informal description given here for determining the stability of a stationary point has its limitations. A more formal and correct analysis involves linearizing the equations describing the system at each stationary point and carrying out an eigen-analysis of the linearized equations. The nature of the eigenvalues and eigenvectors describes the dynamics of the system close to the stationary point. The general solution of the differential equations is

$$\Delta \vec{X}(t) = c_1 e^{\lambda_1 t} \vec{V}_1 + c_2 e^{\lambda_2 t} \vec{V}_2 \qquad (6.11)$$

for the two-dimensional case considered here, and

$$\Delta \vec{X}(t) = \sum_i c_i e^{\lambda_i t} \vec{V}_i \qquad (6.12)$$

for the general case, where $\Delta \vec{X}(t)$ is the time-dependent state of the system relative to the stationary point $\vec{X}^*$:

$$\Delta \vec{X}(t) \equiv \vec{X}(t) - \vec{X}^* \qquad (6.13)$$

The c_i are constants that depend on the initial conditions, the λ_i are the eigenvalues, and the $\vec{V}_i$ the corresponding eigenvectors. A more detailed treatment of this form of analysis is given in chapter 7 in the description of the Lotka-Volterra models of predator-prey populations.

The two stable stationary points in the Griffith model, which are *stable nodes*, each have two negative real eigenvalues, corresponding to directions in which, at long time, the displacement goes to zero. The saddle point has one negative real (stable) and one positive real (unstable) eigenvalue. Other behaviors, not present in this example, include: two positive real eigenvalues, which represent *unstable nodes*; purely imaginary eigenvalues, which represent stationary points that are *centers* and correspond to oscillatory solutions (and although the mathematics is different, oscillatory solutions in nonlinear models correspond to *limit cycles*); and complex solutions with real and imaginary parts that represent stationary points that are *spirals* and correspond to solutions that oscillate while growing (*unstable spiral*, with positive real part) or decaying (*stable spiral*, with negative real part).

Returning to our example, we've drawn the nullclines and characterized the stationary points, which occur at the intersections of the nullclines. We've also indicated the direction of the flow in each segment of each nullcline and in each region between the nullclines. Using these flow directions and stationary points as guides, we can produce a sketch of the phase portrait for the system, as shown in figure 6.4B. Each of the stationary points is shown, along with representative flow lines. A flow line depicts a trajectory as a function of time taken from some initial condition. Because of the uniqueness of solutions to the system of differential equations, trajectory lines do not cross except at stationary points; otherwise, two outcomes could result from the set of initial conditions represented by their crossing point. Stationary points represent unchanging solutions, and so multiple trajectories can terminate at a stationary point.

The phase portrait in figure 6.4B exhibits *multistability* for this dynamical system—it has more than one steady state, in this case, two. The figure shows that sets of initial conditions with relatively low protein and mRNA concentrations all reach the low steady state represented by the stable node at the origin (zero protein and zero mRNA). With a sufficiently high initial protein or mRNA concentration, the system will instead reach the high steady state. There is a particular line (the dashed line passing through the saddle point, called the *separatrix*) that divides the phase portrait into points to the left (which always end in the low steady state) and points to the right (which always end in the high steady state). In fact, no matter what set of initial conditions is selected, the steady-state expression level can never exceed that of the high steady state.

Let's pause to review. In this section, we've followed one particular example involving gene expression to examine how one can characterize the dynamical behavior of nonlinear systems of differential equations used throughout this text. We've seen that we can encapsulate the behavior in terms of a phase portrait. The phase portrait defines the time-dependent dynamics for all possible initial conditions but a single set of parameters. To trace the dynamics, we just find the point on the phase portrait corresponding to a particular initial condition and trace forward in time along the flow line going through that point; often the flow line will end in a stable node or spiral, corresponding to a stationary state. Now, what happens if the parameters change—maybe a rate constant doubles? A new phase portrait governs the changed system, but the new phase portrait is often very similar to the original one. Doubling a rate constant may change the position of a steady state, so the positions of the stationary points might move somewhat, but often the number and character of stationary points, and their general relationships to one another, remain the same, as does the pattern of flow lines. That is, although quantitatively the details of the dynamical and steady-state behavior depend on the parameters, often the qualitative behavior (the numbers and types of steady states, for instance) is relatively insensitive to the exact value of the parameters. Interestingly, for many systems of biological interest, this property holds across wide swaths of parameter space.

Suppose that we printed out phase portraits for small increments that systematically marched through parameter space for a particular system and assembled all of these pages into a flip book. If we then fanned through the pages of the book, it would display a movie. For the most part, the movie would show the stationary points and flow lines smoothly and slowly shifting on the page. But every once in a while, there would be a sharp transition in the movie—a stationary point might change character or disappear, or a new one might appear, so that a pure oscillation might become damped and then overdamped, for example. Each of these sudden transitions represents an interesting division in parameter space and potentially a different set of biological behaviors. In the terminology of nonlinear dynamics, these sudden transitions are called *bifurcations*.

The gene expression system under discussion here has two parameters, a and b, corresponding to effective protein and mRNA degradation rate constants, respectively. Imagine that we keep b fixed and slowly increase a. If we return to our sketch of the nullclines, reproduced in figure 6.5A, we recall that the sigmoidal nullcline is given by the equation $y = x^2/(b(1 + x^2))$, which depends on b but not a, and so remains constant. The linear nullcline is given by the equation $y = ax$, in which a gives the slope of the line passing through the origin. As a increases, so does the slope. The stationary points correspond to the intersections of the nullclines. Thus, as the parameter a increases slowly, the slope rises, and the low stable node remains at the origin, but the other two stationary points become closer together (the saddle point and the high stable node; figure 6.5B). Increasing the parameter a further results in a value for which the upper two stationary points coalesce into a single stationary point, because the linear nullcline is tangent to the sigmoidal nullcline

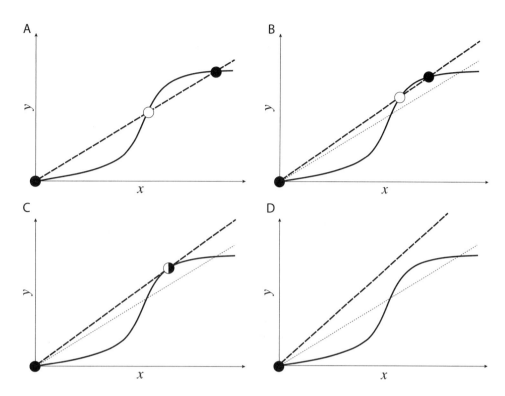

Figure 6.5 Demonstration of a bifurcation. (A) The same nullclines as shown in figure 6.4A, corresponding to a particular choice of parameters a and b with $ab < 1/2$. The nullclines intersect in three stationary points, a low stable node (closed circle at the origin), a saddle point (open circle), and a high stable node (the other closed circle). The remaining panels show the effect of increasing parameter a while parameter b remains fixed, which increases the slope of the linear nullcline while leaving the sigmoidal nullcline fixed. The original position of the linear nullcline is shown as a thin, light line in the remaining panels. (B) As parameter a increases a small amount, the slope of the linear nullcline increases such that the two nullclines still intersect at three points. (C) Increasing parameter a further leads to a value at which the nullclines intersect in only two points. This is the critical value of a at which a bifurcation occurs. (D) All further increases to parameter a result in a phase portrait with just one stationary point.

(figure 6.5C). For larger values of a, there is only one stationary point, at the origin, and just one steady state corresponding to no gene expression (figure 6.5D). That is, if the product of the protein and mRNA degradation rates is too great, then the system can not support any expression at steady state. No matter what initial conditions are set, the system will eventually reach a state at which none of the desired protein or its mRNA is in the cell and none of either is being expressed. The point at which this transition occurs, from three stationary points down to one, is the bifurcation.

Example 6-2 What is the critical value of the parameter a at the bifurcation?

Solution In example 6-1, we saw that the x value for the saddle point and the high stable node are given by the equation $x = \dfrac{1 \pm \sqrt{1-4a^2b^2}}{2ab}$, with the positive sign giving the high stable node and the negative giving the saddle point. These points coalesce when the square-root term is zero, that is, for $2ab = 1$, or $a = \dfrac{1}{2b}$.

Example 6-3 With increasing values of the parameter a, the high steady state moves to the left before disappearing at the bifurcation. What is the minimum value of the protein concentration variable x for the high steady state?

Solution The value of the high steady state will reach its minimum value at the bifurcation, and so we are being asked to find the value of x at the bifurcation. From example 6-2, we know that $x = \dfrac{1 \pm \sqrt{1-4a^2b^2}}{2ab}$ for the two points that coalesce in the bifurcation, which occurs when the square-root term is zero. Thus, the minimum value of x for the high steady state is $x = \dfrac{1}{2ab} = 1$, because $2ab = 1$ at the bifurcation.

This analysis can be depicted in a plot that shows the steady states of the system, represented by the steady-state value of the protein level x_{ss} as a function of the parameter a (for some fixed value of b). Such a plot is sketched in figure 6.6. The two black curves represent the two possible steady states—the high steady state (high stable node) that exists only for values of $a < \dfrac{1}{2b}$ curving down from the vertical axis and the low steady state (low stable node) at no protein expression as the line along the horizontal axis. This system displays a property shared by many dynamical systems known as *hysteresis*, which means that the state the system reaches can depend not only on its current conditions but also on its history. In this case, two possible steady states are attainable for values of $0 < a < \dfrac{1}{2b}$. If we carefully control a, we can drive the system through a process so that for the same value of a, it exhibits different behaviors that depend on its past history. This is illustrated by the gray line in the figure. Starting from the high steady state, as a is slowly increased, the system remains in the high steady state, whose value drops slowly and continuously with increasing a. This continues until the bifurcation at $a = \dfrac{1}{2b}$ is reached, at which point the high steady state disappears and the system drops discontinuously

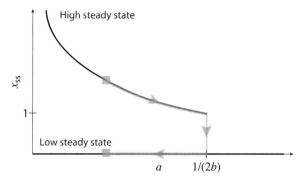

Figure 6.6 Hysteresis diagram showing how steady-state values of x change as a function of the value of parameter a for fixed b. The upper black curve, part of which extends under the gray curve, represents the high steady state, which exists only for $a < \frac{1}{2b}$. The low steady state is represented by the black line along the horizontal axis. The gray lines illustrate a process demonstrating hysteresis, starting from the upper gray square and ending at the lower gray square, both with the same value of a.

into the low steady state. As the value of parameter a returns to its original value, the system remains in the low steady state.

Example 6-4 Consider a cell with R_o integrin receptors attached to a substrate presenting L_o adhesive ligands ($L_o \gg R_o$). A detachment force F is applied to the cell. What is the minimum value of F that can fully detach the cell?

Solution Following the theory of Bell [4], the dissociation rate constant of each integrin-ligand complex can be described by

$$k_{\text{off}} = k_{\text{off},o} e^{\left(\frac{\gamma F}{k_B TC}\right)} \tag{6.14}$$

where $k_{\text{off},o}$ is the basal dissociation rate constant, γ is known as the reactive compliance (in units of distance), k_B is the Boltzmann constant, T is temperature, and C is the number of complexes. (The term $\gamma F/C$ can therefore be thought of as the work performed on each bond.)

Thus, we can write the rate equation for complex number as

$$\dot{C} = k_{\text{on}} L_o R - k_{\text{off}} C \tag{6.15}$$

$$= k_{\text{on}} L_o (R_o - C) - k_{\text{off},o} C e^{\left(\frac{\gamma F}{k_B TC}\right)} \tag{6.16}$$

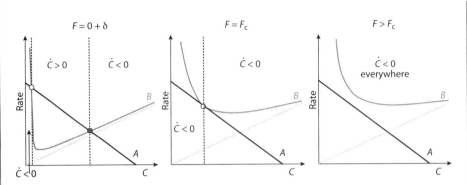

Figure 6.7 Effect of detachment force F on system stability, assessed by plotting individual rate terms from equation 6.16 (i.e., the complex association rate A and the complex dissociation rate B) versus C. The system undergoes a bifurcation at $F = F_c$.

If we define $A = k_{on} L_o (R_o - C)$ and $B = k_{off,o} C e^{\left(\frac{\gamma F}{k_B T C}\right)}$, we can plot these two rate terms as a function of C to better understand the dynamics of complex formation and dissolution as a function of F (figure 6.7).

The intersection points of A and B represent the fixed points of the system, and the gray dotted line in each panel represents the plot of B when $F = 0$, as a point of reference.

When F is only infinitesimally larger than 0 (i.e., $0 + \delta$; left panel), the dissociation behavior at large values of C is similar to the case when $F = 0$, because $k_{off} \approx k_{off,o}$. However, as C approaches zero, B eventually deviates from the dotted line and tends to infinity due to the exponential term. The two intersection points partition this graph into three regions (vertical dashed lines): in the leftmost region, $\dot{C} < 0$, because $A < B$ everywhere in this region; in the middle region, $\dot{C} > 0$; and in the rightmost region, $\dot{C} < 0$. These values of $\dot{C}$ make clear that the low-C fixed point is unstable, and the high-C fixed point is stable.

If F is very large (right panel in figure 6.7), B is sufficiently large at all values of C that it never intersects A, and $\dot{C} < 0$ for all C, ultimately leading to zero complexes and, thus, cell detachment. At a critical force F_c (middle panel), A will be just tangent to B: Below this force, there will be two fixed points (as in the left panel), and above this force, there will be no fixed points (as in the right panel). Thus, F_c represents the minimum force to fully detach the cell.

This critical force can be calculated by noting that A and B in the middle panel have (1) the same value and (2) the same slope at their intersection point. Using these two relationships to solve for F_c it can be shown that

$$F_c = \frac{k_B T R_\circ}{\gamma} \alpha_c \qquad (6.17)$$

where the value of α_c is determined from the relationship $\alpha_c e^{(\alpha_c+1)} = k_{on} L_\circ / k_{off,\circ}$.

6.2 Switches and Thresholds

In many instances, a cellular response can be switchlike: Decisions to proliferate, differentiate, or apoptose are all-or-none. Switchlike dose responses are sometimes referred to as *ultrasensitive*, because the output increases significantly over a narrow range in input signal. Ultrasensitivity can, in principle, be achieved for a pairwise biomolecular interaction via a mechanism such as allostery, but this is dependent on highly nuanced protein structural changes and is not an easily generalizable strategy. By contrast, noncooperative binding events can combine in a specific network structure that engenders a switchlike response; this strategy is more generalizable, because the dose response is a fundamental feature of the network structure and is not dependent on cooperativity of the specific proteins involved. Here we describe some common biomolecular interaction networks capable of achieving such switchlike responses. The figure of merit to assess the degree of switchlike behavior will generally be an effective value of the Hill coefficient, with larger values corresponding to greater ultrasensitivity.

6.2.1 Ultrasensitivity through Homomultimerization

If a protein must self-assemble to carry out its function (as in the Griffith model discussed in section 6.1), this multimerization imparts a nonlinearity in the system that can result in more switchlike activity. To see how this ultrasensitivity arises, let's consider the illustrative case of a transcription factor monomer M that must form a homodimer D before binding to its genomic target G to form a complex C:

$$M + M \rightleftharpoons D; \quad K_d = \frac{[M]^2}{[D]} \qquad (6.18)$$

$$D + G \rightleftharpoons C; \quad K_c = \frac{[D][G]}{[C]} \quad (6.19)$$

Assuming the concentration of genomic targets is negligible compared to the total concentration of monomer, we can substitute the mass balance $[M] \approx [M]_{\text{tot}} - 2[D]$ into equation 6.18 to solve for $[D]$ at equilibrium:

$$[D] = \frac{4[M]_{\text{tot}} + K_d - \sqrt{8[M]_{\text{tot}}K_d + K_d^2}}{8} \quad (6.20)$$

From equation 6.19, we can readily obtain the familiar result for fractional occupancy of G (i.e., $y = [C]/[G]_{\text{tot}}$):

$$y = \frac{[D]}{K_c + [D]} \quad (6.21)$$

Substituting equation 6.20 into equation 6.21 yields

$$y = \frac{4m + 1 - \sqrt{8m + 1}}{8\kappa + 4m + 1 - \sqrt{8m + 1}} \quad (6.22)$$

where $m = [M]_{\text{tot}}/K_d$ and $\kappa = K_c/K_d$. Plotting y versus m for different values of κ yields the curves in figure 6.8, each of which can be closely fit by a Hill equation: $y = m^{n_H}/(EC_{50}^{n_H} + m^{n_H})$. It is particularly noteworthy that, as κ is decreased, the Hill coefficient (n_H) increases but appears to asymptote to a value of 2. A very low κ means that the affinity for dimerization is very weak relative to the DNA-binding affinity of the dimer (i.e., $K_d \gg K_c$).

To understand this apparent upper limit on the Hill coefficient as $\kappa \to 0$, let's consider the situation where dimerization is so weak that the contribution of the dimer species to the overall mass balance is negligible: $[M] \approx [M]_{\text{tot}}$. Thus, from equation 6.18, $[D] = [M]_{\text{tot}}^2/K_d$. Plugging this into equation 6.21, we obtain:

$$y = \frac{[M]_{\text{tot}}^2}{K_c K_d + [M]_{\text{tot}}^2} \quad (6.23)$$

which clearly has a Hill coefficient of 2. This result is a direct consequence of the squared term in the monomer-dimer equilibrium expression.

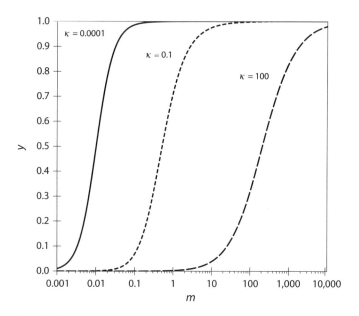

Figure 6.8 The fractional occupancy of a target bound by a homodimer, plotted as a function of total dimensionless monomer concentration (equation 6.22). For $\kappa = 100$, the Hill coefficient is 1.03; for $\kappa = 0.1$, the Hill coefficient is 1.44; and for $\kappa = 0.0001$, the Hill coefficient is 1.95.

Example 6-5 If the functional form of a bacterial repressor is a dimer of dimers (i.e., a homotetramer), determine the Hill coefficient for this DNA-binding isotherm when the protein is predominantly in the monomeric state.

Solution If K_t represents the equilibrium dissociation constant for the binding of two dimers, then the concentration of tetramers, $[T]$, is

$$[T] = \frac{[D]^2}{K_t} = \frac{1}{K_t}\left(\frac{[M]^2}{K_d}\right)^2 \approx \frac{[M]_{\text{tot}}^4}{K_t K_d^2} \tag{6.24}$$

Because the tetramer is the DNA-binding species, the relevant expression for y in this case is

$$y = \frac{[T]}{K_c + [T]} = \frac{[M]_{\text{tot}}^4}{K_c K_t K_d^2 + [M]_{\text{tot}}^4} \tag{6.25}$$

which has a Hill coefficient of 4.

6.2.2 Ultrasensitivity through Molecular Titration

The previous section described ultrasensitivity arising from homomultimerization, but the maximum Hill coefficient attainable is at most equal to the stoichiometry of the multimer complex. Buchler and Louis [5] described another mechanism based on *molecular titration* that generates highly ultrasensitive responses by sequestering the actuating protein P at low concentrations through a high-affinity buffering molecule B.

Interestingly, this particular behavior can actually be achieved using the general binding model that we have already considered

$$P + B \rightleftharpoons S; \quad K_d = \frac{[P][B]}{[S]} \tag{6.26}$$

but now it is not the sequestering complex S that is functional, but rather the free protein P that can drive downstream responses, by binding to a genomic target, for example:

$$P + G \rightleftharpoons C; \quad K_c = \frac{[P][G]}{[C]} \tag{6.27}$$

Again assuming a negligible concentration of genomic targets ($[P] \approx [P]_{tot} - [S]$), we can solve for the concentration of free protein:

$$[P] = \frac{[P]_{tot} - [B]_{tot} - K_d + \sqrt{([P]_{tot} - [B]_{tot} - K_d)^2 + 4[P]_{tot}K_d}}{2} \tag{6.28}$$

Plugging this expression into the fractional occupancy of G ($y = \frac{[C]}{[G]_{tot}} = \frac{[P]}{K_c + [P]}$), we obtain

$$y = \frac{p - \beta - 1 + \sqrt{(p - \beta - 1)^2 + 4p}}{2\kappa + p - \beta - 1 + \sqrt{(p - \beta - 1)^2 + 4p}} \tag{6.29}$$

where $p = [P]_{tot}/K_d$, $\beta = [B]_{tot}/K_d$, and $\kappa = K_c/K_d$.

Plotting y versus p for different values of β and κ (figure 6.9), we note that both a high β and a high κ are needed to observe ultrasensitivity. The high β means that the buffering molecule is at a sufficiently high concentration to titrate away low concentrations of protein, so that it is not available to participate in any downstream reactions until a tipping point is reached. Additionally, the protein should have a strong preference for binding the buffering molecule compared to the genomic target ($\kappa \gg 1$); otherwise this molecular titration effect would similarly be lost.

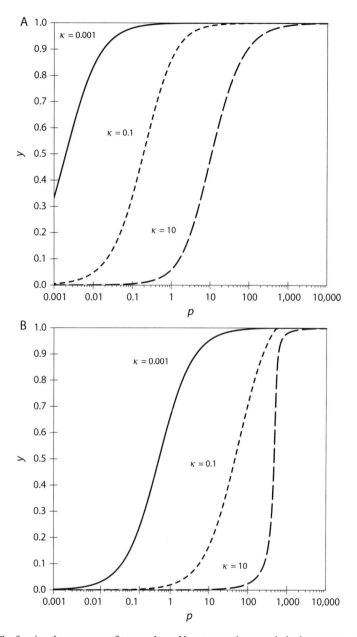

Figure 6.9 The fractional occupancy of a target bound by an actuating protein in the presence of a buffering molecule, plotted as a function of total dimensionless protein concentration (equation 6.29). (A) Response curves for $\beta = 1$ and $\kappa = 0.001, 0.1, 10$. All Hill coefficients are approximately 1. (B) Response curves for $\beta = 500$ with the same κ values. For the κ values less than 1, the Hill coefficients are again approximately 1, but for $\kappa = 10$, the Hill coefficient is 6.3.

6.2.3 Zero-Order Ultrasensitivity

Ultrasensitivity can also be achieved through simple enzymatic networks. Futile cycles are ubiquitous in cell signaling. In such signaling cycles, one enzyme adds a functional group to a substrate, and another enzyme removes this group. This is exemplified by a kinase K and a phosphatase P that act on the same substrate S (figure 6.10). At first blush, these reactions may seem to negate one another, but Goldbeter and Koshland [6] showed that this simple network of three proteins can actually generate a response of activated substrate with unlimited ultrasensitivity. They demonstrated that this ultrasensitivity arises when the substrate concentration is in great excess relative to the K_m values of both enzymes. Under these conditions, the velocity of each reaction is zeroth order with respect to substrate, so Goldbeter and Koshland called this mechanism *zero-order ultrasensitivity*.

For the cycle shown in figure 6.10, let us write appropriate rate equations using the theory developed for enzymes in chapter 4:

$$\frac{d[S]}{dt} = -k_{on,K}[K][S] + k_{off,K}[X_K] + k_{cat,P}[X_P] \tag{6.30}$$

$$\frac{d[S^*]}{dt} = -k_{on,P}[P][S^*] + k_{off,P}[X_P] + k_{cat,K}[X_K] \tag{6.31}$$

$$\frac{d[X_K]}{dt} = k_{on,K}[K][S] - k_{off,K}[X_K] - k_{cat,K}[X_K] \tag{6.32}$$

$$\frac{d[X_P]}{dt} = k_{on,P}[P][S^*] - k_{off,P}[X_P] - k_{cat,P}[X_P] \tag{6.33}$$

where X_K is the bound complex of S with K, and X_P is the bound complex of S^* with P. At steady state, all time derivatives are equal to zero, so we can solve for $[X_K]$ and $[X_P]$ from the last two equations using appropriate mass balances on the enzymes (i.e., $[K]_{tot} = [K] + [X_K]$, and $[P]_{tot} = [P] + [X_P]$):

$$[X_K] = \frac{[K]_{tot}[S]}{K_{m,K} + [S]} \quad [X_P] = \frac{[P]_{tot}[S^*]}{K_{m,P} + [S^*]} \tag{6.34}$$

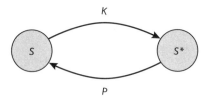

Figure 6.10 A futile cycle in which a substrate S is phosphorylated by a kinase K to generate the state S^*, which is in turn dephosphorylated by a phosphatase P.

where $K_{m,K} = \frac{k_{\text{off,K}} + k_{\text{cat,K}}}{k_{\text{on,K}}}$ and $K_{m,P} = \frac{k_{\text{off,P}} + k_{\text{cat,P}}}{k_{\text{on,P}}}$.

Summing equations 6.30 and 6.32, we obtain the following steady-state result:

$$\frac{d[S]}{dt} + \frac{d[X_K]}{dt} = k_{\text{cat,P}}[X_P] - k_{\text{cat,K}}[X_K] = 0 \tag{6.35}$$

Plugging the expressions from equation 6.34 into 6.35, we obtain

$$\frac{V_{\text{max,P}}[S^*]}{K_{m,P} + [S^*]} = \frac{V_{\text{max,K}}[S]}{K_{m,K} + [S]} \tag{6.36}$$

where $k_{\text{cat,P}}[P]_{\text{tot}} = V_{\text{max,P}}$ and $k_{\text{cat,K}}[K]_{\text{tot}} = V_{\text{max,K}}$.

If we assume that the substrate is in great excess compared to both enzymes, we can neglect the contribution of enzyme-substrate complexes to the mass balance on substrate. Thus,

$$[S]_{\text{tot}} \approx [S] + [S^*] \tag{6.37}$$

Substituting this result into equation 6.36 and rearranging, we obtain

$$\frac{V_{\text{max,K}}}{V_{\text{max,P}}} = \frac{[S^*](K_{m,K} + [S]_{\text{tot}} - [S^*])}{([S]_{\text{tot}} - [S^*])(K_{m,P} + [S^*])} \tag{6.38}$$

Further nondimensionalizing the right-hand side by dividing both the numerator and the denominator by $[S]_{\text{tot}}$ results in

$$\frac{V_{\text{max,K}}}{V_{\text{max,P}}} = \frac{\tilde{S}^*(\tilde{K}_{m,K} + 1 - \tilde{S}^*)}{(1 - \tilde{S}^*)(\tilde{K}_{m,P} + \tilde{S}^*)} \tag{6.39}$$

where the tilde above each quantity represents its nondimensional form (i.e., its dimensional form divided by $[S]_{\text{tot}}$).

The ratio of V_{max} values is proportional to the ratio of the active enzyme concentrations, which can be considered an input to the system. Note that K and P represent active forms of the kinase and phosphatase, respectively. In many instances, the kinase molecules that act on the substrate are present in the cell but inactive until posttranslationally modified (e.g., phosphorylation triggered by some external factor). Thus, an extracellular cue can modulate $\frac{V_{\text{max,K}}}{V_{\text{max,P}}}$ through activation (or deactivation) of existing enzymes.

Figure 6.11 shows how the fraction of phosphorylated substrate concentration changes as a function of this input ratio for different $\tilde{K}_m$ values.

When $\tilde{K}_{m,K} = \tilde{K}_{m,P} = 5$ (or any value significantly larger than 1), equation 6.39 can be reduced to

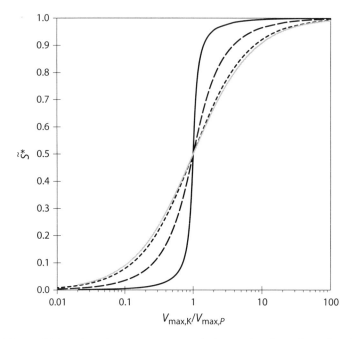

Figure 6.11 Ultrasensitivity arises when enzymes operate in the zeroth-order regime. The solid gray curve is $\tilde{S}^* = (V_{\text{max},K}/V_{\text{max},P})/(1+V_{\text{max},K}/V_{\text{max},P})$ (equation 6.41). The black curves are all plots of equation 6.39 with equal $\tilde{K}_{\text{m},K}$ and $\tilde{K}_{\text{m},P}$ values: 5 for the short-dashed curve, 0.5 for the long-dashed curve, and 0.05 for the solid curve. The solid black curve is highly ultrasensitive, with an effective Hill coefficient n_H of 7.8.

$$\frac{V_{\text{max},K}}{V_{\text{max},P}} \approx \frac{\tilde{S}^*}{1-\tilde{S}^*} \tag{6.40}$$

which can be rearranged to the standard hyperbolic (non-ultrasensitive) response curve, plotted in gray in figure 6.11:

$$\tilde{S}^* = \frac{\frac{V_{\text{max},K}}{V_{\text{max},P}}}{1+\frac{V_{\text{max},K}}{V_{\text{max},P}}} \tag{6.41}$$

Intriguingly, when $\tilde{K}_{\text{m},K} = \tilde{K}_{\text{m},P} = 0.05$, the response becomes highly ultrasensitive. When these K_m values are much less than the total substrate concentration, the underlying reactions are zeroth order with respect to substrate, so this type of response is known as zero-order ultrasensitivity. By modulating the concentrations of active kinase and phosphatase acting on a common substrate, a cell can generate highly switchlike responses without the need for any of the participating components to have inherent cooperativity in their mode of action.

We can further quantify the response dynamics in these cascades and relate them to the Hill coefficient that we defined in chapter 3. For a graded response (Hill coefficient $n_H = 1$), the fold increase in input that is needed to go from a 10% response to a 90% response is 81. More generally, it can be readily shown for a Hill equation that the fold increase in input required is $(81)^{\frac{1}{n_H}}$. For comparison, we can derive a similar result for a futile cycle by determining the fold increase in $V_{\max}$ ratio required to go from $[S^*] = 0.1[S]_{\text{tot}}$ to $[S^*] = 0.9[S]_{\text{tot}}$:

$$\frac{\left(\frac{V_{\max,K}}{V_{\max,P}}\right)_{90\%}}{\left(\frac{V_{\max,K}}{V_{\max,P}}\right)_{10\%}} = \frac{\frac{(0.9)(\tilde{K}_{m,K}+1-0.9)}{(1-0.9)(\tilde{K}_{m,P}+0.9)}}{\frac{(0.1)(\tilde{K}_{m,K}+1-0.1)}{(1-0.1)(\tilde{K}_{m,P}+0.1)}} = \frac{81(\tilde{K}_{m,K}+0.1)(\tilde{K}_{m,P}+0.1)}{(\tilde{K}_{m,K}+0.9)(\tilde{K}_{m,P}+0.9)} \quad (6.42)$$

It can be seen from this result that, in the limit of very large $K_{m,K}$ and $K_{m,P}$, this fold increase asymptotes to 81, which corresponds to a Hill coefficient of 1 and indicates a graded response. However, as these Michaelis constants become vanishingly small compared to the total substrate concentration, this ratio approaches 1, which indicates a perfect step function (because 10% and 90% responses would occur at the same input value when this ratio is unity).

6.2.4 Bistable Switches

As described in section 6.1, a nonlinear system has the potential to be multistable. If this system has two stable states, it is referred to as *bistable* and the observed steady state is determined by the initial conditions of the system. An intuitive requirement for bistable systems is the presence of positive feedback, as we saw in the Griffith model in section 6.1. If the system is initially in a *low* state, the positive feedback is not sufficiently triggered, and the system remains stably in this low state unless some stimulus external to the system is applied to induce a transition to the *high* state. Conversely, if the system is initially in the high state, the positive feedback can provide sufficient activation independent of any stimulus. A perhaps less-obvious requirement for bistability is that there must also be an ultrasensitive step in the network (e.g., if equation 6.7 in the Griffith model had a Hill coefficient of 1 in the first term, the system would not exhibit bistability). Let's examine a minimal positive-feedback network (figure 6.12) to see why this is the case.

Here, A represents a species in the cell—say, a transcription factor—that is synthesized at a basal rate V_A and degraded at a rate $k_{\text{deg}}[A]$. Additionally, A drives its own synthesis in a manner that can be captured by a Hill equation.

Let's first consider this system when $n = 1$, such that the positive-feedback loop is not ultrasensitive. In this scenario, we have

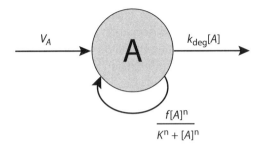

Figure 6.12 A minimal positive-feedback network.

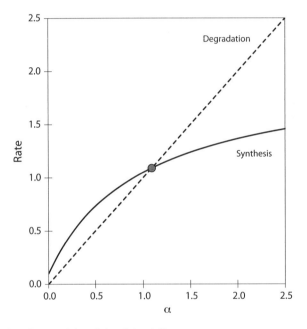

Figure 6.13 Rate plot when $n = 1$ ($\nu = 0.1$ and $\phi = 1.9$).

$$\frac{d[A]}{dt} = V_A + \frac{f[A]}{K + [A]} - k_{\text{deg}}[A] \quad (6.43)$$

To simplify our analysis, we examine the nondimensionalized form of this equation, because it reduces the number of parameters that must be considered:

$$\dot{\alpha} = \frac{d\alpha}{d\tau} = \nu + \frac{\phi\alpha}{1+\alpha} - \alpha \quad (6.44)$$

where $\alpha = [A]/K$, $\tau = k_{\text{deg}}t$, $\nu = V_A/(k_{\text{deg}}K)$, and $\phi = f/(k_{\text{deg}}K)$.

We use simple graphical methods to quickly determine the number of stable fixed points for this system. By overlaying plots of the dimensionless degradation rate (α) versus α and the synthesis rate ($\nu + \frac{\phi\alpha}{1+\alpha}$) versus α, we recognize that the intersection points of these two curves represent the fixed points (i.e., when $d[A]/dt = 0$).

From figure 6.13, it is evident that only one fixed point is possible when $\nu > 0$ (because the synthesis rate curve does not pass through the origin). If the system is perturbed even slightly from its fixed point to a higher α value, figure 6.13 shows that the degradation rate exceeds the synthesis rate above the fixed point, which means that $\dot{\alpha} < 0$, and the system will return to this fixed point. Similarly, if the system is perturbed to a slightly lower α value, then $\dot{\alpha} > 0$ below the fixed point, and the system will return to the same fixed point. Thus, this fixed point is stable, but the system is fundamentally incapable of exhibiting bistability. (Note that even if $\nu = 0$, which would add a fixed point at the origin, this point is unstable, so the system would still only have one stable fixed point.)

If we now consider the system in figure 6.12 when $n = 2$ (e.g., when the transcription factor is a homodimer, as described in section 6.2.1), then the rate plot can, under certain conditions, reveal three fixed points (figure 6.14). In this case, the lowest and highest fixed points are stable, and the intermediate fixed point is

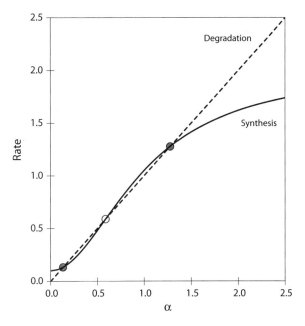

Figure 6.14 Rate plot when $n = 2$ ($\nu = 0.1$ and $\phi = 1.9$).

unstable. This can again be assessed graphically by determining the sign of $\dot{\alpha}$ for small perturbations away from each fixed point.

As illustrated by this simple example, the presence of a positive-feedback loop is not sufficient for the generation of a bistable response; there must also be a nonlinearity in the system. This nonlinearity can arise in various ways, including multimerization, molecular titration, and zero-order ultrasensitivity. Thus, the ultrasensitive motifs described earlier in this section can serve as building blocks for more complex behaviors, such as bistability.

Case Study 6-1 N. A. Shah and C. A. Sarkar. Robust network topologies for generating switch-like cellular responses. *PLOS Comput. Biol.* 7: e1002085 (2011). S. Palani and C. A. Sarkar. Synthetic conversion of a graded receptor signal into a tunable, reversible switch. *Mol. Syst. Biol.* 7: 480 (2011) [7, 8].

From an engineering perspective, it is particularly useful to identify robust network topologies for engendering a specific cellular response. A common problem in biomolecular network engineering is that the parameter values in the system are not all known, so even if one has an idea of how to construct a network that can in principle lead to the desired response, this may not be experimentally observed if the network response is highly dependent on the specific parameter values.

In the study by Shah and Sarkar [7], the authors took a reverse-engineering approach to identify networks that were robust to parameter variation in creating switchlike responses to ligand. They examined all possible three-component network topologies: one node (A) was considered the input (receptor) node, a second node (C) was considered the output node, and a third regulatory node (B) was also included. With just these three nodes and considering all possible connections among the nodes (including autofeedback loops), there were ≈16,000 distinct topologies that maintained connectivity between the input and output nodes. Each of these individual topologies was then analyzed under four different compositional classes: all enzymes (EEE), all transcription factors (TTT), or hybrid networks in which only A is an enzyme (ETT) or only C is a transcription factor (EET). The distinction between enzymes and transcription factors in this framework is that an enzyme changes the activity state of its target, whereas a transcription factor modulates target abundance. For each network, the parameter values were not prescribed but rather randomly chosen within biologically relevant bounds, and then the network was simulated to assess switchlike behavior (both ultrasensitivity ($n_H > 2$) and bistability (fold change in ligand exhibiting bistable behavior > 5)). This random parameter assignment and network simulation was repeated 1,000 times, and the robustness of a particular network topology was defined as the fraction of networks that exhibited a desired switchlike response (figure 6.15).

Notably, hybrid networks (EET and ETT) are markedly more robust than EEE and TTT networks in achieving both ultrasensitivity and bistability. The cause of this disparity is due in part to zero-order ultrasensitivity (both for ultrasensitivity and also as a building block for bistability); hybrid networks are better able to leverage zero-order ultrasensitivity, because this is an enzyme-specific phenomenon that also benefits from the presence of

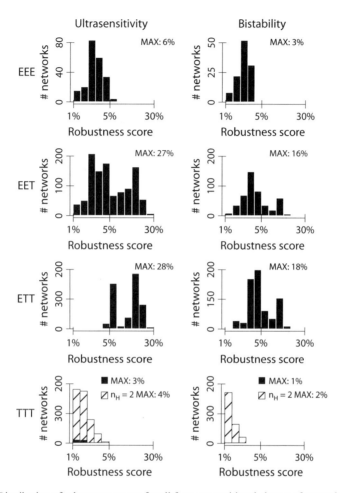

Figure 6.15 Distribution of robustness scores for all four compositional classes of networks considered in the topology search algorithm. Maximum robustness scores for each class are also noted. Throughout the simulations, transcriptional reactions were modeled as noncooperative; however, even when these reactions were modeled with a Hill coefficient of 2, the robustness scores for the TTT compositional class remained low. Figure from reference [7].

transcription factors that can increase substrate abundance so that the necessary reactions can be zeroth order. The network motifs identified agnostically through this computational search also exhibit parallels in natural biological systems, including ultrasensitivity in Yan phosphorylation in *Drosophila* [9], MAPK activation in *Xenopus* oocyte maturation [10, 11], and lineage commitment in erythropoiesis [12].

In the related study by Palani and Sarkar [8], the authors sought to experimentally test one of the most robust designs identified from this computational search (figure 6.16A). Notably, this dual-feedback model does not explicitly include any Hill coefficients greater

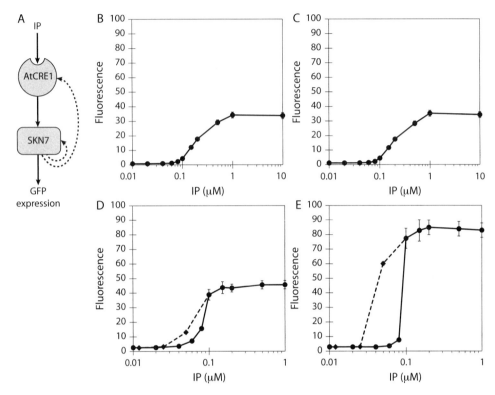

Figure 6.16 (A) Minimal network topology predicted to be robustly ultrasensitive and bistable from computational search algorithm. Experimental implementation of the network in *S. cerevisiae* utilized the synthetic receptor AtCRE1 from *Arabidopsis thaliana*, which signals to endogenous transcription factor SKN7 when bound by isopentenyl adenine (IP). Synthetic response elements were used to introduce positive-feedback synthesis of AtCRE1 and/or SKN7. (B) Behavior of the system without either positive-feedback loop. The response is largely graded ($n_H \approx 2$) and not bistable. (C) Behavior of the system with only AtCRE1 feedback. The response is largely graded ($n_H \approx 2$) and not bistable. (D) Behavior of the system with only SKN7 autofeedback. The response is slightly more ultrasensitive ($n_H \approx 4$) and minimally bistable. (E) Behavior of the full system, incorporating both feedback loops shown in panel A. The response is highly ultrasensitive ($n_H \approx 20$) with a wider bistable window [8].

than 1; however, one of the positive-feedback loops can generate ultrasensitivity that is reinforced by the other positive-feedback loop to create bistability. Computationally, this network exhibited near-maximal robustness for ultrasensitivity (26%) and bistability (13%). It had the added benefit of only requiring two nodes—a receptor input node and a transcription factor output node—for experimental implementation.

The network was constructed in *S. cerevisiae* and immediately exhibited very high ultrasensitivity ($n_H \approx 20$) and a pronounced bistable window when both positive-feedback loops were introduced (figure 6.16E) but not when one or both of the feedback loops was absent (figure 6.16B,C,D), underscoring the importance of incorporating the full topology to achieve the desired response. This approach is particularly notable, because no post hoc tweaking was required to achieve the desired response, suggesting that computational identification of robust network topologies can serve to streamline experiments needed to create desired cellular responses.

6.3 Adaptation

Another phenotype achievable via multiple network topologies is adaptation. Adaptation is the process by which a system responds to a stimulus by varying its output and then returns to approach prestimulus output activity (figure 6.17). In the case of *perfect adaptation*, the system returns to exactly match prestimulus activity. In control theory, this is also termed zero offset, zero steady-state error, or asymptotic tracking. Adaptation aids the maintenance of homeostasis; enables differential response to a wide range of input levels by responding to input deviations rather than absolute values; and limits the duration of a response, which can aid system efficiency. Thus, adaptive behavior is valuable in various settings.

Robust perfect adaptation requires integral control [13, 14]. Yet adaptive biological networks often exhibit topologies for which integral feedback is not immediately evident. Several network topologies are capable of enabling adaptive response and have been evolved in natural systems or engineered into synthetic systems. In particular, perfect or near-perfect adaptation has been demonstrated

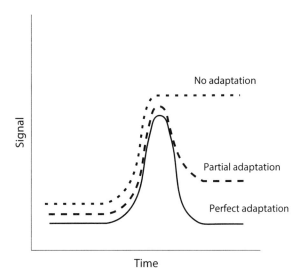

Figure 6.17 System responses with varying levels of adaptation.

from negative-feedback loops, incoherent feedforward systems, state-dependent inactivation systems, and antithetic integral feedback.

6.3.1 Negative Feedback

Adaptation can result from a negative-feedback loop that involves a pair of proteins with an activated A positively driving a constitutively inhibited B, while B inhibits A (figure 6.18). To be adaptive, the feedback from B needs to lag behind the input stimulus. In addition, B needs to be ultrasensitive to changes in A with a steady-state level of B that is close to the threshold for ultrasensitive activation (section 6.2). The response can monotonically decay to the original output activity or exhibit damped oscillatory behavior.

Assuming Michaelis-Menten forms for the activation and inactivation kinetics, the material balances for A and B can be written as

$$\frac{d[A]}{dt} = \frac{k_{IA}[I][A_{\text{inactive}}]}{K_{IA} + [A_{\text{inactive}}]} - \frac{k_{BA}[B][A]}{K_{BA} + [A]} \tag{6.45}$$

$$\frac{d[B]}{dt} = \frac{k_{AB}[A][B_{\text{inactive}}]}{K_{AB} + [B_{\text{inactive}}]} - \frac{k_B[B]}{K_B + [B]} \tag{6.46}$$

where $[I]$ is the concentration of the input molecule, and k_{XY} and K_{XY} are the rate constant and Michaelis-Menten constant, respectively, for the action of X on Y.

Consider that the inactive concentrations of A and B are the balance between the respective total concentrations and active concentrations (i.e., $[A_{\text{inactive}}] = [A]_{\text{total}} - [A]$) and nondimensionalize their active concentrations (i.e., $\theta_A = \frac{[A]}{[A]_{\text{total}}}$):

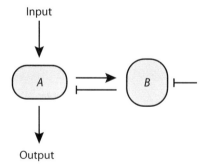

Figure 6.18 A negative-feedback loop in which input activates A, which in turn activates B, which in turn inhibits A.

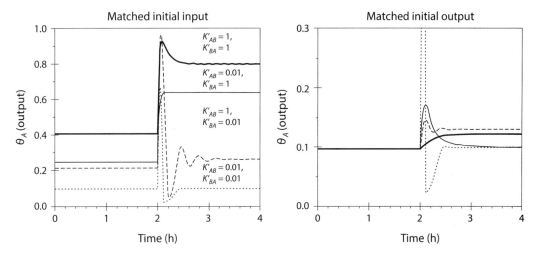

Figure 6.19 The output from a simple negative-feedback system (figure 6.18), for which the input is doubled at $t = 2$ hours, is presented for high and low values of the Michaelis-Menten constants for activation and inhibition. $k'_{IA} = 100$ h^{-1}, $k'_{AB} = 10$ h^{-1}, $k'_{BA} = 100$ h^{-1}, $k'_B = 1$ h^{-1}, $K'_{IA} = 0.01$, $K'_B = 0.01$. The left plot has equal initial $[I]$ for all conditions. The right plot has equal initial $[A]$ for all conditions.

$$\frac{d\theta_A}{dt} = \frac{k'_{IA}[I](1-\theta_A)}{K'_{IA}+(1-\theta_A)} - \frac{k'_{BA}\theta_B\theta_A}{K'_{BA}+\theta_A} \quad (6.47)$$

$$\frac{d\theta_B}{dt} = \frac{k'_{AB}\theta_A(1-\theta_B)}{K'_{AB}+(1-\theta_B)} - \frac{k'_B\theta_B}{K'_B+\theta_B} \quad (6.48)$$

Numerical solution of this system of equations reveals highly differential responses that are dependent on system parameters (figure 6.19). For example, nearly perfect adaptation is observed for highly sensitive activation and inactivation mechanisms (low K'_{AB} and K'_{BA}), whereas minimal adaptation is present when both are insensitive.

A quantitative metric of the degree of adaptation is often called the precision. This term has multiple implementations, one of which is simply the ratio of final output to initial output:

$$precision' = \frac{[A]_{\text{final}}}{[A]_{\text{initial}}} \quad (6.49)$$

for which 1 represents perfect adaptation, but imperfect adaptation (or none whatsoever) can approach zero or infinity. A second implementation calculates the deviation of the activity from the prestimulus condition normalized by the deviation in the input signal and inverts the ratio such that high values correspond to strong adaptation:

$$precision = \frac{\frac{[I]_{\text{final}} - [I]_{\text{initial}}}{[I]_{\text{initial}}}}{\frac{[A]_{\text{final}} - [A]_{\text{initial}}}{[A]_{\text{initial}}}} \qquad (6.50)$$

Example 6-6 Calculate the Michaelis-Menten constants that generate high-precision adaptation for a simple negative-feedback loop (figure 6.18).

Solution This is done by combining equations 6.45, 6.46, and 6.50. As an example, consider the following parameter values: $k_{IA} = 100 \text{ h}^{-1}$, $k_{AB} = 10 \text{ h}^{-1}$, $k_{BA} = 100 \text{ h}^{-1}$, $k_B = 1 \text{ h}^{-1}$, $K_{IA} = 0.01$, $K_B = 0.01$, $K_{AB} = 0.01$, $K_{BA} = 0.01$. Two Michaelis-Menten constants are varied at a time, and then the precision is calculated (figure 6.20).

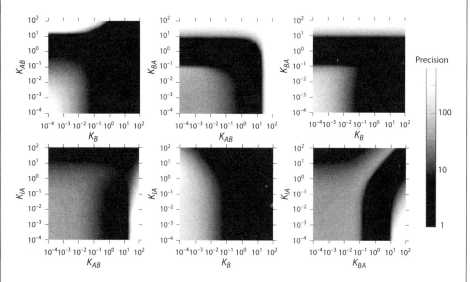

Figure 6.20 The precision (equation 6.50) from a simple negative-feedback system (figure 6.18), for which the input is doubled, is presented for a range of values of the Michaelis-Menten constants. $k_{IA} = 100 \text{ h}^{-1}$, $k_{AB} = 10 \text{ h}^{-1}$, $k_{BA} = 100 \text{ h}^{-1}$, $k_B = 1 \text{ h}^{-1}$, $K_{IA} = 0.01$, $K_B = 0.01$, $K_{AB} = 0.01$, $K_{BA} = 0.01$, unless otherwise indicated.

Case Study 6-2   N. Barkai and S. Leibler. Robustness in simple biochemical networks. *Nature* 387: 913–917 (1997). U. Alon, M. G. Surette, N. Barkai, and S. Leibler. Robustness in bacterial chemotaxis. *Nature* 397: 168–171 (1999). T.-M. Yi, Y. Huang, M. I. Simon, and J. Doyle. Robust perfect adaptation in bacterial chemotaxis through integral feedback control. *Proc. Natl. Acad. Sci. U.S.A.* 97: 4649–4653 (2000) [14–16].

An important feature of biochemical networks is that they must function reliably in cells across changes in cell size, stochastic variation in numbers of individual molecules, and different environments (such as temperature, particularly for organisms that cannot regulate their own temperature). Moreover, for biochemical networks to effectively adapt through evolution, it is necessary that individual point mutations that change a single rate constant in a complex network do not result in the complete loss of network function. In other words, network function is unlikely to be encoded through finely tuning network parameters, because changes in effective rate constants could abrogate function. Instead, at some level, network function should persist across parameter changes. That is, certain fundamental network functions should be robust to at least modest parameter variation. This notion was explored in a pioneering set of papers by Leibler et al. [15,16], who examined the bacterial chemotaxis network, which controls the movement of bacteria swimming toward sources of attractants and away from those of repellents. Yi et al. drew parallels between the control mechanism encoded in the chemotaxis network and integral feedback systems common in modern controls [14].

E. coli motion consists of periods of smooth swimming punctuated by brief periods of tumbling. The organism proceeds in a straight line during smooth swimming, whereas tumbling randomizes the direction for the next straight run. Interestingly, smooth swimming results from counterclockwise rotation of the flagella, and tumbling behavior results from clockwise rotation. The frequency of tumbling is a function of the environment, such that tumbling is less frequent after increases in the local attractant concentration, as would be sensed when swimming up a gradient (figure 6.21). The accumulated evidence from careful study is that *E. coli* executes a biased random walk, whose net effect is to travel up attractant gradients (toward the source of attractant) and down repellent gradients (away from the source of repellent) [17]. Rather than directly measuring the gradient, changes in concentration over time are sensed, assuming movement during that time provides an indirect approximation to gradient information. Bacteria exhibit the same tumbling frequency when exposed to a constant ligand (attractant or repellent) concentration, whether that constant concentration is high or low, because constant concentration represents zero gradient. This perfect adaptation is important for permitting gradient sensing across a large concentration range.

Barkai and Leibler proposed a model of bacterial chemotactic signaling based largely on the known literature [15]. Their model focuses on ligand binding by the transmembrane enzyme-receptor complex (the receptor and the proteins CheA (a histidine kinase) and CheW (an adaptor protein)) as well as two enzymes that modify the complex by methylation (CheR) and demethylation (CheB) of protein side chains. A fundamental assumption of the model is that the receptor exhibits two-state behavior, active and inactive, where the active state is characterized by an active form of CheA that phosphorylates CheY, which increases the propensity of the motor to initiate tumbling. The receptor complex can exist

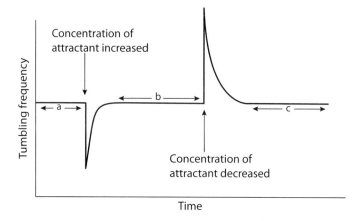

Figure 6.21 A schematic plot of tumbling frequency as a function of time during an experiment in which the concentration experienced by a single *E. coli* bacterium changes abruptly at two time points. An increase in attractant leads initially to a precipitous drop in tumbling frequency; although the concentration does not change further, the tumbling frequency progressively returns to its original value. This is the property of perfect adaptation. Likewise, upon a decrease in attractant, the tumbling frequency increases sharply but then adapts back to its zero-gradient value. Although the concentration of attractant may be different in the time periods marked a, b, and c, the tumbling frequency is the same. The tumbling frequency measures not the concentration but the perceived gradient, which is zero in all three cases.

in a multiplicity of forms involving binding or unbinding of ligand as well as methylation and demethylation of a set of sites through Michaelis-Menten kinetics of the CheR and CheB enzymes, as shown below (figure 6.22).

Importantly, none of the forms of the enzyme-receptor complex in figure 6.22 is explicitly active or inactive. Instead, each form has a probability α of being active in the model. The amount of active receptor in a cell is determined by summing over the amounts of receptor in each of the different forms times the probability of being active. Key assumptions of the model include (1) CheR, the methylation enzyme, acts on active and inactive receptor states and operates at saturation; (2) CheB, the demethylation enzyme, only acts on active receptors; and (3) the previously mentioned assumption of two-state activity. Simulations exhibit perfect adaptation over a wide range of parameter values describing the model, such as the rate constants for the binding and catalytic steps. We examine below how the model produces perfect adaptation without dependence on specific parameter values. Before doing so, it is important to note that not all properties of the model are robust. Some depend strongly on parameter values. One property is the adaptation time, which represents a characteristic time for return to the perfect adaptation value of tumbling activity after a discrete change in ambient ligand concentration. In this model, and in experimental studies, the adaptation time is strongly dependent on aspects of the system, such as rate constants or protein concentrations [15, 16].

Perfect adaptation is a property of the model's topology rather than of specific parameter values. That is, it is a function of the form of the equations rather than the particular numbers used. This property can be shown using a simplified example [15]. To simplify, imagine that only one methylation site exists, and unmethylated enzyme is inactive. If

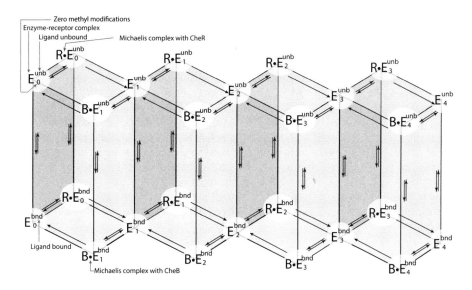

Figure 6.22 The states of the enzyme-receptor complex (E) are shown with their interconverting chemical reactions. In the upper-left corner is the complex with no ligand bound and no methylated groups (E_0^{unb}). Moving to the right, association with CheR produces a Michaelis complex ($R \cdot E_0^{unb}$) that can dissociate or go on to form enzyme-receptor complex with no ligand bound and one methylated group (E_1^{unb}). Further reaction with CheR adds additional methyl groups, passing through a Michaelis complex before each methylation event, for up to a total of four methylation events illustrated here (E_4^{unb}). Each methylated species can also be demethylated by reacting with CheB, again first forming a Michaelis complex and then going on to the demethylated product. Finally, each form of the complex with no ligand bound can undergo reversible ligand binding (vertical arrows).

CheR operates at saturation, then

$$\frac{d[E_1(\text{active})]}{dt} = k_{\text{cat}}^{\text{CheR}}[R]_\circ - \frac{k_{\text{cat}}^{\text{CheB}}[B]_\circ [E_1(\text{active})]}{K_{\text{m}}^{\text{CheB}} + [E_1(\text{active})]} \quad (6.51)$$

The second term on the right-hand side is the Michaelis-Menten expression for enzyme CheB with total concentration $[B]_\circ$, but the substrate is the concentration of active enzyme-receptor complex $[E_1(\text{active})]$. The first term on the right-hand side is the corresponding Michaelis-Menten term for CheR, but because the enzyme operates at saturation (that is, in the limit of high substrate concentration), the dependence on substrate is zeroth order. Perfect adaptation implies that the steady-state activity of enzyme-receptor complex is a constant, independent of ligand concentration, for example. Setting equation (6.51) to zero and solving yields

$$[E_1(\text{active})]_{\text{ss}} = \frac{K_{\text{m}}^{\text{CheB}} k_{\text{cat}}^{\text{CheR}} [R]_\circ}{k_{\text{cat}}^{\text{CheB}} [B]_\circ - k_{\text{cat}}^{\text{CheR}} [R]_\circ} \quad (6.52)$$

which is constant. The value of the constant depends on specific rate parameters and total enzyme concentrations. The fact that it is a constant independent of ligand concentration—which is the essence of perfect adaptation—does not depend on particular parameter values. This demonstrates the fundamental notion that function can be encoded in network topology as opposed to in specific values for network parameters.

A related work analyzed the Barkai-Leibler model. Yi et al. interpreted the model in terms of engineering control systems, recognizing that the biochemical chemotactic network acts like a feedback control circuit [14]. The input to the control circuit is external ligand concentration, and the output is the tumbling activity; perfect adaptation corresponds to a control circuit that at steady state produces the same constant output for any constant input. Yi et al. demonstrated that the role of the biochemical mechanism, largely through the pattern of methylation, is to feed back the time integral of the difference between the instantaneous network output and the target steady-state output [14]. Interestingly, this principle of *integral feedback control* has been a long-established technique in control theory for maintaining a set point (a constant output) in the face of variable input. As we understand more about the principles by which biological systems operate, it is an open question whether we will discover new engineering principles or whether biology will have adopted the same ones developed by humans. Although there are arguments for both sides, this example seems to be one in which biology and engineering have independently adopted the same principle.

6.3.2 Incoherent Feedforward

Adaptation can also result from an incoherent feedforward system in which input activates A and B, with A inhibited by B and B basally inhibited (figure 6.23). The delay between activation and inhibition on A enables response followed by a return to near-initial output.

Material balances for A and B, assuming Michaelis-Menten kinetics as above, yield

$$\frac{d\theta_A}{dt} = \frac{k'_{IA}[I](1-\theta_A)}{K'_{IA}+(1-\theta_A)} - \frac{k'_{BA}\theta_B\theta_A}{K'_{BA}+\theta_A} \tag{6.53}$$

$$\frac{d\theta_B}{dt} = \frac{k'_{IB}[I](1-\theta_B)}{K'_{IB}+(1-\theta_B)} - \frac{k'_B\theta_B}{K'_B+\theta_B} \tag{6.54}$$

6.3.3 Additional Adaptive Topologies

An additional topology that can exhibit adaptation entails a signaling entity with three states: off (but activatable), activated, and inactivated (and unable to respond) (figure 6.24). If new off-state molecules are not introduced (via synthesis or reversion from the inactivated state), the system will be perfectly adaptive for systems with zero basal activity for any parameter values but would require a resource loss of nonfunctional inactivated molecules.

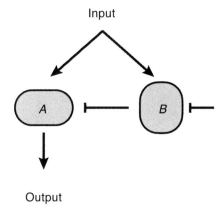

Figure 6.23 An incoherent feedforward system in which input activates both A and B, while B inhibits A.

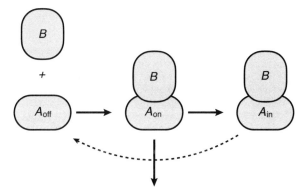

Figure 6.24 A state-dependent inactivation model in which an input of molecule B converts A from an off state to an on state, which is then inactivated. The inactivated state is differentiated from the off state in that the former cannot respond to stimulus, whereas the latter can respond. The inactivated state can be converted to the off state, for example via dissociation of the activating B.

Briat et al. [18] have demonstrated that a topology termed *antithetic integral feedback* enables adaptation; interestingly, adaptation from this topology does not merely tolerate stochasticity but instead requires it.

6.4 Oscillations

We have discussed in this chapter how network responses may be switchlike or adaptive in response to step changes in the cell's environment. Some cellular processes exhibit more exotic oscillatory or pulsatile behavior following a nonoscillatory input. As dynamic measurements have begun to be made at the single-cell level

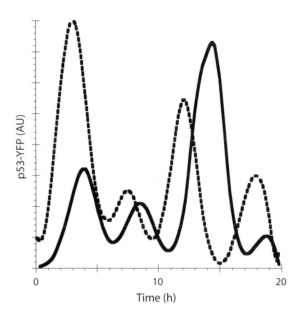

Figure 6.25 Pulsatile expression of p53-YFP in two single human cells after DNA damage occurs at time zero [24]. The apparently random variation in the pulse amplitude and period can obscure the presence of oscillations when measurements are made on a large population of cells.

in asynchronous populations, it is clear that many gene regulatory systems pulse on and off repeatedly rather than following a smooth monotonic trajectory to a new steady state. Such responses have been observed for receptor tyrosine kinase signaling [19], calcium responses [20], inflammatory signaling [21], and many others [22].

A paradigmatic example of such pulsatile signaling is the response of mammalian cells to DNA damage by expressing and degrading the p53 protein (figure 6.25). Do these oscillations matter? Such behavior could either serve a functional role shaped by natural selection, or it may be that these fascinating dynamics are simply an effect rather than a cause. To explore this question, p53 pulses were converted to sustained levels by use of a pharmacological inhibitor of p53 degradation [23]. Loss of pulsatile p53 expression significantly impaired cell recovery from DNA damage and dramatically altered the sets of genes induced by p53 upregulation. So these oscillations can in fact be critical for functional cellular responses.

The p53 protein is multifunctional, inducing transcription of many genes involved in DNA repair. The pulsatile dynamics of p53 allow target genes to respond differentially on the basis of their relative mRNA degradation rates. Different patterns of p53-responsive mRNA accumulation are observed to cluster together according to the particular genes' mRNA stability. The relative time scales of p53 oscillation and mRNA turnover substantially determine the presence or absence of induction or oscillations for each gene [25].

This general phenomenon is well known in the field of process control and signal analysis; it is called *low-pass filtering*. When a system has a capacity for and resistance to the flow of some quantity—be it electrons; fluids; or in this case, mRNA—the dynamic model is essentially the same, exhibiting a characteristic response time τ and a proportional gain K. When such a system is subjected to oscillating input of varying frequency, for frequencies lower than $1/\tau$, the amplitude of the input signal is passed along to the output, but for frequencies higher than $1/\tau$, the output is damped down. Lower-frequency inputs pass through such a system, hence the term *low-pass filter*. Each gene's mRNA degradation rate serves as a low-pass filter to either propagate the p53 pulses or squelch them.

Example 6-7 If transcription of a gene is controlled by a transcription factor whose expression varies sinusoidally, how does the response magnitude depend on the gene's mRNA degradation rate constant?

Solution We first analytically solve the simple gene-expression model from chapter 5

$$\frac{d[mRNA]}{dt} = k_R - (\gamma_R + \mu)[mRNA] \tag{6.55}$$

for the case of the idealized sinusoidal input

$$k_R = 1 + \sin\left(\frac{2\pi}{\tau_{p53}}t - \frac{\pi}{2}\right) \tag{6.56}$$

which yields the following solution:

$$[mRNA] = \tau\left(1 - \left(1 - \frac{1}{\omega^2\tau^2 + 1}\right)e^{-t/\tau} - \frac{\omega\tau\sin(\omega t) + \cos(\omega t)}{\omega^2\tau^2 + 1}\right) \tag{6.57}$$

where $\tau \equiv \frac{1}{\gamma_R + \mu}$ and $\omega \equiv \frac{2\pi}{\tau_{p53}}$.

Solutions to equation 6.57 for various mRNA degradation rates are shown in figure 6.26. Note that the most stable mRNA is significantly induced by p53 pulses, whereas the least stable mRNA remains effectively uninduced, flickering near zero concentration levels. One can see how varying the frequency of an input could effectively select for expression of sets of genes on

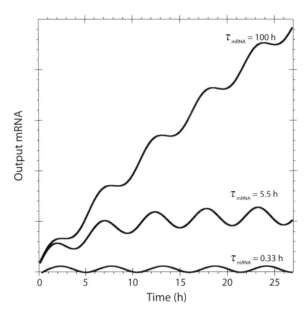

Figure 6.26 The expression kinetics of individual mRNAs act as low-pass filters to p53 fluctuations; rapidly degraded mRNA species do not accumulate in response to more slowly fluctuating p53 levels, while stable species accumulate. The curves are the solution to the simple gene expression model equation 6.57, at various mRNA degradation rates $\tau_{mRNA} = 1/(\gamma_R + \mu)$. $\tau_{p53} = 5.5$ hours, as shown, for example, in figure 6.25.

the basis of their mRNA stability, enabling frequency modulation or FM control.

Pulsed responses have been shown to control downstream gene expression by frequency modulation. An example of this phenomenon is yeast's response to calcium, which leads to rapid cycling of essentially all the Crz1 transcription factor between the cytoplasm and the nucleus [20]. Intriguingly, the duration of time that Crz1 remains in the nucleus is largely independent of calcium level, but the frequency of these nuclear localization bursts correlates directly with calcium level (figure 6.27).

Given the wide variety of processes that exhibit oscillatory responses, is there a general dynamic mechanism that might account for many of them? It turns out that a negative-feedback loop with a time delay may be a very common motif in such oscillatory circuits [26], although a negative-feedback loop controlling a first-order process only exhibits oscillatory instabilities if time delay is introduced to

Network Dynamics

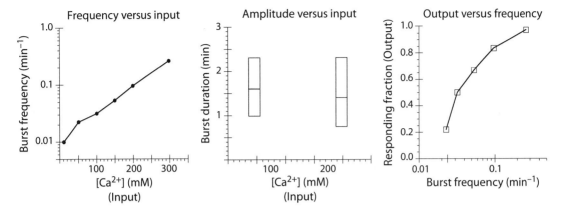

Figure 6.27 Yeast exposed to increasing calcium concentrations vary the frequency of nucleocytoplasmic shuttling (left panel), but not the time spent in the nucleus per burst (middle panel; box plots for median, top and bottom quartiles). Consequently, the fraction of cells responding to varying calcium is a function of the burst frequency (right panel), and is FM controlled.

the system [27]. As we have seen in chapter 5, gene expression introduces such pure time delays during transcription and translation, and it therefore could directly contribute to the presence of oscillatory signaling dynamics.

Case Study 6-3 A. Hoffmann, A. Levchenko, M. L. Scott, and D. Baltimore. **The IκB–NF-κB signaling module: Temporal control and selective gene activation.** *Science* **298**: 1241–1245 (2002) [21].

Nuclear-cytoplasmic transport plays an important role in some cell signaling networks. In this study, Hoffmann et al. studied signaling involving the transcriptional activator NF-κB, which is modulated by the inhibitor IκB (actually three isoforms: IκBα, IκBβ, and IκBε). Binding of NF-κB by IκB leads to localization of the complex in the cytoplasm, preventing transcriptional activation by NF-κB. The NF-κB pathway can be activated (e.g., by treating cells with tumor necrosis factor-α (TNF-α)), which leads to increased activity of the IκB kinase (IKK) complex, degradation of the various IκB isoforms, and release of NF-κB. Free NF-κB enters the nucleus and activates transcription of its targets.

Hoffmann et al. examined the role of the individual isoforms of IκB through a combination of experimental study and computational modeling. Because IκBα expression is upregulated by NF-κB, the pathway forms a negative-feedback loop under stimulatory conditions: IKK degrades IκB, NF-κB translocates to the nucleus and activates transcription of IκBα, which binds NF-κB and reduces transcription of IκBα. Negative-feedback systems can exhibit a range of behaviors, depending on the detailed strengths of various interactions. The most common behaviors are undamped oscillatory, damped oscillatory, and nonoscillatory. (Undamped oscillations continue for multiple cycles with the same amplitude, whereas damped oscillations progress with smaller amplitude for successive cycles.)

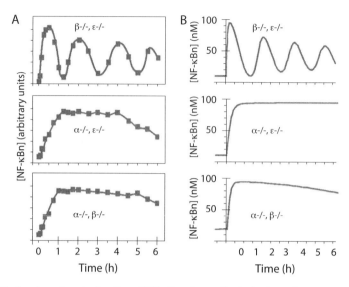

Figure 6.28 Nuclear/cytoplasmic shuttling of NF-κB varies qualitatively, depending on which IκB subunits are expressed, with oscillations present when only the α isoform is expressed. (A) Experimental behavior of cells containing only one of the three IκB isoforms in response to TNF-α. β-/-, ε-/- indicates cells deleted for IκBβ and ε, so they contain only the α isoform. (B) Computed behavior of cells under the same stimulation using models. Figure from reference 21.

Three mouse cell lines were engineered, each containing only one of the three IκB isoforms. Stimulation of each with TNF-α produced very different results: Cells containing only IκBα exhibited damped oscillations of nuclear NF-κB (NF-κBn), whereas cells with only the β or ε isoform showed nonoscillatory behavior, as shown in figure 6.28.

A useful model of this pathway was constructed by individually fitting the trajectories from the deletion mutants, combining the individual models to create a preliminary wild-type model, and then refitting to data from wild-type cells. Interestingly, the wild-type model was significantly different than simply the superposition of the individual models, suggesting regulatory crosstalk. Aspects of the fitted wild-type model were examined both computationally and experimentally to better understand the functional characteristics of this important pathway (figure 6.29). Notably, simulation of short pulses of TNF-α of varying duration produced transient activation of NF-κB of nonvarying duration but with larger amplitudes for longer stimuli. Longer simulated pulses of TNF-α stimulation resulted in NF-κB activation lasting roughly as long as the input pulse. Thus, the model predicts a bimodal response: Short input pulses (less than 1 hour) result in similar-length activation, but longer input stimulation results in activation of duration proportional to the input. Aspects of these predictions were tested experimentally by Hoffmann et al. [21].

An interesting question arises as to why it might be useful to have a bimodal response mechanism in the NF-κB network. One possibility, put forward by Hoffmann et al., is that it could be useful in the expression of two separate classes of genes. The first class responds to short and long input stimulation, and the second responds only to persistent stimulation. To test this notion, the authors examined the expression of some NF-κB regulated genes

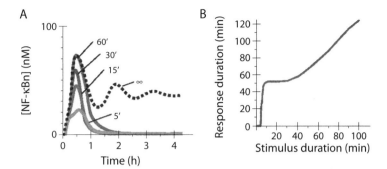

Figure 6.29 (A) Simulated input pulses of durations 5, 15, 30, and 60 minutes all result in an activation of nuclear NF-κB of uniform duration, lasting nearly an hour. The amplitude of activation increased with the duration of the input pulse. A constant activation (infinite-width pulse) resulted in damped oscillatory activation, the first peak of which lasted roughly an hour. (B) Longer simulated input stimuli resulted in activation of duration proportional to the input duration. Figure from reference 21.

and found that IP-10 exhibited the first behavior and RANTES the second. The net result of complex control is a bimodal activation profile that leads to a single pathway capable of activating genes with multiple responses. Integration of computational modeling and experimental measurement was necessary to unravel the complex dynamics of this system.

6.4.1 Cell Cycle

Growing and dividing eukaryotic cells progress through the G_1, S, G_2, and M cell-cycle phases before returning to G_1, giving the superficial impression of a periodically oscillating system. One might therefore expect to find limit cycles occurring in a mechanistic cell-cycle model, but this is not the case. Instead, the cell cycle consists of sequential progression through a series of bistable switches at the G_1–S, G_2–M, and M–G_1 checkpoints. These switches are created by enzymatic phosphorylation/dephosphorylation feedback loops. The Nobel Prize in Physiology or Medicine was awarded in 2001 to Leland Hartwell, Tim Hunt, and Sir Paul Nurse for the discovery of the essential components of this cell-cycle regulatory system. A simplified model of feedback control at the G_2–M checkpoint was instrumental in understanding cell-cycle dynamics (figure 6.30) [28].

6.4.2 Circadian Rhythms

Nearly all life on earth must cope with the daily rhythm of light and dark. Eukaryotes almost universally do so by cyclically varying expression of large sets of genes with a 24-hour period, allowing them to anticipate changes in their environment to provide a survival advantage. These circadian oscillations are autonomous—for example, the leaves of a mimosa plant placed in the dark will nevertheless open during the

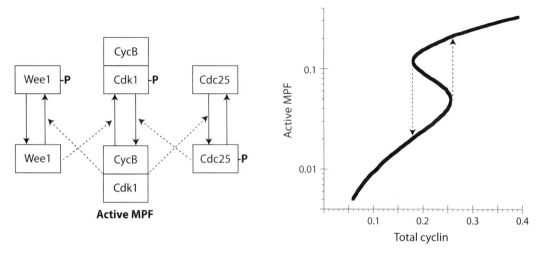

Figure 6.30 The three-enzyme phosphorylation/dephosphorylation network on the left creates a bistable response as shown on the right [28]. Dashed lines indicate enzymatic catalysis of the indicated reaction (e.g., phosphorylated Cdc25 is a phosphatase that catalyzes the removal of a phosphate from Cdk1 complexed with CycB). The dephosphorylated Cdk1/CycB complex is also known as active M-phase promoting factor (MPF). As total cyclin B increases during the S and G2 phases, eventually a threshold is reached and the concentration of MPF jumps severalfold, leading to the cell entering M phase. The system exhibits hysteresis, in that cyclin levels must fall to a threshold value lower than that required for the upward MPF jump, in order to jump down to the lower trajectory (concentrations are scaled as a fraction of total Cdk1).

day and close at night. However, circadian gene circuits also possess light-sensor components, so that the cycles become entrained with seasonal changes in the day's length while not being derailed by sporadic cloud cover. The 2017 Nobel Prize in Physiology or Medicine was awarded to Jeffrey Hall, Michael Rosbash, and Michael Young for uncovering the molecular mechanism of circadian cycling.

Studies with fruit flies were pivotal in identifying transcription-translation feedback loops that oscillate. Levels of PER protein, the product of the *period* gene, were shown to peak in neurons in the middle of the night, and drop to a low point at midday. It turns out that the PER protein represses transcription of its own gene, after forming a heterodimer with the product of the *timeless* gene, TIM. (The fruit fly genetics community is notorious for creative and evocative gene names, such as *hedgehog* and *son of sevenless*.) A simple model of the fruitfly circadian network [29] is shown in figure 6.31. With a simplifying assumption lumping the TIM and PER proteins together, the following model equations were obtained:

$$\frac{d[M]}{dt} = \frac{v_m}{1 + ([P_t](1-q)/(2P_{\text{crit}}))^2} - k_m[M] \qquad (6.58)$$

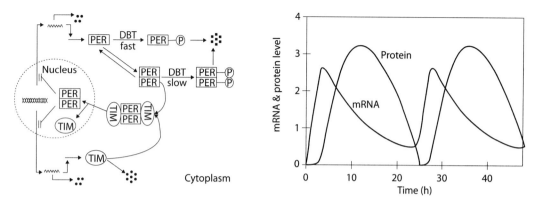

Figure 6.31 A simple mechanistic model of the circadian gene control network (left panel) produces stable oscillations in protein and mRNA levels for the fruit fly *period* gene (right panel) [29].

$$\frac{d[P_t]}{dt} = v_p[M] - \frac{k_{p1}q + k_{p2}}{1 + J_p/[P_t]} - k_{p3}[P_t] \qquad (6.59)$$

where $[M]$ is PER's mRNA, $[P_t]$ is the sum of PER and TIM protein, $q \equiv \frac{2}{1+\sqrt{1+8K_{eq}[P_t]}}$, and the remaining parameters describe transcription and translation rate constants and binding constants. With realistic values for these parameters, the solution to these equations was shown to oscillate with a period of about one day, consistent with experimental observations. Models of circadian gene networks in mammals have incorporated many other gene products and feedback loops, but the same self-inhibitory feedback loop of a PER gene's expression remains at their core [30]. These models have been useful tools for understanding pathophysiology related to dysregulation of the normal circadian rhythms.

Suggestions for Further Reading

U. Alon. *An Introduction to Systems Biology: Design Principles of Biological Circuits*. Chapman & Hall/CRC Mathematical and Computational Biology, 2006.

J. E. Ferrell, Jr. Perfect and near-perfect adaptation in cell signaling. *Cell Systems* 2: 62–67 (2016).

S. H. Strogatz. *Nonlinear Dynamics and Chaos: With Applications to Physics, Biology, Chemistry, and Engineering*, second edition. CRC Press, 2018.

J. J. Tyson, K. C. Chenz, and B. Novak. Sniffers, buzzers, toggles and blinkers: Dynamics of regulatory and signaling pathways in the cell. *Curr. Opin. Cell Biol.* 15: 221–231 (2003).

Problems

6-1 Justify the fact that a three-component system—in which each component can have activating, inhibitory, or null action on the other components—has 138 distinct motifs.

6-2 Consider an elaboration of the model of gene expression of Griffith, equations 6.6 and 6.7, which also incorporates constitutive expression of the gene at a level signified by a new parameter c:

$$\dot{x} = y - ax \qquad (6.60)$$

$$\dot{y} = c + \frac{x^2}{1+x^2} - by \qquad (6.61)$$

Draw the nullclines and phase portrait, and describe the stationary states. For the original model, we described a bifurcation that eliminated the high steady state. For this new system, identify a bifurcation that eliminates the low steady state. Give both a mathematical explanation for the bifurcation and a biological one.

6-3 Consider a predator-prey model represented by

$$\dot{x} = x(x(1-x) - y) \qquad (6.62)$$

$$\dot{y} = y(x - a) \qquad (6.63)$$

where $x \geq 0$ represents the prey populations (as a scaled quantity), $y \geq 0$ the predator population, and $a \geq 0$ is a parameter of the model. Briefly stated, the predator is a species that hunts and eats the prey, whereas the prey eats plants freely available from the land.

   a. Interpret the model biologically. That is, describe how the rate equations in the model correspond to the elements of the biological situation. (Prey live off the land with limited food supply; predators live off the prey.)

   b. Find the stationary points of the model in the biologically relevant quadrant ($x \geq 0$, $y \geq 0$) for $a > 1$.

   c. [*Advanced problem*] Characterize the fixed point at $(1, 0)$ by linearizing about it for $a > 1$. What type of fixed point is it?

   d. What happens to the predator population for $a > 1$? Rationalize in the context of the model.

6-4 Derive equation 6.17 by setting $\alpha_c = \frac{\gamma F_c}{k_B T C_c} - 1$, where C_c is the critical number of complexes at $F = F_c$.

6-5 Consider the following dynamical system inside a cell:

$$\dot{T} = E - \gamma T \tag{6.64}$$

$$\dot{E} = \frac{L_o T}{K + T} - E \tag{6.65}$$

where E represents the dimensionless concentration of an enzyme, T represents the dimensionless concentration of a transcription factor, and γ and K are positive constants. A small-molecule ligand added to the system at dimensionless concentration L_o (constant) increases the rate of enzyme synthesis in a manner directly proportional to the ligand concentration.

a. Determine the fixed points of the system (T^*, E^*).
b. If $L_o > \gamma K$, sketch the phase portrait. Which fixed points are stable, and which are unstable?
c. If $L_o < \gamma K$, sketch the phase portrait. Which fixed points are stable, and which are unstable?
d. Plot T^* versus L_o (include negative values for mathematical completeness, even though such values are of course irrelevant for biological variables and parameters). Label the stable and unstable branches on the plot. There is a bifurcation in this system, but this one entails a change in the stability of existing fixed points rather than a change in the number of fixed points. At what critical value of L_o does this bifurcation occur?

6-6 Consider a simple negative-feedback loop (figure 6.18) with the following parameters:

$$k_{IA} = 100 \text{ h}^{-1}$$
$$k_{AB} = 10 \text{ h}^{-1}$$
$$k_{BA} = 100 \text{ h}^{-1}$$
$$k_B = 1 \text{ h}^{-1}$$
$$K_{IA} = K_B = K_{AB} = K_{BA} = 0.01$$

a. Plot the output signal with a series of stepwise changes in input (+0.1 units of signal every hour for 3 hours, followed by −0.1 units of signal every hour for 3 hours) from an initial input of 0.1 units.
b. Generate a series of plots in which you vary the sensitivity of B and the speed of B.

6-7 Consider a simple incoherent feedforward system (figure 6.23) with the following parameters:

$$k_{IA} = 50 \text{ h}^{-1}$$
$$k_{IB} = 10 \text{ h}^{-1}$$
$$k_{BA} = 100 \text{ h}^{-1}$$
$$k_B = 10 \text{ h}^{-1}$$
$$K_{IA} = K_{IB} = K_B = K_{BA} = 1$$

a. Plot the output signal with a series of stepwise changes in input (+1 unit of signal every hour for 3 hours followed by −1 unit of signal every hour for 3 hours) from an initial input of 0.1 units.

b. Generate a series of plots in which you vary the sensitivity of B and the speed of B.

6-8 Derive equation 6.57, which describes mRNA dynamics under sinusoidal transcription.

References

1. J. J. Tyson and B. Novák. Functional motifs in biochemical reaction networks. *Annu. Rev. Phys. Chem.* 61: 219–240 (2010).
2. S. H. Strogatz. *Nonlinear Dynamics and Chaos: With Applications to Physics, Biology, Chemistry, and Engineering*, second edition. CRC Press (2018).
3. J. S. Griffith. Mathematics of cellular control systems. II. Positive feedback to one gene. *J. Theor. Biol.* 20: 209–216 (1968).
4. G. I. Bell. Models for the specific adhesion of cells to cells. *Science* 200: 618–627 (1978).
5. N. E. Buchler and M. Louis. Molecular titration and ultrasensitivity in regulatory networks. *J. Mol. Biol.* 384: 1106–1119 (2008).
6. A. Goldbeter and D. E. Koshland, Jr. An amplified sensitivity arising from covalent modification in biological systems. *Proc. Natl. Acad. Sci. U.S.A.* 78: 6840–6844 (1981).
7. N. A. Shah and C. A. Sarkar. Robust network topologies for generating switch-like cellular responses. *PLOS Comput. Biol.* 7: e1002085 (2011).
8. S. Palani and C. A. Sarkar. Synthetic conversion of a graded receptor signal into a tunable, reversible switch. *Mol. Syst. Biol.* 7: 480 (2011).
9. G. J. Melen, S. Levy, N. Barkai, and B.-Z. Shilo. Threshold responses to morphogen gradients by zero-order ultrasensitivity. *Mol. Syst. Biol.* 1: 2005.0028 (2005).
10. J. E. Ferrell. The biochemical basis of an all-or-none cell fate switch in *Xenopus* oocytes. *Science* 280: 895–898 (1998).
11. W. Xiong and J. E. Ferrell. A positive-feedback-based bistable "memory module" that governs a cell fate decision. *Nature* 426: 460–465 (2003).

12. S. Palani and C. A. Sarkar. Positive receptor feedback during lineage commitment can generate ultrasensitivity to ligand and confer robustness to a bistable switch. *Biophys. J.* 95: 1575–1589 (2008).

13. B. A. Francis and W. M. Wonham. The internal model principle of control theory. *Automatica* 12: 457–465 (1976).

14. T.-M. Yi, Y. Huang, M. I. Simon, and J. Doyle. Robust perfect adaptation in bacterial chemotaxis through integral feedback control. *Proc. Natl. Acad. Sci. U.S.A.* 97: 4649–4653 (2000).

15. N. Barkai and S. Leibler. Robustness in simple biochemical networks. *Nature* 387: 913–917 (1997).

16. U. Alon, M. G. Surette, N. Barkai, and S. Leibler. Robustness in bacterial chemotaxis. *Nature* 397: 168–171 (1999).

17. H. C. Berg and E. M. Purcell. Physics of chemoreception. *Biophys. J.* 20: 193–219 (1977).

18. C. Briat, A. Gupta, and M. Khammash. Antithetic integral feedback ensures robust perfect adaptation in noisy biomolecular networks. *Cell Syst.* 2: 15–26 (2016).

19. J. G. Albeck, G. B. Mills, and J. S. Brugge. Frequency-modulated pulses of ERK activity transmit quantitative proliferation signals. *Cell* 49: 249–261 (2013).

20. L. Cai, C. K. Dalal, and M. B. Elowitz. Frequency-modulated nuclear localization bursts coordinate gene regulation. *Nature* 455: 485–491 (2008).

21. A. Hoffmann, A. Levchenko, M. L. Scott, and D. Baltimore. The IκB–NF-κB signaling module: Temporal control and selective gene activation. *Science* 298: 1241–1245 (2002).

22. J. H. Levine, Y. Lin, and M. B. Elowitz. Functional roles of pulsing in genetic circuits. *Science* 342: 1193–1200 (2013).

23. J. E. Purvis, K. W. Karhohs, C. Mock, E. Batchelor, A. Loewer, and G. Lahav. p53 dynamics control cell fate. *Science* 336: 1440–1444 (2012).

24. J. Stewart-Ornstein, H. W. Cheng, and G. Lahav. Conservation and divergence of p53 oscillation dynamics across species. *Cell Syst.* 5: 410–417 (2017).

25. J. R. Porter, B. E. Fisher, and E. Batchelor. p53 pulses diversify target gene expression dynamics in an mRNA half-life-dependent manner and delineate co-regulated target gene subnetworks. *Cell Syst.* 2: 272–282 (2016).

26. N. A. M. Monk. Oscillatory expression of Hes1, p53, and NF-κB driven by transcriptional time delays. *Curr. Biol.* 13: 1409–1413 (2003).

27. D. E. Seborg, T. F. Edgar, D. A. Mellichamp, and F. J. Doyle III. *Process Dynamics and Control*, fourth edition. Wiley (2016).

28. B. Novák and J. J. Tyson. Numerical analysis of a comprehensive model of M-phase control in *Xenopus* oocyte extracts and intact embryos. *J. Cell Sci.* 106: 1153–1168 (1993).

29. J. J. Tyson, C. I. Hong, C. D. Thron, and B. Novák. A simple model of circadian rhythms based on dimerization and proteolysis of PER and TIM. *Biophys. J.* 77: 2411–2417 (1999).

30. J. H. Abel and F. J. Doyle III. A systems theoretic approach to analysis and control of mammalian circadian dynamics. *Chem. Eng. Res. Des.* 116: 48–60 (2016).

7

Population Growth and Death Models

Population, when unchecked, increases in a geometrical ratio. Subsistence increases only in an arithmetical ratio. A slight acquaintance with numbers will show the immensity of the first power in comparison of the second.
—Thomas Robert Malthus, *An Essay on the Principle of Population*

Death, like generation, is a secret of Nature.
—Marcus Aurelius, *Meditations*

Growth, reproduction, and death are central preoccupations of both biological systems and the people who study them. In this chapter, we consider the kinetic description of these processes for populations of cells as well as multicellular organisms. The earliest kinetic descriptions of population growth originated in the actuarial examination of the growth and mortality statistics of human populations, and the statistical distributions discovered in this way are often empirically applicable to cellular and animal populations. For example, descriptions of microbial growth play a central role in bioprocess design and operation. In another example, design and analysis of therapeutic regimens in cancer and infectious disease rely on quantitative descriptions of the proliferation and death of immune cells and tumors. Studies of ecosystem dynamics also often employ empirical models of individual birth, growth, and death. Surprisingly, these population growth models are often equally applicable to cell cultures or populations of individual organisms, even humans. Consequently, we consider population models ranging from bacterial fermentors to human epidemics.

Many of the models discussed in this chapter differ from those considered elsewhere in this book in that they lack grounding in molecular mechanisms. Given the common need for a mathematical representation of cell growth, however, statistical descriptions of population dynamics can be quite useful. As the molecular mechanisms of cellular differentiation and programmed cell death are elucidated, new mechanistic models must reproduce the well-established overall growth and mortality dynamics that these empirical models describe so well.

7.1 Typical Growth Curves

On introduction to a new environment, a cell population typically goes through a series of distinct *growth phases*, each characterized by different kinetics (figure 7.1). In the first, *lag phase*, cells adapt to their new environment by changing gene expression patterns and metabolic capabilities. Once physiologically adapted to their new environs, cells undergo a period of rapid expansion. During this *exponential growth phase* (or log phase), the cell population increases exponentially until reaching some upper limit—either exhaustion of critical external resources or accumulation of toxic growth by-products. At this point, the population enters *stationary phase*, during which cells often induce stress-resistance programs to weather subsequent periods of hardship. In some cases, cells begin to die off, with kinetics that are also exponential, though decreasing.

We now consider several purely empirical mathematical descriptions of growth, starting with ideal exponential growth and then moving on to some common representations for resource-limited growth.

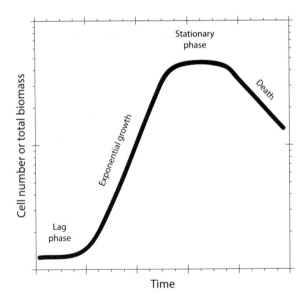

Figure 7.1 A typical microbial growth curve following introduction to a new environment. After a lag phase of metabolic adaptation, cell number and mass increase exponentially during this log phase (note the logarithmic ordinate scale). Once nutrients are consumed, or toxins have accumulated, cells enter a stationary phase and cease growth. In some instances, a phase of cell death follows, during which viable cell number declines exponentially.

7.1.1 Exponential Growth

Consider a population of cells that synchronously divide after every time increment τ_D (the *doubling time*). Such a population grows in discrete jumps, so the number of cells present after i generations is 2^i times the starting number (figure 7.2). However, in most cases, cell populations grow asynchronously, with the time between divisions for each cell varying randomly around the population average τ_D. There are various experimental methods for driving initially synchronized cell division, but such populations generally lapse into asynchronous growth due to random variations in the generation time from cell to cell (figure 7.3).

Now consider an asynchronously dividing population. In the limit of a large number of cells, one can as a practical matter treat the cell number as a continuous variable, N. We can derive the growth law for such a population by direct analogy to compound interest in finance. In one doubling time τ_D, the cell population effectively pays 100% *interest* by increasing the cell number by 100%. In a synchronous population, this *interest payment* is paid only at precisely $\Delta t = \tau_D$ steps. In an asynchronous population, such *payments* are made continuously. If one divides each doubling time τ'_D (where $'$ distinguishes the asynchronous case) into n equal time increments, then the population N increases by a factor $(1 + 1/n)$ during each time

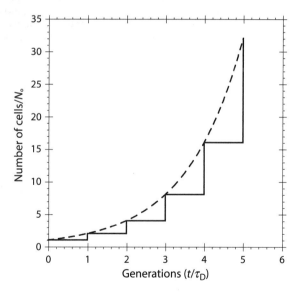

Figure 7.2 Growth curves for a synchronized population (solid line) and an asynchronous population (dashed line). The synchronous population exhibits discrete values of $2^i N_o$ cells, where $i = \lfloor t/\tau_D \rfloor$. The asynchronous population varies smoothly as $N_o e^{t/\tau_D}$.

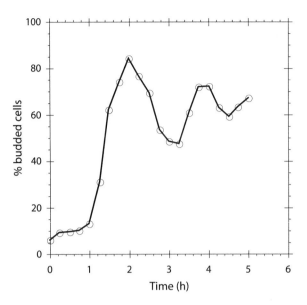

Figure 7.3 Loss of synchronization in a growing yeast population. The yeast *Saccharomyces cerevisiae* grows by asymmetric budding, so newborn cells can be isolated by their smaller size by centrifugal elutriation to obtain a population of synchronized newborn cells. Subsequent rounds of cell division can then be tracked by counting the fraction of cells with buds, as plotted here. Following one fairly simultaneous wave of division at 2 hours, random variation in subsequent generation times leads to an overall averaging such that by the third generation, the fraction of budded cells in the population is essentially constant [1].

increment, or $(1+1/n)^{t/(\tau'_D/n)}$ for the whole series of $t/(\tau'_D/n)$ increments. At the end of time t, the population will be

$$N = N_\circ \left(1 + \frac{1}{n}\right)^{\frac{t}{\tau'_D/n}} \tag{7.1}$$

The continuous function is derived in the limit of a large number of infinitesimally small time increments:

$$N = \lim_{n \to \infty} N_\circ \left[\left(1 + \frac{1}{n}\right)^n\right]^{\frac{t}{\tau'_D}} \tag{7.2}$$

$$= N_\circ e^{\frac{t}{\tau'_D}} \tag{7.3}$$

Notably, because of compounding within the time increments, the actual doubling time (τ_D) is less than the initial nominal asynchronous τ'_D:

$$\tau_D(N = 2N_o) = \ln(2)\tau'_D \qquad (7.4)$$

An alternative derivation also leads to an exponential growth function identical to equation 7.3. If cells are treated as a pseudochemical species, a mass-action rate law can be written to describe the increase in cell number. A population unconstrained by limiting resources will often grow in proportion to the number of individuals present at that instant. The mathematical statement of this observation is

$$\frac{dN}{dt} = \mu N \qquad (7.5)$$

where N is the number of individuals, and μ is a constant known as the *specific growth rate*, which has units of reciprocal time. With initial condition $N = N_o$ at $t = 0$, equation 7.5 is easily solved to yield

$$N = N_o e^{\mu t} \qquad (7.6)$$

This functional form results in a period of rapid expansion, the exponential growth phase. During exponential growth, the number of individuals doubles with each passing increment of doubling time τ_D:

$$N(t = \tau_D) = 2N_o \qquad (7.7)$$

$$\tau_D = \frac{\ln 2}{\mu} \qquad (7.8)$$

When cells in an exponentially growing population maintain a constant average size and composition, it is called balanced growth, and the total cellular biomass X (units of cell mass per volume) is proportional to the number of individuals, such that $X = X_o e^{\mu t}$. These units are commonly used in bioreactor design.

Thomas Robert Malthus was the first to explore the characteristics and consequences of unfettered exponential growth in *An Essay on the Principle of Population* (1798), and consequently, such human population growth is often called Malthusian. Examining the statistics of population growth under a variety of societal conditions, he concluded that "population when unchecked goes on doubling itself every twenty-five years, or increases in a geometrical ratio." (This is about as mathematical as the *Essay* gets—Malthus was primarily concerned with the sociopolitical consequences of population growth. He terms his efforts "an investigation of the causes that have hitherto impeded the progress of mankind towards happiness.")

> **Example 7-1** Express Malthus's observation about the doubling of an unchecked human population in mathematical form.
>
> **Solution**
>
> $$\mu = \frac{\ln 2}{\tau_D} \quad (7.9)$$
>
> $$= \frac{\ln 2}{25 \text{ yr}} \quad (7.10)$$
>
> $$= 0.028 \text{ yr}^{-1} \quad (7.11)$$
>
> Therefore,
>
> $$N = N_\circ e^{0.028t} \quad (7.12)$$
>
> where $N \equiv$ the population size at time t (years), and $N_\circ \equiv$ the population at time zero.

7.1.2 Subexponential Growth Kinetics

Logistic equation The dire predictions of unlimited exponential growth made by equation 7.6 run counter to our experience at every scale. Bacterial cultures do not fill earth's biosphere in a matter of days, and, at least to date, human populations do not grow exponentially without limit. A simple model to describe this limitation on Malthusian growth was first proposed by Verhulst in 1838 [2] and independently rediscovered by Pearl and Reed in 1920 [3]. The *logistic* model posits that a given environmental setting has a finite *carrying capacity*, beyond which further growth of the organism in question is not supportable. (Pearl [4] attributes the term *logistique* to Verhulst, who did not use it in his original 1838 publication but coined it in a later paper [5].)

A simple linear decrease in growth rate as a population approaches its carrying capacity is provided by the following expression:

$$\frac{dN}{dt} = \mu \left(1 - \frac{N}{K}\right) N \quad (7.13)$$

The variable K is the carrying capacity for the environment in which the population is growing, because it defines the population at which the growth rate reaches zero. This model also features a negative growth rate when the population is greater than K, which is thus the steady state of the system for any nonzero initial population. However, neither Verhulst nor Pearl and Reed made any pretense of a mechanistic

Population Growth and Death Models

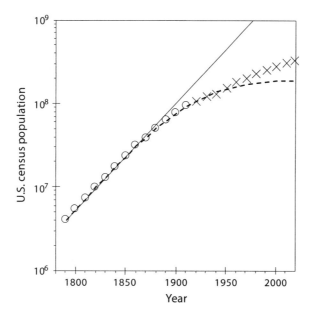

Figure 7.4 Growth in the U.S. population as assessed in the census performed every ten years since 1790. In 1920, Pearl and Reed [3] formulated the logistic equation (originally developed by Verhulst in 1838) to predict subexponential growth due to a finite "carrying capacity" of the land. The solid line reflects Malthusian exponential growth extrapolated from the years 1790–1850, and the dashed line represents Pearl and Reed's fit parameters to the logistic equation for data through 1910. The purely empirical logistic equation extrapolated rather accurately for the next four decades, after which growth exceeded its predictions. It should also be pointed out that from 1790 to 1850, the population was in essentially exponential growth, indicative of growth without limiting resources.

basis for this choice of the functional form. It is merely a convenient representation that fits the available data.

The solution to this equation for $N = N_o$ at $t = 0$ is

$$N = \frac{K}{1 + \left(\frac{K}{N_o} - 1\right) e^{-\mu t}} \qquad (7.14)$$

Note that if $N_o \ll K$, then at early times, $N \approx N_o e^{\mu t}$, or simple exponential growth. However, as $t \to \infty$, $N \to K$. Equation 7.14 is sigmoidal in N versus t and is symmetrical about its inflection point at $N = K/2$.

The population data that inspired Pearl and Reed's rediscovery of the logistic rate form are shown in figure 7.4. As Verhulst did, Reed and Pearl extrapolated rather optimistically from a slight deviation below Malthusian exponential growth to a prediction of a stable maximum population level. They went so far as to predict an asymptotic U.S. population of 1.97×10^8 individuals, at which level, "it will be necessary ... to import nearly or quite one-half of the calories necessary for that population." That is, "unless our agricultural production radically increases," which clearly has been the case, given that the 2010 U.S. census population was 3.09×10^8, and the country is a net agricultural exporter. Nevertheless, the logistic equation predicted the subsequent 90 years of population growth with far greater accuracy than would have been expected for Malthusian exponential growth—a particularly impressive extrapolation for what is, in essence, a curve fit to a mathematically convenient functional form. Consider also some mechanistic details absent from this model for the growth of a human population: gender, age, and death. This logistic model for growth of a sexually reproducing population should only be considered a mathematically convenient representation, not expected to extrapolate accurately outside the initial data set used to curve-fit the parameters.

Gompertz growth equation Another function widely used to empirically describe sigmoidal growth curves bears the name of Benjamin Gompertz, although his original work dealt with mortality rather than growth (section 7.4.2). It was first noted in passing (in a book review [6]) that the specific growth rate of a population appears to decrease by a constant *proportion* with each passing increment of time. This dependence leads to an exponentially decreasing specific growth rate (by contrast with the logistic equation 7.14, which predicts a constant linear decrease in specific growth rate with population size). The analogy to exponentially increasing death rate was later drawn [7], and this formula has since been termed "Gompertzian growth kinetics." The Gompertz growth law is thus written as follows:

$$\frac{dN}{dt} = \mu e^{-\alpha t} N \qquad (7.15)$$

With the initial condition $N = N_o$ at $t = 0$, this can be solved to yield

$$N = N_o e^{\frac{\mu}{\alpha}(1 - e^{-\alpha t})} \qquad (7.16)$$

Let's consider the behavior of equation 7.16 at short and long times. At short times ($t \to 0$), it can be shown that $N \to N_o e^{\mu t}$, which is simple Malthusian exponential growth. As $t \to \infty$, $N \to N_o e^{\frac{\mu}{\alpha}}$, a constant maximum value for the population size. Note that, in contrast to the logistic model, the Gompertz model assumes that N_o is below the carrying capacity, because the rate in equation 7.15 can never be negative and its steady-state value is not independent of N_o (as K is in the logistic model), but instead is N_o times a number greater than one. The Gompertz growth law has been particularly widely used for the description of tumor cell populations, which is appropriate, given that the starting cell population of a tumor is typically below the carrying capacity.

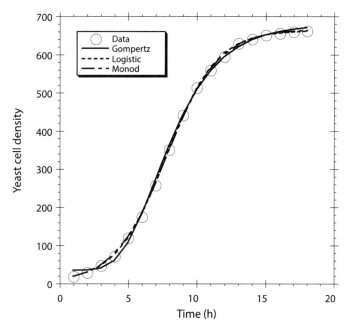

Figure 7.5 Fit of sigmoidal growth laws (Gompertz, Verhulst-Pearl-Reed logistic, and Monod) to microbial growth data. Yeast was grown in shake-flask cultures [8] and exhibited a sigmoidal increase in cell mass versus time. The Monod growth rate law describes depletion of a critical nutrient and is described in section 7.2.1 (equation 7.36). These three models cannot be distinguished on the basis of statistical goodness-of-fit. However, only the Monod model incorporates a mechanistic description of the cause of growth limitation (nutrient depletion).

There is no strong rationale for choosing between the Gompertz and the logistic growth equations to describe the growth of a given population. Both expressions require fitting three parameters (N_o, α, and μ for the Gompertz equation; N_o, K, and μ for the logistic). It can be shown that the two models yield the same steady states (carrying capacities) when $\alpha = \frac{\mu}{\ln(K/N_o)}$, and the subtle differences in the shapes of the two sigmoidal expressions are generally insufficient to usefully discriminate between their agreement with a particular data set (see, for example, figure 7.5).

Allometric scaling laws Compared to larger creatures, smaller animals "live fast and die young." Max Kleiber [9] noted in 1932 that an animal's basal metabolic rate scales as the 3/4 power of the body mass (figure 7.6), an exponent of enigmatic origin. Some have argued that a 2/3 power dependence on mass should be expected, as respiration and heat loss depend on surface area. However, the 3/4 value is better supported by many experimental studies, and a theoretical examination of the scaling properties of fluid dynamics in circulatory and respiratory networks in animals has been shown to lead to a prediction of a 3/4 exponent [10].

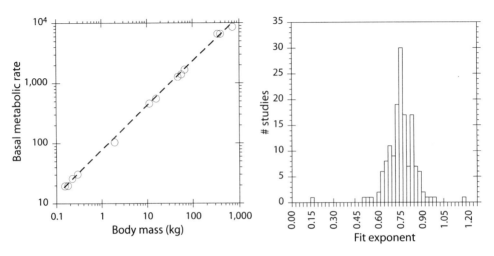

Figure 7.6 Organismal metabolic rate scales as the 3/4 power of body mass. Left panel, a log-log plot of basal metabolic rate versus body mass indicates that the former scales as the 0.74 power of the latter [9]. The data points represent, from smallest to largest: dove, rat (female, male), pigeon, hen, dog (female, male), sheep, woman, man, steer, cow, and steer (different data set). In the right panel, the least squares-fit exponent for the power law relating body mass and biological rates is summarized for 146 independent studies [12]. The mean exponent determined is 0.74 ± 0.11.

Scaling relationships between biological rates and body mass are termed *allometric* laws and have been useful in extrapolating from animal studies to predict effects of drugs and toxins in humans. They also lead to two commonly used and closely related growth-law forms: von Bertalanffy, and West-Brown-Enquist. These laws describe the accumulation of biomass X as the balance between two rate processes: synthesis and degradation. In the von Bertalanffy derivation [11], synthesis rate is proportional to biomass to the 2/3 power (i.e., the organism's surface area), and degradation is directly proportional to biomass:

$$\frac{dX}{dt} = k_1 X^{2/3} - k_2 X \tag{7.17}$$

which can be integrated to give, in dimensionless form

$$\hat{X} = \left(1 - \left(1 - \rho^{-1/3}\right) e^{-\tau}\right)^3 \tag{7.18}$$

where nondimensional biomass $\hat{X} \equiv X/X_{max}$, overall growth ratio $\rho \equiv X_{max}/X_o$, and dimensionless time $\tau = \alpha t$, with $\alpha \equiv k_2/3$ and $X_{max} = (k_1/k_2)^3$.

The West-Brown-Enquist derivation [13] instead uses a 3/4 exponent for the synthetic term:

$$\frac{dX}{dt} = k_1 X^{3/4} - k_2 X \tag{7.19}$$

leading to

$$\hat{X} = \left(1 - \left(1 - \rho^{-1/4}\right) e^{-\tau}\right)^4 \tag{7.20}$$

where nondimensional biomass $\hat{X} \equiv X/X_{\max}$, overall growth ratio $\rho \equiv X_{\max}/X_\text{o}$, and dimensionless time $\tau = \alpha t$, with $\alpha \equiv k_2/4$ and $X_{\max} = (k_1/k_2)^4$.

In these same nondimensional variables, with $\alpha \equiv \mu/\ln \rho$, the Gompertz growth law (equation 7.16) is

$$\hat{X} = \rho^{-e^{-\tau}} \tag{7.21}$$

How well do equations 7.18, 7.20, and 7.21 describe experimental growth data? In figure 7.7, all three curves are fit to published data on tumor cell growth in cell culture, animals, and human patients. Each equation has three fitted parameters ($X_{\max}$, X_o, and α), and because the fits are of indistinguishable quality, the same symbol (a dashed line) is used for all three. The three growth laws are also curve-fit to the mass trajectories of a variety of vertebrate animals in figure 7.8. Again, the fit is excellent in every case, with all three growth laws capturing the growth curves equally well.

A stringent test of any model is to determine its predictive capability, which is performed retrospectively in figure 7.9. Each of the three growth models was fit to the data accounting for approximately the first half of the experimentally observed growth (open circles). The model was extrapolated with these parameters for the second half of the observed growth experiment, for comparison to the actual experimental points (solid circles). None of the growth models reliably predicted the second half of the growth curves. The von Bertalanffy and West-Brown-Enquist predictions were very similar, as might be expected, given the similarity of their derivations and rate forms.

At first glance, it is not clear why the Gompertz growth law (equation 7.21) should give fits similar to the von Bertalanffy and West-Brown-Enquist laws. In dimensionless form, these three equations are numerically similar for the same value of the parameter ρ, as shown in figure 7.10. The agreement is better as the overall growth ratio ρ decreases (left panel versus right panel).

Why might these three growth laws exhibit such numerically similar behavior? It can be shown that all three laws are members of one common algebraic family, with the Gompertz law as a limiting case (personal communication, R. Braatz). Let's take a look at the nondimensional differential equation forms:

von Bertalanffy:

$$\frac{d\hat{X}}{d\tau} = 3(\hat{X}^{2/3} - \hat{X}) \tag{7.22}$$

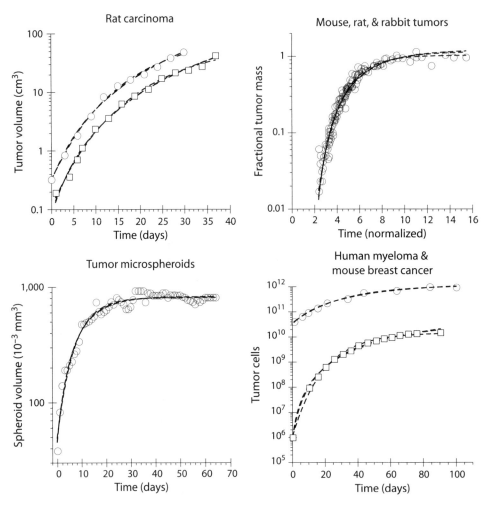

Figure 7.7 The growth of tumor cells in animals, humans, and cell culture are equivalently well described by the von Bertalanffy, West-Brown-Enquist, and Gompertz rate forms (equations 7.18, 7.20, and 7.21, respectively). The nonlinear least squares curve fit of each rate form to the data is shown, and the three curves essentially overlay one another. (It should be noted, however, that in the primary references, only one model was originally fit to the data in each case.) The rat carcinoma data is from Norton et al. [14]. The mouse, rat, and rabbit tumor data represent pooled data from 19 examples of 12 different tumor types transplanted into the three species [15]. The time coordinate was normalized by the doubling time immediately before the inflection point, which was set to time zero, and tumor size is expressed as a fraction of asymptotic size. The growth of tumor microspheroids was measured in cell culture [16] and shown to be consistent with the West-Brown-Enquist rate form [17]. Human myeloma cell number was calculated in patients by the production of a tumor clonal IgG [18]. One million cells of spontaneous C3H mouse breast cancer were transplanted at time zero [19].

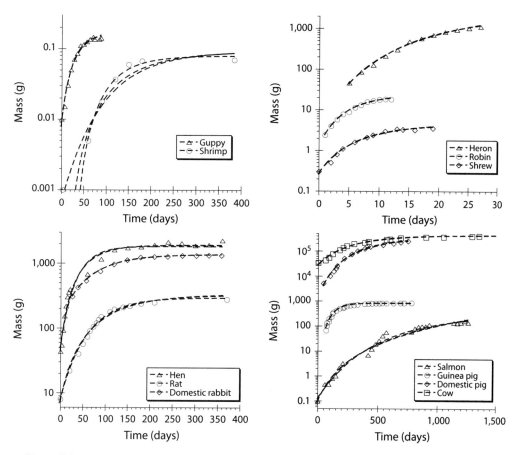

Figure 7.8 The growth trajectory of a variety of vertebrate animals is equally well described by the von Bertalanffy, Gompertz, and West-Brown-Enquist rate forms (least squares fit curves for each of the three growth equations are plotted, which overlap closely in most cases). This particular collection of primary data from a variety of studies was shown to be consistent with the West-Brown-Enquist growth expression [13].

West-Brown-Enquist:

$$\frac{d\hat{X}}{d\tau} = 4(\hat{X}^{3/4} - \hat{X}) \qquad (7.23)$$

Gompertz (substituting $\ln \hat{X} = -e^{-\tau} \ln \rho$ from equation 7.21):

$$\frac{d\hat{X}}{d\tau} = -\hat{X} \ln \hat{X} \qquad (7.24)$$

Note that the von Bertalanffy and West-Brown-Enquist growth laws are both of the following form:

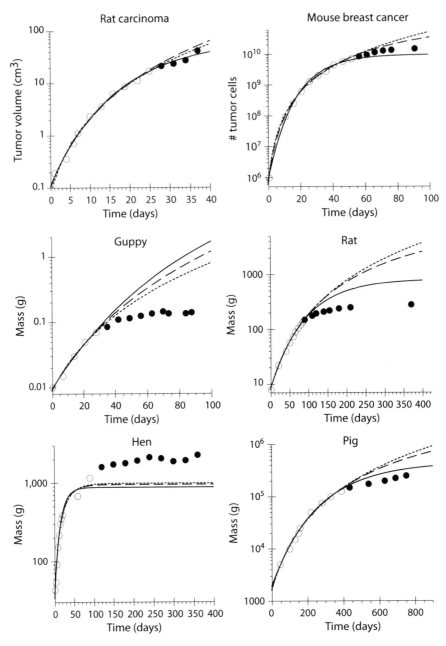

Figure 7.9 Empirical allometric relationships do not reliably forecast future growth trajectories. The von Bertalanffy (short-dashed line), Gompertz (solid line), and West-Brown-Enquist (long-dashed line) equations were least squares fit to data points from the first portion of several of the growth curves in figures 7.7 and 7.8. In each case, the equation was fit to those data points representing cumulative growth of less than half of the highest size attained (open circles). The subsequent growth data (closed circles) is plotted to test the forecasting power of the three growth laws. No consistent trend of over- or underprediction is found, and the three rate laws performed roughly equivalently in accuracy. The take-home lesson is that extrapolating a curve-fit empirical relationship, even one that fits the existing data points very closely, should not be expected to provide accurate predictions. The data in this figure are a subset of the datasets shown in figures 7.7 and 7.8. In those previous figures, the existence of parameter sets that fit the full range of the growth curve is demonstrated. However, these are not the parameters that best fit the first half of the data.

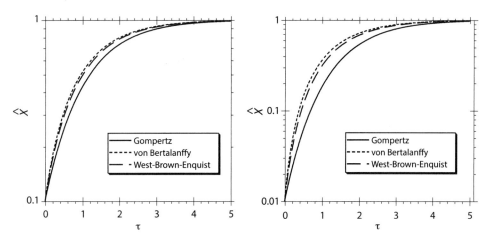

Figure 7.10 Comparison of the dimensionless forms of the Gompertz, von Bertalanffy, and West-Brown-Enquist growth curves illustrates the numerical similarities among these relationships. The majority of datasets in figures 7.7 and 7.8 are approximately within the growth range $\rho = 10$, corresponding to the left panel of this figure. The right panel corresponds to $\rho = 100$.

$$\frac{d\hat{X}}{d\tau} = n(\hat{X}^{(n-1)/n} - \hat{X}) \quad (7.25)$$

Let's examine the limit of this form as $n \to \infty$ (applying l'Hôpital's rule to obtain equation 7.28):

$$\lim_{n \to \infty} n(\hat{X}^{(n-1)/n} - \hat{X}) = \hat{X} \lim_{n \to \infty} n(\hat{X}^{-1/n} - 1) \quad (7.26)$$

$$= \hat{X} \lim_{n \to \infty} \frac{(\hat{X}^{-1/n} - 1)}{1/n} \quad (7.27)$$

$$= \hat{X} \lim_{n \to \infty} \frac{1/n^2 \cdot \hat{X}^{-1/n} \ln \hat{X}}{-1/n^2} \quad (7.28)$$

$$= -\hat{X} \ln \hat{X} \lim_{n \to \infty} \hat{X}^{-1/n} \quad (7.29)$$

$$= -\hat{X} \ln \hat{X} \quad (7.30)$$

And so these three laws are special cases of equation 7.25, with $n = 3$ for von Bertalanffy, $n = 4$ for West-Brown-Enquist, and $n \to \infty$ for Gompertz. As a practical matter, though, how numerically similar are these formulas? Figure 7.11 indicates that the numerical agreement is generally good, in particular for the final two doublings, as $\hat{X} \to 1$.

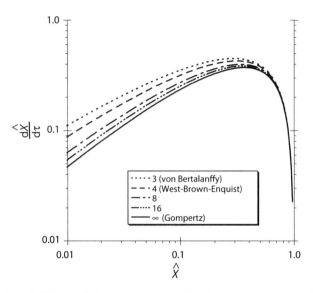

Figure 7.11 Despite their different origins, the von Bertalanffy, West-Brown-Enquist, and Gompertz growth laws are just instances of equation 7.25 with different values of the parameter n.

So, what are the conclusions of this perhaps overly exhaustive consideration of three different growth laws? One of the most important lessons is that consistency is not proof: Although a model's predictions should match the available experimental data, one must always remember that if a model matches the data, it is only one among an infinite set of alternative models that also do so. We considered here only three alternative representations, each of which has its practitioners and adherents. All three can be fit to diverse data sets with three fitted parameters (figures 7.7 and 7.8). However, as shown in figure 7.9, the predictive performance of the three models is similar—and is not strong.

For what purpose, then, would one employ these growth laws? Reflecting on the discussion of modeling motivations in chapter 1, improved mechanistic understanding of the growth process does not arise from the models. (The Gompertz law is purely an empirical fit, and the differing synthetic rate forms of the von Bertalanffy and West-Brown-Enquist laws do not result in distinguishable consistency with the available data.) However, also remember that a condensed mathematical representation of a dynamic process can be useful (or even necessary) for design and analysis, even in the absence of mechanistic understanding. These growth laws have been extensively used in this spirit in such diverse fields as ecology and cancer therapy.

Table 7.1 Parameters for generalized Savageau growth equations.

Growth law	α_1	g_{11}	g_{12}	β_1	h_{11}	h_{12}	α_2	g_{21}	g_{22}	β_2	h_{21}	h_{22}
Exponential	μ	1	0	0	0	0	0	0	0	0	0	0
Logistic	μ	1	0	$-\mu/K$	2	0	0	0	0	0	0	0
Theta-logistic	μ	1	0	$-\mu/K^\theta$	$1+\theta$	0	0	0	0	0	0	0
von Bertalanffy	k_1	2/3	0	$-k_2$	1	0	0	0	0	0	0	0
West-Brown-Enquist	k_1	3/4	0	$-k_2$	1	0	0	0	0	0	0	0
Gompertz	1	1	1	0	0	0	0	0	0	$-\alpha$	0	1
Lotka-Volterra	b	1	0	$-s$	1	1	$-d$	0	1	es	1	1

Generalized growth equations Savageau has pointed out that all commonly used growth equations in the literature are special cases of a single general form [20]:

$$\frac{dX_1}{dt} = \alpha_1 X_1^{g_{11}} X_2^{g_{12}} + \beta_1 X_1^{h_{11}} X_2^{h_{12}} \qquad (7.31)$$

$$\frac{dX_2}{dt} = \alpha_2 X_1^{g_{21}} X_2^{g_{22}} + \beta_2 X_1^{h_{21}} X_2^{h_{22}} \qquad (7.32)$$

where X_i is either total biomass or a component of it. This form emerges when one assumes that (1) the relevant rate forms for cellular reactions obey power laws (i.e., rates are proportional to the reactant raised to some power); (2) most cellular reactions are much faster than overall growth rates, and so reach a quasi-steady state; and (3) the overall observed growth rate is dominated by one or two slow reactions. The particular exponents in equations 7.31 and 7.32 that yield the growth equations discussed in this chapter are shown in table 7.1. The theta-logistic model is an extension of the logistic growth law that is frequently used in ecological studies of animal population dynamics to better fit empirical data [21–23]. The Lotka-Volterra model of predator-prey interactions is considered later in this chapter in section 7.5.4.

7.2 Limitations on Growth of Cell Cultures

Exponential growth generally ends due to depletion of resources (nutrients or growth factors), accumulation of waste by-products, or specific cell-cell signaling.

7.2.1 Consumption of Nutrients

Microbial growth medium can generally be formulated such that all components except one (the limiting substrate S) are in excess for maximal growth rate (i.e., further increases in any component other than S do not increase the specific growth rate μ). The carbon source, such as glucose, is generally chosen as the limiting substrate.

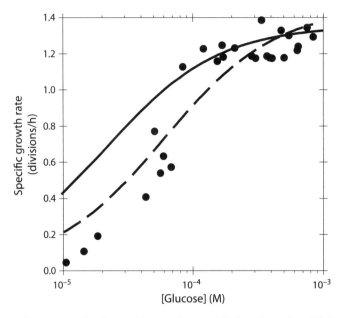

Figure 7.12 Data from Monod for the specific growth rate of *Escherichia coli* at 37°C as a function of glucose concentration [24]. Glucose concentration is plotted on a logarithmic scale to expand the portion of the curve that most influences the fit of the parameter K_s. The solid curve is Monod's hand-drawn fit with the reported parameters $\mu_{max} = 1.35$ divisions/h and $K_s = 2.2 \times 10^{-5}$ M. The nonlinear least squares fit to the data is shown as a dashed line, with $\mu_{max} = 1.5$ divisions/h and $K_s = 6.2 \times 10^{-5}$ M. Neither fit captures the relationship in the data particularly well. However, equation 7.33 is such a convenient two-parameter approximation to the data that it is almost universally used to describe the dependence of cell growth rates on a limiting growth substrate's concentration.

How might the specific growth rate μ decrease as the concentration of limiting substrate S decreases? Of course, cellular growth is the sum of a large number of complex interacting chemical reactions, and so a simple rate form can be, at best, a gross empirical simplification. By analogy with the Michaelis-Menten enzymatic rate form and the Langmuir-Hinshelwood rate form for surface-catalyzed reactions, Monod proposed that a hyperbolic functional form was consistent with bacterial growth rate dependence on carbon source concentration:

$$\mu = \frac{\mu_{max}[S]}{K_s + [S]} \tag{7.33}$$

Examination of typical data from which Monod drew this conclusion (figure 7.12) indicates that this rate form is not convincingly consistent with the experimental relationship between S and μ. Nevertheless, the hyperbolic rate form of equation 7.33 is so easy to work with that Monod growth kinetics are by far the most common model of microbial growth.

The parameter K_s is generally much smaller than initial glucose concentrations in batch culture. For example, for *E. coli* with glucose as the limiting substrate, $K_s = 2-4$ mg/L [25,26], and for *S. cerevisiae* with glucose as the limiting substrate, $K_s = 25-180$ mg/L. For many batch growth media, $[S]_o \geq 20$ g/L, so that $K_s \ll [S]$ and $\mu \approx \mu_{max}$ is approximately constant during most of the growth curve, until the great majority of substrate has been consumed.

An analytical solution to growth kinetics governed by the Monod equation (equation 7.33) can be obtained. The increase in biomass is generally proportional to the consumption of the limiting substrate, such that

$$X - X_o = Y_{X/S}([S]_o - [S]) \tag{7.34}$$

where the *yield coefficient* $Y_{X/S}$ is a constant. Using this relationship to substitute for $[S]$ in the Monod expression for specific growth rate μ, we have

$$\frac{dX}{dt} = \frac{\mu_{max}(Y_{X/S}[S]_o + X_o - X)}{(K_s Y_{X/S} + Y_{X/S}[S]_o + X_o - X)} X \tag{7.35}$$

Integrating from 0 to time t and X_o to X, one obtains

$$\left(1 + \frac{Y_{X/S}[S]_o + X_o}{K_s Y_{X/S}}\right) \ln \frac{X}{X_o} - \ln\left(1 - \frac{X - X_o}{Y_{X/S}[S]_o}\right) = \frac{Y_{X/S}[S]_o + X_o}{K_s Y_{X/S}} \mu_{max} t \tag{7.36}$$

This is the relationship that was fit to the yeast growth data in figure 7.5, for comparison to the Verhulst-Pearl-Reed logistic and Gompertz growth laws.

7.2.2 Growth-Factor Dose Responses

The expression *growth factor*, although conceivably applicable to any chemical substance that promotes growth, is used primarily to describe proteins or peptides that bind to cellular receptors that subsequently signal cells to grow.

Animal cells grown in culture typically grow more rapidly when their medium is supplemented with fetal bovine serum, which contains a mixture of growth factors. The dependence of growth rate on serum content can generally be approximately described by a Monod-type functional form, as shown in figure 7.13. Dependence of cell culture growth rate on individual factors also often approximates a Monod-type form (figure 7.13 [27]).

The simplest expectation for a growth response curve for a single growth factor would be that it would mimic the curve of growth factor binding to cell-surface receptors—which is approximately true in some cases, as in the response to epidermal growth factor (EGF) shown in figure 7.14. The mitogenic response tracks the binding curve closely.

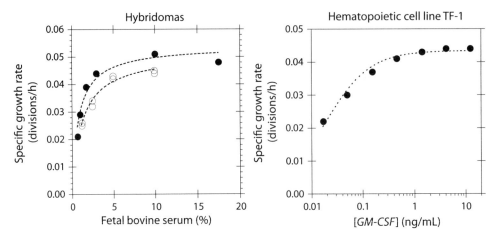

Figure 7.13 (Left panel) The dependence of hybridoma specific growth rate on serum is reasonably well described by a Monod rate form. Open circles, data from Ozturk and Palsson [28]; closed circles, data from Glacken [29]. (Right panel) Growth rate of human hematopoietic cell line TF-1 as a function of initial concentration of granulocyte-macrophage colony-stimulating factor (GM-CSF) in the growth medium (data from reference [27]). The dotted lines in both panels are least squares fits to equation 7.33.

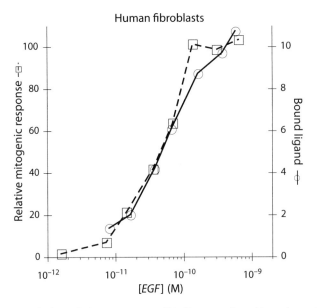

Figure 7.14 Similarity of mitogenic dose response and binding curve for epidermal growth factor (EGF) on the cell surface [30]. The mitogenic response (DNA synthesis by radiolabeling) at steady state was measured following 22 hours of exposure to ligand. The binding curves are not equilibrium isotherms but steady-state surface ligand values in the presence of endocytic trafficking of both the receptor and the receptor-ligand complexes (see chapter 5).

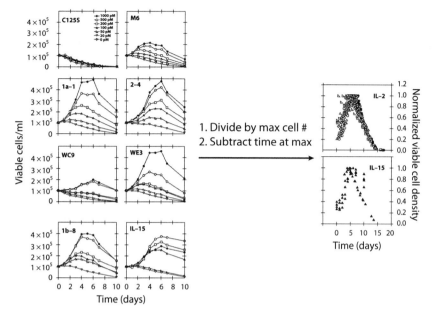

Figure 7.15 A simple growth/death switch may underlie apparently complex growth responses. On the left, the growth of an IL-2-dependent human T cell line was measured following pulse exposures to IL-2, IL-15, or a variety of IL-2 mutants with varying receptor-binding affinity [33]. Despite the seemingly wide variety of responses, normalizing the cell densities by the maximum and then centering the time axis at the time of maximum cell density reveals that the growing cultures had very similar doubling times (≈ 2 days), and once the culture began to die off, the half-time for death was ≈ 4 days.

In other cases, the growth decision appears binary, once a minimum threshold of integrated signal has accumulated. The response of T cells to the growth factor IL-2 appears to be of this type [31], with cell-to-cell random variation in the number of receptors smoothing out the apparent dose-response relationship between growth and IL-2. This phenomenon has been generalized into a hypothesis that a minimum ligand-complex signaling threshold is required to drive self-renewal divisions versus differentiation in many cell types [32]. A simple binary renew/death decision was shown in the response of an IL-2-dependent T cell line (figure 7.15) to a variety of kinetic mutants of IL-2.

Many growth factors stimulate cells via homodimerization of two receptor subunits on the cell surface [34]. As shown in chapter 3, such crosslinking equilibria exhibit a maximum with respect to soluble ligand concentration; consequently, the dose responses to many growth factors similarly exhibit such a maximum (figure 7.16). The activity inhibition at high growth-factor concentrations in such plots may never be realized under physiological conditions. Instead it is an artifact of the extremely high concentrations in some in vitro assays and indicates the importance of receptor crosslinking in the stimulated biological activity being measured.

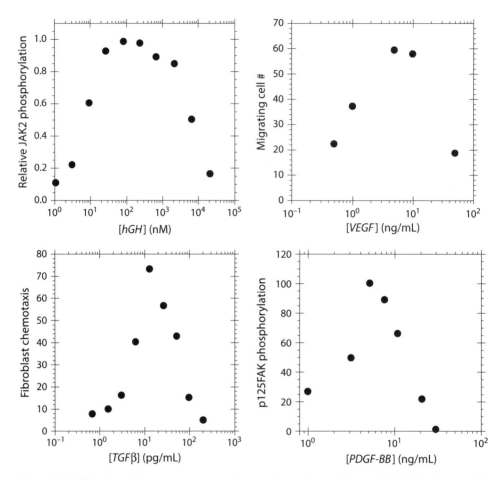

Figure 7.16 Bell-shaped dose response curves for a number of growth factors. Such behavior typically indicates a maximum in growth-factor-mediated receptor crosslink formation at the maximum of the curve. JAK2 phosphorylation versus human growth hormone (hGH) [35]; migration versus vascular endothelial growth factor (VEGF) [36]; chemotaxis versus transforming growth factor β (TGFβ) [37]; tyrosine phosphorylation versus platelet-derived growth factor isoform BB (PDGF-BB) [38].

7.2.3 Toxic Metabolite Accumulation

As cells metabolize growth substrates, they produce waste by-products that are toxic in sufficient concentration. In a multicellular organism, waste is carried away by bodily fluids and excreted, and in natural ecosystems, waste diffuses or is swept away by flow. However, in contained volumes with high cell densities, as is often the case for microbial or animal cell bioreactors, these wastes present a limitation on increasing cell density to desired high levels.

Table 7.2 Inhibitory concentrations of metabolic by-products.

Species	Metabolite	Range of IC_{50}
S. cerevisiae [39,40]	Ethanol	100–900 mM
Clostridium thermosaccharolyticum [41]	Ethanol	900 mM
E. coli [42]	Acetate	100 mM
Mouse hybridoma [43]	Ammonia	2–10 mM
Mouse hybridoma [43]	Lactate	20–55 mM

Typical inhibitory concentrations of several metabolites are shown in table 7.2, represented as IC_{50}, the inhibitory concentration at which growth is reduced by 50%. Ethanol is a product of anaerobic fermentation in yeast and occurs in aerated cultures when a fermentable carbon source is present in excess of the respiratory capacity. When glucose is fed to bacterial cultures in excess of their oxidative respiratory capacity, it is converted to weak acids, predominantly acetate, which are toxic due to decoupling of the intracellular pH gradient relative to the extracellular fluid. Animal cells in culture produce ammonia and lactate, which on accumulation slow cell growth.

When designing and operating bioreactors, it is useful to have a condensed mathematical representation of the effects of environmental parameters on cell growth. Consequently, metabolic by-product growth inhibition has been represented by different mathematical forms chosen for convenience and not mechanistically based. Examples of these functional forms and the data they represent are shown in figure 7.17.

To a first approximation, the effects of different growth inhibitors and limiting nutrients can be treated as independent in bioreactor design. For example, the effects of glucose, glutamine, ammonium, and lactate on the growth rate of hybridomas in continuous culture were found to fit reasonably well [44] to the following functional form:

$$\mu = \mu_{max} \left(\frac{[Gluc]}{K_{Gluc} + [Gluc]} \right) \left(\frac{[Gln]}{K_{Gln} + [Gln]} \right) \left(\frac{K_{NH4+}}{K_{NH4+} + [NH_4^+]} \right) \left(\frac{K_{Lac}}{K_{Lac} + [Lac]} \right)$$

(7.37)

with $\mu_{max} = 1.5$ day^{-1}, $K_{Gluc} = 0.15$ mM, $K_{Gln} = 0.15$ mM, $K_{NH4+} = 20$ mM, and $K_{Lac} = 140$ mM.

A similar empirical relationship was found by other investigators [29]:

$$\mu = \mu_{max} \left(\frac{[Gln]}{K_{Gln} + [Gln]} \right) \left(\frac{[FCS]}{K_{FCS} + [FCS]} \right) \left(\frac{K_{NH4+}}{K_{NH4+} + [NH_4^+]^2} \right) \left(\frac{K_{Lac}}{K_{Lac} + [Lac]^2} \right)$$

(7.38)

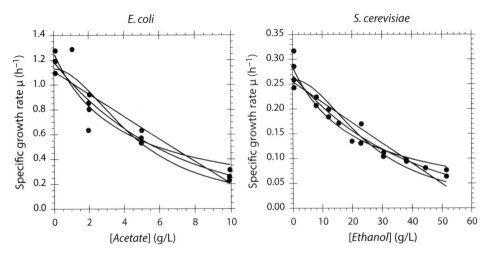

Figure 7.17 Four different rate forms are plotted for two inhibitory metabolic by-products (acetate and ethanol), with parameters fit by nonlinear least squares: $e^{-k[I]}$, $(1-k[I])$, $\frac{K_I}{K_I+[I]}$, or $\frac{K_I^2}{K_I^2+[I]^2}$, where $[I]$ is the concentration of the by-product. The scatter in the data is sufficient to obscure any meaningful differences among the alternative functional forms representing inhibition. Data on acetate inhibition of *E. coli* growth from Luli and Strohl [42]; data on ethanol inhibition of yeast growth from Aiba et al. [40].

where FCS is fetal calf serum and $\mu_{max} = 1.5$ day^{-1}, $K_{FCS} = 1.6\%$, $K_{Gln} = 0.15$ mM, $K_{NH4+} = 45$ mM^2, and $K_{Lac} = 12{,}000$ mM^2.

7.3 Bioreactors

Humankind has harnessed cellular biosynthetic potential for millennia by growing cells in vessels, beginning with alcoholic fermentation and leading to pharmaceutical production and replacement-tissue generation. Empirical models of cell growth play a central role in the design and operation of such bioreactors.

7.3.1 Cell Growth as a Pseudochemical Reaction

It is often important in bioprocess design and operation to predict, or infer from indirect experimental measurements, the extent of conversion of nutrients into cellular biomass and extracellular products. Cells synthesize more cells by drawing carbon, hydrogen, oxygen, and nitrogen (as well as sulfur, phosphorus, and other trace elements) from their surroundings. This process can be framed as a pseudochemical reaction:

$$\text{Carbon source} + O_2 + \text{Nitrogen source} \rightarrow \text{Biomass} + \text{By-products} + H_2O + CO_2 \tag{7.39}$$

For the case of a culture using ammonia as the nitrogen source, equation 7.39 can be written as follows:

$$CH_aO_b + \alpha O_2 + \beta NH_3 \rightarrow \delta CH_cO_dN_e + \varepsilon CH_fO_gN_h + \zeta H_2O + \eta CO_2 \quad (7.40)$$

Cellular biomass is represented in equation 7.40 as the chemical pseudospecies $CH_cO_dN_e$. Data on the elemental composition of a wide variety of microbial cells display the following ranges [25,45]: hydrogen, $c = 1.64$ to 2.00 (1.80 ± 0.09); oxygen, $d = 0.27$ to 0.60 (0.49 ± 0.07); and nitrogen, $e = 0.10$ to 0.26 (0.20 ± 0.04). Consequently, a good estimate for the elemental composition of microbial cellular biomass is reflected by the empirical chemical formula $CH_{1.8}O_{0.5}N_{0.2}$.

Of course, cells also contain numerous other components not accounted for here. For example, 3–10% of biomass, called ash, remains as a residue of combustion and is composed primarily of inorganic oxides. Furthermore, this empirical chemical formula for biomass should be viewed only as a crude approximation, because cellular composition can vary considerably with changing growth media and conditions.

For a given situation where the carbon source and by-products are specified (i.e., a–h in equation 7.40 are known), one can solve for the coefficients $\alpha, \beta, \delta, \varepsilon, \zeta,$ and η by performing atomic balances on carbon, hydrogen, oxygen, and nitrogen:

$$C: 1 = \delta + \varepsilon + \eta \quad (7.41)$$

$$H: a + 3\beta = \delta c + \varepsilon f + 2\zeta \quad (7.42)$$

$$O: b + 2\alpha = \delta d + \varepsilon g + \zeta + 2\eta \quad (7.43)$$

$$N: \beta = \delta e + \varepsilon h \quad (7.44)$$

Unfortunately, this provides only four equations for specifying the six unknown variables $\alpha, \beta, \delta, \varepsilon, \zeta,$ and η. One additional relationship often obtained experimentally is the *respiratory quotient* (RQ), defined as the ratio of carbon dioxide production to oxygen consumption:

$$RQ = \frac{\eta}{\alpha} \quad (7.45)$$

The parameter RQ is commonly used as an output variable to be maintained by feedback control of a nutrient feed stream in microbial fermentation.

An additional relationship for specification of the variables in equations 7.41–7.44 can be obtained by balancing the available electrons in each of the species of equation 7.40. The degree of reduction γ of a particular molecule is defined as the number of electrons available for oxidation and is straightforwardly calculated by summing the following values: 4 for each carbon, 1 for each hydrogen,

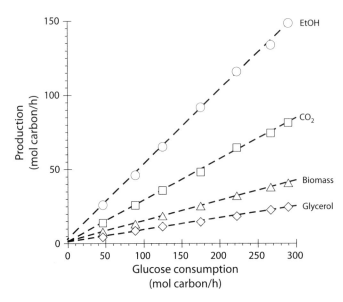

Figure 7.18 Conversion of glucose into ethanol (EtOH), carbon dioxide, biomass, and glycerol during anaerobic continuous culture of the yeast *S. cerevisiae*. The dashed lines are linear least squares fits, which intersect close to the origin. Consequently, across this full range of glucose consumption rates, a proportionality constant, or yield coefficient, is applicable for the molar conversion of carbon for each product (0.50 for EtOH, 0.28 for CO_2, 0.14 for biomass, and 0.08 for glycerol). Note that these fractions sum to 1.0, a circumstance referred to as a "closed material balance," indicating that the major carbon-containing products have been quantitatively accounted for [25].

−2 for each oxygen, and −3 for each nitrogen in the chemical formula. By this algorithm, $\gamma_{NH_3} = \gamma_{H_2O} = \gamma_{CO_2} = 0$. The average cellular composition represented by the empirical formula $CH_{1.8}O_{0.5}N_{0.2}$ yields a biomass degree of reductance $\gamma_b = 4.2$. Performing an available electron balance on equation 7.40 gives the following relationship:

$$(4 + a - 2b) - 4\alpha = 4.2\delta + \varepsilon(4 + f - 2g - 3h) \tag{7.46}$$

The degree of reductance balance is redundant with the collective elemental balances. The balances, RQ, and one other piece of information can be used to specify the coefficients α, β, δ, ε, ζ, and η.

To the extent that equation 7.40 provides an adequate description of cell growth, one can expect that the amount of biomass, extracellular products, and CO_2 produced will be proportional to the mass of growth substrate consumed—in fact, often a good approximation, as shown in figure 7.18.

In given growth conditions, the proportionality constant between products and feeds is referred to as a yield coefficient:

$$Y_{A/B} = \frac{\Delta A}{\Delta B} \qquad (7.47)$$

where A is a product (e.g., biomass, ethanol, glycerol, carbon dioxide, ATP, heat), and B is a feed (e.g., glucose, other carbon sources, oxygen). Yield coefficients are often expressed on a mass-per-mass basis, but to spare confusion, it is a good policy to always explicitly provide the units when presenting yield coefficients.

Yield coefficients have been collected for aerobic cultures of a variety of microbes with glucose as the carbon source [46]. The average glucose yield coefficient is $Y_{X/S} = 0.6 \pm 0.1$ (grams biomass produced)/(grams glucose consumed), with a range of 0.32–0.80 for 27 different published values. The average yield coefficient for biomass formation from oxygen consumption (with glucose as the carbon source) is $Y_{X/O_2} = 1.9 \pm 0.7$ (grams biomass produced)/(grams oxygen consumed), with a range of 0.53–3.57 for 27 independent reports. The yield coefficient $Y_{X/S}$ decreases for substrates with decreasing degree of reductance γ_S (figure 7.19).

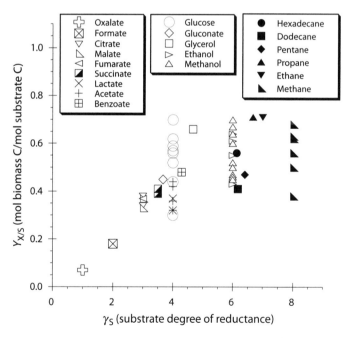

Figure 7.19 Approximately half of the carbon atoms from most growth substrates are incorporated into biomass. Efficiency is lower for substrates with lower degrees of reductance [47].

Example 7-2 Yeast is grown with glucose and ammonia to generate a recombinant protein ($CH_{1.5}O_{0.3}N_{0.3}$). The respiratory quotient is 0.2. Determine the reaction stoichiometry if the yield coefficient for biomass from glucose is 0.6 g/g.

Solution From equation 7.40, we have

$$C_6H_{12}O_6 + \alpha O_2 + \beta NH_3 \rightarrow \delta CH_{1.8}O_{0.5}N_{0.2} + \varepsilon CH_{1.5}O_{0.3}N_{0.3}$$
$$+ \zeta H_2O + \eta CO_2 \quad (7.48)$$

We can then write the elemental balances:

$$C: 6 = \delta + \varepsilon + \eta \quad (7.49)$$
$$H: 12 + 3\beta = 1.8\delta + 1.5\varepsilon + 2\zeta \quad (7.50)$$
$$O: 6 + 2\alpha = 0.5\delta + 0.3\varepsilon + \zeta + 2\eta \quad (7.51)$$
$$N: \beta = 0.2\delta + 0.3\varepsilon \quad (7.52)$$

with additional constraints for the respiratory quotient

$$\frac{\eta}{\zeta} = 0.2 \quad (7.53)$$

and the yield coefficient

$$\frac{\delta (24.6 \text{ g/mol})}{1(180 \text{ g/mol})} = 0.6 \quad (7.54)$$

This series of equations can be written in matrix form:

$$\begin{bmatrix} 0 & 0 & 1 & 1 & 0 & 1 \\ 0 & -3 & 1.8 & 1.5 & 2 & 0 \\ -2 & 0 & 0.5 & 0.3 & 1 & 2 \\ 0 & -1 & 0.2 & 0.3 & 0 & 0 \\ 0 & 0 & 0 & 0 & 0.2 & -1 \\ 0 & 0 & 24.6 & 0 & 0 & 0 \end{bmatrix} \times \begin{bmatrix} \alpha \\ \beta \\ \delta \\ \varepsilon \\ \zeta \\ \eta \end{bmatrix} = \begin{bmatrix} 6 \\ 12 \\ 6 \\ 0 \\ 0 \\ 108 \end{bmatrix} \quad (7.55)$$

The solution to this system of equations reveals the following stoichiometry:

$$C_6H_{12}O_6 + 0.39 \, O_2 + 1.2 \, NH_3 \rightarrow 4.4 \, CH_{1.8}O_{0.5}N_{0.2}$$
$$+ 1.0 \, CH_{1.5}O_{0.3}N_{0.3} + 3.1 \, H_2O + 0.61 \, CO_2 \quad (7.56)$$

Population Growth and Death Models

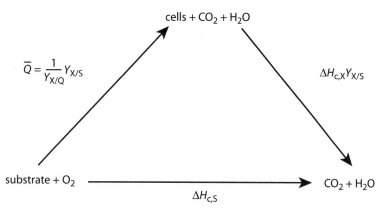

Figure 7.20 Metabolic heat generation can be evaluated via analysis of substrate consumption and combustion.

As with all conversion of energy from one form to another, the process of substrate metabolism is not perfectly efficient and so leads to the loss of some fraction of the available energy as heat. In bioreactors with high cell density, this heat represents a significant practical issue, because it must be removed to avoid overheating and consequent cell death. A yield coefficient for cell mass produced per heat released can be defined as follows:

$$Y_{X/Q}(\text{g biomass/kcal}) \equiv \frac{\Delta X}{Q_{\text{metab}}} \qquad (7.57)$$

where Q_{metab} is the heat released (kcal) due to metabolic inefficiency. The released heat per gram of substrate ($\bar{Q}$, kcal/g substrate) can be calculated by subtracting the heat of combustion of biomass from that of the substrate (figure 7.20):

$$\bar{Q} = \frac{1}{Y_{X/Q}} Y_{X/S} = \Delta H_{c,S} - \Delta H_{c,X} Y_{X/S} \qquad (7.58)$$

where $\Delta H_{c,i}$ is the heat of combustion of component i. The heat of combustion of biomass ranges from 5 to 6 kcal/(gram dry weight), with an average of 5.4. Thus, if one knows the heat of combustion of the growth substrate and the yield coefficient, the heat liberated per biomass growth can be calculated from equation 7.57. For a variety of growth substrates, it has been found that approximately 110 kcal of heat are evolved per mol of O_2 consumed in aerobic metabolism (figure 7.21 [48]). Therefore, a crude estimate for heat evolution can be obtained as $Y_{X/Q}(\text{g biomass/kcal}) = 0.3 Y_{X/O_2}(\text{g biomass/g oxygen})$.

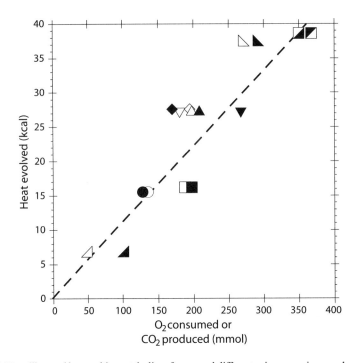

Figure 7.21 Heat liberated by aerobic metabolism for several different microorganisms and carbon sources. The dotted line represents the relationship 110 kcal per mol O_2 consumed. Open symbols are heat produced per mol CO_2 produced, and closed symbols are heat produced per mol O_2 consumed. The following symbols correspond to microbe and carbon source, respectively: circles, *Candida intermedia* and glucose; squares, *Aspergillus niger* and glucose; diamonds, *Bacillus subtilis* and glucose; downward triangles, *Bacillus subtilis* and molasses; upward triangles, *E. coli* and glucose; leftward wedges, *Candida intermedia* and molasses; rightward wedges, *Bacillus subtilis* and soybean meal medium; half-squares, *Aspergillus niger* and molasses [48].

Direct measurement of yield coefficients is complicated by the fact that the carbon source is metabolized to provide not only mass and energy for cell growth but also energy for maintenance (e.g., maintaining transmembrane pH, ionic, and osmotic gradients). Maintenance substrate-consumption requirements are generally expressed in terms of a constant m_S (mass substrate consumed/mass cell dry weight (CDW)/h). A survey of 59 published measurements for m_S for microbial monocultures showed a mean value of 0.055 ± 0.048 (g substrate/g CDW/h), with a range of 0.003–0.28 [46]. The maintenance requirement for oxygen consumption, m_{O_2}, was found to range from 0.0025 to 0.31 (g O_2/g CDW/h), with a mean of 0.052 ± 0.054 for 53 observations.

Assuming that the amount of energy required per unit of biomass growth is constant, one can lump together the requirements of substrate consumption for growth energy and anabolic metabolism (excluding consumption for maintenance) into one term, ΔS_{growth}:

$$\Delta S_{\text{growth}} = \frac{1}{Y_{X/S}} \Delta X \qquad (7.59)$$

In a batch culture, substrate consumption for maintenance requirements is given by the integral

$$\Delta S_{\text{maintenance}} = m_S \int_0^t X(+')dt' \qquad (7.60)$$

The apparent yield coefficient is perturbed by consumption of substrate for maintenance requirements as follows:

$$Y_{X/S,\text{apparent}} = \frac{\Delta X}{\Delta S_{\text{growth}} + \Delta S_{\text{maintenance}}} \qquad (7.61)$$

Rearranging gives

$$\frac{1}{Y_{X/S,\text{apparent}}} = \frac{1}{Y_{X/S}} + \frac{m_S \int_0^t X(+')dt'}{\Delta X} \qquad (7.62)$$

For gross estimation purposes, it is often useful to have rough estimates for the growth parameters described in this section. In table 7.3, consensus parameter estimates for the two workhorse microbes *E. coli* and *S. cerevisiae* have been put together from the most commonly used bioprocess engineering texts (see Suggestions for Further Reading at the end of the chapter).

7.3.2 Steady-State Monod Chemostat

A *chemostat* is a microbial bioreactor operated continuously at a steady state. Consider a well-mixed aerated vessel holding a growing microbial population in a volume V. Fresh growth medium is added to the vessel at a constant rate F (volume/time), and culture fluid is withdrawn at the same rate F, so V is constant (figure 7.22). The growth limiting substrate S is fed at concentration $[S]_o$.

Table 7.3 Consensus growth parameter estimates for yeast and bacteria.

	$\mu_{\max}$ (h^{-1})	K_s (mg/L)	$Y_{X/S}$ (g/g)	Y_{X/O_2} (g/g)	$Y_{X/Q}$ (g CDW/g)	m_S g/(g CDW · h)	m_{O_2} g/(g CDW · h)
E. coli (37°C, glucose)	0.9	3	0.4	0.4	0.12	0.057	0.014
S. cerevisiae (30°C, glucose)	0.6	25	0.5	0.9	0.12	0.017	0.019

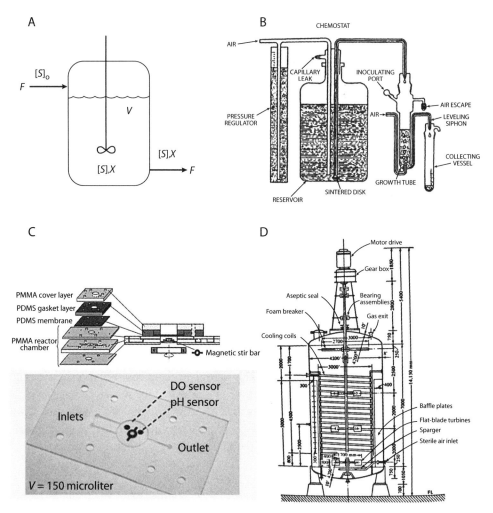

Figure 7.22 Microbial bioreactors have been designed and operated at volumes that span nine orders of magnitude. (A) Steady-state chemostat schematic, identifying variables in the model. (B) Diagram of the first reported chemostat [49]. Agitation was provided by bubbles rising through the culture volume, and *E. coli* were grown with tryptophan as the limiting substrate. (C) 150 μL working volume chemostat applicable for continuous chemostat culture of *E. coli* and *S. cerevisiae*, with operating parameters (pH, dissolved oxygen (DO), and optical density) demonstrated to be identical to those for that a 0.5 L instrumented fermentor (from reference [50]). (D) Schematic diagram of a 100,000 L fermentor used for penicillin production. Linear dimensions are in mm, and the overall height is 14 m [51].

Accumulation = Biomass inflow − Biomass outflow + Biomass growth (7.63)

$$V\frac{dX}{dt} = FX_{in} - FX + \mu XV \quad (7.64)$$

The feed stream is usually cell-free ($X_{in} = 0$) for convenience of operation, and the reactor vessel is initially seeded with the microbial culture of interest before beginning flow through the reactor. When the system reaches steady state ($dX/dt = 0$), the production of new cells by growth will equal the removal of biomass X by efflux:

$$\mu XV = FX \quad (7.65)$$

Simplifying, and substituting in the *dilution rate* $D \equiv F/V$, which has units of inverse time, results in

$$D = \mu \quad (7.66)$$

A remarkable aspect of the steady-state biomass material balance equation 7.66 is that the left-hand side reflects fluidics (culture volume and flow rates), whereas the right-hand side reflects biology (specific growth rate)—because at steady state, by definition, the cells grow at a rate specified by the reactor operating conditions. The ability to mechanically specify a biological state is a particular advantage of chemostat experiments—the precise and reproducible operator specification of cellular physiology. This advantage is offset somewhat by the technical demands of operating an instrumented sterile bioreactor for extended periods without contamination by adventitious microbes.

We can further specify the operating conditions by constructing a steady-state material balance on the growth substrate S:

Accumulation = Flow in − Flow out − Cell consumption of substrate (7.67)

$$0 = F[S]_o - F[S] - \frac{1}{Y_{X/S}}\mu XV \quad (7.68)$$

Substituting equation 7.65 into equation 7.68, one obtains

$$X = Y_{X/S}([S]_o - [S]) \quad (7.69)$$

However, we have not yet determined the value of $[S]$, which will be less than $[S]_o$ due to cell consumption of substrate. Substituting Monod dependence of specific growth rate on substrate concentration $\mu([S])$ into equation 7.66 gives

$$D = \frac{\mu_{max}[S]}{K_s + [S]} \quad (7.70)$$

which can be rearranged to give

$$[S] = \frac{K_s D}{\mu_{max} - D} \quad (7.71)$$

Substituting equation 7.71 into equation 7.69 results in

$$X = Y_{X/S} \left([S]_o - \frac{K_s D}{\mu_{max} - D} \right) \quad (7.72)$$

Equations 7.71 and 7.72 completely specify the chemostat conditions $(X, [S])$ as a function of the operating conditions $(D, [S]_o)$ and cellular growth parameters (μ_{max}, K_s). Note that equation 7.72 formally allows the potential for physically unrealizable negative values for biomass X, unless $[S]_o - \frac{K_s D}{\mu_{max} - D} \geq 0$. This constraint can be rearranged to provide the maximum dilution rate D_{max} that allows a nonzero steady-state biomass:

$$D_{max} = \frac{\mu_{max}[S]_o}{K_s + [S]_o} \quad (7.73)$$

When $D \geq D_{max}$, the chemostat is in *washout*, because the cells cannot grow rapidly enough to keep up with the efflux, and the steady-state solution is $X = 0$.

Let's nondimensionalize these results to examine their qualitative behavior. Define the dimensionless dilution rate, substrate concentration, and biomass concentration as follows:

$$\hat{D} \equiv \frac{D}{D_{max}} \quad (7.74)$$

$$\hat{S} \equiv \frac{[S]}{[S]_o} \quad (7.75)$$

$$\hat{X} \equiv \frac{X}{Y_{X/S}[S]_o} \quad (7.76)$$

The dimensionless forms of equations 7.71 and 7.72 are

$$\hat{S} = \frac{\alpha \hat{D}}{1 + \alpha - \hat{D}} \quad (7.77)$$

$$\hat{X} = 1 - \hat{S} \quad (7.78)$$

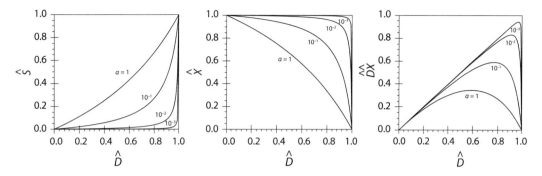

Figure 7.23 Steady-state Monod chemostat behavior as a function of dimensionless dilution rate $\hat{D}$.

where

$$\alpha \equiv \frac{K_s}{[S]_o} \quad (7.79)$$

The dependence of biomass and substrate concentrations on dilution rate is illustrated in figure 7.23. Recall from table 7.3 that the Monod parameter K_s is typically much smaller than the initial concentration of the limiting growth medium substrate $[S]_o$, such that $\alpha \ll 1$. For small values of α, substrate concentration is almost zero until $D \to D_{max}$, but the biomass concentration is near its maximal value $Y_{X/S}[S]_o$ for most dilution rates until $D \to D_{max}$.

If the biomass itself is the desired product, or if the desired product is produced at a rate proportional to biomass concentration, at what dilution rate should a chemostat be operated? The overall rate of biomass production is FX/V, or DX. By inspection of the right panel of figure 7.23, the productivity DX would be expected to increase with increasing D, right up to a point near D_{max}, at which $DX \to 0$. Consequently, one would expect there to be a value of the dilution rate D_{opt} at which the productivity DX would be maximal. The optimal dilution rate can be determined by taking the derivative of productivity with respect to D and setting it to zero. In other words, $D = D_{opt}$ at $\frac{d(DX)}{dD} = 0$. With a bit of algebra, the following value for the optimal dilution rate is obtained:

$$D_{opt} = \mu_{max}\left(1 - \sqrt{\frac{K_s}{K_s + [S]_o}}\right) \quad (7.80)$$

Because $K_s \ll [S]_o$ in most cases, $D_{opt} \approx \mu_{max}$. However, operating a real-world chemostat at this point could be practically problematic, because a small system fluctuation could result in washout of the culture.

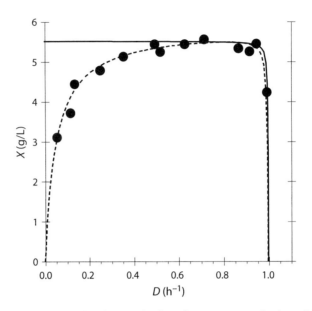

Figure 7.24 Continuous chemostat data for growth of *Aerobacter aerogenes* in glycerol (10 g/L feed). The solid line is for the Monod chemostat model with $K_s = 0.01$ g/L, $Y_{X/S} = 0.55$, and $\mu_{max} = 1.0$ h^{-1}. The dashed line is for the chemostat model incorporating a term for maintenance consumption of substrate, with the least squares fit parameters to equation 7.82 as: $K_s = 0.025$ g/L, $Y_{X/S,true} = 1.7$, $\mu_{max} = 1.0$ h^{-1}, and $m = 0.086$ g glycerol/g biomass/h. Data from references 52 and 25.

The Monod chemostat model breaks down when the chemostat is run at very slow dilution rates. Equation 7.72 predicts that $X \to Y_{X/S}[S]_o$ as $D \to 0$; however, in actual practice, biomass concentration drops as the feed rate approaches zero, as shown in figure 7.24. One simple explanation for this phenomenon would be that at low flow rates, biomass levels are limited by the substrate consumption requirement for cellular maintenance. Incorporating the maintenance term into the substrate balance equation gives

$$\frac{d[S]}{dt} = D([S]_o - [S]) - \frac{1}{Y_{X/S}}\mu X - mX \tag{7.81}$$

or, setting the derivative to zero at steady state:

$$X = \frac{D([S]_o - [S])}{m + \frac{D}{Y_{X/S}}} \tag{7.82}$$

The biomass balance is unchanged by the maintenance requirement and results again in equation 7.72.

An observed yield coefficient as a function of dilution rate can be derived as follows:

$$Y_{X/S,obs} = \frac{X}{[S]_o - [S]} \tag{7.83}$$

$$= \frac{DY_{X/S}}{D + mY_{X/S}} \tag{7.84}$$

Analogously to equation 7.62, we can rewrite this as

$$\frac{1}{Y_{X/S,obs}} = \frac{1}{Y_{X/S}} + \frac{m}{D} \tag{7.85}$$

7.3.3 Fed-Batch Fermentors

To obtain very high microbial cell densities (50–150 grams cell dry weight per liter) in a bioreactor, a proportionally large amount of carbon source must be fed: $\Delta X = Y_{X/S}\Delta[S]$, or 5–15% sugar by weight. However, extremely high concentrations of sugar raise liquid viscosities to impractical levels and are toxic to microorganisms due to overflow metabolic by-product toxicity (section 7.2.3) and osmotic stress. Rather than start with a very high $[S]_o$, a practical strategy is to add carbon source to the reactor at a steady and defined rate without drawing off any liquid stream. This approach is called fed-batch fermentation [53].

In a fully instrumented fermentor, the carbon-source feed rate can be used as the manipulated variable in a feedback control loop to maintain a given setpoint as an indirect measure of cellular metabolic rate. That setpoint may be the dissolved oxygen concentration, pH, or the calculated respiratory quotient (RQ). If such instrumentation is not available, one can instead calculate a feeding rate to maintain a constant specific growth rate μ. By selecting a sufficiently low specific growth rate (e.g., 0.10–0.15 h^{-1} for *E. coli*), the formation of inhibitory metabolic by-products can be minimized.

Example 7-3 What volumetric flow rate F of carbon source at concentration $[S]_o$ should be fed as a function of time to maintain a constant specific growth rate μ? At time zero, biomass concentration is $X = X_o$, and the liquid volume in the reactor is $V = V_o$.

Solution If total reactor biomass grows exponentially, then

$$VX = V_o X_o e^{\mu t} \tag{7.86}$$

As the reactor volume increases, the total amount of substrate must increase to maintain the constant substrate concentration requisite for constant specific growth rate:

$$\frac{d}{dt}(V[S]) = F[S]_o - \frac{1}{Y_{X/S}}\frac{d}{dt}(VX) \qquad (7.87)$$

Consider steady substrate concentration and volumetric change equivalent to feed ($dV/dt = F$):

$$F([S]_o - [S]) = \frac{1}{Y_{X/S}}\mu XV \qquad (7.88)$$

In most scenarios, substrate is fed at high concentration. Thus, $[S]_o \gg [S]$. Simplifying and applying equation 7.86, we have

$$F[S]_o = \frac{\mu XV}{Y_{X/S}} = \frac{\mu X_o V_o e^{\mu t}}{Y_{X/S}} \qquad (7.89)$$

Writing a volumetric balance gives

$$\frac{dV}{dt} = F \qquad (7.90)$$

$$= \frac{\mu X_o V_o e^{\mu t}}{[S]_o Y_{X/S}} \qquad (7.91)$$

which on integration yields

$$V = V_o\left(1 + \frac{X_o}{[S]_o Y_{X/S}}(e^{\mu t} - 1)\right) \qquad (7.92)$$

Substituting this expression for V into equation 7.86 gives the biomass concentration as a function of time:

$$X = \frac{X_o e^{\mu t}}{1 + \frac{X_o}{[S]_o Y_{X/S}}(e^{\mu t} - 1)} \qquad (7.93)$$

Note that although the cells are growing at a constant specific growth rate μ, cell density does not increase exponentially, because the increasing liquid volume in the reactor dilutes the cell mass (the denominator term in equation 7.93). Note also that equation 7.93 is functionally identical to logistic growth (equation 7.14) with a carrying capacity $K \equiv [S]_\circ Y_{X/S}$. In fact, the logistic equation has been found to be particularly suitable for a description of mammalian cell fed-batch growth cultures [54].

7.3.4 Product-Formation Kinetics

When biomass is not itself the product of interest, it is useful for bioprocess design purposes to obtain a simple kinetic description of product accumulation. In microbial fermentation, the two general categories of product are those formed during cell growth and those formed independently of cell growth. The former category of growth-associated products includes those that arise directly from constitutive enzymatic processes linked to cellular energy metabolism and central carbon metabolism—for example, ethanol. For such products, the specific rate of accumulation is simply the culture-specific growth rate ($\mu = \frac{1}{X}\frac{dX}{dt}$) times the constant amount of product produced per unit biomass (α):

$$\frac{1}{X}\frac{d[P]}{dt} = \frac{d[P]}{dX}\left(\frac{1}{X}\frac{dX}{dt}\right) = \alpha\mu \qquad (7.94)$$

The biosynthesis of secondary metabolites, such as many antibiotics, is often induced once cells enter stationary phase and can be modeled by the simple relationship

$$\frac{1}{X}\frac{d[P]}{dt} = \beta \qquad (7.95)$$

Some products appear to have mixed kinetics, with both a growth-associated and a growth-independent component:

$$\frac{1}{X}\frac{d[P]}{dt} = \alpha\mu + \beta \qquad (7.96)$$

This type of product-formation kinetics was first described by Luedeking and Piret [55] for lactic acid fermentation by the bacterium *Lactobacillus acidophilus* (figure 7.25).

Monoclonal antibodies are often produced industrially by secretion from Chinese hamster ovary (CHO) cells. Although the viable cell count peaks and then drops during batch cultivation, antibody product accumulation in the culture supernatant

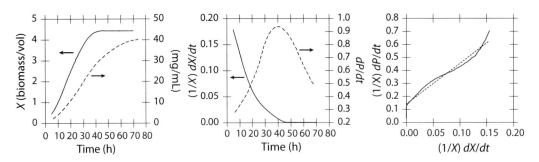

Figure 7.25 The product acetic acid is formed by both growth-associated and growth-independent kinetics in fermentations of *Lactobacillus acidophilus* [55]. In the left panel, cell accumulation and product accumulation are shown. In the middle panel, the specific growth rate and rate of product formation are plotted. In the right panel, the specific rate of product formation is plotted versus the specific biomass growth rate (solid line). A linear curve fit (dashed line) gives the slope of β and the intercept of α in equation 7.96.

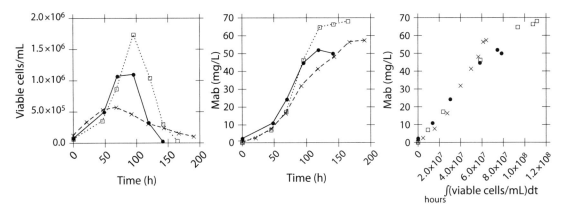

Figure 7.26 Monoclonal antibody (Mab) production kinetics by a hybridoma line under varying conditions (data from reference [57]). Three independent experiments are represented by three different symbols (closed circle, open square, cross). In all three experiments, the viable cell number reaches a maximum and declines midway through the experiment (left panel), while Mab continues to accumulate (center panel). Plotting Mab concentration versus the integral of viable cell number yields a straight line for all three experiments (right panel), with slope equal to the specific Mab productivity as per equation 7.97.

continues during the decline in cell viability (figure 7.26). Assuming that antibody secretion is not growth associated and integrating equation 7.95, one obtains:

$$[P] = \beta \int_0^t X dt \qquad (7.97)$$

where X represents viable cell number. A plot of antibody concentration versus viable cell integral should produce a straight line with the slope equal to β. This relationship appears to be closely followed in actual cell culture conditions for antibody production (figure 7.26 [56,57]). Typical optimized productivities for antibody production are approximately 100 pg/cell/day [58].

When the product itself is a growth inhibitor, one must account for this growth inhibition when formulating a kinetic description of the bioprocess. Hoppe and Hansford [39] showed that for ethanol production by the yeast *S. cerevisiae*, growth-associated ethanol production led to a specific growth-rate function that was well described by the following relationship:

$$\mu = \mu_{max} \left(\frac{[S]}{K_s + [S]} \right) \frac{K_P}{K_P + Y_{P/S}([S]_f - [S])} \quad (7.98)$$

where $[S]$ is the glucose substrate concentration in the bioreactor, and $[S]_f$ is the feed glucose concentration. The denominator term on the far right substitutes an ethanol concentration predicted by growth-associated kinetics. For their data set, the following parameters were estimated: $\mu_{max} = 0.64 \text{ h}^{-1}$, $K_s = 3.3$ g/L, $K_P = 5.2$ g/L, and $Y_{P/S} = 0.43$ g/g.

7.4 Cell and Organismal Death

The two general types of mortality kinetics correspond to death by injury or death by senescence (colloquially, by "natural causes").

7.4.1 Death by Injury

Single-target kinetics Under some conditions, a constant fraction of individuals in a population perish in each successive time increment, such that the rate of decrease in the population is proportional to the number of individuals still alive:

$$\frac{dN}{dt} = -\gamma N \quad (7.99)$$

Under such conditions, the number of surviving individuals decreases exponentially:

$$N = N_o e^{-\gamma t} \quad (7.100)$$

Some examples of exponential death kinetics are shown in figure 7.27.

A constant *hazard function* γ as in equation 7.99 can result from environmental exposure to predation, parasitism, toxins, radiation, heat, or other insults. As we will see in chapter 9, this system corresponds to a Poisson random process, with a constant probability of a lethal event γdt during each time increment from t to $t + dt$. Because there is no dependence of γdt on history (i.e., γ is a constant), it is the

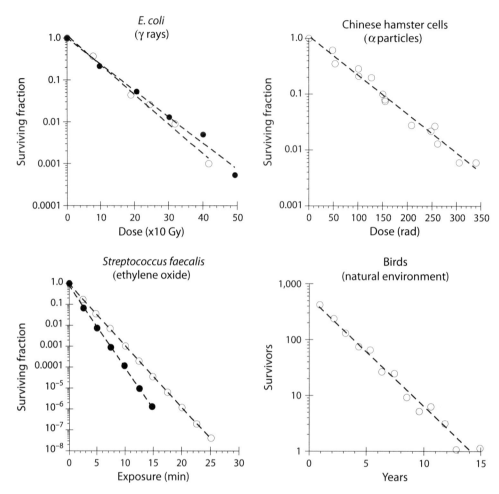

Figure 7.27 Surviving fraction decreases exponentially (equation 7.100) when single random events lead to the death of an individual in a population. This description applies to many situations, with four examples shown here. In the upper-left panel, *E. coli* cells became unviable on exposure to γ irradiation [59]. In the upper-right panel, Chinese hamster cells were killed by exposure to α particles [60]. In the lower-left panel, two strains of *Streptococcus faecalis* were sterilized by exposure to ethylene oxide [61]. In the lower-right panel, banded fledglings of the bird *Vanellus vanellus* were released, and then surviving birds were counted in successive years [62].

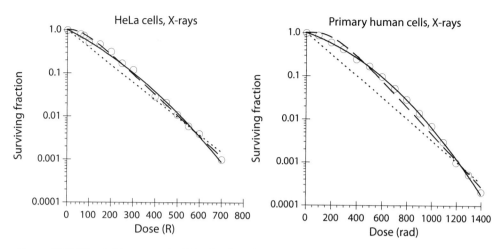

Figure 7.28 Threshold-type survival kinetics for eukaryotic cells exposed to low linear energy-transfer ionizing radiation from X-rays. The dotted line represents the least squares fit to the single-hit model (equation 7.101). The dashed line is the least squares fit to the multitarget single-hit model (equation 7.102). The solid line is the least squares fit to the linear-quadratic model (equation 7.106). Although both the multitarget single-hit and linear-quadratic models are consistent with the data set on the left [66], the linear-quadratic model is most consistent with the data set on the right [64]. (Note that there appears to be less scatter in the data set on the right, allowing this subtle distinction to be made.)

probability of an isolated individual event as opposed to an accumulation of events. Consequently, this form of kinetics is known as a single-target or single-hit model.

Ionizing radiation: Multiple-hit and linear-quadratic models Radiobiologists have found that the primary cause of cell death from ionizing radiation exposure is double-stranded DNA breaks. Haploid microorganisms with only one copy of each gene often show single-hit death kinetics, leading to an exponential dependence of survival on radiation dose (e.g., *E. coli* in figure 7.27). By contrast, diploid microbial or animal cells exhibit a threshold-type survival curve as shown in figure 7.28. Such survival curves have a shoulder at low dose followed by an approximately exponential decline, consistent with a requirement for accumulated damage before cells begin to die. (However, α particles have high linear energy transfer, and a single particle track through the nucleus generally kills even a diploid cell, leading to single-hit death kinetics as shown for diploid CHO cells in figure 7.27.)

Two theories have been developed to describe threshold-type survival curves: target theory and the linear-quadratic model. Multitarget single-hit target theory was developed by Lea [63] based on the following assumptions: (1) there are n targets in each cell, (2) each target has the same probability density p of being hit, and (3) the cell dies when all targets are hit. The survival curve for each target obeys single-hit kinetics, such that the probability that a given target has *not* been hit as a function of radiation dose D is

$$S = e^{-pD} \qquad (7.101)$$

Thus, the probability that all n targets have been hit is $(1 - e^{-pD})^n$. Consequently, the survival curve for the multitarget single-hit theory is

$$S = 1 - (1 - e^{-pD})^n \qquad (7.102)$$

It can be shown that in the limit of large pD, S in equation 7.102 approaches ne^{-pD}. Thus, extrapolating backward from the logarithmic decline at high dosage D, the intercept at zero dose is n, the effective number of targets that must be hit to kill a cell.

An alternative theory that has seen wider acceptance is called both the molecular model and the linear-quadratic model [64]. In this representation, two different mechanisms are assumed to produce double-strand breaks: (1) both strands are broken by one hit and (2) two independent hits cause separate strand breaks in sufficiently close proximity to result in an overall double-strand break. Incorporating constants to account for the probability of strand-break repair and the number of vulnerable sites, and letting $f_1 \equiv$ the fraction of the dose acting through mechanism (1), the average number of double-strand breaks in a given cell by mechanism (1) is

$$n_1 = C_1(1 - e^{-k_1 f_1 D}) \qquad (7.103)$$

and the number of double-strand breaks by mechanism (2) is

$$n_2 = C_2(1 - e^{-k_2(1-f_1)D})^2 \qquad (7.104)$$

We follow the published treatment here, although it might be more correct to replace the use of f_1 with an independent treatment of each mechanism causing a double-strand break. By Poisson statistics, the probability of survival of a given cell is then directly related to the total average number of double-strand breaks $(n_1 + n_2)$:

$$S = e^{-C_3(n_1 + n_2)} \qquad (7.105)$$

For small values of $k_2(1 - f_1)D$ and $k_1 f_1 D$, a truncated Taylor series linearization can be employed to simplify this equation to the following final form, which is called the linear-quadratic model for its algebraic dependence on dose D:

$$S = e^{-\alpha D - \beta D^2} \qquad (7.106)$$

The consistency of the single-hit, multitarget single-hit, and linear-quadratic models with two different data sets for mammalian cell killing by X-rays is shown in figure 7.28. The linear-quadratic model is somewhat more consistent with the data and is generally the base-case model for describing animal-cell killing by ionizing radiation. More detailed models that account for the kinetics of enzymatic DNA repair and the rate of dose delivery have been developed [65].

7.4.2 Death by Senescence

Exponential death kinetics reflects a process where the proportion of individuals that die per time increment is a constant value. However, experience suggests that the risk of death increases with age due to a general accumulation of damage and a lessening capability of an organism to maintain and repair itself. For example, the fraction of 90-year-old humans that will perish in a given year is expected to be higher than the fraction of 50-year olds.

Actuarial data on human mortality from the eighteenth century are shown in figure 7.29. Indeed, the number of surviving individuals does not form a straight line on a semilogarithmic plot, as would be expected for the exponential decay resulting from a constant rate constant γ. Instead the curves are concave downward, suggesting an acceleration in the death rate. A semilogarithmic plot reveals that γ is an exponentially increasing function of time, such that

$$\gamma \equiv -\frac{1}{N}\frac{dN}{dt} = \gamma_\circ e^{\alpha t} \qquad (7.107)$$

This relationship was deduced by Benjamin Gompertz in 1825 and remains to this day the point of departure for any description of the kinetics of mortality by

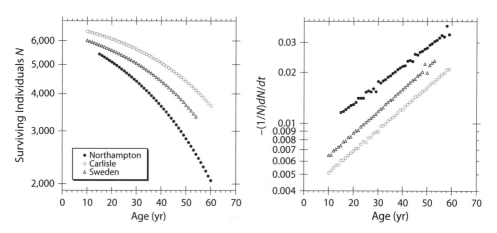

Figure 7.29 The actuarial data from which Benjamin Gompertz deduced equation 7.107 in 1825 [67].

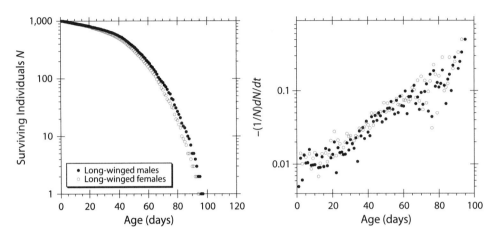

Figure 7.30 Mortality data for *Drosophila* exhibit the same functional form as those for humans. The data shown here were taken a century after Gompertz's observations, and the similarity to human mortality curves was remarked on by the authors as indicative of some potential similarity in the underlying processes [68].

senescence, or old age. The immediate and continuing practical value of equation 7.107 was in the construction of actuarial tables to match insurance premiums with the risk of death. However, such a simple statistically validated empirical relationship additionally places a strong constraint on any potential mechanistic model of senescence.

As it happens, Gompertz death kinetics are surprisingly general: The same exponential increase in the apparent death rate constant γ is observed for other organisms (e.g., the fruit fly *Drosophila* in figure 7.30, and even the baker's yeast *S. cerevisiae* in figure 7.31).

Rapid exponential acceleration of the specific death rate γ carries with it the implication of an effectively limited maximum lifespan due to senescence, even in the absence of risk from external injury. As it happens, the Gompertz relationship breaks down for the very oldest members of a population, as shown in figure 7.32. One plausible explanation for such a mortality rate plateau is a mathematical consequence of random population variation of the parameters of equation 7.107 [70, 71].

Extreme value theory and senescence kinetics The empirically observed kinetics of senescence are consistent with the statistics of *extreme value distributions*, which have been extensively exploited for failure analysis in nonbiological engineering [73–75]. At first glance, this analogy seems to hold promise for mechanistic insight into the process of senescence. However, as we shall see, extreme value distribution

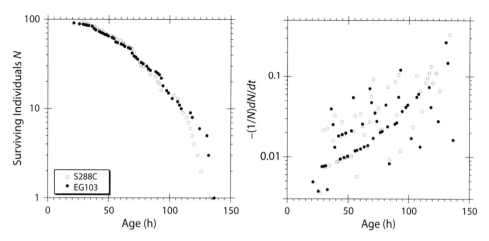

Figure 7.31 Lifespan distribution for two strains of the budding yeast *S. cerevisiae*. Unlike bacteria, which divide by binary fission, yeast cells divide asymmetrically, and a mother cell can be distinguished from a daughter cell, because the mother cell has a *bud scar* where the separation occurs. Mortality, in this case, was determined by the ability of a cell to exclude the dye Phloxine B. An alternative measure of lifespan in yeast is the number of daughter cells produced, which for different strains under different conditions can be an average of 10–30 with a maximum of 30–60. For the data shown here, cells always stopped producing daughter cells prior to loss of viability. Because there are far fewer individuals observed than in the experiments of figures 7.29 and 7.30, there is much more noise in the derivative estimate; clearly, the same trend of increasing $(-1/N)dN/dt$ versus time is observed for yeast cells as for humans and flies [69].

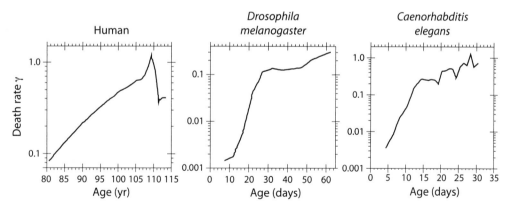

Figure 7.32 Plateau in death rates at advanced ages. In the left panel, death rates from ages 80–122 for human females accumulated for 14 countries for 1950–1997. In the middle panel, death rates for a cohort of 6,338 fruit flies. In the right panel, death rates for a population of 550,000 nematodes. In each case, following 1–2 logs of Gompertzian exponential increase in death rate with time, a plateau death rate appears to be reached [72].

statistics are a good limiting-case approximation for many completely unrelated phenomena.

Imagine drawing a series of n random numbers from an arbitrary continuous probability distribution. Now consider the maximum value from among the random numbers drawn. What is the probability that this maximum assumes a particular value? (Similar findings will hold for the minimum value, which for simplicity we will not consider further.) The rather extraordinary result of extreme value distribution theory is that in the limit of large sample size n, the maximum value will tend to one of only three possible probability distributions: the Weibull, Gumbel/Gompertz, or Frechet distribution. (In the nonbiological literature the term "Gumbel" distribution is most often used, but we will use the term "Gompertz" for consistency with the biological literature.)

A given probability distribution will converge to only one of these three classes of extreme value distribution (Weibull, Gompertz, or Frechet). However, because many different probability distributions converge to each of them, one *cannot obtain* the original probability distribution from the identity or parameters of the extreme value distribution. Thus, for example, it is impossible to identify a mechanism for the cause of death simply by demonstrating that mortality follows a Gompertz distribution, because infinite alternative causes would lead to identical Gompertz mortality kinetics. For survival and reliability analyses, the Weibull and Gompertz distributions find the most common usage.

Cast in terms of surviving fraction $F_G(t)$ and hazard function $\gamma_G(t)$, the Gompertz distribution is

$$F_G(t) = e^{-\frac{\gamma_\circ}{\alpha}(e^{\alpha t}-1)} \qquad (7.108)$$

and

$$\gamma_G(t) = \gamma_\circ e^{\alpha t} \qquad (7.109)$$

The corresponding surviving fraction $F_W(t)$ and hazard function $\gamma_W(t)$ for the Weibull distribution bounded by a lower value of $t=0$ are

$$F_W(t) = e^{-\left(\frac{t}{\tau}\right)^\beta} \qquad (7.110)$$

and

$$\gamma_W(t) = \frac{\beta}{\tau}\left(\frac{t}{\tau}\right)^{\beta-1} \qquad (7.111)$$

The applicability of the survival functions F_G and F_W to a variety of nonbiological extreme value data sets is illustrated in figure 7.33. For the most part, both functions provide equivalent fits to the data. The variety of processes examined is strikingly broad and spans both human-made and natural systems: car lifetime, mechanical failures, precipitation, wind, or car speed. The common thread is that in each case an extreme value from a large random sample is represented. In the case of daily maximum car speed at a given position, the fastest car from a sample of many cars passing that point on a given day is measured. In the mechanical failure case, the strength of a critical flaw from amongst many microscopic flaws in either a chain or an airplane window are represented. Notice that for none of the datasets is the sample size or underlying probability distribution for the random variable known—however, the generally reasonable fit of F_G or F_W to the data implies that a large number of samples from some underlying but unspecified random process has been taken.

Whole-organism mortality curves are often consistent, over at least a portion of the lifespan, with the Gompertz hazard function (equation 7.109). For example, the three cases in figure 7.32 have intervals wherein the semilog plot of γ versus lifespan is linear (80–100 years for the human data, 12–24 days for *Drosophila*, and 5–13 days for *Caenorhabditis elegans*). A different set of survival data for *C. elegans* is fit to both Gompertz and Weibull distributions in the top two panels of figure 7.34. Clearly, neither of the best-fit curves is satisfactory, as the experimental hazard function appears to have two regimes: one with Gompertz behavior, and a second period of approximately constant hazard rate γ.

A set of *Drosophila* survival data distinct from that in figure 7.32 is shown in the bottom two panels of figure 7.34. In this instance, the hazard function γ appears to be fit closely by Gompertz kinetics, but less well by Weibull. The differences are subtle when viewed on the predicted survival curves (lower-left panel, dotted versus solid line) but are highlighted when examining the predicted hazard functions (lower right panel). Note the striking contrast between the *Drosophila* survival curve in figure 7.32 and that in figure 7.34; it is unclear whether differences in the strains or in the handling conditions produce the significantly different survival curves in these two experiments. Both Gompertz and Weibull distributions also provide reasonable fits to survival kinetics for a variety of human diseases, two examples of which are shown in figure 7.35.

The following take-home messages apply to these survival curve analyses:

1. A single survival distribution will not necessarily apply across the full lifespan range for an organism.
2. Gompertz senescence kinetics often apply over some subset of the lifespan interval.

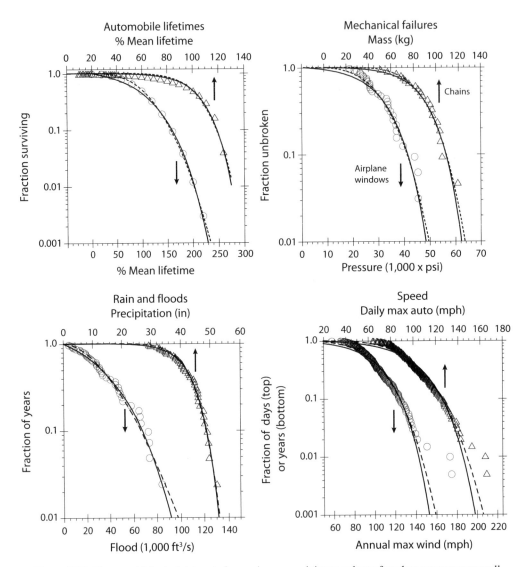

Figure 7.33 Many nonbiological data sets for maximum or minimum values of random processes are well described by Gompertz (solid lines) or Weibull (dashed lines) survival curves. The automobile survival data were taken from registration histories in 1935 (circles; [76]) or 1980 (triangles; [72]). For both cases, the lifetime is expressed as a percentage of the mean lifetime. Data sets for chain breakage, daily maximum car speed, annual maximum wind speed, and annual precipitation were taken from reference [74], airplane window failure data from reference [75], and annual maximum flood data from reference [77].

Population Growth and Death Models

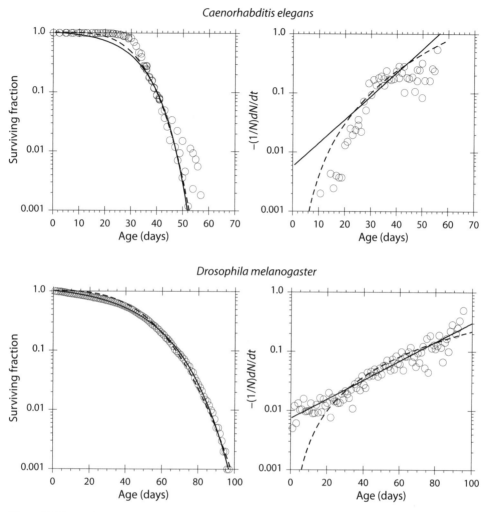

Figure 7.34 Consistency of survival curves with Gompertz (solid line) or Weibull (dashed line) distributions. In the top panels, survival data for *Caenorhabditis elegans* [78] are shown. Neither simple curve captures the trend in the data over the full time range. In the bottom panels, a survival data set for *Drosophila melanogaster* [68] is most consistent with Gompertz kinetics over the full time range.

3. No inference of senescence mechanism can be unambiguously supported by a survival curve, because the extreme value distribution for many different probability densities collapses to either Gompertz or Weibull forms.
4. However, any correct mechanistic description of senescence must produce Gompertz or Weibull survival dependencies to be consistent with the great majority of data sets.

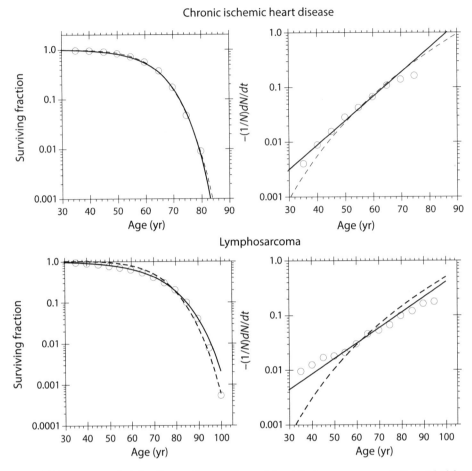

Figure 7.35 Human survival data for two diseases, from a data set for Japanese women [79]. Survival from chronic ischemic heart disease is perhaps somewhat more consistent with a Weibull distribution (dashed lines), whereas lymphosarcoma survival follows Gompertz senescence kinetics (solid lines) more closely.

Case Study 7-1   B. J. Tolkamp, M. J. Haskell, F. M. Langford, D. J. Roberts, and C. A. Morgan. Are cows more likely to lie down the longer they stand? *Appl. Anim. Behav. Sci.* 124: 1–10 (2010) [80].

The Ig Nobel Prizes have been awarded annually since 1991 to "reward research projects that first make people laugh, and then make them think." The 2013 Ig Nobel Probability Prize was awarded to this paper, which addressed what seems at first to be a frivolous issue of bovine preferences. However, measurements of cow standing and lying times are of practical value in veterinary practice, and in fact, the expected trends are counterintuitive.

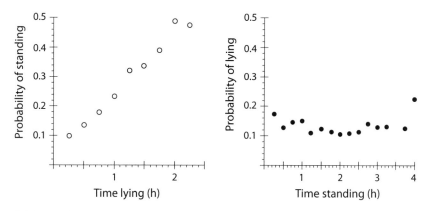

Figure 7.36 Data for cow transitions from the lying to the standing position, and vice versa. The probability of a transition in the ensuing 15 minutes is plotted against the duration of time in the particular pose.

The issue can be posed as two questions: (1) Is a standing cow more likely to lie down the longer it has been standing? (One might expect the answer to be yes, if cows get tired like humans do.) (2) Is a lying cow more likely to stand up the longer it has been lying down? (Such is not necessarily the experience with napping humans.)

Through a combination of visual observations and biosensor measurements, the authors took data on 10,814 lying episodes and obtained the results plotted in figure 7.36. The standing probability from the lying position increases linearly with time, thereby being very consistent with a Gompertzian extreme value distribution. By contrast, the probability of lying down is relatively history independent in the time frame observed and is therefore close to a Poisson process. It can be generally gratifying when complex multifactorial biological processes such as this unexpectedly obey such a simple relationship. One must also admit a grudging respect for bovine stoicism and work ethic, which is not in keeping with prevailing cultural stereotypes for cattle.

Cellular replicative senescence When normal cells are removed from a multicellular organism, they can be coaxed to grow and multiply for only a limited number of generations ($\approx$40–80) in primary cell culture before entering a state called replicative senescence. Leonard Hayflick first demonstrated this phenomenon with human fibroblasts (figure 7.37), and hence this limit has come to be called the Hayflick limit. (The persistence required to establish the validity of this phenomenon can be appreciated by realizing that the experiments portrayed in figure 7.37 required sterile cultures to be maintained and passaged for more than 7 months.)

Careful experiments with cryopreservation, mixed cultures, and nuclear transfer indicated that the Hayflick limit is defined not in terms of clock or calendar

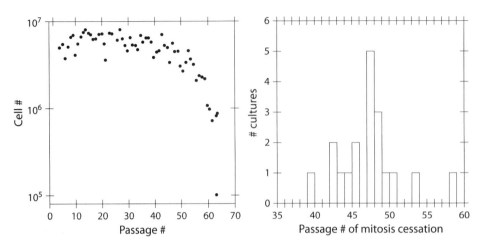

Figure 7.37 In the left panel, cells at each 2:1 passage of a human fetal primary cell culture. When mitosis ceases in the culture, cell number drops by successive dilutions. The right panel shows the distribution of numbers of passages before cessation of mitosis. The average number of passages is 47 ± 4 [81].

time but rather as a cell-division counter that resides in the nucleus. Substantial evidence indicates that telomere shortening during DNA replication serves as the cell's *replicometer*.

Telomeres are repetitive stretches of DNA at the ends of linear chromosomes. Replication of the lagging strand of a DNA double helix involves the annealing of Okazaki fragments primed by short RNA primers. However, at the $3'$ end of telomeres, there is no priming site beyond the end of the DNA, and so each round of replication results in a deletion of some portion of the telomere (figure 7.38). Many microbes and viruses possess circular genomes to topologically evade such end truncations and the progressive genomic degradation that ensues, whereas others rebuild their chromosome tips with an enzyme called telomerase. Telomere shortening has been observed directly in primary cell culture, at a pace of approximately 50 bp/doubling (figure 7.39).

Telomere shortening not only occurs in vivo but is also predictive of replicative lifespan in primary culture (figure 7.40). Furthermore, overexpression of human telomerase extends in vitro lifespan [85]. Given the clear evidence for telomere shortening predicting the number of cell divisions in vitro, it is tempting to speculate that it plays a similar role in vivo. However, so far, such evidence is only circumstantial. As examples, telomere length in blood samples correlates with overall survival [86]; the fraction of skin fibroblasts with measurable telomere damage in baboons grows exponentially with age [87]; and peripheral blood mononuclear cell telomere length was found to be shorter in women subjected to severe life stressors [88].

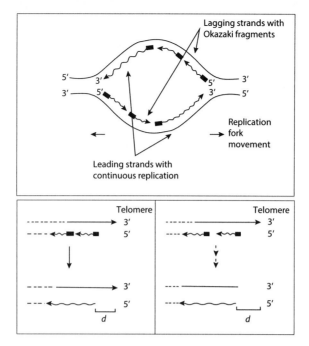

Figure 7.38 DNA replication results in progressive truncations at chromosomal termini with each round of replication [82]. The top panel depicts a replication bubble at which faithful replication of DNA is achieved; this occurs continuously on one strand and piecewise on the other, given the constrained 5′-to-3′ directionality of the DNA polymerase. The bottom panels depict the consequences of this directionality at the ends of linear DNA. Because the RNA primer required for initiation of replication must anneal to the available template but is subsequently degraded, there is necessarily a deletion at the very end of the strand (bottom-left panel) that is then propagated to the other strand, resulting in progressive shortening of the double-stranded DNA.

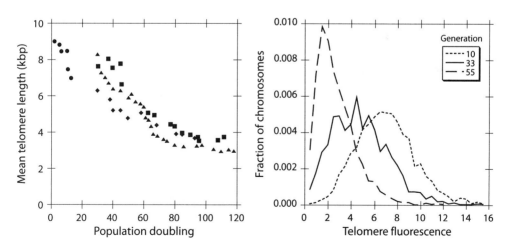

Figure 7.39 Primary human fibroblasts were passaged and telomere shrinkage was measured by restriction digestion (left panel [83]) or quantitative fluorescence in situ hybridization (right panel [84]). In general, ≈50 bp/doubling are lost.

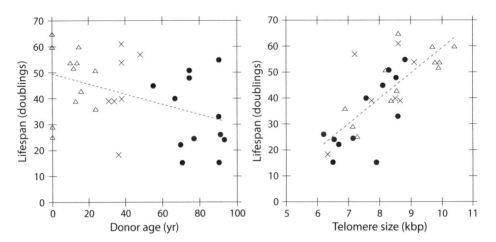

Figure 7.40 The age of 31 different human donors does not predict the replicative potential in primary cell culture of donated fibroblasts (left panel, $R^2 = 0.2$). Telomere size better predicted in vitro lifespan, consistent with its serving a role as a counter of cell divisions prior to senescence (right panel, $R^2 = 0.6$) [89].

Case Study 7-2 M. Z. Levy, R. C. Allsopp, A. B. Futcher, C. W. Greider, and C. B. Harley. Telomere end-replication problem and cell aging. *J. Mol. Biol.* 225: 951–960 (1992) [82].

In this paper, the authors used discrete accounting of telomere deletion events to determine whether the constraints on telomere shortening processes are consistent with the empirically observed behavior of the cellular *replicometer*. The assumptions of their model are:

1. All telomeres behave identically following incomplete replication, although initial lengths may vary.
2. Generation time is constant.
3. Overhangs from single-strand deletions are not degraded.
4. Deletions on either or both strands of a telomere are a "deletion event."
5. Cells enter senescence when one telomere is reduced to a critical length.

The first and second assumptions greatly simplify the analysis, and there is no evidence contrary to them to justify complicating the model. The third and fourth assumptions define the number of telomere deletion events possible per DNA replication. The fifth and final assumption has subsequently been supported by experiments indicating that the length of the shortest rather than the average telomere drives senescence [90].

All possible telomere deletion combinations on a single chromosome for four generations of replication are enumerated in figure 7.41. The probability that a particular telomere has d deletions following n generations of replication is

Population Growth and Death Models

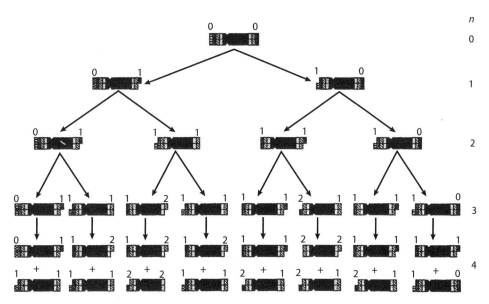

Figure 7.41 The complete four-replication lineage of telomere deletions for a given chromosome. The number above each telomere is the number of deletions d. Figure from reference 82.

$$p(d,n) = \begin{cases} \dfrac{\binom{n+1}{2d}}{2^n} & \text{if } d \leq \dfrac{n+1}{2} \\ 0 & \text{if } d > \dfrac{n+1}{2} \end{cases} \quad (7.112)$$

To find the probability that a given telomere has fewer than the critical number of deletions, d_c, after n replications, one sums over all values $d < d_c$:

$$P(d_c, n) = \sum_{i=0}^{d_c-1} p(d_i, n) \quad (7.113)$$

If at the beginning of the experiment, k telomeres are short enough to reach d_c during the period of interest, then the fraction of dividing cells is simply the probability of k failures to reach d_c:

$$F(d_c, n) = P(d_c, n)^k \quad (7.114)$$

The effect of the parameters k and d_c on the predicted fraction of viable cells is shown in figure 7.42. The predicted viability curve is not a strong function of the parameter k, the number of telomeres with initial lengths short enough to lead to d_c deletions in the number of generations n considered (left panel, figure 7.42). This is fortunate, because this parameter is somewhat arbitrary: Its primary effect on the predicted senescence curve is just to sharpen the transition.

However, the generation of senescence is a strong function of the critical telomere deletion number d_c (right panel, figure 7.42). It should be noted that telomeres need

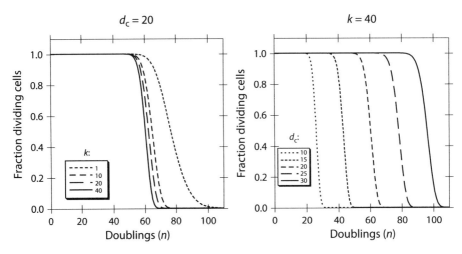

Figure 7.42 Parameter sensitivity of equation 7.114. The number of telomeres k vulnerable to critical deletion serves to sharpen the transition to senescence, and the critical deletion number d_c directly determines the number of doublings before senescence, as expected.

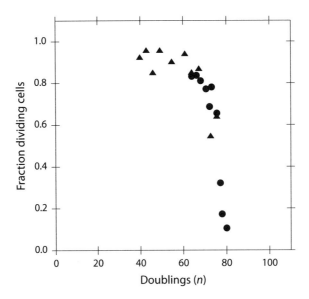

Figure 7.43 The fraction of dividing cells in primary cultures of human fibroblasts, demonstrating cellular replicative senescence at a Hayflick limit of 70–80 [92]. These data are qualitatively consistent with the simple probabilistic model for telomere shortening (equation 7.114).

not shrink to zero length to lead to senescence (left panel, figure 7.39). In fact, it does not appear to be gross mechanical breakdown of the chromosomal ends that leads to senescence, but instead the status of telomere-binding protein factors that sends a signal to cells to stop dividing [91].

The authors compared their model predictions to a particular data set for human fibroblasts, as shown in figure 7.43. The behavior is qualitatively similar to that of the model (i.e., a sharp decrease in viable cell number at 40–80 generations) with a value of $d_c \approx$ 20–25.

The conclusion that can be drawn from this modeling exercise is that the telomere shortening mechanism for signaling maximum division number is consistent with the observed kinetics of primary cell culture senescence.

7.4.3 Cell-Cell Death Signaling

Cells may sense the approach to full occupancy of their volumetric niche not only by depletion of nutrients or growth factors or accumulation of waste products, but also sometimes by specific signals sent by their companion cells. These signals may be soluble or result from cell-cell contact: Killer lymphocytes carrying Fas ligands can directly signal to a target cell to undergo apoptosis by ligating the Fas receptor and triggering caspase-mediated apoptosis. The effect of T-cell/T-cell contacts mediating cell death can be mathematically described with a death rate that is second-order in T cell number [93].

7.5 Compartmental Growth Models

We have so far considered populations with a single type of individual. How can population heterogeneity be modeled? A simple first-level mathematical model for the growth of mixed populations is to treat different cell types as discrete homogeneous *compartments*, with the transitions between compartments considered as reactions obeying mass-action kinetics (i.e., with the rate proportional to the *concentration* of cells or individuals in the compartment). We have demonstrated the utility of this approach in analyzing protein trafficking dynamics (chapter 5), but this simplified description has also proven useful for such divergent systems as cell cycle progression, infectious disease dynamics, predator-prey interactions, stem cell differentiation, and tumorigenesis.

7.5.1 Cell Cycle

Before the detailed molecular machinery that controls cell division was elucidated, simple empirical models of progression from division to division were formulated. More mechanistic models of the cell cycle can be constructed that incorporate the regulatory role of cyclins, M-phase promoting factors, Cdc25, and other kinases and transcription factors [94].

Case Study 7-3 J. A. Smith and L. Martin. Do cells cycle? *Proc. Natl. Acad. Sci. U.S.A.* 70: 1263–1267 (1973) [95].

One generally imagines the periodic behavior of the cell cycle as having a regular *clock-like* mechanism underlying it. Smith and Martin asked whether the experimental data could instead be interpreted as including a stage of random length, during which continuous progress toward division is not made. In their parlance, this would be distinct from a pure cycle, because cyclical cell growth would require "a continuous progression through a chain of events leading to division." To test this intriguing idea, they constructed a simple model in which cell growth consists of two alternating states. The B phase consists of progressive behavior through S, G_2, and M phases, and possibly contains progressive portions of G_1. This is interspersed with the A state, in which there is no progressive movement toward cell division; instead, the cell waits in this state until it transitions to the B phase, with a constant transition probability P. The model is sketched in the top portion of figure 7.44. Experimental observations using radionucleotides provide a basis for the model by showing that S and G_2 phases have a relatively narrow distribution of durations within a cell type, but that G_1 duration has a wide distribution both within and across cell types.

Imagine tracking a population of cells, and whenever a cell divides, recording its age upon division. For this purpose, consider the age as measured from the previous division leading to that cell. Smith and Martin consider a plot of the fraction of cells remaining in interphase α (i.e., that have not entered mitosis yet) as a function of cell age. If each cell spent the same amount of time in interphase, the plot would be square (part A of the left panel in figure 7.44); no cells would divide until reaching the age at which they all divide. Part B of the same panel shows the curve resulting from normally distributed interphase durations, and part C shows a logarithmic ordinate version. The negative slope of the curve, $-\frac{d\log\alpha}{dt} = -\frac{d\alpha}{\alpha\, dt}$, gives the relative rate of entering mitosis, which in part C is shown to increase continuously with time. Smith and Martin suggest that this is the expected result for a cyclic process with variation in cycle time (and we will explore this idea more in the problems). Part D is the plot for the model with a constant B-phase duration (T_B) and constant transition probability P; it shows no division until age T_B, followed by constant-rate negative slope P on the logarithmic plot.

Results of experiments carried out with time-lapse imaging on a variety of cell cultures in exponential growth are shown in the right-hand panel of figure 7.44. These curves all clearly match Part D, which was constructed from the probabilistic model, rather than the progressive model in Part C. (The sharp transition in Part D is somewhat smoother in the experimental data, which can be explained by variation in T_B.) Thus, the authors conclude that cells don't strictly cycle because there is not a pure "regularly recurring succession of events."

Although a mathematical expression of this model was not presented in this paper, it is, in essence, a first-order process followed by a pure time delay. The Smith-Martin model is expressed mathematically for lymphocyte labeling experiments, including cell death terms, in equation 7.115 [96].

Population Growth and Death Models

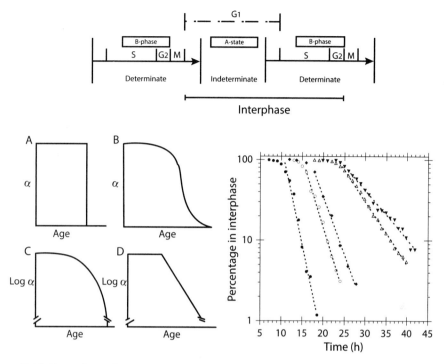

Figure 7.44 Comparison of expected results from alternative conceptual models of cell cycle progression [95].

7.5.2 In Vivo Cell Population Models

The complex interconversions among cell types during development, infections, and tumorigenesis can often be described to a first approximation using compartmental models.

Tracking cell growth in vivo How can one track the population dynamics of a particular cell type in the complex milieu of other cells that exists in an intact organism? Two broad categories of labeling experiments are used: (1) label a cell cohort and reintroduce it to the organism; or (2) label all cells in the organism for a pulse period, then isolate the subpopulation of interest at time points during a chase period.

In the first category of approaches, labeling a cell population with an inert fluorescent tracer dye has been commonly applied to track immune cell dynamics in vivo. The dye typically used is carboxyfluorescein diacetate succinimidyl ester (CFSE), which diffuses freely across cell membranes and is processed by intracellular esterases to a fluorescent fluorescein product, which then covalently

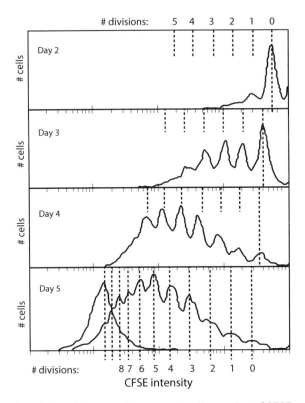

Figure 7.45 Progressive dilution of the trapped fluorescent reaction product of CFSE can be used with flow cytometry to count the number of divisions of B cells following activation [97]. Note that the CFSE intensity scale is logarithmic, so that the successive 2-fold reductions in intensity give evenly spaced peaks.

labels free amines on cellular proteins. With each cell division, the cellular fluorescence drops by a factor of approximately two, and so the number of cells that have undergone a given number of divisions can be counted directly (figure 7.45; [97]).

Quantitative analysis of curves such as those in figure 7.45 is not straightforward, requiring a mathematical model to describe the kinetics of cell division and death. In the most rigorous form, such models account for the approximately constant delay time to progress through the S, G_2, and M phases by using the Smith-Martin model [96]. The analytical solution of the delay-ODE equations for this model is rather complex and must be evaluated numerically.

Letting $A_i \equiv$ the number of cells in the indeterminate A phase (figure 7.44) that have doubled i times, and $B_i \equiv$ the number of cells in the determinate B phase that have doubled i times, then the total number of cells that have doubled i times $N_i \equiv A_i + B_i$. The solution is as follows:

$$A_0(t) = A_0(0) e^{-(\lambda_0 + d_0)t} \tag{7.115}$$

$$A_i(t) = \frac{2^i \lambda^{i-1} \lambda_0 A_0(0) e^{(\lambda_0 + d_A - d_B)\tilde{\Delta}_i} e^{-(\lambda_0 + d_0)t}}{i!} \times$$
$$H(t - \tilde{\Delta}_i)\Gamma(i, 0, (t - \tilde{\Delta}_i)(\lambda - \lambda_0)) \qquad (7.116)$$

$$B_0(t) = \lambda_0 \int_0^{\Delta_0} A_0(t - \tau) e^{-d_B \tau} d\tau \qquad (7.117)$$

$$B_i(t) = \lambda \int_0^{\Delta} A_i(t - \tau) e^{-d_B \tau} d\tau \qquad (7.118)$$

for $i = 1, \ldots, \infty$. Parameters for this set of equations are: λ_0 and λ, the specific activation rate constants for transition from the A phase to the B cell cycle phase for undivided and previously divided cells, respectively; d_A and d_B, the specific death rates of cells in the A and B cell cycle phases, respectively; d_0, the specific death rate of nondivided cells; $H(t - \tau)$, the Heaviside step function with delay τ; Δ_0, the length of the B cell cycle phase for as yet undivided cells; Δ, the length of the B cell cycle phase for all previously divided cells; $\tilde{\Delta}_i \equiv (i - 1)\Delta + \Delta_0$; and $\Gamma(a, z_0, z_1) = \int_{z_0}^{z_1} e^{-t} t^{a-1} dt$, the incomplete gamma function. These parameters can be fit to experimental CFSE data to obtain physiologically reasonable estimates [96].

The pulse-chase method for following cell populations in vivo is illustrated in the following case study.

Case Study 7-4 H. Mohri, A. S. Perelson, K. Tung, R. M. Ribeiro, B. Ramratnam, M. Markowitz, R. Kost, A. Hurley, L. Weinberger, D. Cesar, M. K. Hellerstein, and D. D. Ho. Increased turnover of T lymphocytes in HIV-1 infection and its reduction by antiretroviral therapy. *J. Exp. Med.* **194**: 1277–1287 (2001) [98].

A common perturbation used to probe the kinetics of synthesis and degradation of a component is to label it with a metabolic radiolabel for a brief period of time (the *pulse*) and then measure the concentration of labeled species as a function of time following cessation of the label (the *chase*). In this paper, Mohri and colleagues use pulse-chase isotopic labeling of DNA in whole human beings to track the growth and death of T cell populations in healthy and HIV-infected individuals. The issue at hand was whether HIV causes increased T cell production and destruction or, alternatively, if it decreases the body's production of new T cells. To radiolabel all DNA synthesized in the subjects, they were hospitalized for seven days, placed on a diet with minimal carbohydrates, and intravenously administered deuterium-labeled glucose. The deuterium label was metabolized and incorporated into freshly synthesized DNA during the pulse time period. Blood samples were taken, and $CD4^+$ and $CD8^+$ T lymphocytes were independently isolated by flow cytometric sorting. Total DNA in the sorted sample was analyzed for deuterium

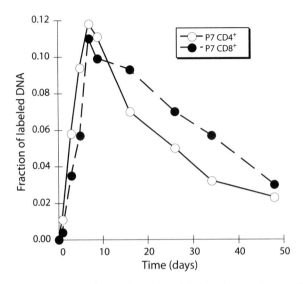

Figure 7.46 Fraction of labeled DNA in CD4$^+$ and CD8$^+$ T lymphocytes from an HIV-positive patient during a 7-day radiolabeling and 43-day chase experiment [98].

incorporation by mass spectrometry. A sample data set from one HIV-positive patient is shown in figure 7.46.

A simple model was constructed to describe these dynamics, shown schematically in figure 7.47.

There are three kinetic regimes to be described: before, during, and after the labeling pulse with deuterated glucose.

Before:

$$\frac{dT}{dt} = s + (p-d)T \qquad (7.119)$$

where T is the total T cell count, s is the source rate for new T cells differentiated from precursors, p is the specific proliferation rate constant, and d is the specific death rate.

During the labeling pulse:

$$\frac{dL}{dt} = s_L + pU + (p-d)L \qquad (7.120)$$

$$\frac{dU}{dt} = s_U - dU \qquad (7.121)$$

where L is the concentration of labeled T cell DNA, and U is the concentration of unlabeled T cell DNA. Note that because we are accounting for the DNA rather than the cells, a cell replication event leaves U constant (from the template DNA strand) while producing one equivalent of newly replicated L (i.e., $U \to U + L$). The constant terms s_L and s_U reflect an assumption that the rate of introduction of labeled and unlabeled DNA into the T cell compartment via differentiation of precursors is constant. In the absence of more detailed

Population Growth and Death Models

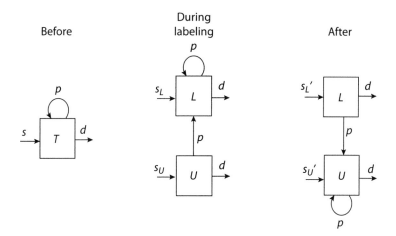

Figure 7.47 Compartmental models of labeled (L) and unlabeled (U) T cell dynamics, during three distinct time periods of a pulse-chase metabolic labeling experiment in human subjects [98].

measurements of hematopoiesis, this approximation is the simplest available.

After (chase period):

$$\frac{dL}{dt} = s'_L - dL \qquad (7.122)$$

$$\frac{dU}{dt} = s'_U + (p-d)U + pL \qquad (7.123)$$

Note that the hematopoietic source terms s'_L and s'_U are also held constant but distinct from the pulse period. Labeled template DNA strands are preserved upon replication ($L \rightarrow U + L$).

The measured parameter is the fraction of labeled DNA in a sample. Assuming that the total DNA in the T cell compartment is constant over the labeling pulse, $L(t) + U(t) = L(0) + U(0)$. Because there is no labeled DNA at time zero, $L(t) = U(0) - U(t)$. During the labeling pulse, the fraction of labeled DNA $f_{L,\text{pulse}}(t) = L(t)/[U(t) + L(t)]$ can be rewritten as $f_{L,\text{pulse}}(t) = [U(0) - U(t)]/U(0)$, or $f_{L,\text{pulse}}(t) = 1 - U(t)/U(0)$. Solving equation 7.121 with the initial condition $U(t) = U(0)$ at $t = 0$ gives the following solution:

$$f_{L,\text{pulse}}(t) = \left(1 - e^{-dt}\right)\left(1 - \frac{\hat{s}_U}{d}\right) \qquad (7.124)$$

where $\hat{s}_U \equiv s_U/U(0)$ is the specific source rate for unlabeled cells.

Assuming a constant amount of DNA during the chase period, as we did during the pulse, we can solve equation 7.122 for $L(t)$ to obtain $f_{L,\text{chase}}(t) = L(t)/U(0)$, with the initial condition $L(t_e) = f_{L,\text{pulse}}(t_e)U(0)$ at $t = t_e$:

$$f_{L,\text{chase}}(t) = \left(f_{L,\text{pulse}}(t_e) - \frac{\hat{s}'_L}{d}\right)e^{-d(t-t_e)} + \frac{\hat{s}'_L}{d} \qquad (7.125)$$

where $\hat{s}'_L \equiv s'_L/U(0)$ is the specific source rate for labeled cells during the chase.

Note that in equations 7.124 and 7.125, the exponential time dependence involves only the parameter d, or the specific death rate; the specific proliferation rate p does not appear in these equations. Therefore, nonlinear least squares fitting of the labeled DNA fraction during the pulse and chase periods provides a direct estimate of the specific death rate d but not of the proliferation rate constant p. An estimate of the proliferation parameter is obtained by assuming that the T cell compartment is at steady state before the labeling experiment, such that $\hat{s} + p = d$. Assuming continued steady-state T cell growth during the pulse period, $\hat{s}_U + \hat{s}_L + p = d$. However, from equation 7.124 we can estimate $\hat{s}_U$ but not $\hat{s}_L$. This allows us to state the limit:

$$p > d - \hat{s}_U \tag{7.126}$$

The source rate $\hat{s}'_L$ for labeled cells during the chase period is estimated by fitting equation 7.125. This source term corresponds to DNA from labeled precursor cells that become CD4$^+$ or CD8$^+$ T cells during the chase period. Because there are more labeled precursor cells following the pulse than at its initiation, $\hat{s}'_L > \hat{s}_L$. Consequently, the following holds:

$$p < d - (\hat{s}_U + \hat{s}'_L) \tag{7.127}$$

The authors took the mean of the upper limit in equation 7.127 and the lower limit in equation 7.126 as an estimate for the proliferation rate p.

The parameter fits reported in table I of the paper, unfortunately, do not uniformly reproduce the putative model curves of figure 3 of the paper; in fact, in three cases, the fitted source rate $\hat{s}_U > d$, which is inconsistent with the steady state approximation. However, the reported death rates d do generally reproduce the data curves. An important caveat is that the measured dynamics are only those of T cells that replicated their DNA during the pulse period; resting T cells would not have been labeled and so would be invisible to the measurement. Consequently, the measured dynamics are not for the total T cell population, but just for those that replicated during the labeling period.

The mean parameters estimated for healthy and HIV-positive individuals are shown in table 7.4. The average death rate for labeled CD4$^+$ T cells in HIV-infected patients is approximately 3-fold higher than normal, but this is compensated for (in the short term of this experiment) by a 3-fold higher rate of production of CD4$^+$ T cells by hematopoiesis. The rate constants for CD8$^+$ T cells are less altered by HIV infection, consistent with the specific effect of HIV on the CD4$^+$ subset.

Table 7.4 Best-fit parameters for death and source terms (day^{-1}) in pulse-chase experiments.

Category	CD4$^+ d$	CD4$^+ \hat{s}_U$	CD8$^+ d$	CD8$^+ \hat{s}_U$
Normal control individuals	0.044 ± 0.011	0.033 ± 0.011	0.043 ± 0.017	0.040 ± 0.017
HIV$^+$ patients	0.129 ± 0.033	0.123 ± 0.037	0.050 ± 0.009	0.031 ± 0.009

In summary, the kinetic analysis reported in this paper is consistent with the hypothesis that HIV infection increases turnover rates for CD4$^+$ T cells, rather than decreasing the production of these lymphocytes by hematopoiesis.

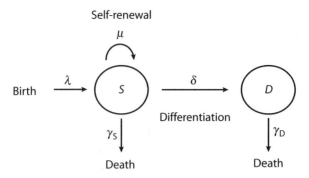

Figure 7.48 A simplified kinetic description of stem cell dynamics. The compartment S represents stem cells, and the compartment D represents differentiated cells.

Stem cell dynamics Each cell in a multicellular organism generally undergoes one of three transitions from a quiescent state: replication, apoptosis, or differentiation. Dysregulation of this decision process lies at the heart of noninfectious diseases, such as cancer and autoimmunity. The significance of apoptosis can be appreciated by realizing that in its absence, an 80-year-old human would have two tons of bone marrow and lymph nodes, and a gut 16 km long [99]! While apoptosis sculpts tissue by removing cells, stem cells are a continuous source of new replacement cells.

A simplified description of stem cell dynamics is shown in figure 7.48. Translating this schematic diagram into differential equations yields the following:

$$\frac{dS}{dt} = \lambda + (\mu - \gamma_S - \delta)S \qquad (7.128)$$

$$\frac{dD}{dt} = \delta S - \gamma_D D \qquad (7.129)$$

In healthy tissue, these processes reach a steady state. The cell compartment levels, found by setting the derivatives to zero, are:

$$S_{ss} = \frac{\lambda}{\delta + \gamma_S - \mu} \qquad (7.130)$$

$$D_{ss} = \frac{\delta}{\gamma_D} S_{ss} \qquad (7.131)$$

Clearly, the parameters λ, μ, δ, γ_S, and γ_D are not constants but will depend on the tissue environment: cell density, extracellular matrix composition, and the concentrations of nutrients and growth factors. It is generally found experimentally that differentiated cells are in considerable excess over stem cells, indicating that $\gamma_D \ll \delta$ (i.e., the specific death rate of differentiated cells is much less than the specific stem cell differentiation rate).

Models of this type have been used to consider the possible processes leading to tumorigenesis [100]. For example, can a reduction in the apoptosis rate for differentiated cells lead to unbounded tumor growth? Examination of equation 7.131 shows that reducing the value of γ_D increases the steady-state differentiated cell population D_{ss} to a higher value, but it does not lead to unbounded growth. (Of course, this is a necessary consequence of our definition of the differentiated cell compartment D as non-self-replicating.) Such amplified numbers of differentiated cells in the absence of unbounded growth is a hallmark of many premalignant tissue formations.

By contrast, a reduction in the rate of apoptosis in the stem cell compartment γ_S, in the absence of any change in the other parameters, leads to exponential growth of the stem cell population with a specific growth rate $(\mu - \gamma_S - \delta)$. (A corresponding decrease in the rate of differentiation δ also leads to exponential growth.) More generally, if $\mu > \gamma_S + \delta$, the stem cell population grows exponentially. The significance of stem cells in cancer progression is therefore clear: Therapies that preferentially target the differentiated phenotype (i.e., increase γ_D) without shifting stem cell dynamics such that $\mu \leq \gamma_S + \delta$ cannot terminate exponential tumor growth.

When a stem cell divides to produce one differentiated cell incapable of further replication and one stem cell still capable of dividing, it is termed an *asymmetric cell division*. Perfect asymmetric cell division occurs when $\mu = \delta$, such that the loss of one cell to the differentiated state is perfectly compensated by the self-renewal of one stem cell. Note that this mathematical condition can be achieved by at least two very different biological events: (1) half the cells dividing to form new stem cells and half the cells differentiating without replication and (2) every cell division producing one differentiated and one stem cell. The second mechanism is predominant during tissue development, but stem cells are also capable of replenishing their pool by symmetric division under some circumstances, and excess symmetric stem cell division may also lead to cancer [101].

Asymmetric division can be described in the framework of figure 7.48 by considering there to be no cell death or new stem cell introduction during the time period of interest (i.e., $\gamma_S = \gamma_D = \lambda = 0$), and $\mu = \delta$. Equations 7.128 and 7.129 are then considerably simplified:

$$\frac{dS}{dt} = 0 \qquad (7.132)$$

$$\frac{dD}{dt} = \delta S_\circ \qquad (7.133)$$

Integrating these equations leads to $S = S_\circ$ and $D = D_\circ + \delta S_\circ t$, so that the total cell population $S + D$ grows linearly with time.

Case Study 7-5   H. Quastler and F. G. Sherman. Cell population kinetics in the intestinal epithelium of the mouse. *Exp. Cell Res.* **17**: 420–438 (1959) [102].

In this paper, the kinetics of cell replication in the gut of mice were studied. No explicit mathematical models were used, as was the case for the Smith and Martin cell division model discussed previously; rather, pulse-chase experiments were performed and the results analyzed in terms of an abstract quantitative model. To quote the authors on this point:

> The analysis given here is crude. There exist general mathematical methods by which problems of the type considered can be solved accurately; but before using them one wants to be certain that the amount of information in the data is commensurate with the amount of information one wishes to extract.... Our data are not good enough for more than the crude approximation methods used.

Based on morphological examination of structures in the gut, the authors formulated a schematic model shown in figure 7.49. Progenitor cells (in effect, stem cells) were observed to be the only cells to undergo mitosis in the crypt structure. Differentiated *functional* or *columnar* cells move steadily up the villi, to be sloughed off when they reach the end.

To visibly mark cells during DNA synthesis, tritiated thymidine was administered intraperitoneally, and incorporation into DNA was detected by film autoradiography of tissue slices. Labeling of any given cell was found to reach its maximum value by 10 minutes, but only those cells that were synthesizing DNA were labeled. Cells in mitosis can be identified by their morphology. By varying the delay time between label injection and animal sacrifice, the length of the different cell cycle phases could be inferred (figure 7.50). A cell that has completed DNA synthesis goes through a G_2 growth phase prior to mitosis (figure 7.44); the observed delay before labeled mitoses appear represents

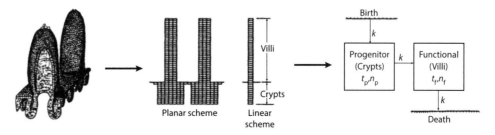

Figure 7.49 A schematic model for cell division in colon crypts and villi [102].

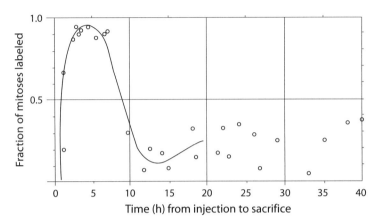

Figure 7.50 The fraction of mitotic cells whose DNA is labeled with tritiated thymidine as a function of the waiting time between label injection and sacrifice of the animal [102].

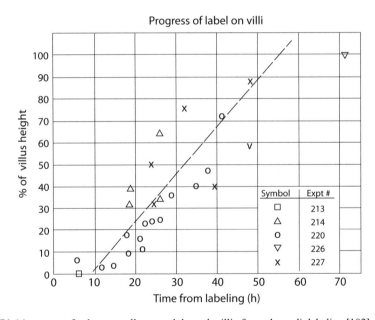

Figure 7.51 Movement of columnar cells upward through villi after pulse radiolabeling [102].

the minimum length of this phase (0.5–1.0 h). From 1 to 8 hours, essentially all mitoses are labeled; thus, the S phase was estimated to be approximately 7.5 hours in duration. The proportion of labeled to total cells should be the same as the proportion of S phase to the entire cell cycle. Because 40% of the cells in the crypt were found to be labeled, it was therefore estimated that the overall cell cycle is approximately 19 hours.

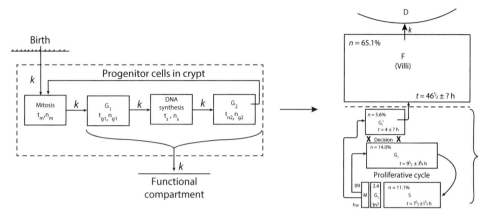

Figure 7.52 A compartmental model of the cell cycle for stem cells, differentiation to villi cells, and movement of villi cells and their subsequent sloughing off and death [102].

To crudely track the functional columnar cells as they move upward through the villi, they were labeled by intraperitoneal injection of ^{32}P inorganic phosphate, and the distance of the labeled front was measured (figure 7.51). It is found that cells progress to the full length of the villi in approximately 50 hours.

A summary of the kinetic parameters estimated by these approximate methods is shown in figure 7.52. It is somewhat remarkable how this detailed, if quantitatively approximate, understanding of the cellular dynamics was extracted by simple examination of two pulse radiolabeling experiments. The progenitor cell cycle, differentiation, and migration of differentiated cells are all represented. In subsequent years, as the molecular details of the differentiation process have begun to be elucidated, and as imaging capabilities have progressed, numerous more detailed mathematical models of this process have been formulated (reviewed in Van Leeuwen et al. [103]).

7.5.3 Epidemics and Infections

Kinetic modeling has played a central role in understanding and responding to outbreaks of infectious diseases in human populations. In such models, individuals are classified on the basis of their susceptibility to infection and infection status. A general classification scheme for such models is shown in figure 7.53. Individuals fall into five different classes: M, infants born with passive immunity against the pathogen of interest via maternal antibodies; S, susceptible individuals lacking immunity; E, individuals exposed and infected with the pathogen but not yet infectious themselves; I, infectious individuals; and R, individuals who have recovered from infection and have immunity against the pathogen. The type of model used is classified according to the particular classifications included: for example, an SIR model does not include the M or E classes.

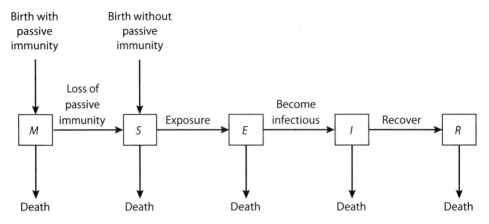

Figure 7.53 Commonly used categories of individuals in epidemiological models. Adapted from reference [104].

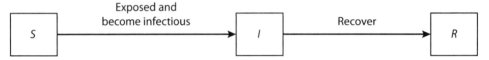

Figure 7.54 The simple *SIR* epidemic model.

SIR models were the first type developed to describe epidemics [105]. As with all models, the degree of complexity chosen for incorporation should be matched with the objectives for the modeling exercise. The simplest *SIR* models (figure 7.54) miss numerous details but capture some of the simple, central, qualitative aspects of epidemics.

Let's consider a simple means to describe the rate of infection of susceptible individuals S by exposure to infectious individuals I, in a population of size $N = S + I + R$ (where N does not significantly change over the time of interest). Let $\beta \equiv$ the number of interactions per unit time that a given individual experiences, with I/N the fraction of such interactions that occur with an infected individual. The overall rate of infectious interactions per unit time is therefore $\beta(I/N)S$.

Let's now consider the simplest means to describe recovery from the disease. In the *SIR* model, infected individuals immediately become infectious (i.e., there is no class E). If there is a constant probability γ per unit time of recovery, then the rate of recovery of infectious individuals is γI. Incorporating these rate expressions for infection and recovery into balances on S, I, and R yields

$$\frac{dS}{dt} = -\beta IS/N \qquad (7.134)$$

$$\frac{dI}{dt} = \beta IS/N - \gamma I \tag{7.135}$$

$$\frac{dR}{dt} = \gamma I \tag{7.136}$$

However, note that the equation for R is not independent of the first two, because $S + I + R = N$, a constant. It is also useful to nondimensionalize the variables as fractions by dividing by N: $s \equiv S/N$, $i \equiv I/N$. Time will be nondimensionalized by the mean time $1/\gamma$ before an infected individual recovers: $\tau \equiv \gamma t$. Making these substitutions yields

$$\frac{ds}{d\tau} = -\sigma is \tag{7.137}$$

$$\frac{di}{d\tau} = \sigma is - i \tag{7.138}$$

where $\sigma \equiv \beta/\gamma$ is called the contact number. σ is the average number of potentially infectious contacts an infected individual makes prior to recovering from the infection: β is the number of such contacts per unit time, and $1/\gamma$ is the average duration of infection; therefore, their product is the number of contacts. Numerical solutions to equations 7.137 and 7.138 are shown in figure 7.55. The general features of these curves mimic those of actual epidemics (figure 7.56).

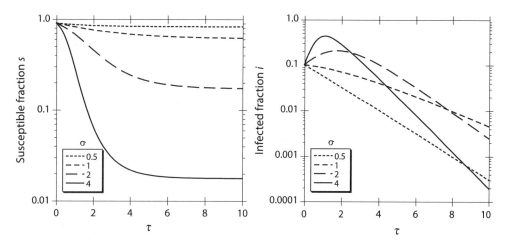

Figure 7.55 Trajectories of susceptible (s) and infected (i) fractions in the nondimensionalized *SIR* model of equations 7.137 and 7.138. For values of the contact number $\sigma > 1$, the infected fraction increases, goes through a maximum, and then decreases.

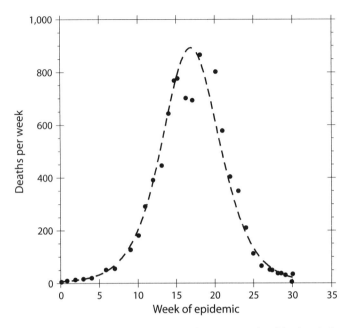

Figure 7.56 Consistency of the *SIR* model with data from an actual epidemic of plague in Bombay, 1905–1906 [105]. Because 80–90% of reported cases were fatal, the number of deaths in a week (dots) is approximately equal to the number of infected individuals. The dashed line is an approximate analytical solution to equations 7.137 and 7.138, with parameters chosen to produce solutions resembling the data. The study's authors are careful to point out: "A close fit is not to be expected, and deductions as to the actual values of the various constants should not be drawn. It may be said, however, that the calculated curve, which implies that the rates did not vary during the period of the epidemic, conforms roughly to the observed figures" [105].

A key prediction of this simple *SIR* epidemic model is the "threshold theorem": if $\sigma \cdot s(0) \leq 1$, then $i \to 0$ as $t \to \infty$; if $\sigma \cdot s(0) > 1$, then i increases to a maximum before declining to zero as $t \to \infty$. (This theorem can be proven by solving for possible solution trajectories s versus i, dividing equation 7.137 by equation 7.138, and solving analytically.) The product $\sigma \cdot s(0)$ is also known as the basic reproduction ratio R_o in the literature and is a central parameter in predicting the growth or extinction of epidemics [106].

The practical significance of the threshold theorem is apparent: If a vaccination program can reduce the susceptible fraction such that $s(0) < 1/\sigma$, then no epidemic with increasing numbers of infections will occur. Another interesting prediction of the *SIR* model is that the epidemic becomes exhausted before all susceptible individuals have become infected (i.e., note in the left panel of figure 7.55 that s does not go to zero as $t \to \infty$).

To better capture specific features of disease transmission, mechanistic improvements are often incorporated into epidemic models [107]. The benefit of improved realism is traded off against the requirement for an increasing number of model parameter estimates and less facile analysis of model solutions. The distribution of individual age is often incorporated, because obviously, the social interactions and susceptibility to disease vary markedly from infancy through senior citizenship. The first-order recovery function γI used in the simple SIR model implies an exponential waiting time for recovery (see chapter 9), whereas often the time to recovery more closely resembles a simple delay. The infection term implies uniform mixing of individuals, whereas in actuality, people are often spatially segregated and interact with a restricted number of other people. Finally, a deterministic model structure cannot capture events driven by the randomness of small numbers of events. Stochastic models are required to describe epidemics that may be extinguished by fluctuations in small numbers of infected individuals.

By analogy to epidemics in human or animal populations, the progress of an infection in the cells of an individual organism can be described by mass-action kinetic compartmental models. A paradigmatic example of this approach is presented in the following case study on viral dynamics in humans.

Case Study 7-6 A. S. Perelson, A. U. Neumann, M. Markowitz, J. M. Leonard, and D. D. Ho. HIV-1 dynamics in vivo: Virion clearance rate, infected cell life-span, and viral generation time. *Science* 271: 1582–1586 (1996) [108].

This paper describes a classic kinetic analysis of HIV clearance in AIDS patients following administration of a protease inhibitor. Important aspects of the virus's replication were elucidated that guided future therapeutic designs (the paper has been cited thousands of times).

HIV infection proceeds through three distinct phases (figure 7.57): (1) an initial acute infection with flu-like symptoms, (2) a quasi-steady-state phase with slow declines in $CD4^+$ T cell number, and (3) initiation of AIDS symptoms following depletion of $CD4^+$ T cells to sufficiently low levels. Before this picture of the disease's progression became clear, however, the kinetics of the middle phase dominated the perception of HIV, which was considered to be a "slow" virus given the often decade-long period of negligible adverse symptoms.

Once potent inhibitors of HIV protease became available, the kinetic response to specific perturbations revealed that the quasi-steady state phase is the result of very small differences in unexpectedly large rates of virus production and clearance [110]. In this paper, important features of the HIV life cycle were elucidated through kinetic descriptions of the response to inhibitor perturbations. Treatment with the HIV protease inhibitor ritonavir causes freshly synthesized HIV virions to be noninfectious. Measurements of total and infectious HIV virions in a patient over the first week following the initiation of protease inhibitor treatment are shown in figure 7.58.

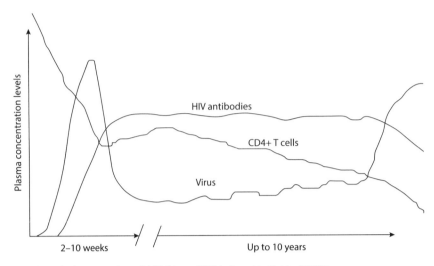

Figure 7.57 Typical progression of AIDS in an HIV-infected individual [109].

The total viral load drops more slowly after what appears to be a delay, whereas the infectious virus load drops more rapidly and with less, if any, noticeable delay. One might expect there to be some delay while the drug is absorbed (it is taken orally), distributes through the bloodstream, and enters cells. However, this pharmacokinetic delay should apply equally to both measurements. The slower total viral mRNA decay was expected, because active virus present when the protease inhibitor was dosed can still infect T cells. Infected T cells also continue to produce noninfectious virions until their own deaths.

Based on previous results by these authors demonstrating the rapid turnover of HIV and infected cells [110], the following model is used to describe the dynamics of T cell infection prior to treatment:

$$\frac{dT^*}{dt} = kVT - \delta T^* \qquad (7.139)$$

$$\frac{dV}{dt} = N\delta T^* - cV \qquad (7.140)$$

where $T \equiv$ uninfected target T cells; $V \equiv$ free HIV virions; $T^* \equiv$ HIV-infected T cells; $N \equiv$ the number of virions produced per infected cell during its lifetime; $\delta \equiv$ the rate constant for infected cell lysis with coincident release of virions; $c \equiv$ the specific rate constant for clearance of free virions from circulation (by an unspecified mechanism); and $k \equiv$ the rate constant for T cell infection by free virions, assuming that infection can be described by a mass-action rate form.

Once patients begin treatment with the HIV protease inhibitor ritonavir, the following equations were used to describe the dynamics of the system:

$$\frac{dT^*}{dt} = kV_1T - \delta T^* \qquad (7.141)$$

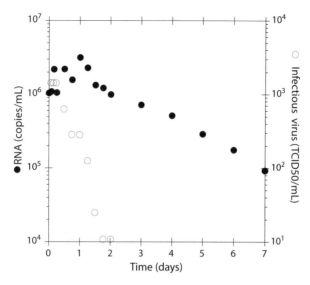

Figure 7.58 Response to treatment with the HIV protease inhibitor ritonavir, as measured by two different methods. The open circles represent a measure of infectious virions, and the closed circles represent total virions, whether infectious or inactivated due to synthesis during protease inhibition.

$$\frac{dV_{NI}}{dt} = N\delta T^* - cV_{NI} \qquad (7.142)$$

$$\frac{dV_I}{dt} = -cV_I \qquad (7.143)$$

where $V_I \equiv$ infectious virions present at the beginning of treatment, and $V_{NI} \equiv$ non-infectious virions produced in the presence of the inhibitor. For simplicity, it is assumed that the drug was dosed at a sufficiently high concentration to completely block protease processing required for virion activation.

Choosing to focus on the dynamics of virus clearance in the first week following drug administration, the authors assumed that the uninfected T cell concentration T remains constant over the period of interest. This allowed an analytical solution for the total virion concentration $V = V_I + V_{NI}$ to be obtained:

$$V = V_\circ e^{-ct} + \frac{cV_\circ}{c-\delta}\left(\frac{c}{c-\delta}(e^{-\delta t} - e^{-ct}) - \delta t e^{-ct}\right) \qquad (7.144)$$

The plasma RNA data for five patients treated with ritonavir were fit to equation 7.144 by nonlinear least squares with the parameters $V_\circ$, c, and δ. (A pharmacokinetic delay of 2–6 hours was either taken from infectivity measurements or also fit.) The parameter estimates and 68% confidence limits obtained by a stochastic bootstrap method are plotted in the right-hand panel of figure 7.59. The fit to the plasma mRNA data is good (figure 7.60). Strong support for the model's assumptions is provided by the curve-fit value for infectious

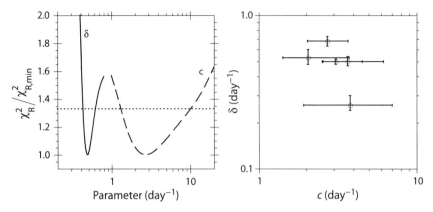

Figure 7.59 Parameter estimation for nonlinear least squares fitting of equation 7.144 to the plasma HIV mRNA data for five treated patients. In the left panel, the confidence limits for one patient's data (patient 107 in the paper) are determined by the method outlined in chapter 3 (the value of $F_{0.32}(12,3) = 1.3$ was obtained from figure 3.39). In the right panel, the reported confidence limits on the viral clearance rate c and the cellular clearance rate δ are shown for all five patients.

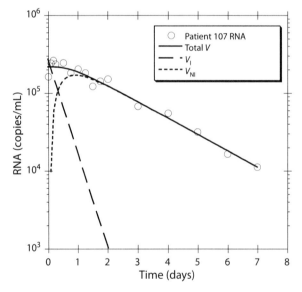

Figure 7.60 Fit of equation 7.144 to one patient's data (patient 107 in the paper). The time scale for infectious virus decay is consistent with the independently measured rate of decay (figure 7.58).

virus clearance half-life of 0.3–0.5 days, consistent with the independently measured values (open circles in figure 7.58).

For the five patients studied, the average virion clearance rate constant is $c = 3.07 \pm 0.64$ day^{-1}, and the average death rate constant for infected cells is $\delta = 0.49 \pm 0.13$ day^{-1}. The average lifetime for a system undergoing first-order decays such as this is the reciprocal of the rate constant (chapter 9). Therefore, the average lifetime in circulation for a virus is $1/3.07 = 0.33$ day or 8 hours, and the average lifetime of an infected cell is $1/0.49 = 2$ days.

Of particular interest is the number of virions replicated per day in an individual, because each error-prone replication cycle has the possibility of creating a drug-resistant mutant. At the quasi-steady state prior to drug treatment, viral production $N\delta T^*$ is equal to viral clearance rate cV. The average initial viral load prior to treatment for these patients was 2.2×10^5 virions per mL. Using the estimated blood volume for each patient, a range of virion production rates of 0.4–32.1×10^9 per day was calculated, with an average of 10^{10} virions produced per day per individual.

This phenomenal rate is a veritable juggernaut of diversity in the virus population. Retroviruses such as HIV have highly error-prone polymerases, with an error rate of 3.4×10^{-5} per base pair per replication cycle for HIV-1, which has a genome of 10^4 base pairs. Each possible point mutation will occur in an average of $10^{10} \times 3.4 \times 10^{-5} = 340{,}000$ virions every day. The probability of any particular double mutation is the square of the individual rate, or 1.2×10^{-9} (if the mutational events are independent). Consequently, an average of ten virions are produced every day with any given possible double mutation. This is why single-agent drugs against HIV fail so rapidly: Several mutations usually exist that confer resistance to a drug with only one or two base changes from the wild-type sequence. Combination drug therapy can only be evaded by ultra-low-probability infectious many-base-change mutants such that resistant variants are too unlikely to be sampled even among the 10^{10} candidates created every day in a patient.

7.5.4 Growth of Mixed Populations

When two different species of organisms share a common environment, the ways in which they influence each other's growth and reproduction can be classified into several distinct types. Let's consider the general model:

$$\frac{dX_1}{dt} = f_1(X_1, X_2) \tag{7.145}$$

$$\frac{dX_2}{dt} = f_2(X_1, X_2) \tag{7.146}$$

At a particular steady state, an approximate linearized model can be constructed to examine the local dynamics of the mixed population. The partial derivatives of the growth function are

$$a_{i,j} = \frac{\partial f_i}{\partial X_j} \tag{7.147}$$

Table 7.5 Classification of mixed population growth types.

		Sign of $a_{i,j}$		
		−	0	+
Sign of $a_{j,i}$	−	Competition	Amensalism	Predation
	0	Amensalism	Neutralism	Commensalism
	+	Predation	Commensalism	Mutualism

The influence of organism j on the growth of species i depends on the sign of $a_{i,j}$. The terms used to describe the different classifications are given in table 7.5.

In *amensalism*, the growth of one species is impeded by the presence of the other. An example of this is the production of antibiotics by microbes that suppress the growth of surrounding microbes, or the sequestering of critical nutrients. In *commensalism*, one organism benefits from the presence of the other, either because a key nutrient is produced as a by-product or because a growth-impeding feature is removed. *Mutualism* is also often called symbiosis and occurs when each organism benefits from the presence of the other. We will consider the remaining two conditions, *competition* and *predation*, in more detail.

Competition Volterra [111] considered a simple theory for two species competing for a common resource, assuming that each species obeyed an extended logistic-type growth function as follows:

$$\frac{dX_1}{dt} = (\mu_1 - \gamma_1(h_1X_1 + h_2X_2))X_1 \tag{7.148}$$

$$\frac{dX_2}{dt} = (\mu_2 - \gamma_2(h_1X_1 + h_2X_2))X_2 \tag{7.149}$$

where μ_i is a specific growth rate constant, γ_i describes the impact of nutrient depletion on species i, and h_i relates to the carrying capacity of species i. Multiplying the first equation by γ_2/X_1 and the second equation by γ_1/X_2, and subtracting the second from the first, yields

$$\frac{\gamma_2}{X_1}\frac{dX_1}{dt} - \frac{\gamma_1}{X_2}\frac{dX_2}{dt} = \mu_1\gamma_2 - \mu_2\gamma_1 \tag{7.150}$$

Integrating with respect to time yields

$$\frac{X_1^{\gamma_2}}{X_2^{\gamma_1}} = Ce^{(\mu_1\gamma_2 - \mu_2\gamma_1)t} \tag{7.151}$$

Population Growth and Death Models

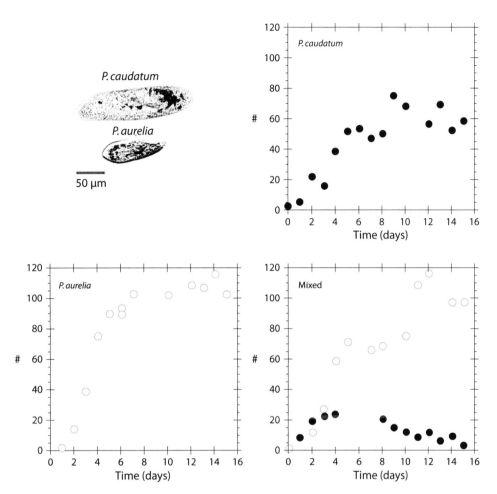

Figure 7.61 Gause's experimental confirmation of Volterra's competitive exclusion principle. Cultures of two species of paramecia were maintained alone or mixed, and were fed every day a fresh supply of the bacterium *Bacillus pyocyaneus*. When both species were in the same culture volume, the quantity of *P. caudatum* steadily declined toward zero. Data are from the evocatively titled paper "Experimental analysis of Vito Volterra's mathematical theory of the struggle for existence" [112]. Paramecium images are from the Protist Information Server [115].

where C is a constant of integration dependent on the initial conditions. Thus, so long as $\mu_1 \gamma_2 \neq \mu_2 \gamma_1$, the proportion of one of the two species with respect to the other will drop to zero approximately exponentially. At steady state, only one of the two species will persist. This concept, called the competitive exclusion principle, was later confirmed experimentally by Gause (figure 7.61 [112]). Levin [113] has cited the following contribution from Theodor Geisel [114] as capturing the essence of this principle well:

And NUH is the letter I use to spell Nutches
Who live in small caves, known as Nitches, for hutches.
These Nutches have troubles, the biggest of which is
The fact there are many more Nutches than Nitches.
Each Nutch in a Nitch knows that some other Nutch
Would like to move into his Nitch very much.
So each Nutch in a Nitch has to watch that small Nitch
Or Nutches who haven't got Nitches will snitch.

Predation The interactions of predators and prey constitute a key topic in the field of ecology. One might imagine a delicate balance exists between a pair of species, one of which is the food source for the other. For example, it seems that growth of the predator species could overwhelm the prey species, reducing its population to zero; or, at the opposite extreme, growth in the prey species could fuel essentially uncontrolled growth of the predator species. These unstable scenarios, although potentially plausible, contradict our sense of long-term coexistence of not only pairs of predator and prey species but also of larger-scale networks in which one species preys on those lower on the food chain and is preyed on by those higher up. Relatively simple models of potentially complex systems reveal key elements of their long-term stability properties.

Early models of the coexistence behavior of predators and their prey were developed independently by Lotka [116] and by Volterra [117] (and are therefore often referred to as Lotka-Volterra models). In the absence of predators, the prey is considered to reproduce exponentially without bound (equation 7.5); in the absence of prey, the predator population dies off in a first-order process. With coexistence, encounters between predator and prey are modeled as second-order processes that augment the predator population and diminish that of the prey. Following the notation of Boccara [118], we write

$$\frac{dH}{dt} = bH - sHP \tag{7.152}$$

$$\frac{dP}{dt} = -dP + esHP \tag{7.153}$$

where H is the population density of prey, P is the population density of predator, b is the birth rate of prey, d is the death rate of predator, s represents the searching/hunting efficiency of predators killing prey, and e gives the influence of killed prey on new predator births. As written, the equations contain four parameters (b, d, s, and e) and are thus difficult to fully explore. The following four substitutions produce a dimensionless system of equations with only a single parameter, ρ [118]:

Population Growth and Death Models

$$h = \frac{esH}{d}; \quad p = \frac{sP}{b}; \quad \tau = \sqrt{bd}\, t; \quad \rho = \sqrt{\frac{b}{d}} \qquad (7.154)$$

The new system of equations is

$$\frac{dh}{d\tau} = \rho h(1-p) \qquad (7.155)$$

$$\frac{dp}{d\tau} = -\frac{1}{\rho} p(1-h) \qquad (7.156)$$

It is useful to plot sample trajectories to examine their behavior. Figure 7.62 shows the results of integrating equations 7.155 and 7.156 for $\tau \in [0, 25]$, $\rho = 2$, and $(h_0, p_0) = (4, 1)$. The trajectories in the left panel show regular periodic behavior, with sharp rises and declines in each population. First the prey population rises, then the predator population follows; next, the prey population drops sharply, then the predator population follows again. This is reasonable, in that a low predator population provides a slow enough feeding rate to allow growth of the prey population. Increasing the prey population provides a larger food base for the predator, and so its population grows. More predation overcomes the increased prey population and it decreases. Finally, fewer prey causes a reduction in the predator population, and the cycle begins again. The cyclic behavior of predator and prey repeats exactly in each cycle, as shown by the perfect coincidence of repeated cycles in the right panel of figure 7.62. Changes to the initial population (h_0, p_0) result in oscillatory cycles as well, although the trajectories follow a different and nonintersecting path, as shown in figure 7.63. Changes to the dimensionless parameter ρ modify the relative amplitudes of the predator and prey cycles, with the two amplitudes being the same for $\rho = 1$.

A useful mechanism for analyzing systems of this sort is to apply now-standard approaches from the field of nonlinear dynamics (chapter 6; [119]). One first identifies the stationary points by setting equations 7.155 and 7.156 to zero, which produces two solutions: $(h, p) = (0, 0)$ and $(1, 1)$. Next, one examines the behavior of the system of equations in the neighborhood of each stationary point by performing a linear expansion about each stationary point. This is achieved by translating the system, so the new origin is at the stationary point, and keeping the first term of a Taylor series.

For the stationary point at $(h, p) = (0, 0)$, the translation to the new origin is trivial: $(\hat{h}, \hat{p}) = (h, p)$, and the expansion is produced by

$$\frac{d\hat{h}}{d\tau} \approx \frac{d}{dh}\left(\frac{dh}{d\tau}\right)\bigg|_{(h=0,\, p=0)} h + \frac{d}{dp}\left(\frac{dh}{d\tau}\right)\bigg|_{(h=0,\, p=0)} p \qquad (7.157)$$

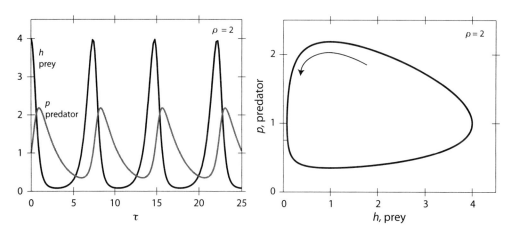

Figure 7.62 Integration of the dimensionless Lotka-Volterra model for predator-prey interactions. Equations 7.155 and 7.156 were numerically integrated for $\tau \in [0, 25]$, $\rho = 2$, and $(h_0, p_0) = (4, 1)$. Notice that the predator and prey populations oscillate with a regular period and amplitude. In each cycle, the prey population rises first, followed by a rise in the predator population; then there ensues a drop, first in prey, and then in predator. (Left panel) Trajectories show values of h and p as a function of dimensionless time τ. (Right panel) The same pair of trajectories is now shown as a plot of h versus p, with the arrow indicating the direction of motion.

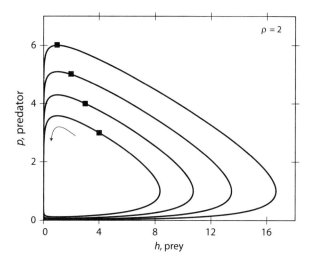

Figure 7.63 The dimensionless Lotka-Volterra model for predator-prey interactions, equations 7.155 and 7.156, were numerically integrated for a variety of initial conditions (h_0, p_0) with $\rho = 2$. Oscillations of predator and prey populations occur for all initial conditions, but the path of each trajectory depends strongly on initial conditions (indicated by a square on each trajectory).

$$\frac{d\hat{p}}{d\tau} \approx \frac{d}{dh}\left(\frac{dp}{d\tau}\right)\bigg|_{(h=0,\,p=0)} h + \frac{d}{dp}\left(\frac{dp}{d\tau}\right)\bigg|_{(h=0,\,p=0)} p \qquad (7.158)$$

which can be more compactly represented in matrix form:

$$\begin{pmatrix} \frac{d\hat{h}}{d\tau} \\ \frac{d\hat{p}}{d\tau} \end{pmatrix} = \begin{pmatrix} \rho & 0 \\ 0 & -\frac{1}{\rho} \end{pmatrix} \begin{pmatrix} h \\ p \end{pmatrix} \qquad (7.159)$$

Performing the same operation for the stationary point at $(h,p)=(1,1)$ produces the matrix equation

$$\begin{pmatrix} \frac{d\hat{h}}{d\tau} \\ \frac{d\hat{p}}{d\tau} \end{pmatrix} = \begin{pmatrix} 0 & -\rho \\ \frac{1}{\rho} & 0 \end{pmatrix} \begin{pmatrix} h \\ p \end{pmatrix} \qquad (7.160)$$

A general solution to this form of equation, $\dot{\vec{X}} = A\vec{X}$ (where A is a matrix), can be obtained by substituting $\vec{X}(t) = e^{\lambda t}\vec{V}$:

$$\dot{\vec{X}} = A\vec{X} \qquad (7.161)$$
$$\lambda e^{\lambda t}\vec{V} = Ae^{\lambda t}\vec{V} \qquad (7.162)$$
$$(A - \lambda I)\vec{V} = 0 \qquad (7.163)$$

with I being the identity matrix. Specializing to the case of a 2×2 matrix, as for the instances described above, let

$$A = \begin{pmatrix} a & b \\ c & d \end{pmatrix} \qquad (7.164)$$

so equation 7.163 can be represented as

$$\begin{pmatrix} a-\lambda & b \\ c & d-\lambda \end{pmatrix} \vec{V} = 0 \qquad (7.165)$$

for which the nontrivial solution is

$$\det\begin{bmatrix} a-\lambda & b \\ c & d-\lambda \end{bmatrix} = 0 \qquad (7.166)$$

or

$$(a-\lambda)(d-\lambda) - bc = 0 \qquad (7.167)$$
$$\lambda^2 - (a+d)\lambda + (ad-bc) = 0 \qquad (7.168)$$

$$\lambda^2 - \text{Tr}(A)\lambda + \text{Det}(A) = 0 \tag{7.169}$$

whose solutions are

$$\lambda_1 = \frac{\text{Tr}(A) + \sqrt{\text{Tr}^2(A) - 4 \cdot \text{Det}(A)}}{2}, \quad \lambda_2 = \frac{\text{Tr}(A) - \sqrt{\text{Tr}^2(A) - 4 \cdot \text{Det}(A)}}{2} \tag{7.170}$$

where $\text{Tr}(A) = a + d$ and $\text{Det}(A) = ad - bc$ are the trace and determinant of the matrix A, respectively. If the eigenvalues are distinct ($\lambda_1 \neq \lambda_2$), then the eigenvectors $\vec{V}_1$ and $\vec{V}_2$ are linearly independent and solutions of the form

$$\vec{X}(t) = c_1 e^{\lambda_1 t} \vec{V}_1 + c_2 e^{\lambda_2 t} \vec{V}_2 \tag{7.171}$$

are valid for the linearized system.

The form of the solution for the system of equations linearized about $(0, 0)$ is

$$\begin{pmatrix} \hat{h}(\tau) \\ \hat{p}(\tau) \end{pmatrix} = c_1 e^{\rho \tau} \begin{pmatrix} 1 \\ 0 \end{pmatrix} + c_2 e^{-\frac{\tau}{\rho}} \begin{pmatrix} 0 \\ 1 \end{pmatrix} \tag{7.172}$$

and it can be shown that this solution, which corresponds to that of a saddle point, is valid in a region sufficiently close to the fixed point at $(0, 0)$ [119]. The solution corresponding to the linearization about the fixed point at $(1, 1)$ is

$$\begin{pmatrix} \hat{h}(\tau) \\ \hat{p}(\tau) \end{pmatrix} = c_1 e^{i\tau} \begin{pmatrix} i\rho \\ 1 \end{pmatrix} + c_2 e^{-i\tau} \begin{pmatrix} -i\rho \\ 1 \end{pmatrix} \tag{7.173}$$

For equation 7.173, c_1 and c_2 are complex conjugates of one another, and the oscillatory solutions correspond to the cyclic behavior shown in figures 7.62 and 7.63. This is a linearization of the actual dimensionless Lotka-Volterra mechanism, and it is theoretically possible that the higher-order terms in the full equation will upset the oscillatory behavior [119]. That is not the case here, and other approaches exist to guarantee oscillatory behavior of nonlinear systems (chapter 6). The linearized solution is a series of nested ellipses centered at $(1,1)$; the nonlinear terms (involving $h \times p$) in equations 7.155 and 7.156 are responsible for the nonelliptical shapes of the cycles in figure 7.63. This solution is called neutrally stable, because multiple oscillating trajectories exist for a range of initial conditions. This differs from situations in which altered initial conditions lead to solutions that converge over time to the same eventual path, known as a *limit-cycle* solution.

What about the solution in equation 7.172? This represents a formal class of solutions that pass through the origin. The number of prey h increases without bound, while the number of predators p decreases exponentially. This represents a collapse of what could be a self-sustaining ecosystem. (It is actually remarkable that the model predicts a sustainable population model over the wide range of conditions that it does.) Moreover, the uncontrolled growth is unphysical, but it is expected from the model, because the only control on the prey population is the presence of predator. An absence of predator and any finite number of prey should lead to this result, and the unphysical outcome is due to the simplicity of the model. Indeed, the types of limitations on growth introduced earlier in this chapter are appropriate to add to the core Lotka-Volterra oscillating model.

It is interesting to examine perturbations to predator-prey systems in the context of simple models. For example, imagine that the predator is hunted in an attempt to reduce its numbers. What will be the effect on its population and that of the prey? Likewise, if disease were suddenly to kill off a large number of prey, how does that propagate to the predator population, and what are the long-term effects on both species? The results of exploring these questions with the Lotka-Volterra model are explored in figure 7.64. Interestingly, the fixed point remains unmoved, and the perturbation moves the system to a new oscillatory solution that is concentric with the

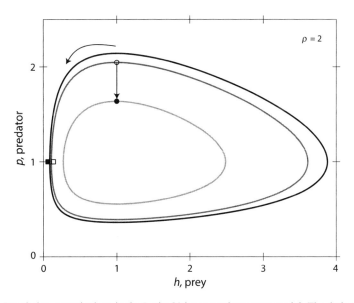

Figure 7.64 Population perturbations in the Lotka-Volterra predator-prey model. The dark gray path represents a stable population trajectory. The open circle represents the time at which the predator population was instantaneously reduced by 20%, to represent purposeful culling of the population, and the closed circle shows the new stable trajectory followed by the population. The open square represents the initial population at a time when disease suddenly removes 20% of the prey. Interestingly, reducing the number of individuals in a species when near its oscillator peak reduces the amplitude of oscillations of both populations by about the same average; a similar reduction near the oscillatory trough actually increases the amplitude of oscillation of both species, again by about the same average.

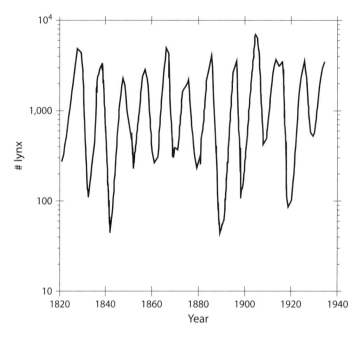

Figure 7.65 The variation in the lynx population in northwest Canada displays oscillations with a period of approximately 10 years. The abscissa covers the years 1821–1934, and the ordinate is on a logarithmic scale [121].

original solution (that is, the old and new solutions do not cross). A reduction in the predator (or prey) population near its peak leads to a new oscillatory solution, with smaller amplitude swings in both predator and prey populations; however, a reduction in the predator (or prey) population near its trough actually leads to increased magnitude oscillations of both populations.

The core Lotka-Volterra model is remarkably useful in representing the fundamental oscillatory behavior of predator-prey populations. Indeed, analysis of population records corresponding to hunter and hunted displays the types of oscillations inherent in the core model. One example is taken from an analysis of records for the Hudson Bay Company, which report on the numbers of animals captured by trapping in the Mackenzie River district of northwest Canada [120]. The population of lynx, shown in figure 7.65, indicates a relatively regular oscillation with somewhat uneven amplitude and a period of close to 10 years. A full accounting of the oscillations would require an understanding and tracking of the hierarchy of predators, prey, and competitors in the region, and significant work has been done to get at some of this information.

The simplicity of the Lotka-Volterra mechanism is useful for studying the fundamental reasons underlying oscillatory population behavior, but the model lacks important features. For example, it predicts that in the absence of predator, the population of prey will increase without bound. A logistic or Gompertz-type growth equation has been used to correct this behavior. The equations also suggest that a predator population will increase more with increasing prey, without limit. A more realistic treatment provides for saturation behavior, in which a predator can only hunt a maximum number of prey and produce a corresponding maximum increase in the predator population. Further refinements are possible in the area of the model describing how increases in prey lead to increases in predator. Although increased prey will lead almost immediately to fewer predator deaths, the effect on new predator births will be delayed by at least the gestation period. The use of delay differential equations is one way to simulate time delays in such systems. Other additions to the model are illustrated in Boccara [118], one of which is

$$\frac{dH}{dt} = r_H H \left(1 - \frac{H}{K}\right) - \frac{a_H P H}{b+H} \qquad (7.174)$$

$$\frac{dP}{dt} = \frac{a_P P H}{b+H} - cP \qquad (7.175)$$

where the prey population has a finite carrying capacity K; the effectiveness of predators saturates with sufficient prey (the $\frac{1}{b+H}$ terms); and hunting that incrementally decreases the prey population increases the predator population by a factor of a_P/a_H as much, which is termed a *functional response*.

Suggestions for Further Reading

J. E. Bailey and D. E. Ollis. *Biochemical Engineering Fundamentals*, second edition. McGraw-Hill, 1986.

H. W. Blanch and D. S. Clark. *Bioreaction Engineering Principles*, third edition. Marcel Dekker, 1997.

M. L. Shuler, F. Kargi, and M. DeLisa. *Bioprocess Engineering: Basic Concepts*, third edition. Prentice-Hall, 2017.

S. H. Strogatz. *Nonlinear Dynamics and Chaos: With Applications to Physics, Biology, Chemistry, and Engineering*, second edition. CRC Press, 2018.

J. Villadsen, J. Nielsen, and G. Lidén. *Bioreaction Engineering Principles*, third edition. Springer, 2011.

Problems

7-1 The objective of this exercise is to compare the volumetric productivity of a steady-state chemostat to that of a batch reactor. The batch operating time is the time for exponential biomass growth from X_o to X plus a turnaround time t_{turn}. Show that the ratio of volumetric biomass productivity for a chemostat versus a batch reactor is approximately $\ln \frac{X}{X_o} + \mu t_{turn}$.

7-2 Show that for Gompertzian growth kinetics (equation 7.16), as $t \to 0$, $N \to N_o e^{\mu t}$. (*Hint:* Expand the $e^{-\alpha t}$ term in a Taylor series, and truncate beyond linear terms in t.)

7-3 You are trying to kill a tumor with beta radiation.

a. What radiation dose is necessary to kill 99.99% of the tumor cells?

b. What radiation dose is necessary if you use a fractionated therapy once a day for seven days?

You may assume a linear quadratic model with $\alpha = 0.09\,\text{Gy}^{-1}$ and $\beta = 0.05\,\text{Gy}^{-2}$. The doubling time for the tumor is one week. (Contributed by K. J. Davis.)

7-4 The age at death for a collection of fossilized *Tyrannosaurus rex* skeletons was estimated [122], and the survivorship curve for an initial hypothetical cohort of 1,000 is presented in table 7.6.

Table 7.6 *T. rex* survival table.

Age (yr)	Surviving #
2	385
4	372
6	360
8	336
9	314
11	304
12	284
13	266
14	249
15	217
16	181
17	164
18	125
19	108
20	71
21	54
23	36
28	18

For ages 2–28 years, what type of death kinetics are these data consistent with? What does this imply about the types of cause of death for *T. rex* in this age range? How does this contrast with the causes of death for *T. rex* under 2 years of age?

7-5 Bacterial contamination in mammalian cell culture can be a significant problem without proper precautions. Derive the relationship to estimate the time it would take a single bacteria cell to double until the bacterial cell population size reaches that of the mammalian cell population size, in terms of the initial mammalian cell population size ($N_{o,m}$) and doubling times of each cell type. Assume exponential growth for each population. How long would this time be for a flask initially containing 3 million mammalian cells, with doubling times of 20 hours and 30 minutes for the mammalian cells and bacteria cells, respectively? (Contributed by B. Belmont.)

7-6 The data in table 7.7 were obtained for a batch hybridoma culture producing an IgG1 of interest. What is the productivity of the cells, in pg antibody/cell/day? What volume of reactor would need to be designed to produce 10^5 kg/yr of this antibody?

Table 7.7 Antibody production kinetics.

Time (h)	Viable cells/mL	IgG1 (mg/L)
0	1.19×10^5	0.3
21	3.32×10^5	2.5
46	5.28×10^5	7.7
68	5.68×10^5	16.3
92	4.61×10^5	31.8
118	3.03×10^5	41.3
143	2.37×10^5	48.0
166	1.52×10^5	56.3
190	1.00×10^5	57.2

7-7 In the Smith-Martin model of cell replication, the fraction of cells that have not yet divided is considered as a function of cell age (that is, the time since the division that created that cell). An essential feature of the model is that the relative rate for entering mitosis, $-\frac{1}{\alpha}\frac{d\alpha}{dt}$, increases with age. For simplicity, imagine a culture of young cells that are all the same age. Let $f(t)$ describe the distribution of times at which the original cells first leave interphase. Compute and plot the relative rate for entering mitosis as a function of age using each of the following for $f(t)$:

a. $f(t) = \begin{cases} 1/a & \text{for } t_0 < t < t_0 + a \\ 0 & \text{otherwise} \end{cases}$

b. $f(t) = A_0 \exp(-k(t-t_0))$

c. $f(t) = \begin{cases} \dfrac{1}{a^2}(t-t_0) + \dfrac{1}{a} & \text{for } t_0 - a < t < t_0 \\ -\dfrac{1}{a^2}(t-t_0) + \dfrac{1}{a} & \text{for } t_0 < t < t_0 + a \\ 0 & \text{otherwise} \end{cases}$

7-8 Equations 7.174 and 7.175 can be nondimensionalized using

$$h = \frac{H}{H^*}; \quad p = \frac{P}{P^*}; \quad \tau = r_H t; \quad k = \frac{K}{H^*}; \quad \beta = \frac{b}{H^*}; \quad \gamma = \frac{c}{r_H} \qquad (7.176)$$

(where (H^*, P^*) is a nontrivial stationary point) to give

$$\frac{dh}{d\tau} = h\left(1 - \frac{h}{k}\right) - \frac{\alpha_h p h}{\beta + h} \qquad (7.177)$$

$$\frac{dp}{d\tau} = \frac{\alpha_p p h}{\beta + h} - \gamma p \qquad (7.178)$$

subject to the conditions

$$\alpha_h = \left(1 - \frac{1}{k}\right)(\beta + 1) \qquad (7.179)$$

and

$$\alpha_p = \gamma(\beta + 1) \qquad (7.180)$$

with fixed point $(h, p) = (1, 1)$.

Solve for the Jacobian (derivative matrix) of this system at the fixed point. Using a value of $1 < k < 2 + \beta$, plot a series of trajectories starting from different initial conditions. Do the same for $k > 2 + \beta$. This system has a Hopf bifurcation at $k = 2 + \beta$ and displays qualitatively different behaviors on either side of this condition. (Adapted from reference [118].)

7-9 The main tourist attraction at Furano, Hokkaido, is its beautiful lavender garden, an expanse covered with artfully arranged wildflowers. Unfortunately, the flower stems are the favorite food of the borer; the flower and borer populations fluctuate cyclically in accordance with Lotka-Volterra's predator-prey equations. To boost the wildflower level for the tourists, the director wants to fertilize the garden, so that the flower growth will outrun that of the borer. Assume that the fertilizer would boost the wildflower growth rate (in the absence of borers) by 25%. What do you think of this proposal? What if we genetically modified the flower so that it is slightly toxic to borers and decrease the growth rate of borers by 25%? (Contributed by Y. Matsumoto.)

7-10 Estes et al. [123] tagged and monitored sea otters in two different populations, Clam Lagoon and Kuluk Bay. One key difference between the populations is the presence of killer whales as a predator in Kuluk Bay. Based on the data shown in table 7.8, describe how you would model each case, and determine the appropriate model parameter(s) for the population in Kuluk Bay. (Contributed by K. Bernick.)

Table 7.8 Sea otter survival rates.

Time (months)	Kuluk Bay	Clam Lagoon
0	100	100
6	60	100
12	33	90
25	5	68

7-11 Obtain an analytical solution for i in terms of s to the *SIR* epidemic model (equations 7.137 and 7.138), assuming that the recovered population is zero at time zero. Show that the infectious population will be monotonically decreasing only if $\sigma \cdot s(0) \leq 1$. How would one find the fixed point(s) of the solution for a given set of parameters? Use this method to find the fixed points for a system in which $s(0) = 0.4$ and $\sigma = 3$. (Contributed by J. Spangler.)

7-12 *Saccharomyces cerevisiae* ferments sugar to ethanol in a closed batch system. If the initial sugar and cell concentrations are 100 g/L and 2 g/L, respectively, what is the final ethanol concentration achieved? Assume a gram of sugar yields 0.5 grams of biomass metabolized. Plot the time profiles of the sugar, cell, and ethanol concentrations. (Contributed by C. Zopf).

7-13 You are trying to determine the maintenance requirement of a cell culture in a chemostat reactor as well as the actual biomass yield coefficient $Y_{X/S}$. You run a chemostat at different flow rates with 20 g/L substrate in the entering stream and measure the substrate concentration and cell concentration in the outgoing stream, as shown in table 7.9. Determine m and $Y_{X/S,\text{actual}}$. (Contributed by D. V. Liu.)

Table 7.9 Chemostat biomass and substrate levels.

D (1/h)	X (g cell/L)	$[S]$ (g substr/L)
0.1	3.10	14.20
0.2	4.12	14.56
0.3	4.29	15.12
0.4	4.56	15.23
0.5	4.75	15.30
0.6	4.89	15.34
0.7	4.98	15.38
0.8	5.03	15.43

7-14 A food-processing plant has a 100 L/h waste stream containing 15 g/L glucose. They wish to produce yeast biomass from this stream, to be used as a supplement for cattle feed. The growth characteristics of the yeast are plotted in figure 7.66.

a. What volume of bioreactor will be required?

b. How much biomass will be produced per day?

c. Due to changes upstream in the plant, the waste stream flow rate increases to 400 L/h (still 15 g/L glucose). Using the bioreactor of the volume calculated in part a, how much biomass can now be produced per day?

7-15 Develop a simple structured model for bacterial growth on two nutrients, A and B, where the bacteria prefer to utilize A. Use the following assumptions: (1) The rate of growth on A is proportional to the concentration of a constitutive intracellular enzyme, which is produced at a constant rate per unit biomass and degraded by a first-order process; (2) The rate of growth on B is proportional to the concentration of an inducible intracellular enzyme, which is produced at a rate that decreases with increasing concentration of A and approaches a constant as the concentration of A goes to zero, per unit biomass. This enzyme is also degraded by a first-order process. In addition to the aforementioned enzyme dependencies, the specific rates of growth also depend on their respective nutrients, A and B, in a Michaelis-Menten manner.

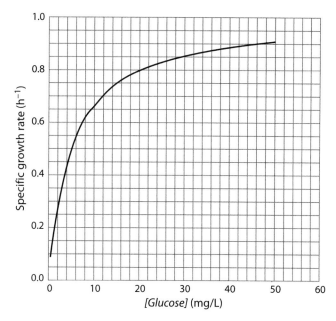

Figure 7.66 Relationship between glucose concentration and specific growth rate for yeast.

Write the appropriate equations for growth in a well-mixed batch reactor. For balanced growth, develop an expression for the dependence of specific growth rate on the concentrations of A and B, using the additional assumption that the enzyme degradation rates are rapid compared to cell growth rates.

7-16 A chemostat (feed sugar concentration = 10 g/L) is operated at a variety of dilution rates, and the steady-state biomass concentration as a function of dilution rate is plotted in figure 7.67. Assuming that the cells follow Monod growth kinetics, what would the maximum biomass productivity be (g biomass/L/h) if the sugar concentration were increased to 50 g/L?

7-17 a. Plot the Lotka-Volterra predator-prey versus time and phase plots using the following parameters, and comment on the steady-state solution:
b (birth rate of prey): 6
s (hunting efficiency of predator): 0.3
d (death rate of predator): 2
e (influence of killed prey on predator birth rate): 0.2
Initial prey population: 15
Initial predator population: 10

b. The Lotka-Volterra model does not account for resource limitations arising from competition within a species group. To include such competition,

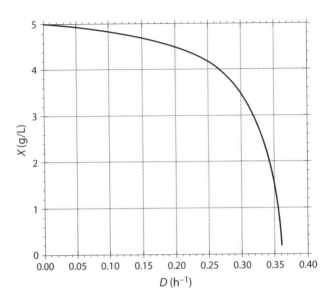

Figure 7.67 Biomass versus dilution rate in a chemostat.

a quadratic expression is subtracted from each differential equation. The expressions, αX^2 and βY^2, represent the rate of death of prey and predators, respectively, as a result of interspecies competition. Produce similar plots as in part a using this new model, and comment on the steady-state solution when $\alpha = 0.005$ and $\beta = 0.02$. These parameters represent the probability of a fatal interaction between same-species individuals. (Contributed by D. Weingeist.)

7-18 A nongrowth-associated product is to be produced in two chemostats operated in series, with the effluent from the first reactor being fed to the second, which has no other feed streams entering. The first reactor volume is half that of the second. What dilution rate in the first stage will maximize the rate of production of the nongrowth-associated product? Prepare a graph of productivity ($D[P]$) for each stage. Compare the maximum productivity obtainable in the two-stage reactor with that obtainable in a one-stage chemostat that has the same total volume as the sum of both reactors in the two-stage system. What reactor configuration would you recommend for this type of product? The parameter values for this problem are: $\mu_{max} = 0.4$ h^{-1}, $K_s = 0.05$ g/L, $Y_{X/S} = 0.5$ g/g; $[S]_o = 10$ g/L; $Y_{P/S} = 0.1$ g/g; $\beta = 0.1$ h^{-1}.

7-19 Consider a culture of bacteria with a composition consistent with the empirical biomass formula $CH_{1.7}O_{0.46}N_{0.18}$, or MW$_B$=23.6 g/mol, growing aerobically with glucose ($C_6H_{12}O_6$) as the carbon source. The measured yield coefficient for glucose is $Y_{X/S} = 85$ g biomass/mol glucose. The measured yield coefficient for oxygen is $Y_{X/O_2} = 39$ g biomass/mol O_2.

a. Show that the measured yield coefficients are stoichiometrically consistent with each other.

b. A batch culture of this organism initially containing 0.01 g of biomass inoculum and 20 mmol glucose is incubated overnight and observed by optical density measurements to have stopped growing. The total biomass is calculated to be 1.0 g. Estimate the remaining glucose, and speculate as to the reasons for cessation of growth. (Contributed by J. Haugh.)

7-20 A bacterial culture is grown using pyruvate ($C_3H_4O_3$) as the carbon source. The nitrogen source is ammonium salts. The empirical biomass formula is $CH_{1.8}O_{0.5}N_{0.17}$, or $MW_B = 24.2$ g/mol.

a. Based on this information alone, estimate the maximum theoretical yield of biomass per mol of pyruvate (g DCW/mol).

This culture is carried out aerobically, and based on measurements of carbon dioxide in the headspace and dissolved in the culture, you determine that 45 mmol of CO_2 is liberated for every g DCW biomass produced.

b. Estimate the actual biomass yield per mol pyruvate consumed.

c. Speculate as to any discrepancy between the yield coefficients from parts a and b. (Contributed by J. Haugh.)

7-21 Under substrate-limited conditions, a microorganism exhibits the following growth rate relationship:

$\mu(h^{-1}) = 0.7[S]/(0.1 + [S])$, [S] in g/L; $Y_{X/S}$ (g DCW/g) = 0.40

The available growth medium contains 10 g/L substrate.

a. A 100 L batch bioreactor is inoculated with 1.0 g DCW of biomass. Estimate the maximum cell density achieved, and the approximate time required to achieve it after exponential growth resumes.

b. As an alternative, a chemostat of this microorganism is operated with the same growth medium. Estimate the dilution rate for maximum steady-state biomass productivity.

c. Calculate and compare the overall biomass productivities in g DCW/L/h of the two scenarios from parts a and b. What other consideration(s) will decrease the batch process productivity further? (Contributed by J. Haugh.)

7-22 Yeast grown with glucose as the carbon source generally metabolizes glucose primarily via glycolysis, producing ethanol as a by-product. Once glucose is completely consumed, the ethanol by-product is then consumed by aerobic respiration. This shift in growth regimes is often called diauxic growth. A simple structured model has been constructed that describes this effect well [124, 125].

The assumptions of this model are as follows. The length of the G_1 phase of the cell cycle depends on the availability of the limiting substrate; the length of the division phase (the sum of G_2, M, and S phases) is assumed to be independent of substrate. The cell mass is considered to be composed of two parts: mass A, which carries out substrate uptake and energy production; and mass B, which carries out reproduction and division. B is converted to A at a constant rate, whereas A consumes substrate and produces B at a variable rate. Thus the total biomass is $X = A + B$.

The following pseudo-reactions describe the model:

$$A + a_1[S] \xrightarrow{r_A} 2B + a_2[E] + [CO_2] \qquad (7.181)$$

$$[O_2] + A + a_3[E] \xrightarrow{r_B} 2B + [CO_2] \qquad (7.182)$$

$$B \xrightarrow{r_C} A \qquad (7.183)$$

where $[S]$ = glucose concentration, and $[E]$ = ethanol concentration.

The rates of glycolysis, respiration, and cell mass interconversion are given as

$$r_A = k_1 A \frac{[S]}{K_s + [S]} \qquad (7.184)$$

$$r_B = k_2 A \frac{[E]}{K_e + [E]} \qquad (7.185)$$

$$r_c = KB \qquad (7.186)$$

where parameter estimates from experiments are as follows: $k_1 = 5.0$ h^{-1}; $K = 0.51$ h^{-1}; $k_2 = 0.30$ h^{-1}; $K_s = 0.5$ g/L; $K_e = 0.02$ g/L; $a_1 = 6.67$; $a_2 = 2.80$; $a_3 = 2.22$.

a. Consider a batch growth experiment with the following initial conditions at time zero: $[E] = 0$, $[S] = 10$ g/L, A = 0.01 g/L, and B = 0.01 g/L. Simulate this experiment using the structured model above, and predict the specific growth rate μ before depletion of glucose.

b. If the product of interest in this fermentation is ethanol, at what time should the fermentation be stopped?

c. Repeat parts a and b, except with initial sugar concentrations $[S] = 1$ g/L.

d. Repeat part c with $[S] = 0.1$ g/L.

e. Plot (maximum ethanol concentration)/(initial glucose concentration) versus (initial glucose concentration). What sugar concentration should be used to most efficiently convert sugar to ethanol? (Contributed by J. Haugh.)

7-23 You are producing bacteria with a specific growth rate of 0.0236 min^{-1} in a fed-batch fermentor using glucose as a substrate. If the initial concentration of glucose inflow to the fermentor is 25.0 g/L and the system is run for 50 minutes, what is the final concentration of bacteria if the initial population is 5.0 mg/L? If the initial fermentor volume was 1 L, what is the total yield of biomass? (Contributed by L. Llemke.)

7-24 You are culturing yeast cells, which double every 3 hours, in a 1 L pilot reactor. As you remove a tiny aliquot of culture to determine the number of cells in the reactor (0.3 million), you accidentally introduce 1,000 bacteria into the reactor. The bacteria, which double every 25 minutes, secrete a short-lived toxin that inhibits yeast growth. As such, you can describe the yeast growth rate as follows:

$$\mu = \begin{cases} \mu_o(1 - \beta E) & E < 1/\beta \\ 0 & E \geq 1/\beta \end{cases}$$

in which μ_o is the uninhibited growth rate constant, β is a constant related to the strength of inhibition ($\beta = 1$ L/g bacteria), and E is the cell density of bacteria. Note that these bacteria enter stationary phase at a cell density of 2 g/L.

a. Derive an expression for the yeast cell density as a function of time.

b. Calculate the steady-state yeast cell density.

c. Calculate the steady-state yeast cell density if the contamination were a single bacterium and occurred when the yeast density was initially 10× higher (i.e., 3 million yeast cells in the reactor).

References

1. L. H. Johnston and A. L. Johnson. Elutriation of budding yeast. *Methods Enzymol.* 283: 342–350 (1997).
2. P.-F. Verhulst. Notice sur la loi que la population poursuit dans son accroissement. *Correspondance Mathématique et Physique* 10: 113–121 (1838).
3. R. Pearl and L. J. Reed. On the rate of growth of the population of the United States since 1790 and its mathematical representation. *Proc. Natl. Acad. Sci. U.S.A.* 6: 275–288 (1920).
4. R. Pearl. The growth of populations. *Q. Rev. Biol.* 2: 532–548 (1927).

5. P.-F. Verhulst. Recherches mathématiques sur la loi d'accroissement de la population. *Nouv. mem. de l'Acad. Roy. des Sci. et Belles-Lett. de Bruxelles* 18: 1–38 (1845).

6. S. Wright. Book review. *J. Am. Stat. Assoc.* 21: 494 (1926).

7. C. P. Winsor. The Gompertz curve as a growth curve. *Proc. Natl. Acad. Sci. U.S.A.* 18: 1–8 (1932).

8. T. Carlson. Über Geschwindigkeit und Grösse der Hefevermehrung in Würze. *Biochemische Zeitschrift* 57: 313–334 (1913).

9. M. Kleiber. Body size and metabolism. *Hilgardia* 6: 315–353 (1932).

10. G. B. West and J. H. Brown. The origin of allometric scaling laws in biology from genomes to ecosystems: Towards a quantitative unifying theory of biological structure and organization. *J. Exp. Biol.* 208: 1575–1592 (2005).

11. L. von Bertalanffy. Quantitative laws in metabolism and growth. *Q. Rev. Biol.* 32: 217–231 (1957).

12. R. H. Peters. *The Ecological Implications of Body Size*. Cambridge University Press (1986).

13. G. B. West, J. H. Brown, and B. J. Enquist. A general model for ontogenetic growth. *Nature* 413: 628–631 (2001).

14. L. Norton, R. Simon, H. D. Brereton, and A. E. Bogden. Predicting the course of Gompertzian growth. *Nature* 264: 542–545 (1976).

15. A. K. Laird. Dynamics of growth in tumors and in normal organisms. *Natl. Cancer Inst. Monogr.* 30: 15–28 (1969).

16. R. Chignola, A. Schenetti, G. Andrighetto, E. Chiesa, R. Foroni, S. Sartoris, G. Tridente, and D. Liberati. Forecasting the growth of multicell tumour spheroids: Implications for the dynamic growth of solid tumours. *Cell Prolif.* 33: 219–229 (2000).

17. C. Guiot, P. G. Degiorgis, P. P. Delsanto, P. Gabriele, and T. S. Deisboeck. Does tumor growth follow a "universal law"? *J. Theor. Biol.* 225: 147–283 (2003).

18. P. W. Sullivan and S. E. Salmon. Kinetics of tumor growth and regression in IgG multiple myeloma. *J. Clin. Invest.* 51: 1697–1708 (1972).

19. H. E. Skipper. Kinetics of mammary tumor cell growth and implications for therapy. *Cancer* 28: 1479–1499 (1971).

20. M. A. Savageau. Growth of complex systems can be related to the properties of their underlying determinants. *Proc. Natl. Acad. Sci. U.S.A.* 76: 5413–5417 (1979).

21. B.-E. Sæther, S. Engen, and E. Matthysen. Demographic characteristics and population dynamical patterns of solitary birds. *Science* 295: 2070–2073 (2002).

22. M. E. Gilpin and F. J. Ayala. Global models of growth and competition. *Proc. Natl. Acad. Sci. U.S.A.* 70: 3590–3593 (1973).

23. R. M. Sibly, D. Barker, M. C. Denham, J. Hone, and M. Pagel. On the regulation of populations of mammals, birds, fish, and insects. *Science* 309: 607–610 (2005).

24. J. Monod. The growth of bacterial cultures. *Annu. Rev. Microbiol.* 3: 371–394 (1949).

25. J. Nielsen, J. Villadsen, and G. Lidén. *Bioreaction Engineering Principles*, second edition. Springer (2003).

26. D. S. Clark and H. W. Blanch. *Biochemical Engineering*. CRC Press (1995).

27. M. A. Chaudhry, B. D. Bowen, C. J. Eaves, and J. M. Piret. Empirical models of the proliferative response of cytokine-dependent hematopoietic cell lines. *Biotechnol. Bioeng.* 5: 348–358 (2004).

28. S. S. Ozturk and B. Ø. Palsson. Growth, metabolic, and antibody production kinetics of hybridoma cell culture: 2. Effects of serum concentration, dissolved oxygen concentration, and medium pH in a batch reactor. *Biotechnology* 7: 481–494 (1991).

29. M. W. Glacken. *Development of mathematical descriptions of mammalian cell culture kinetics for the optimization of fed-batch bioreactors*. D.Sc. thesis (Massachusetts Institute of Technology, 1987).

30. D. J. Knauer, H. S. Wiley, and D. A. Lauffenburger. Relationship between epidermal growth factor receptor occupancy and mitogenic response. Quantitative analysis using a steady state model system. *J. Biol. Chem.* 259: 5623–5631 (1984).

31. D. A. Cantrell and K. A. Smith. The interleukin-2 T-cell system: A new cell growth model. *Science* 224: 1312–1316 (1984).

32. P. W. Zandstra, D. A. Lauffenburger, and C. J. Eaves. A ligand–receptor signaling threshold model of stem cell differentiation control: A biologically conserved mechanism applicable to hematopoiesis. *Blood* 96: 1215–1222 (2000).

33. B. M. Rao, I. Driver, D. A. Lauffenburger, and K. D. Wittrup. High-affinity CD25-binding IL-2 mutants potently stimulate persistent T cell growth. *Biochemistry* 44: 10,696–10,701 (2005).

34. R. M. Stroud and J. A. Wells. Mechanistic diversity of cytokine receptor signaling across cell membranes. *Science STKE* 231: re7 (2004).

35. K. H. Pearce Jr., B. C. Cunningham, G. Fuh, T. Teeri, and J. A. Wells. Growth hormone binding affinity for its receptor surpasses the requirements for cellular activity. *Biochemistry* 38: 81–89 (1999).

36. S. Rousseau, F. Houle, J. Landry, and J. Huot. p38 MAP kinase activation by vascular endothelial growth factor mediates actin reorganization and cell migration in human endothelial cells. *Oncogene* 15: 2169–2177 (1997).

37. A. E. Postlethwaite, J. Keski-Oja, H. L. Moses, and A. H. Kang. Stimulation of the chemotactic migration of human fibroblasts by transforming growth factor beta. *J. Exp. Med.* 165: 251–256 (1987).

38. S. Rankin and E. Rozengurt. Platelet-derived growth factor modulation of focal adhesion kinase (p125FAK) and paxillin tyrosine phosphorylation in Swiss 3T3 cells. Bell-shaped dose response and cross-talk with bombesin. *J. Biol. Chem* 269: 704–710 (1994).

39. G. K. Hoppe and G. S. Hansford. Ethanol inhibition of continuous anaerobic yeast growth. *Biotechnol. Lett.* 4: 39–44 (1982).

40. S. Aiba, M. Shoda, and M. Nagatani. Kinetics of product inhibition in alcohol fermentation. *Biotechnol. Bioeng.* 10: 845–864 (1968).

41. S. Baskaran, H.-J. Ahn, and L. R. Lynd. Investigation of the ethanol tolerance of *Clostridium thermosaccharolyticum* in continuous culture. *Biotechnol. Prog.* 11: 276–281 (1995).

42. G. W. Luli and W. R. Strohl. Comparison of growth, acetate production, and acetate inhibition of *Escherichia coli* strains in batch and fed-batch fermentations. *Appl. Environ. Microbiol.* 56: 1004–1011 (1990).

43. S. S. Ozturk, M. R. Riley, and B. Ø. Palsson. Effects of ammonia and lactate on hybridoma growth, metabolism, and antibody production. *Biotechnol. Bioeng.* 39: 418–431 (1992).

44. W. M. Miller, H. W. Blanch, and C. R. Wilke. A kinetic analysis of hybridoma growth and metabolism in batch and continuous suspension culture: Effect of nutrient concentration, dilution rate, and pH. *Biotechnol. Bioeng.* 32: 947–965 (1988).

45. B. Atkinson and F. Mavituna. *Biochemical Engineering and Biotechnology Handbook*. Nature Press (1983).

46. J. J. Heijnen and J. A. Roels. A macroscopic model describing yield and maintenance relationships in aerobic fermentation processes. *Biotechnol. Bioeng.* 23: 739–763 (1981).

47. J. A. Roels. Application of macroscopic principles to microbial metabolism. *Biotechnol. Bioeng.* 22: 2457–2514 (1980).

48. C. L. Cooney, D. I. C. Wang, and R. I. Mateles. Measurement of heat evolution and correlation with oxygen consumption during microbial growth. *Biotechnol. Bioeng.* 11: 269–281 (1969).

49. A. Novick and L. Szilard. Description of the chemostat. *Science* 112: 715–716 (1950).

50. Z. Zhang, N. Szita, P. Boccazzi, A. J. Sinskey, and K. F. Jensen. A well-mixed, polymerbased microbioreactor with integrated optical measurements. *Biotechnol. Bioeng.* 93: 286–296 (2006).

51. J. E. Bailey and D. F. Ollis. *Biochemical Engineering Fundamentals*. McGraw-Hill (1986).

52. D. Herbert. Some principles of continuous culture. *Recent Prog. Microbiol.* 7: 381–396 (1959).

53. L. Yee and H. W. Blanch. Recombinant protein expression in high cell density fed-batch cultures of *Escherichia coli*. *Nat. Biotechnol.* 10: 1550–1556 (1992).

54. C. T. Goudar, K. Joeris, K. B. Konstantinov, and J. M. Piret. Logistic equations effectively model mammalian cell batch and fed-batch kinetics by logically constraining the fit. *Biotechnol. Prog.* 21: 1109–1118 (2005).

55. R. Luedeking and E. L. Piret. Transient and steady states in continuous fermentation. *J. Biochem. Microbiol. Tech. Eng.* 1: 393–412 (1959).

56. S. S. Ozturk and B. Ø. Palsson. Growth, metabolic, and antibody production kinetics of hybridoma cell culture: 2. Effects of serum concentration, dissolved oxygen concentration, and medium pH in a batch reactor. *Biotechnol. Prog.* 7: 481–494 (1991).

57. J. M. Renard, R. Spagnoli, C. Mazier, M. F. Salles, and E. Mandine. Evidence that monoclonal antibody production kinetics is related to the integral of the viable cells curve in batch systems. *Biotechnol. Lett.* 10: 91–96 (1988).

58. M. E. Reff. High-level production of recombinant immunoglobulins in mammalian cells. *Curr. Opin. Biotechnol.* 4: 573–576 (1993).

59. S. Iwanami and N. Oda. Theory of survival of bacteria exposed to ionizing radiation. I. X and gamma rays. *Radiat. Res.* 102: 46–58 (1985).

60. E. J. Hall, W. Gross, R. F. Dvorak, A. M. Kellerer, and H. H. Rossi. Survival curves and age response functions for Chinese hamster cells exposed to x-rays or high LET alpha-particles. *Radiat. Res.* 52: 88–98 (1972).

61. K. Kereluk, R. A. Gammon, and R. S. Lloyd. Microbiological aspects of ethylene oxide sterilization. II. Microbial resistance to ethylene oxide. *Appl. Microbiol.* 19: 152–156 (1970).

62. W. K. Kraak, G. L. Rinkel, and J. Hoogenheide. Oecologische bewerking van de Europese ringgegevens van de Kievit (*Vanellus vanellus* (L.)) *Ardea* 29: 151–157 (1940).

63. D. E. Lea. *Actions of Radiations on Living Cells*. Cambridge University Press (1955).

64. K. H. Chadwick and H. P. Leenhouts. A molecular theory of cell survival. *Phys. Med. Biol.* 18: 78–87 (1973).

65. E. L. Alpen. *Radiation Biophysics*. Academic Press (1990).

66. T. T. Puck and P. I. Marcus. Action of x-rays on mammalian cells. *J. Exp. Med.* 103: 653–666 (1956).

67. B. Gompertz. On the nature of the function expressive of the law of human mortality, and on a new mode of determining the value of life contingencies. *Phil. Trans. Royal Soc. London* 115: 513–583 (1825).

68. R. Pearl and S. L. Parker. Experimental studies on the duration of life. I. Introductory discussion of the duration of life in *Drosophila*. *Am. Nat.* 55: 481–509 (1921).

69. N. Minois, M. Frajnt, C. Wilson, and J. W. Vaupel. Advances in measuring lifespan in the yeast *Saccharomyces cerevisiae*. *Proc. Natl. Acad. Sci. U.S.A.* 102: 402–406 (2005).

70. J. S. Weitz and H. B. Fraser. Explaining mortality rates in patients. *Proc. Natl. Acad. Sci. U.S.A.* 98: 15,383–15,386 (2001).

71. J. W. Vaupel and A. I. Yashin. The deviant dynamics of death in heterogeneous populations. *Sociol. Methodol.* 15: 179–211 (1985).

72. J. W. Vaupel, J. R. Carey, K. Christensen, T. E. Johnson, A. I. Yashin, N. V. Holm, I. A. Iachine, V. Kannisto, A. A. Khazaeli, P. Liedo, et al. Biodemographic trajectories of longevity. *Science* 280: 855–860 (1998).

73. R. A. Fisher and L. H. C. Tippett. Limiting forms of the frequency distribution of the largest or smallest member of a sample. *Math. Proc. Cambridge Philos. Soc.* 24: 180–190 (1928).

74. E. Castillo, A. S. Hadi, N. Balakrishnan, and J. Sarabia. *Extreme Value and Related Models with Applications in Engineering and Science*. Wiley (2004).

75. C. Croarkin and P. Tobias. *NIST/SEMATECH e-Handbook of Statistical Methods*. NIST (2003). http://www.itl.nist.gov/div898/handbook/.

76. R. Pearl and J. R. Miner. Experimental studies on the duration of life. XIV. The comparative mortality of certain lower organisms. *Q. Rev. Biol.* 10: 60–79 (1935).

77. E. J. Gumbel and N. Goldstein. Analysis of empirical bivariate extremal distributions. *J. Am. Stat. Assoc.* 59: 794–816 (1964).

78. J. R. Vanfleteren, A. De Vreese, and B. P. Braeckman. Two-parameter logistic and Weibull equations provide better fits to survival data from isogenic populations of *Caenorhabditis elegans* in axenic culture than does the Gompertz model. *J. Gerontol., Ser. A* 53: B393–B403 (1998).

79. D. A. Juckett and B. Rosenberg. Comparison of the Gompertz and Weibull functions as descriptors for human mortality distributions and their intersections. *Mech. Ageing Dev.* 69: 1–31 (1993).

80. B. J. Tolkamp, M. J. Haskell, F. M. Langford, D. J. Roberts, and C. A. Morgan. Are cows more likely to lie down the longer they stand? *Appl. Anim. Behav. Sci.* 124: 1–10 (2010).

81. L. Hayflick. The limited in vitro lifetime of human diploid cell strains. *Exp. Cell Res.* 37: 614–636 (1965).

82. M. Z. Levy, R. C. Allsopp, A. B. Futcher, C. W. Greider, and C. B. Harley. Telomere end- replication problem and cell aging. *J. Mol. Biol.* 225: 951–960 (1992).

83. C. M. Counter, A. A. Avilion, C. E. LeFeuvre, N. G. Stewart, C. W. Greider, C. B. Harley, and S. Bacchetti. Telomere shortening associated with chromosome instability is arrested in immortal cells which express telomerase activity. *EMBO J.* 11: 1921–1929 (1992).

84. U. M. Martens, E. A. Chavez, S. S. S. Poon, C. Schmoor, and P. M. Lansdorp. Accumulation of short telomeres in human fibroblasts prior to replicative senescence. *Exp. Cell Res.* 256: 291–299 (2000).

85. A. G. Bodnar, M. Ouellette, M. Frolkis, S. E. Holt, C. P. Chiu, G. B. Morin, C. B. Harley, J. W. Shay, S. Lichtsteiner, and W. E. Wright. Extension of life-span by introduction of telomerase into normal human cells. *Science* 279: 349–352 (1998).

86. R. Cawthon, K. Smith, E. O'Brien, A. Sivatchenko, and R. Kerber. Association between telomere length in blood and mortality in people aged 60 years or older. *Lancet* 361: 393–395 (2003).

87. U. Herbig, M. Ferreira, L. Condel, D. Carey, and J. M. Sedivy. Cellular senescence in aging primates. *Science* 311: 1257 (2006).

88. E. S. Epel, E. H. Blackburn, J. Lin, F. S. Dhabhar, N. E. Adler, J. D. Morrow, and R. M. Cawthon. Accelerated telomere shortening in response to life stress. *Proc. Natl. Acad. Sci. U.S.A.* 101: 17,312–17,315 (2004).

89. R. C. Allsopp, H. Vaziri, C. Patterson, S. Goldstein, E. V. Younglai, A. B. Futcher, C. W. Greider, and C. B. Harley. Telomere length predicts replicative capacity of human fibroblasts. *Proc. Natl. Acad. Sci. U.S.A.* 89: 10,114–10,118 (1992).

90. M. T. Hemann, M. Strong, L. Y. Hao, and C. W. Greider. The shortest telomere, not average telomere length, is critical for cell viability and chromosome stability. *Cell* 107: 67–77 (2001).

91. J. Karlseder, A. Smogorzewska, and T. de Lange. Senescence induced by altered telomere state, not telomere loss. *Science* 295: 2446–2449 (2002).

92. C. B. Harley and S. Goldstein. Retesting the commitment theory of cellular aging. *Science* 207: 191–193 (1980).

93. R. E. Callard, J. Stark, and A. J. Yates. Fratricide: A mechanism for T memory-cell homeostasis. *Trends Immunol.* 24: 370–375 (2003).

94. A. Csikász-Nagy, D. Battogtosk, K. C. Chen, B. Novák, and J. J. Tyson. Analysis of a generic model of eukaryotic cell-cycle regulation. *Biophys. J.* 90: 4361–4379 (2006).

95. J. A. Smith and L. Martin. Do cells cycle? *Proc. Natl. Acad. Sci. U.S.A.* 70: 1263–1267 (1973).

96. R. J. De Boer and A. S. Perelson. Estimating division and death rates from CFSE data. *J. Comput. Appl. Math.* 184: 140–164 (2005).

97. P. D. Hodgkin, J. H. Lee, and A. B. Lyons. B cell differentiation and isotype switching is related to division cycle number. *J. Exp. Med.* 184: 277–281 (1996).

98. H. Mohri, A. S. Perelson, K. Tung, R. M. Ribeiro, B. Ramratnam, M. Markowitz, R. Kost, A. Hurley, L. Weinberger, D. Cesar, M. K. Hellerstein, and D. D. Ho. Increased turnover of T lymphocytes in HIV-1 infection and its reduction by antiretroviral therapy. *J. Exp. Med.* 194: 1277–1287 (2001).

99. G. Melino. The Sirens' song. *Nature* 412: 23 (2001).

100. I. P. M. Tomlinson, M. R. Novelli, and W. F. Bodmer. The mutation rate and cancer. *Proc. Natl. Acad. Sci. U.S.A.* 93: 14,800–14,803 (1996).

101. S. J. Morrison and J. Kimble. Asymmetric and symmetric stem-cell divisions in development and cancer. *Nature* 441: 1068–1074 (2006).

102. H. Quastler and F. G. Sherman. Cell population kinetics in the intestinal epithelium of the mouse. *Exp. Cell Res.* 17: 420–438 (1959).

103. I. M. M. Van Leeuwen, H. M. Byrne, O. E. Jensen, and J. R. King. Crypt dynamics and colorectal cancer: Advances in mathematical modelling. *Cell Prolif.* 39: 157–181 (2006).

104. H. W. Hethcote. The mathematics of infectious diseases. *SIAM Rev.* 42: 599–653 (2000).

105. W. O. Kermack and A. G. McKendrick. A contribution to the mathematical theory of epidemics. *Proc. R. Soc. London, Ser. A* 115: 700–721 (1927).

106. J. M. Heffernan, R. J. Smith, and L. M. Wahl. Perspectives on the basic reproductive ratio. *J. R. Soc. Interface* 2: 281–293 (2005).

107. N. M. Ferguson, M. J. Keeling, W. J. Edmunds, R. Gani, B. T. Grenfell, R. M. Anderson, and S. Leach. Planning for smallpox outbreaks. *Nature* 425: 681–685 (2003).

108. A. S. Perelson, A. U. Neumann, M. Markowitz, J. M. Leonard, and D. D. Ho. HIV-1 dynamics in vivo: Virion clearance rate, infected cell life-span, and viral generation time. *Science* 271: 1582–1586 (1996).

109. A. S. Perelson and P. W. Nelson. Mathematical analysis of HIV-1 dynamics in vivo. *SIAM Rev.* 41: 3–44 (1999).

110. D. D. Ho, A. U. Neumann, A. S. Perelson, W. Chen, J. M. Leonard, and M. Markowitz. Rapid turnover of plasma virions and CD4 lymphocytes in HIV-1 infection. *Nature* 373: 123–126 (1995).

111. V. Volterra. Variazioni e fluttuazioni del numero di individui in specie animali conviventi. *Mem. Acad. Lincei* 2: 31–13 (1926).

112. G. F. Gause. Experimental analysis of Vito Volterra's mathematical theory of the struggle for existence. *Science* 79: 16–17 (1934).

113. S. Levin. Community equilibria and stability, and an extension of the competitive exclusion principle. *Am. Nat.* 104: 413–423 (1970).

114. Dr. Seuss [Theodor Geisel]. *On Beyond Zebra*. Random House (1955).

115. Protist information server. http://protist.i.hosei.ac.jp/.

116. A. J. Lotka. *Elements of Physical Biology*. Williams and Williams (1925).

117. V. Volterra. Fluctuations in the abundance of a species considered mathematically. *Nature* 118: 558–560 (1926).

118. N. Boccara. *Modeling Complex Systems*. Springer (2004).

119. S. H. Strogatz. *Nonlinear Dynamics and Chaos: With Applications to Physics, Biology, Chemistry, and Engineering*. second edition. CRC Press (2018).

120. C. Elton and M. Nicholson. The ten-year cycle in numbers of the lynx in Canada. *J. Anim. Ecol.* 11: 215–244 (1942).

121. M. G. Bulmer. A statistical analysis of the 10-year cycle in Canada. *J. Anim. Ecol.* 43: 701–718 (1974).

122. G. M. Erickson, P. J. Currie, B. D. Inouye, and A. A. Winn. Tyrannosaur life tables: An example of nonavian dinosaur population biology. *Science* 313: 213–217 (2006).

123. J. A. Estes, M. T. Tinker, T. M. Williams, and D. F. Doak. Killer whale predation on sea otters linking oceanic and nearshore ecosystems. *Science* 282: 473–476 (1998).

124. A. H. E. Bijkerk and R. J. Hall. A mechanistic model of the aerobic growth of *Saccharomyces cerevisiae*. *Biotechnol. Bioeng.* 19: 267–296 (1977).

125. N. B. Pamment, R. J. Hall, and J. P. Barford. Mathematical modeling of lag phases in microbial growth. *Biotechnol. Bioeng.* 20: 349–381 (1978).

8

Coupled Transport and Reaction

A good man always knows his limitations.
—Clint Eastwood (as Harry Callahan in *Magnum Force* [1973], written by Harry Fink and Rita Fink)

I think knowing what you cannot do is more important than knowing what you can.
—Lucille Ball

How quickly can two biomolecules bind? Is there a "speed limit," and if there is, what determines it? In this chapter, we consider binding processes in biological systems when mass-action kinetics may not be applicable. If one or both of the partners in a binding reaction are attached to a membrane or to a surface, then the observed rates may be dominated by slow diffusive processes rather than by the rate of molecular collisions in a well-mixed solution. An exact mathematical description of such cases of coupled reaction and diffusion would require the solution of partial differential equations to account for spatial concentration gradients. However, we will show that simplified approximations can be applied by subdividing the volume of interest into compartments and by describing the rate of transport between the compartments as a pseudo-mass-action rate, known as a mass-transfer coefficient. We will also consider how temperature, salt ions, and the presence of other macromolecules influence observed binding rates.

8.1 Diffusion, Collision, and Binding

8.1.1 Transport Effects on Binding

For reactions occurring in aqueous media, it is useful to separate binding into two consecutive processes—one for the physical transport of protein and ligand into the vicinity of each other, and a second one for the chemical binding that ensues. The overall rate for a sequence of steps is limited by the slowest. Thus, an important question when considering the rates of biological binding events is whether they are gated by the physical approach of molecules, rather than the complementary chemical interactions formed at the interface once they are in proximity to each other. For instance, if our goal were to engineer cells with altered metabolic properties, attempts to change reaction rates by altering chemical complementarity or enzymatic turnover numbers might be fruitless if reactions were instead limited by the rate of physical association by diffusion. Here we find that many binding reactions are not limited by diffusion, but that certain circumstances can conspire to create diffusion-limited binding conditions.

Up until this point, we have imagined the association of two molecules to be a single process with rate $k_{on}[P][L]$. This is an extraordinarily useful abstraction, not only for its simplicity but also for its general applicability. Nevertheless, thinking mechanistically, there are some rather complex and subtle processes involved in binding that may not always be treated appropriately with this simple mass-action expression. Molecules need to find each other in solution and approach closely enough to influence each other. Intermolecular forces between binding partners will influence their motion, as will collisions with solvent molecules. For a collision to occur between P and L, solvent molecules must move out of the way. A collision is likely to occur before the binding partners are properly aligned for binding, and realignment may be required before the bound complex can form.

We do not examine each of these subtleties in this chapter, but we will first examine the approach of binding partners to each other's neighborhoods. This turns out to have general applicability and gives a useful framework for thinking about other effects on binding. Ultimately, we should be able to connect this more detailed picture with the abstracted one. More than just an exercise, this connection between more mechanistic and more phenomenological perspectives on binding provides additional insight into binding processes and leads to an understanding of how ligands bind to cells whose surfaces are covered by binding proteins. Moreover, this treatment reveals circumstances in which binding properties are substantially altered due to regions where ligand is depleted near binding sites, particularly near cell surfaces coated with large numbers of binding proteins.

Diffusion is the dominant physical transport process by which molecules in solution approach one another in binding reactions. It corresponds to the random motion of individual molecules driven by their inherent kinetic energy. In solution, molecules are constantly colliding with one another and with solvent, changing direction, and heading off toward another collision. Although the resulting motion is random, if the initial distribution of molecules is not uniform, then diffusion acts as a bulk phenomenon to produce a net change in distribution. For example, opening a sealed container of food releases odorant molecules that can diffuse to fill a room. If the distribution of molecules is already uniform, then diffusion still occurs but does not lead to net bulk transport. Instead, if one imagines tagging a single molecule and following its motion, diffusion describes its average progress around the room.

Because of its stochastic nature, diffusion cannot be used to predict the trajectories of individual molecules; however, mathematical relationships can describe the statistical behavior of diffusing molecules, which include the rate of bulk transfer for cases of nonuniform molecular distributions (such as a gradient) and the time-dependent probability distribution for the position of a tagged molecule relative to its initial position.

Diffusing particles are well described by Fick's first and second laws. Fick's first law simply says that the flux of material is related to the magnitude of the local concentration gradient. Stated mathematically, Fick's first law in one dimension is

$$J_x = -D\frac{\partial [C]}{\partial x} \tag{8.1}$$

where J_x is the x-component of the flux of material expressed as a number of molecules per unit area per unit time (such as mol m^{-2} s^{-1}), the diffusivity D is a coefficient that characterizes the process (units of m^2 s^{-1}), and $[C]$ is the concentration of solute. The solute concentration gradient $\partial [C]/\partial x$ has units mol m^{-4}.

Fick's first law describes the bulk transport of material and states that the net flux of material acts to oppose a concentration gradient by a linear relationship. In the absence of a concentration gradient, there is no net transport of solute through diffusion. The above one-dimensional statement is generalized to three dimensions as

$$\vec{J} = -D\vec{\nabla}[C] \tag{8.2}$$

where the one-dimensional gradient $\partial [C]/\partial x$ is replaced by the three-dimensional gradient $\vec{\nabla}[C]$. The proportionality constant in the linear relationship is the diffusivity. Typical values for D are shown in table 8.1.

In the absence of an experimentally measured value, or to scale diffusivity from one temperature to another, the Stokes-Einstein equation predicts the diffusivity of a spherical particle to be

$$D = \frac{k_B T}{6\pi \eta r} \tag{8.3}$$

where k_B is the Boltzmann constant (1.38×10^{-23} J/K), T is the temperature in kelvins, η is the solution viscosity (Pa·s; water has viscosity $\approx 1 \times 10^{-3}$ Pa·s at 20°C), and r is the radius of the spherical particle (in meters). This relationship predicts an inverse 1/3 power dependence of D on molecular weight

Table 8.1 Typical biomolecular diffusivities.

Molecule	Molecular weight (kDa)	Environment	$D\ (\times 10^{-7}\ \text{cm}^2/\text{s})$
NO	0.03	Phosphate-buffered saline [1]	300
O_2	0.032	Blood [2]	170
Glucose	0.18	Tissue [2]	80
GFP	27	Water [3]	8.7
GFP	27	Cytoplasm [3]	2.8
GFP	27	Mitochondria [3]	2.5
GFP	27	Endoplasmic reticulum [3]	0.75
DNA (100 bp)	63	Phosphate-buffered saline [4]	1.8
DNA (100 bp)	63	Cytoplasm [4]	0.3
BSA	66	Phosphate-buffered saline [5]	12.5
BSA	66	Human tumors [5]	9.1
BSA	66	Tumor xenograft in mice [5]	1.2
IgG	150	Four tumor xenografts [6]	0.9–2.0
DNA (500 bp)	314	Phosphate-buffered saline [4]	0.6
DNA (500 bp)	314	Cytoplasm [4]	0.02
IgM	900	Phosphate-buffered saline [5]	5.6
IgM	900	Human tumors [5]	4.2
IgM	900	Tumor xenograft in mice [5]	0.77

for globular proteins in solution, which is consistent with experimental results. Nonspherical macromolecules and proteins in gels or tissues generally exhibit a stronger molecular-weight (MW) dependence, with diffusivity proportional to $(\text{MW})^\alpha$, where $-1 < \alpha < -0.3$.

Example 8-1 What is the predicted diffusivity of hemoglobin in water at 20°C? Assume that hemoglobin is spherical and has a partial specific volume of 0.749 mL/g.

Solution Because hemoglobin has MW $\approx$ 65,000 g/mol, the volume per molecule is

$$V = \frac{0.749\ \text{mL}}{\text{g}} \cdot \frac{65{,}000\ \text{g}}{\text{mol}} \cdot \frac{\text{mol}}{6.02 \times 10^{23}\ \text{molecule}} \cdot \frac{\text{m}^3}{10^6\ \text{mL}}$$

$$= 8.1 \times 10^{-26}\ \frac{\text{m}^3}{\text{molecule}}$$

Thus, the predicted radius r of a spherical hemoglobin molecule is $\sqrt[3]{\frac{3V}{4\pi}} \approx 2.7$ nm.

Using equation 8.3, we find that

$$D = \frac{k_B T}{6\pi \eta r}$$

$$= \frac{1}{6\pi} \cdot \frac{1.38 \times 10^{-23} \text{ J}}{\text{K}} \cdot \frac{293 \text{ K}}{1} \cdot \frac{1}{10^{-3} \text{ Pa} \cdot \text{s}} \cdot \frac{1}{2.7 \times 10^{-9} \text{ m}} \cdot \frac{\text{Pa} \cdot \text{m}^3}{\text{J}}$$

$$= 8.0 \times 10^{-11} \frac{\text{m}^2}{\text{s}}$$

This value corresponds very well to the experimentally measured diffusivity of hemoglobin ($D = 7.0 \times 10^{-11}$ m^2/s).

Figure 8.1 provides a basis for deriving Fick's second law, first using the one dimensional case for simplicity. Fick's second law describes the change in concentration with time resulting from diffusional processes. Consider the change in concentration $\delta[C]$ in the box over the brief time interval δt due to the sum of the particle fluxes across the walls (each of whose area is A) of the volume element. Applying conservation of mass to the volume element gives

$$A \delta x \delta[C] = [J_x(x, t) - J_x(x + \delta x, t)] A \delta t \tag{8.4}$$

$$\approx \left\{ J_x(x, t) - \left[J_x(x, t) + \frac{\partial J_x}{\partial x} \delta x \right] \right\} A \delta t \tag{8.5}$$

$$\approx -\frac{\partial J_x}{\partial x} A \delta x \delta t \tag{8.6}$$

That is, the change in the amount of material in the volume element $A \delta x \delta[C]$ is equal to the amount that flows in (flux in $\times$ A $\times$ δt) minus the amount that flows out

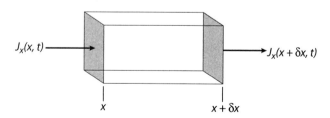

Figure 8.1 A small volume element is shown of width δx with diffusional flux inward through the left wall given by $J_x(x, t)$ and outward through the right wall by $J_x(x + \delta x, t)$. The area of the left and right walls is A.

(flux out $\times A \times \delta t$) in the given time interval. Taking the limit $\delta t \to 0$ to produce an instantaneous time derivative, and $\delta x \to 0$ so the approximate equality becomes an exact equality, gives the continuity equation

$$\frac{\partial [C]}{\partial t} = -\frac{\partial J_x}{\partial x} \tag{8.7}$$

which, when combined with Fick's first law (assuming constant diffusivity), gives

$$\frac{\partial [C]}{\partial t} = D\frac{\partial^2 [C]}{\partial x^2} \tag{8.8}$$

This is Fick's second law in one dimension and can be generalized to the three-dimensional form:

$$\frac{\partial [C]}{\partial t} = D\nabla^2 [C] \tag{8.9}$$

This equation describes how a nonuniform solute concentration distribution will change with time through diffusion. The operator ∇^2 is called the Laplace operator, or the Laplacian, and corresponds to the second-derivative operator. Equation 8.9 indicates that the concentration changes in time by diffusion only when its Laplacian is nonzero. That is, if the gradient is constant (constant first derivative and zero second derivative), there will be no local concentration change, because the amount flowing in will equal the amount flowing out.

We will see that Fick's first and second laws are powerful tools that can be applied to gain an understanding of the combination of diffusion, collision, and binding steps.

It is worth noting that there are other transport phenomena that, like diffusion, can bring molecules together. Because they tend not to be dominant in general molecular binding events in biology, we will state them but not develop them further here. Convection results in the movement of molecules via the bulk flow of solutions. If both binding partners are free in solution, convection produces a similar movement of both and thus tends not to generate relative motion that would contribute to their association. If one of the binding partners is immobilized on a surface and the other is free in solution, then convection can be an important component of molecular association. A useful tool for quantifying the relative importance of diffusion and convection is the Péclet number Pe:

$$Pe = \frac{v\ell}{D} \tag{8.10}$$

where v is the velocity of the flow, and D is the diffusion coefficient. The characteristic length scale ℓ can be in either the direction parallel or perpendicular to the flow, yielding either an axial or a radial Péclet number, respectively. When this dimensionless quantity is much greater than 1, convection effects dominate the motion of molecules relative to diffusive motion. In cells, there is generally no convection to speak of; however, some macromolecules are vectorially transported along cytoskeletal structures, as considered in example 8-3 below.

Example 8-2 A surface plasmon resonance biosensor device has a flow channel that is 50 μm high, 500 μm wide, and 2,400 μm long [7]. If an analyte flows at 50 μL/min through the channel, compare the time scale for diffusion across the channel height to the time scale for convective transport through the device.

Solution The fluid velocity is the volumetric flow rate divided by the cross-sectional area:

$$v = \frac{50 \text{ μL/min}}{50 \times 500 \text{ μm}^2} \times \frac{10^9 \text{ μm}^3}{\text{μL}} \times \frac{1 \text{ min}}{60 \text{ s}} = 3.3 \times 10^4 \text{ μm/s} \quad (8.11)$$

Because the channel is 2,400 μm long in the direction of flow, a molecule entering the device with this velocity exits in a fraction of a second if it does not bind to the surface. A typical diffusivity for a protein analyte in water is 10^{-6} cm²/s (table 8.1), or 100 μm²/s. Therefore the Péclet number is

$$Pe = \frac{v\ell}{D} \quad (8.12)$$

$$= \frac{3.3 \times 10^4 \text{ μm/s} \times 50 \text{ μm}}{100 \text{ μm}^2/\text{s}} \quad (8.13)$$

$$= 1.7 \times 10^4 \quad (8.14)$$

Clearly, convective transport through the device is much more rapid than the rate of diffusion across the height of the device.

Example 8-3 Secretory vesicles move on actin cables in a yeast cell due to myosin motors at an average speed of 3 μm/s [8]. If a typical cytoplasmic diffusivity for a protein is 30 μm²/s (table 8.1), will a protein on the surface of a vesicle come into contact with other cytoplasmic proteins by a process dominated more by diffusion or cytoskeletal transport? A yeast cell is typically 5 μm in diameter.

> **Solution**
>
> $$Pe = \frac{v\ell}{D} \tag{8.15}$$
>
> $$= \frac{3 \text{ μm/s} \times 5 \text{ μm}}{30 \text{ μm}^2/\text{s}} \tag{8.16}$$
>
> $$= 0.5 \tag{8.17}$$
>
> Because $Pe < 1$, diffusive transport is somewhat more rapid than motor-driven cytoskeletal transport, although not substantially so.

A third transport mechanism is molecular motion driven by electric fields. This can become important as a modulating factor on diffusion in certain circumstances (e.g., in the vicinity of charged membrane surfaces). The interested reader can obtain a more detailed treatment of these effects in Alan J. Grodzinsky's text *Fields, Forces, and Flows in Biological Systems* [9].

8.1.2 Two-Step Binding Processes

Here we frame the simple binding reaction

$$L + P \underset{k_{\text{off}}}{\overset{k_{\text{on}}}{\rightleftharpoons}} C \tag{8.18}$$

as consisting of two separate and reversible steps

$$L + P \underset{k_-}{\overset{k_+}{\rightleftharpoons}} L^* + P^* \underset{k_{-1}}{\overset{k_1}{\rightleftharpoons}} C \tag{8.19}$$

where $L^* + P^*$ represents a conceptual intermediate in which the ligand and protein have approached one another but are not yet engaged chemically (L^* represents a ligand in the vicinity of at least one receptor, and P^* represents a protein receptor in the vicinity of at least one ligand; figure 8.2).

It is informative to apply the quasi-steady-state approximation (QSSA) to L^*, which results in a useful relationship between the overall forward (k_{on}) and reverse (k_{off}) rate constants and the individual constants for the separate physical (k_+ and k_-) and chemical (k_1 and k_{-1}) processes. The dynamic material balance for production and loss of L^* by the mechanism in equation 8.19 is

$$\frac{d[L^*]}{dt} = k_+[L][P] - k_-[L^*][P^*] - k_1[L^*][P^*] + k_{-1}[C] \tag{8.20}$$

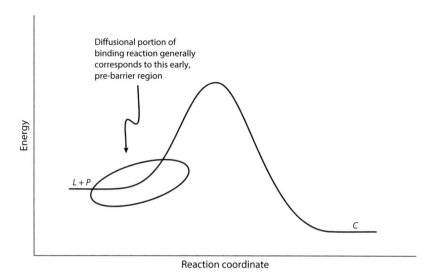

Figure 8.2 The binding reaction profile for ligand (*L*) and binding protein (*P*) to form a complex (*C*) is illustrated schematically. Generally, diffusion corresponds to the early, pre-barrier portions of the association reaction. This pre-barrier region corresponds to the physical approach of the two molecules in the vicinity of each other and precedes the chemical interaction between the binding partners. In certain circumstances that will be explored in this chapter, the physical-association steps become rate limiting.

which we set equal to zero to apply the QSSA. Rearrangement gives

$$[L^*] = \frac{k_+[L][P] + k_{-1}[C]}{(k_1 + k_-)[P^*]} \tag{8.21}$$

for the steady-state concentration of L^*. The overall reaction rate for the mechanism in equation 8.19 can be monitored by the overall rate of product formation:

$$\frac{d[C]}{dt} = k_1[L^*][P^*] - k_{-1}[C] \tag{8.22}$$

$$= \frac{k_1 k_+[L][P] + k_1 k_{-1}[C]}{k_1 + k_-} - k_{-1}[C] \tag{8.23}$$

$$= \frac{k_+ k_1[L][P]}{k_1 + k_-} + \frac{k_1 k_{-1}[C]}{k_1 + k_-} - \frac{(k_1 k_{-1} + k_- k_{-1})[C]}{k_1 + k_-} \tag{8.24}$$

$$= \frac{k_+ k_1[L][P]}{k_1 + k_-} - \frac{k_- k_{-1}[C]}{k_1 + k_-} \tag{8.25}$$

Using the simplified overall mechanism of equation 8.18, the overall reaction rate is

$$\frac{d[C]}{dt} = k_{\text{on}}[L][P] - k_{\text{off}}[C] \tag{8.26}$$

which, by comparison with equation 8.25, provides values for k_{on} and k_{off}:

$$k_{\text{on}} = \frac{k_+ k_1}{k_1 + k_-} \tag{8.27}$$

$$k_{\text{off}} = \frac{k_- k_{-1}}{k_1 + k_-} \tag{8.28}$$

These relationships are consistent with intuition. Namely, when the physical transport step parameterized by k_- is sufficiently fast (i.e., $k_- \gg k_1$), then the overall rates are dominated by the chemical steps: $k_{\text{on}} \to k_1$ (k_+/k_-) and $k_{\text{off}} \to k_{-1}$. Note that the k_+/k_- term in the k_{on} limit is unitless and, as will be shown later in this chapter, turns out to be equal to 1 for common cases of interest.

Likewise, the overall equilibrium dissociation constant can be obtained:

$$K_{\text{d}} = \frac{k_{\text{off}}}{k_{\text{on}}} = \frac{k_- k_{-1}}{k_+ k_1} \tag{8.29}$$

Note that this overall K_{d} is the product of the effective equilibrium dissociation constants of the transport (k_-/k_+) and chemical (k_{-1}/k_1) steps.

Case Study 8-1 A. D. Vogt and E. Di Cera. Conformational selection or induced fit? A critical appraisal of the kinetic mechanism. *Biochemistry* 51: 5894–5902 (2012) [10].

The previous section describes two-step binding that first depends on the physical proximity of a ligand and protein receptor, but conceptually analogous two-step processes can also arise when the protein undergoes a conformational change either before or after binding. These mechanisms are referred to as *conformational selection* and *induced fit*, respectively. The reaction scheme for conformational selection can be written as

$$P^* \underset{k_{-r}}{\overset{k_r}{\rightleftharpoons}} P \underset{k_{\text{off}}}{\overset{k_{\text{on}}[L]}{\rightleftharpoons}} C \tag{8.30}$$

and the scheme for induced fit can be written as

$$P^* \underset{k_{\text{off}}}{\overset{k_{\text{on}}[L]}{\rightleftharpoons}} C^* \underset{k_{-r}}{\overset{k_r}{\rightleftharpoons}} C \tag{8.31}$$

where P^* and P represent two different conformations of the protein, and C^* and C represent two different conformations of the complex.

Based on a rapid-equilibrium approximation for binding, these two schemes had been thought to be mutually exclusive in their dependence on the observed rate constant for approach to equilibrium, k_{obs}, as a function of ligand concentration: This plot would monotonically decrease for conformational selection but monotonically increase for induced fit. In this paper, the authors relaxed the assumption of rapid-equilibrium binding and showed that, in certain parameter regimes, k_{obs} can actually increase as a function of ligand concentration in the conformational selection scheme, thus confounding the underlying mechanism. However, a decrease in k_{obs} as a function of ligand concentration would unambiguously indicate binding via conformational selection.

For the reaction scheme shown in equation 8.30, the corresponding system of equations can be written as

$$\begin{bmatrix} d[P^*]/dt \\ d[P]/dt \\ d[C]/dt \end{bmatrix} = \begin{bmatrix} -k_r & k_{-r} & 0 \\ k_r & -(k_{-r}+k_{on}[L]) & k_{off} \\ 0 & k_{on}[L] & -k_{off} \end{bmatrix} \begin{bmatrix} [P^*] \\ [P] \\ [C] \end{bmatrix} \quad (8.32)$$

There are two nonzero eigenvalues of the rate-constant matrix in equation 8.32:

$$-\lambda_{1,2} = \frac{k_{-r}+k_r+k_{off}+k_{on}[L] \pm \sqrt{(k_{off}+k_{on}[L]-k_{-r}-k_r)^2+4k_{-r}k_{on}[L]}}{2} \quad (8.33)$$

The contribution of the larger eigenvalue, $-\lambda_1$, generally can't be measured by standard experimental techniques given the rapidity of its kinetics, but the smaller eigenvalue, $-\lambda_2$, often can be determined by methods such as stopped-flow spectrometry (chapter 2). It is therefore $-\lambda_2$ that relates to the observed kinetics and k_{obs}.

Since $k_{obs} = -\lambda_2$, it is informative to examine how $-\lambda_2$ changes as a function of ligand concentration, which is most easily seen by analysis at the extrema (i.e., when $[L] = 0$ and as $[L] \to \infty$). The results can be divided into two regimes: $k_{off} > (k_{-r}+k_r)$ and $k_{off} < (k_{-r}+k_r)$. When $k_{off} > (k_{-r}+k_r)$, $k_{obs} = k_{-r}+k_r$ at $[L] = 0$ and $k_{obs} \to k_r$ as $[L] \to \infty$; thus, k_{obs} always decreases as a function of $[L]$ in this regime. When $k_{off} < (k_{-r}+k_r)$, $k_{obs} = k_{off}$ at $[L] = 0$ and $k_{obs} \to k_r$ as $[L] \to \infty$; thus, k_{obs} can either increase or decrease as a function of $[L]$ in this regime. Simulations of k_{obs} as a function of $[L]$ are shown in figure 8.3.

By performing a similar analysis for the induced-fit scheme described by equation 8.31, it can be shown that k_{obs} always increases as a function of $[L]$. Thus, if k_{obs} decreases with respect to $[L]$, the mechanism of binding must be conformational selection. Notably, if k_{obs} increases with respect to $[L]$, the mechanism of binding cannot be uniquely determined because, although this behavior would be consistent with an induced-fit model, it is also consistent with conformational selection when $k_r > k_{off}$.

The authors performed stopped-flow fluorescence experiments to measure k_{obs}, the observed kinetics of binding to thrombin by four ligands (figure 8.4): p-aminobenzamidine (PABA), K^+, Na^+, and H-D-Val-Pro-Arg-p-nitroanilide (VPR). Free thrombin has been

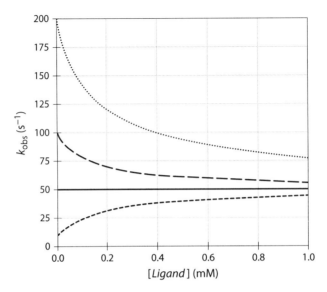

Figure 8.3 Simulated dependence of $-\lambda_2 = k_{obs}$ as a function of $[L]$ using equation 8.33. In all simulations, $k_r = 50$ s^{-1}, $k_{-r} = 150$ s^{-1}, and $k_{on} = 10^6$ M^{-1} s^{-1}. From top to bottom in the plot, $k_{off} = 300$ s^{-1} (dotted), 100 s^{-1} (long dash), 50 s^{-1} (solid), and 10 s^{-1} (short dash). As $[L] \to \infty$, all curves approach k_r. For $k_{off} < (k_{-r} + k_r)$ (bottom three curves), $k_{obs} = k_{off}$ at $[L] = 0$; for $k_{off} > (k_{-r} + k_r)$ (top curve), $k_{obs} = (k_{-r} + k_r)$ at $[L] = 0$.

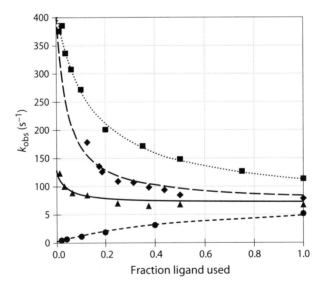

Figure 8.4 Experimental dependence of k_{obs} as a function of $[L]$ for different ligands to thrombin, measured by stopped-flow fluorescence. From top to bottom, the ligands are p-aminobenzamidine (PABA; squares), K$^+$ (diamonds), Na$^+$ (triangles), and H-D-Val-Pro-Arg-p-nitroanilide (VPR; circles). All experiments were performed in 5 mM Tris and 0.1% PEG 8000 (pH 8.0) at 15°C; the total ionic strength was kept constant at 400 mM using choline chloride. Lines are best fits to the data and were used to extract the rate constants.

shown to crystallize in two conformations, one in which the active site is accessible to inhibitors such as PABA, and the other in which this active site is sterically blocked, providing strong experimental support for the conformational selection mechanism in this enzyme. The stopped-flow measurements in this study reveal a rich set of kinetic behaviors across both ligands and concentrations but are fully consistent with the general conformational selection mechanism (figure 8.3). Notably, if the experiments had been carried out only with VPR and rapid-equilibrium binding had been assumed, the data might have been incorrectly interpreted as evidence of an induced-fit model due to the increase in k_{obs} as a function of ligand concentration.

The development of a general mathematical framework and the use of multiple ligands in experimental kinetic measurements not only provided robust support for the conformational selection mechanism in thrombin but also enabled quantification of the ligand-independent protein interconversion parameter values, k_r (≈ 70 s^{-1}) and k_{-r} (≈ 340 s^{-1}), as well as the binding parameters, k_{on} and k_{off}, for some of the ligands. Specifically, when $k_{off} < (k_{-r} + k_r)$, as is the case for the Na$^+$ and VPR ligands, the y-intercept equals k_{off}, so the kinetic parameters k_{off} and k_{on} can be unambiguously extracted from these curve fits: for Na$^+$, $k_{on} = 3.2 \times 10^4$ M^{-1} s^{-1} and $k_{off} = 130$ s^{-1}; for VPR, $k_{on} = 1.7 \times 10^7$ M^{-1} s^{-1} and $k_{off} = 2.4$ s^{-1}.

8.1.3 Smoluchowski Diffusion Limit

A detailed treatment of the process of diffusion followed by collision and eventual binding was reported by Polish physicist Marian Smoluchowski in 1917 [11]. The similarity between the steady-state assumption results presented above and the more detailed analysis is remarkable. Let us model the ligand and protein binding as the association of two sticky spherical objects. We fix the spherical protein P at the origin and examine solutions to the steady-state diffusion equation for the concentration of ligand L as a function of distance from protein (figure 8.5). We restrict our analysis to the case where there is no long-range interaction (such as electrostatics) between protein and ligand. We look at solutions under a few different conditions to build up a conceptual understanding of the behavior of these models and to gain insight into real molecular and cellular systems.

Let us initially consider the case for which diffusion is the slow step in the reaction. Imagine that, as soon as a ligand reaches the contact distance r_o from a protein, the ligand disappears and the protein is free to bind another ligand right away. This is actually a good representation for certain very fast, enzymatically catalyzed chemical reactions (in which the ligand is a reactant, the protein is an enzyme, and the reaction instantaneously converts reactant into product), and it is helpful in formulating general binding reactions as well. The flux of ligand molecules for our configuration is given by Fick's first law (equation 8.2)

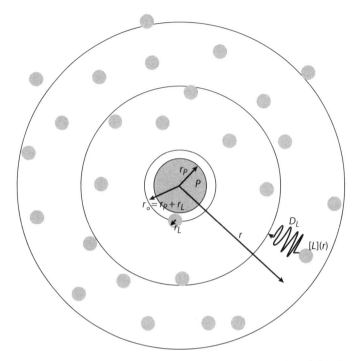

Figure 8.5 Reaction geometry for protein P of radius r_P immobilized at the origin of coordinates in a field of diffusing ligand molecules L (small shaded spheres) of radius r_L and with diffusion coefficient D_L. Collisions between P and L occur when their center-to-center distance approaches $r_o = r_P + r_L$. The protein and ligand are treated as spherically symmetric. The steady-state diffusion equation for this geometry can be solved to give the ligand concentration as a function of distance from the origin, $[L](r)$. Because we ultimately care about the case in which P and L are both free to diffuse, the appropriate diffusion coefficient is the mutual diffusion between the two, $D = D_L + D_P$.

$$\vec{J}_L = -D\vec{\nabla}[L] \qquad (8.34)$$

where we have used the mutual diffusion coefficient $D = D_L + D_P$ for relative diffusion between ligand and protein. The system is spherically symmetric, and so are its boundary conditions. Thus, the flux J_L has only a radial component:

$$J_L = -D\frac{\partial [L]}{\partial r} \qquad (8.35)$$

Consider the net flux of ligand across the boundaries of a thin spherical shell from radius r to $r + \delta r$. At steady state, the net flux into the thin shell must be zero; otherwise, ligand concentration would build up or deplete in the shell. This is expressed using Fick's second law (equation 8.9) at steady state,

$$\frac{\partial [L]}{\partial t} = D\nabla^2 [L] = 0 \tag{8.36}$$

Substituting the Laplacian ∇^2 for spherically symmetric systems gives the steady-state diffusion equation to be solved:

$$D\frac{1}{r^2}\frac{\partial}{\partial r}\left(r^2 \frac{\partial [L]}{\partial r}\right) = 0 \tag{8.37}$$

This equation must be solved together with appropriate boundary conditions on the ligand concentration. For the case under consideration, the ligand concentration far from the receptor protein should be the bulk ligand concentration, which we express as $[L](r) \to [L]_\infty$ for $r \to \infty$. Also, because ligand reacts as soon as it touches protein, $[L] = 0$ for $r = r_o$. This case is often called a perfectly absorbing boundary condition. The solution to this steady-state diffusion equation is

$$[L](r) = [L]_\infty \left(1 - \frac{r_o}{r}\right) \quad \text{for } r \geq r_o \tag{8.38}$$

which produces a flux

$$J_L(r) = \frac{-D[L]_\infty r_o}{r^2} \tag{8.39}$$

One property of this solution is that the total flow rate through a spherical shell centered on P (i.e., the product of flux and transport area) is constant, independent of the shell radius—otherwise, ligand would build up or deplete with time and the steady-state condition would be violated. This total flow rate inward through any shell is $4\pi r_o D[L]_\infty$, which is the overall rate of transport of ligand from bulk (modeled as $r \to \infty$ here) to collision. This quantity has units corresponding to a rate of collisions:

$$\text{collision rate per protein} = \left(4\pi r_o \text{ cm} \cdot D \text{ cm}^2 \text{ s}^{-1}\right) \cdot \left([L]_\infty \text{ mol cm}^{-3}\right) \tag{8.40}$$

$$= 4\pi r_o D[L]_\infty \text{ mol s}^{-1} \tag{8.41}$$

This rate of collisions is associated with the k_+ step from equation 8.19, because the chemical step is infinitely fast in this scenario and thus not rate determining. Therefore,

$$k_+ = 4\pi r_o D \tag{8.42}$$

It is beneficial to briefly examine another extreme case that will provide information on k_-, which describes the other transport step. Using the same problem configuration, diffusion away from the protein can be studied by releasing ligand from the protein surface at some fixed concentration $[L]_o$ in solvent in an otherwise empty solution. The appropriate boundary conditions are $[L] = [L]_o$ at $r = r_o$ and $[L](r) \to 0$ as $r \to \infty$. The solution to equation 8.37 with these new boundary conditions is

$$[L](r) = [L]_o \left(\frac{r_o}{r}\right) \quad (8.43)$$

with flux

$$J_L(r) = \frac{+D[L]_o r_o}{r^2} \quad (8.44)$$

giving a total outward flow rate of $4\pi r_o D[L]_o$. Associating this with the k_- reaction gives

$$k_- = 4\pi r_o D \quad (8.45)$$

which is identical to our expression for k_+. Thus, the ratio $k_-/k_+ = 1$ in this case, and equation 8.29 indicates that the overall equilibrium constant is equal to the chemical equilibrium constant and is unaffected by diffusion. Thus, the rate parameters for the diffusional encounter are the same as for the diffusional separation of binding partners lacking long-range interactions. It should not be surprising that long-range attraction or repulsion between binding partners breaks the symmetry to favor binding (for attraction) or unbinding (for repulsion) [12].

In actual binding processes, there is likely to be some delay between the approach of protein and ligand and the subsequent binding event. Conformational and orientational requirements for either or both binding partners, as well as for the solvent, are a potential cause. The framework developed here can be used to address this situation as well. The boundary condition with $[L](r) \to [L]_\infty$ for $r \to \infty$ can be applied again. However, at the protein surface, not all ligand molecules bind instantaneously. Instead, the rate of binding is limited to be equal to the binding rate described by k_1 in equation 8.19. Thus, the second boundary condition is that the transport rate incident on the protein surface should match the chemical reaction rate at that surface:

$$4\pi r_o^2 D \vec{\nabla}[L](r)\Big|_{r_o} = 4\pi r_o^2 D \frac{\partial [L](r)}{\partial r}\Big|_{r_o} = k_1 [L](r_o) \quad (8.46)$$

where the notation $[L](r)$ is again used to indicate the ligand concentration at a given radius, and the notation $|_r$ indicates that a function is evaluated at a given radius.

Note that binding of ligand by the protein is expected to deplete the concentration of ligand in the neighborhood of the protein below its bulk value. Thus, $[L](r_o) < [L]_\infty$. If the binding rate is equivalently expressed by an overall rate constant times a bulk concentration $k_{on}[L]_\infty$, and by a local rate constant and a local concentration $k_1[L](r_o)$, the depletion of ligand locally at the protein surface implies that the local rate constant must be greater than the overall one: $k_1 > k_{on}$. This case is called a partially absorbing boundary condition.

The solution to this steady-state diffusion problem is

$$[L](r) = [L]_\infty \left[1 - \frac{k_1}{k_1 + 4\pi D r_o} \cdot \left(\frac{r_o}{r}\right) \right] \text{ for } r \geq r_o \tag{8.47}$$

which has the property that the ligand concentration decreases from the bulk value $[L]_\infty$ in a continuous fashion. At the protein surface, where the chemical reaction occurs, the value is given by the expression

$$[L](r_o) = [L]_\infty \left(\frac{4\pi D r_o}{k_1 + 4\pi D r_o} \right) \tag{8.48}$$

At steady state, the overall rate of product formation must match from both the one-step (equation 8.18) and the two-step (equation 8.19) mechanisms. Thus,

$$k_{on}[L]_\infty = k_1[L](r_o) \tag{8.49}$$

$$= k_1[L]_\infty \left(\frac{4\pi D r_o}{k_1 + 4\pi D r_o} \right) \tag{8.50}$$

$$k_{on} = \frac{4\pi D r_o k_1}{k_1 + 4\pi D r_o} \tag{8.51}$$

which produces a result for the overall forward rate constant in terms of the diffusion and chemical steps. Equations 8.42 and 8.45 allow us to associate the $4\pi D r_o$ term with both k_+ and k_- and to rewrite equation 8.51 as

$$\frac{1}{k_{on}} = \frac{1}{k_+} + \frac{1}{k_1} \tag{8.52}$$

The reciprocal of a rate constant can be thought of as a measure of the *resistance* to binding, and equation 8.52 states that the overall binding resistance is the sum of the resistance due to the diffusion step and the resistance due to the binding step. This conceptually matches the idea from electrical circuits, which is that the overall resistance of two elements in series is the sum of their individual resistances. Moreover, in the extreme, where one resistance is much larger than the other, the overall association resistance is essentially equal to the larger resistance; that is, the slower step dominates the binding kinetics.

Example 8-4 Estimate k_+ for a typical protein-protein binding interaction, and compare it to a representative $k_{on} \approx 10^5$ M^{-1} s^{-1}.

Solution We use equation 8.42: $k_+ = 4\pi D r_o$. Protein diffusion coefficients are roughly 10^{-7} cm^2 s^{-1} (see table 8.1), and a reasonable value for a protein radius is $r_o = 3$ nm:

$$k_+ = 4\pi D r_o \tag{8.53}$$

$$= 4\pi (D_P + D_L) r_o \tag{8.54}$$

$$\approx 4\pi (2 \times 10^{-7} \text{ cm}^2 \text{ s}^{-1})(3 \text{ nm}) \tag{8.55}$$

$$\approx (7.5 \times 10^{-6} \text{ cm}^2 \text{ nm s}^{-1})$$

$$\times \left(\frac{1 \text{ cm}}{10^7 \text{ nm}}\right)\left(\frac{1 \text{ L}}{1{,}000 \text{ cm}^3}\right)\left(\frac{6.02 \times 10^{23}}{\text{mol}}\right) \tag{8.56}$$

$$\approx 4.5 \times 10^8 \text{ M}^{-1} \text{ s}^{-1} \tag{8.57}$$

This example shows that the diffusional k_+ is often a few orders of magnitude larger than the observed k_{on}, which means that protein-protein interactions are generally not diffusion limited (i.e., $k_{on} \approx k_1$).

Comparing equation 8.51 to equation 8.27 draws a parallel between the $4\pi D r_o$ in the numerator of equation 8.51 and k_+, and that in the denominator with k_-. The ligand flux is given by

$$J_L(r) = \frac{-D[L]_\infty}{r^2} \cdot \left(\frac{k_1 r_o}{k_1 + 4\pi D r_o}\right) \tag{8.58}$$

$$= \frac{-D[L]_\infty}{r^2} \cdot \left(\frac{k_1 r_o}{k_1 + k_-}\right) \tag{8.59}$$

At the protein surface this ligand flux, which equals the binding flux for this steady-state case, is

$$J_L(r_o) = \frac{-D[L]_\infty}{r_o} \cdot \left(\frac{k_1}{k_1 + k_-}\right) \tag{8.60}$$

which has an intuitive interpretation. The $-D[L]_\infty/r_o$ term is the incident flux for the fully absorbing or diffusion-only case, and the term in parentheses, $\frac{k_1}{k_1+k_-}$, can be associated with a binding probability, where the denominator indicates the sum of the propensity to bind (k_1) and to dissociate without binding (k_-), and the numerator is just the binding propensity. Thus, the reaction rate can be expressed as a collision rate times a reaction probability per collision.

8.1.4 Surfaces: Local Depletion and Recapture

How does immobilization of one of the binding partners affect the rate of protein-ligand binding? We can generalize the analysis from the previous section to this case to determine when diffusive limitations might occur. Imagine that a cell expresses a large number N of receptor molecules on its surface. In this case, ligand molecules diffuse to the surface of the cell and engage with one of a large number of receptors. It is appropriate to replace k_1, which represents the chemical rate for a ligand binding to one receptor in its neighborhood, with an expression that is roughly $k_{1,\text{cell}} = Nk_1$, to represent the now approximately N receptors locally available for the ligand to bind.

This permits rewriting equation 8.51 as

$$k_{\text{on,cell}} = \frac{4\pi Dr_\circ k_{1,\text{cell}}}{4\pi Dr_\circ + k_{1,\text{cell}}} \tag{8.61}$$

$$= \frac{4\pi Dr_\circ Nk_1}{4\pi Dr_\circ + Nk_1} \tag{8.62}$$

$$k_{\text{on,cell}}^{\text{normalized}} = \frac{4\pi Dr_\circ k_1}{4\pi Dr_\circ + Nk_1} \tag{8.63}$$

where D now refers to the mutual diffusion coefficient for the ligand and cell (which is effectively just that of the ligand), $r_\circ$ is the center-to-center distance between cell and ligand when bound (which is effectively just the radius of the cell), and $k_{\text{on,cell}}^{\text{normalized}}$ is the rate constant for ligand binding to cells normalized by the number of receptors per cell.

In the extreme case where each cell tethers one receptor, this case differs from the protein-ligand one only by the changed diffusional step due to the increased size of the cell compared to the protein. However, when the cell displays many receptor molecules on its surface, a very different phenomenon takes place. As the number of receptors per cell increases, the magnitude of Nk_1 relative to the diffusion rate $4\pi Dr_\circ$ increases. Eventually, for a sufficiently large number of receptors per cell, ligand binding becomes diffusion limited. This is illustrated in figure 8.6, which shows the decreasing normalized cell binding rate constant for increasing receptor density. This decrease is due to the depletion of ligand from the neighborhood of the cell at steady state due to the large number of available binding sites. This phenomenon has been observed experimentally (figure 8.7) [13, 14].

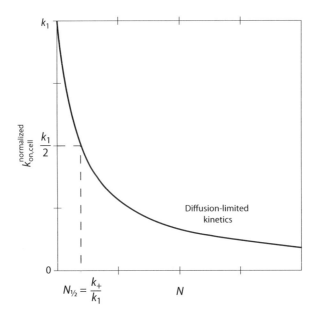

Figure 8.6 Plot of equation 8.63, indicating that the maximum normalized binding rate constant of ligand to receptor-coated cell is k_1, the chemical binding rate of a single ligand to a single receptor. As the receptor density increases, the observed binding rate decreases below this value. When the number of receptors per cell is $N_{\frac{1}{2}} = k_+/k_1$, the diffusion and capture rate constants are of the same magnitude, and the reaction is partially diffusion limited.

Example 8-5 The experimentally determined value for the diffusion-limited binding rate in figure 8.7 is $4\pi D r_o = 3.4 \times 10^{13}$ M^{-1} s^{-1} molecules/cell. If the radius of the cell is 4 µm, estimate the diffusivity of the ligand.

Solution Careful tracking of units is necessary here. First, the units of the diffusion-limited cell binding rate constant $k_+ = 4\pi D r_o$ should be converted so that the number of receptors per cell cancels with the moles of soluble ligand:

$$k_+ = \frac{3.4 \times 10^{13} \text{ L} \cdot \text{molecule}}{\text{mol} \cdot \text{s} \cdot \text{cell}} \times \frac{\text{mol}}{6.02 \times 10^{23} \text{ molecule}}$$
$$= 5.6 \times 10^{-11} \text{ L}/(\text{s} \cdot \text{cell}) \qquad (8.64)$$

Converting volume to µm³, $k_+ = 5.6 \times 10^4$ µm³/(s cell). Therefore, $D = k_+/(4\pi r_o) = 5.6 \times 10^4$ µm³/(s cell)/$(4\pi)/(4$ µm/cell$) = 1100$ µm²/s, or 1.1×10^{-5} cm²/s. Compared to the diffusivities given in table 8.1, this value is plausible for the small-molecule ligand (2,4- dinitrophenol, MW = 184 g/mol) of this example.

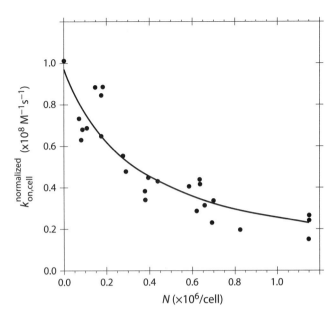

Figure 8.7 Experimental validation of a reduced binding rate at cell surfaces due to local depletion [13]. An IgE antibody against a small molecule (2,4-dinitrophenol) was tethered at varying levels to the surface of rat basophilic leukemia cells by an anti-IgE receptor, and the binding rate constant was measured. The nonlinear least-squares parameter fit to equation 8.63 gives $4\pi D r_\circ = 3.4 \times 10^{13}$ M^{-1} s^{-1} molecules/cell and $k_1 = 9.8 \times 10^7$ M^{-1} s^{-1}. This corresponds to a half-maximum $N_{\frac{1}{2}}$ of 3.5×10^5 molec/cell.

Case Study 8-2 S. Y. Shvartsman, H. S. Wiley, W. M. Deen, and D. A. Lauffenburger. Spatial range of autocrine signaling: Modeling and computational analysis. Biophys. J. 81: 1854–1867 (2001) [15].

An autocrine signaling loop involves cellular secretion of a ligand, which then binds back to a cognate receptor on the same cell, triggering a downstream response. A synthesized and secreted ligand is initially localized near the cell surface and can either bind back to a receptor on its parent cell (thus completing the autocrine signaling loop) or diffuse away. Such autocrine signaling loops are found throughout biology (e.g., T cell proliferation induced by self-secreted cytokines in response to antigen stimulation), but their operational characteristics are difficult to tease out experimentally. In this paper, Shvartsman et al. examine how specific molecular and cellular parameters regulate the likelihood of completing an autocrine signaling loop.

The governing equations for soluble ligand, cell-surface receptor, and cell-surface complex, respectively, are:

$$\frac{\partial [L]}{\partial t} = D\left[\frac{\partial^2 [L]}{\partial r^2} + \frac{2}{r}\frac{\partial [L]}{\partial r}\right] \tag{8.65}$$

$$\frac{d[R_s]}{dt} = -k_{on}[R_s][L](r_{cell}) + k_{off}[C] + s - k_c[R_s] \qquad (8.66)$$

$$\frac{d[C]}{dt} = k_{on}[R_s][L](r_{cell}) - k_{off}[C] - k_e[C] \qquad (8.67)$$

with the following boundary conditions:

$$D\frac{\partial [L](r_{cell}, t)}{\partial r} = -q + k_{on}[R_s][L](r_{cell}) - k_{off}[C] \qquad (8.68)$$

$$[L](\infty, t) = 0 \qquad (8.69)$$

where $[L](r_{cell})$ is the ligand concentration at the cell surface, D is the diffusion coefficient of the ligand, k_{on} is the association rate constant, k_{off} is the dissociation rate constant, s is the synthesis rate of new cell-surface receptors, k_c is the constitutive endocytotic rate constant for free cell-surface receptors, k_e is the ligand-induced endocytotic rate constant for complexes, and q is the steady flux of cell-secreted ligands.

The authors calculated the steady-state probability of ligand recapture, $P_{cap} = 1 - f/q$, where $f = -D\, \partial[L](r_{cell})/\partial r$ is the steady flux of ligand away from the cell surface (i.e., f/q is the probability of a cell-secreted ligand diffusing away from the cell). Solving for the steady cell-surface receptor and complex concentrations as a function of the ligand concentration at the cell surface and substituting these expressions into the boundary value problem for ligand, the following implicit expression for P_{cap} is obtained:

$$Da = \frac{P_{cap}Au}{\gamma} + \frac{P_{cap}}{\delta(1 - P_{cap})} \qquad (8.70)$$

where $Da = k_{on}r_{cell}s/(k_c D)$ is the dimensionless Damköhler number quantifying ligand binding relative to transport; $Au = qr_{cell}/(DK_d)$ is the dimensionless Autocrine number that is the ratio of ligand concentration at the cell surface in the absence of cell-surface receptors to the equilibrium dissociation constant ($K_d = k_{off}/k_{on}$); $\gamma = k_c/k_{off}$; and $\delta = k_e/(k_e + k_{off})$. Dimensionless numbers such as Da and Au can be useful tools for estimating the relative magnitude of rate processes in complex systems, as illustrated further in the following section.

The probability of recapture therefore depends on the nested parameters in equation 8.70. Figure 8.8A shows specifically how P_{cap} varies as a function of cell-surface receptor number in the absence of ligand (receptor density is s/k_c, so receptor number per cell is $4\pi r_{cell}^2 s/k_c$), and ligand secretion rate (ligand secretion flux is q, so ligand secretion rate Q is $4\pi r_{cell}^2 q$). As the cell-surface receptor number is increased, while the ligand secretion rate is held constant, the probability of ligand capture increases, because there are more sinks available for binding the secreted ligand. By contrast, as the ligand secretion rate is increased for a fixed cell-surface receptor number, the probability of ligand capture decreases, because proportionally fewer ligand binding events will occur. Note that the cell-surface receptor number is directly proportional to Da, and that the ligand secretion rate is directly proportional to Au, so the dynamics of the system can also be captured

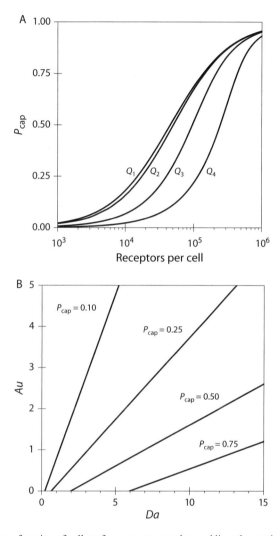

Figure 8.8 (A) P_{cap} as a function of cell-surface receptor number and ligand secretion rate Q ($Q_1 = 0.0167$ molecules/s/cell, $Q_2 = 1.67$ molecules/s/cell, $Q_3 = 16.7$ molecules/s/cell, and $Q_4 = 66.7$ molecules/s/cell). Other model parameters are: $\gamma = 0.1$, $\delta = 0.5$, $k_{off} = 1.67 \times 10^{-3}$ s^{-1}, $k_{on} = 1.67 \times 10^9$ cm^3/mol/s, $k_c = 1.67 \times 10^{-4}$ s^{-1}, $k_e = 1.67 \times 10^{-3}$ s^{-1}, $r_{cell} = 5 \times 10^{-4}$ cm, $D = 10^{-8}$ cm^2/s. (B) Constant lines of P_{cap} as a function of the Damköhler number (Da) and the Autocrine number (Au), which are respectively proportional to cell-surface receptor number and ligand secretion rate. $Q = 1$ molecule/s/cell and other model parameters are as in (A).

by plotting lines of constant P_{cap} as a function of these two nondimensional variables (figure 8.8B).

The authors go on to show in stochastic spatial models that the distances over which such autocrine signaling loops can operate span several orders of magnitude, from submicron to tens of microns, which in turn regulates how a cell samples (and responds to) its environment via the secretion of autocrine ligands. Such models also reveal the complex interconnection between molecular and cellular parameters and spatial sensing. For example, the distance over which autocrine signaling occurs varies as a function of endocytic rate constants, which, at first blush, may seem unrelated. However, the endocytic rate constant affects the expression level of cell-surface receptors, which influences the probability of ligand recapture, which in turn allows for larger excursions of the ligand before eventual recapture by the parent cell.

Overexpression of cell-surface receptors (via increased synthesis rate and/or a decreased endocytotic rate constant) can play a significant role in disease progression. For example, in many cancers, growth factor receptors are overexpressed, allowing more efficient recapture of self-secreted growth factors, thus facilitating uncontrolled proliferation.

8.2 Scaling Analyses of Tissue Penetration

A common scenario is the diffusion of some solute—a nutrient such as oxygen or glucose, or a drug such as an antibody or small molecule—from the bloodstream through tissue. Such solutes are generally of interest because of their reactivity in the tissue, and they may be depleted by such reactions as they percolate between cells through a combination of diffusion and flow. Under some circumstances, the rates of diffusion into the tissue and consumption by reaction may balance at steady state, such that the solute is consumed before reaching distances farthest from a blood vessel. Such circumstances are naturally of particular interest: Tissues not receiving nutrients will die, and cancer cells not reached by a drug will not be killed.

Let us consider a simple model of this process of reaction and diffusion, with the intention of identifying a figure of merit that will tell us how far into tissue a diffusing front of a reactive molecule will penetrate at steady state. This analysis recapitulates a classical model published by Ernest Thiele in 1939 [16].

Consider a sphere of tissue of radius R, bathed in some reactant S (such as oxygen or glucose) that can diffuse through the tissue, with some reaction consuming it along the way. Of course, the tissue is not actually spatially homogenous—it contains cellular substructures and the extracellular matrix. Nevertheless, it is often possible to approximate the process of diffusion through tissue as obeying Fick's first law in a pseudohomogenous medium, with an effective diffusivity D:

$$J_r = -D\frac{d[S]}{dr} \tag{8.71}$$

where r is the radial distance from the center of the sphere.

Let's assume that the reactant S is consumed by a first-order reaction:

$$-r_S = k[S] \tag{8.72}$$

Now let's examine the process of diffusion and reaction occurring in a fictitious thin shell in the tissue sphere, by constructing a *steady-state shell balance*. Consider a thin spherical slice of thickness Δr spanning from radius r to $r + \Delta r$:

$$J_r \cdot (4\pi r^2)|_r - J_r \cdot (4\pi r^2)|_{r+\Delta r} + r_S \cdot 4\pi r^2 \Delta r = 0 \tag{8.73}$$

This balance simply states that at steady state, the difference between the amount of substrate that diffuses into and out of this shell must be equal to the amount consumed by reaction. We now shrink the size of this shell indefinitely, so as to convert to a continuous differential equation, by dividing by $4\pi \Delta r$ and taking the limit as $\Delta r \to 0$:

$$-\frac{d(J_r \cdot r^2)}{dr} - k[S]r^2 = 0 \tag{8.74}$$

Substituting in Fick's law for the J_r term and differentiating yields the following second-order ODE:

$$\frac{d^2[S]}{dr^2} + \frac{2}{r}\frac{d[S]}{dr} - \frac{k}{D}[S] = 0 \tag{8.75}$$

with the boundary conditions that $[S] = [S]_o$ at $r = R$ and that $[S]$ is finite everywhere. As is often the case, it is helpful to nondimensionalize the variables in this equation, which will lead to a dimensionless group that directly captures the relative magnitude of reaction and diffusion time scales in the system. Let's define the dimensionless radius $\rho \equiv r/R$, and the dimensionless substrate concentration $s \equiv [S]/[S]_o$, such that both range from 0 to 1 in magnitude. These substitutions lead to the dimensionless form of equation 8.75:

$$\frac{d^2 s}{d\rho^2} + \frac{2}{\rho}\frac{ds}{d\rho} - \phi^2 s = 0 \tag{8.76}$$

where the Thiele modulus ϕ^2 is defined as

$$\phi^2 \equiv \frac{kR^2}{D} \tag{8.77}$$

With the boundary conditions that $s = 1$ at $\rho = 1$ and that s is finite everywhere, equation 8.76 can be analytically solved to obtain

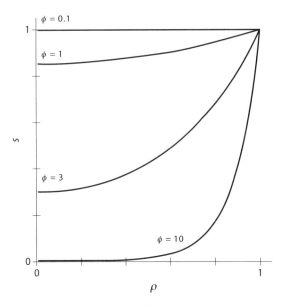

Figure 8.9 Dimensionless solutions to the reaction-diffusion problem in spherical geometry, for a first-order reaction, as a function of the Thiele modulus.

$$s = \frac{\sinh(\phi\rho)}{\rho\sinh(\phi)} \qquad (8.78)$$

where the hyperbolic sine function $\sinh \equiv (e^x - e^{-x})/2$.

Let's take a look at how the features of this solution vary with ϕ^2, as shown in figure 8.9. When $\phi \ll 1$, then $s \approx 1$ throughout the sphere of tissue; in other words, the substrate is at the same level as the surface throughout the tissue, and there is no apparent limitation in diffusive delivery of the substrate to reactive sites anywhere in the sphere. By contrast, when $\phi \gg 1$, note that s drops rapidly as one moves inward from the surface (i.e., $\rho = 1$), indicating that the substrate is consumed more rapidly by reaction than it is replenished by diffusion, leading to a dead zone in the interior of the tissue, with effectively no substrate present.

It is often the case that the most important information is whether the desired reaction in the tissue is diffusion-limited for a given set of parameters. That being the case, the simple expedient of calculating a relevant Thiele modulus, such as equation 8.77, can allow one to assess a base case and can also indicate in what way manipulable parameters must be changed to overcome a diffusive limitation. This approach has been applied to determine what antibody dosage is required to saturate tumor spheroids [17]; to show how growth-factor diffusion and receptor-mediated consumption leads to a morphogen gradient during development [18]; to design immobilized enzyme reactors [19]; and to understand oxygen limitations of engineered tissue implants [20].

Case Study 8-3   G. M. Thurber and K. D. Wittrup. Quantitative spatiotemporal analysis of antibody fragment diffusion and endocytic consumption in tumor spheroids. *Cancer Res.* 68: 3334–3341 (2008) [17].

A primary goal of treating cancer with an antibody drug is to bind all of the specific target receptors everywhere in either a bulk tumor or a micrometastatic cluster of cancer cells. It is therefore troubling when in many cases, detailed examination of antibody microdistribution indicates considerable heterogeneity, with some areas of tumor appearing to accumulate very little or no antibody. This maldistribution appears to be exacerbated for higher affinity antibodies, leading to the coining of the expression *binding site barrier* to describe the phenomenon.

In the present case study, the authors tested the hypothesis outlined in the previous section: Endocytic consumption may balance out diffusive transport in some cases, leading to a failure to fully saturate tumor tissues at steady state. A simplified experimental system was established that lacked the complexities of vascularized tissue architecture in actual tumors but captured the processes of diffusion, binding, and membrane turnover in a tractable in vitro system.

Some tumor cell lines can be coaxed to grow in clumps by suspending them in hanging droplets, such that they settle together into a spheroid of radius R. Such a clump of cells can then be attached as a hemisphere to a microscope slide, and the penetration of fluorescently labeled antibodies into the spheroid can be tracked by two-photon microscopy.

The authors noted that below a certain external bulk concentration, the fluorescent antibody front penetrated to an intermediate depth and never reached the center of the spheroid, potentially having reached a steady state between diffusion and endocytic consumption. To test this hypothesis, a variant of the Thiele modulus shown in equation 8.77 was derived for this particular problem:

$$\phi^2 \equiv \frac{k_e R^2 ([Ag]/\varepsilon)}{D(K_d + [Ab]_{\text{surf}})} \tag{8.79}$$

Each of the parameters in equation 8.79 was experimentally estimated independently, together with its standard error (table 8.2).

Varying concentrations of fluorescent antibody ($[Ab]_{\text{surf}}$) were incubated with spheroids and allowed to penetrate to a steady-state depth, which was then determined by microscopy (figure 8.10). Gratifyingly, the data points lay right along the theoretically predicted curve of saturation versus Thiele modulus ϕ^2 and were well within the experimental error envelope.

To test the model further, spheroids incubated at 22°C were found to be far more uniformly penetrated than at 37°C, a difference attributed to the near absence of endocytic consumption at the somewhat cooler temperature. The central theory was further confirmed in tumor uptake studies in mice [21].

Table 8.2 Parameters important for antibody spheroid penetration.

Parameter	Estimate	Error
Cell density	7.23×10^8 cells/mL	$\pm 1.87 \times 10^8$
Ag (antigen) per cell	3.88×10^5 #/cell	$\pm 4.7 \times 10^3$
Ag concentration [Ag]	4.68×10^{-7} M	$\pm 1.27 \times 10^{-7}$
Void fraction ε	0.1537	± 0.011
Diffusion coefficient D	33.1 $\mu m^2/s$	± 20.7
Endocytosis rate constant k_e	1.23×10^{-5} s^{-1}	$\pm 3.06 \times 10^{-6}$

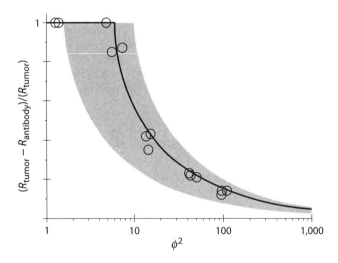

Figure 8.10 Experimentally determined depth of antibody penetration into tumor cell spheroids, compared to model predictions. Circles are data points obtained by two-photon fluorescence microscopy. The solid line is the theoretical solution to the reaction and diffusion model (equation 8.78). The gray shaded area is $\pm$ standard deviation from the theoretical prediction due to uncertainty in the experimentally measured variables.

8.3 Compartmental Models

8.3.1 The Mass-Transfer Coefficient

As shown in section 8.1, we can use Fick's laws to relate the diffusion coefficient (D) to rate constants for transport (k_+ and k_-) for pure diffusion in a simple spherical geometry. In systems involving more complex geometries and modes of transport (e.g., convection or flow), we can still relate the transport flux to a concentration gradient via a *mass-transfer coefficient*:

$$\text{transport flux} = \text{mass-transfer coefficient} \times \text{concentration difference} \quad (8.80)$$

$$J_L = k_t \times ([L] - [L]_\infty) \quad (8.81)$$

A mass-transfer coefficient has units of m s^{-1} and is a function of all properties of the system that impact transport lumped together (e.g., density, viscosity, diffusivity, and convection). This formalism is particularly useful because it approximates the complex analysis of convective and diffusive transport with a simple and familiar mass-action-type rate form, such as those we have utilized throughout this book.

In the simplest case, let us consider purely diffusive steady-state transport of ligand from a bulk concentration of $[L]_\infty$ to a sphere of radius r_o at which $[L] = 0$ (i.e., ligand immediately reacts at the sphere surface). We know from equation 8.39 that the steady-state flux at $r = r_o$ is $-D[L]_\infty/r_o$. Comparing this result to equation 8.81, we see that $k_t = D/r_o$. More generally, we can write $k_t = D/\ell$, where ℓ represents a characteristic length of the system of interest.

If we now consider convective transport in the form of an apparent rate constant k_c, we can define the Sherwood number (Sh) as

$$Sh \equiv \frac{k_c}{D/\ell} \qquad (8.82)$$

Sh is a dimensionless constant representing the ratio of convective mass transfer to diffusive mass transfer. Thus, when $Sh \to 1$, diffusion dominates, and when $Sh \gg 1$, convection dominates. For a number of geometries, Sh can be theoretically determined as a function of the Reynolds number (Re) and the Schmidt number (Sc), which thus connects k_c of the ligand to the physical properties and motion of the fluid. In other cases, useful empirical relationships have been determined by curve fitting power-law relationships among these numbers to experimental data.

Case Study 8-4 D. G. Myszka, X. He, M. Dembo, T. A. Morton, and B. Goldstein. Extending the range of rate constants available from BIACORE: Interpreting mass transport–influenced binding data. *Biophys. J.* 75: 583–594 (1998) [7].

As discussed in chapter 3, surface plasmon resonance (SPR) devices such as Biacore can be used to measure association and dissociation rate constants by flowing ligand or buffer over an immobilized protein, and monitoring real-time changes in mass near the protein surface. However, when the rates of transport of free ligand from and to the bulk are not much faster than the rates of ligand binding and dissociation to protein on the sensor chip, the rates of mass transfer must be explicitly considered. For example, as the protein density on the chip increases, the binding rates increase relative to the rates of mass transfer, eventually reaching a point where transport effects may become rate limiting and cannot be neglected. Therefore, at sufficiently high protein densities, fitting surface plasmon data with the equations derived in chapter 3 (i.e., the rapid-mixing model, in which the ligand concentration is assumed to be spatially uniform throughout the entire flow cell) leads to erroneous values of the true association and dissociation rate constants.

To more accurately simulate the kinetics of association and dissociation in Biacore under a range of conditions, Myszka et al. used a finite element method to solve a partial differential equation (PDE) model of the system that, in addition to the true protein-ligand binding kinetics, explicitly considers the flow velocity, ligand diffusivity, and channel geometry, all of which play a role in transporting the ligand from the bulk to the sensor chip surface. In the limit of negligible mass-transfer limitations, the simulated data from this full PDE model can be accurately fit using the rapid-mixing model. However, a better model is needed for estimating the true values of k_{on} and k_{off} when mass transport must be considered (e.g., at high densities of immobilized protein).

To this end, the authors considered a simple two-compartment ODE model in which the flow chamber is virtually divided into two compartments, each of which has a spatially uniform concentration of ligand. It is assumed that the ligand concentration in the outer compartment (i.e., the one distal from the chip surface) is constant and equal to the input ligand concentration. Of course, this artificial partitioning into two compartments is a convenient fiction, as no such physical boundary line actually exists. A necessary condition for this approximation to be useful is that the data are consistent with the model for physically realistic parameter values.

The ligand concentration [L] in the inner compartment (i.e., the one proximal to the chip surface) and the complex concentration [C] on the chip surface are described by the following equations, respectively:

$$V_i \frac{d[L]}{dt} = S\left[-k_{on}[L]([R]_T - [C]) + k_{off}[C] + k_t([L]_{bulk} - [L])\right] \quad (8.83)$$

$$\frac{d[C]}{dt} = k_{on}[L]([R]_T - [C]) - k_{off}[C] \quad (8.84)$$

where V_i is the volume of the inner compartment, S is the surface area of the flow chamber, $[R]_T$ is the total concentration of protein on the chip, k_{on} is the true association rate constant, k_{off} is the true dissociation rate constant, and k_t is the mass-transfer coefficient describing transport of ligand from the outer compartment (constant concentration $[L]_{bulk}$) to the inner compartment. As might be expected, k_t is a function of the flow velocity, ligand diffusivity, and channel geometry, which are explicitly considered in the PDE model.

The full PDE model was used to simulate several data sets with a range of association rate constants. At low values of the association rate constant, the rapid-mixing model and two-compartment model were both able to accurately estimate the kinetic binding constants. However, at high values of the association rate constant, the rapid-mixing model was unable to properly fit the kinetic data and obtain accurate parameter estimates, whereas the two-compartment model not only provided a good fit to the data but also generated surprisingly high-accuracy estimates of the true rate constants for association and dissociation (figure 8.11). These results also indicate that the rapid-mixing model cannot simply generate a very good fit by erroneously tuning the estimates of the association and dissociation rate constants: The goodness of fit is still poor, because mass-transfer limitations fundamentally change the characteristic shapes of the association and dissociation phases.

In SPR experiments designed to measure true binding kinetics, mass-transfer limitations are unwanted artifacts that confound the results. Although it is, of course, preferable to

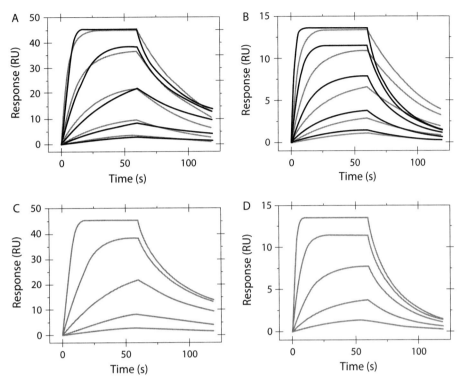

Figure 8.11 Comparison of rapid-mixing model (panels A, B) and two-compartment model (panels C, D) in fitting simulated data with mass-transfer limitations. The black curves represent simulated data from the full PDE model, and the gray curves represent global fits to the two data sets (the two sets of curves overlay completely in panels C and D). The left panels (A, C) have 50 RU maxima and right panels (B, D) have 15 RU maxima. The two-compartment model is able to accurately estimate the true association and dissociation rate constants ($k_{on} = 1 \times 10^8$ M^{-1} s^{-1}, and $k_{off} = 8 \times 10^{-2}$ s^{-1}) used in the PDE model to simulate the data [7].

set up such an experiment to render mass-transfer limitations negligible (e.g., low protein immobilization density and high flow rate), this is not always practical, because there are limits to what can be detected on the chip and how much ligand one can afford to use in an experiment. Additionally, for protein-ligand interactions with very rapid binding kinetics, it may not be possible to set up the experiment to avoid mass-transfer limitations. Thus, it is valuable to employ models that can parse transport effects and yield accurate parameter estimates for the binding rate processes of interest.

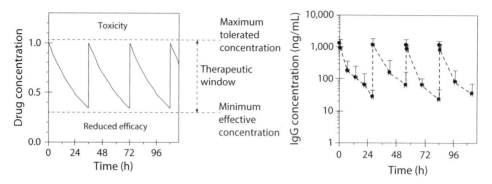

Figure 8.12 (Left panel) Model data for the concentration of drug dosed every 36 hours and exhibiting exponential clearance. (Right panel) Experimental data [22] for patient plasma concentration of IPH2101 antibody dosed repeatedly at 7.5 µg/kg.

8.3.2 Physiologically Based Compartmental Pharmacokinetic Models

Compartmental models can also be applied to describe physiological distribution of a drug throughout a patient's or experimental animal's body. Analysis of pharmacokinetics pertains to the processes by which a drug is administered, distributed, metabolized, and eliminated from the body and its compartments. Pharmacokinetics is often summarized as the effect of the body on the drug, in contrast to pharmacodynamics, which is the effect of the drug on the body.

A common aim of tracking drug concentrations is to maximize the time spent in the therapeutic window, that is, a concentration range above the minimum effective concentration and below the maximum tolerated concentration (figure 8.12). Yet this is an oversimplification, as functional strength—both beneficial and detrimental—varies continuously as a function of concentration (i.e., some effect is still obtained below the minimum effective concentration, and detrimental toxicity is observed below the maximum tolerated dose). Thus, the therapeutic window can provide a useful simple perspective to begin consideration of pharmacokinetics, though a more thorough consideration of molecular impact over the continuum of concentration enables better insight.

8.3.3 Two-Compartment Model

The concentration of administered drug in the plasma ($[C_p]$) after intravenous injection often exhibits biphasic exponential clearance:

$$[C_p](t) = Ae^{-\alpha t} + Be^{-\beta t} \qquad (8.85)$$

This observation is consistent with a two-compartment model, in which the molecule can transfer between the plasma and a general tissue compartment, in addition to being systemically cleared from the plasma (figure 8.13).

Coupled Transport and Reaction

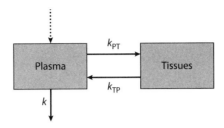

Figure 8.13 Two-compartment model in which drug is input into plasma (instantaneously, continuously, or both), transports between plasma and tissues (which are lumped into a single compartment for simplicity), and clears from plasma. Rate constants for each transport process are indicated.

Balances can be written on the molecule in plasma and in tissue:

$$V_p \frac{d[C_p]}{dt} = k_{TP}[C_t]V_t - k[C_p]V_p - k_{PT}[C_p]V_p \quad (8.86)$$

$$[C_p](t=0) = [C_p]_o \quad (8.87)$$

$$V_t \frac{d[C_t]}{dt} = k_{PT}[C_p]V_p - k_{TP}[C_t]V_t \quad (8.88)$$

$$[C_t](t=0) = 0 \quad (8.89)$$

In these equations, $[C_p]$ is the plasma concentration of the molecule, $[C_t]$ is the tissue concentration, V_t is the volume of the tissue compartment, and V_p is the volume of the plasma compartment. It can be shown that this system of equations is consistent with biphasic exponential clearance (equation 8.85) for the following relationships of the mechanistic rate constants to the empirical rate constants:

$$k = \frac{\alpha\beta(\phi+1)}{\phi\beta+\alpha} \quad (8.90)$$

$$k_{PT} = \frac{\phi(\beta-\alpha)^2}{(\phi+1)(\phi\beta+\alpha)} \quad (8.91)$$

$$k_{TP} = \frac{\phi\beta+\alpha}{\phi+1} \quad (8.92)$$

where $\phi = A/B$. Notably, more complex models are also consistent with this biphasic observation. But as noted in chapter 1, the best model complexity is dependent on the depth of information available and the complexity of the desired information.

The mechanistic rate constant k can be related to additional experimental observations. Consider an initial dose, D_o, supplied at time $t = 0$. The total amount of

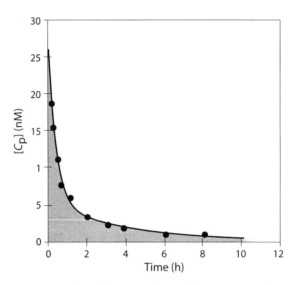

Figure 8.14 The plasma concentration of digoxin is shown following a 1 mg intravenous injection [23]. The integral of this $[C_p](t)$ plot, or AUC, requires estimation of the kinetics beyond 10 hours.

drug in the body, $D(t)$, can be written as the sum of drug in the plasma and tissue:

$$D = [C_p]V_p + [C_t]V_t \quad (8.93)$$

Dynamically, we have

$$\frac{dD}{dt} = V_p\frac{d[C_p]}{dt} + V_t\frac{d[C_t]}{dt} = -k[C_p]V_p \quad (8.94)$$

$$\int_{D_o}^{0} dD = \int_{0}^{\infty} -k[C_p]V_p dt \quad (8.95)$$

$$D_o = kV_p \int_{0}^{\infty} [C_p]dt \quad (8.96)$$

The integral of the plasma concentration is often referenced as its graphical interpretation (figure 8.14), that is, the *area under the curve* (*AUC*):

$$AUC = \int_{0}^{\infty} [C_p]dt \quad (8.97)$$

Thus, by integrating a collection of plasma concentrations over time, the plasma clearance rate constant can be calculated:

$$k = \frac{D_\circ}{V_p \cdot AUC} \tag{8.98}$$

Because the volume of accessible tissue, V_t, is often not known, and the concentration(s) in the tissue is (are) often heterogeneous and challenging to measure, a mathematical construct called the *volume of distribution*, V_D, was created. V_D is the ratio of total drug to plasma concentration, or the effective accessible volume of drug if all accessible locations contained drug at the plasma concentration:

$$V_D = \frac{D}{[C_p]} = \frac{[C_p]V_p + [C_t]V_t}{[C_p]} \tag{8.99}$$

Measurements of V_D are useful to determine initial dosages needed to achieve a desired plasma concentration. But with distribution and clearance, the concentration can drop, often quickly, to suboptimal values. One solution is to provide a continuous input. The rate of change in total drug concentration is described by

$$\frac{dD}{dt} = R - kD \tag{8.100}$$

in which R is the rate of continuous input (moles or mass per time). This can be rearranged, recalling equation 8.99 and assuming constant V_D:

$$\frac{d}{dt}([C_p]V_D) = R - k[C_p]V_D \tag{8.101}$$

$$\frac{d[C_p]}{dt} = \frac{R}{V_D} - k[C_p] \tag{8.102}$$

$$[C_p] = \frac{R}{kV_D}\left(1 - e^{-kt}\right) + [C_p]_\circ e^{-kt} \tag{8.103}$$

This expression can be rearranged to isolate the steady-state term and the dynamic term:

$$[C_p] = \frac{R}{kV_D} + \left([C_p]_\circ - \frac{R}{kV_D}\right)e^{-kt} \tag{8.104}$$

$$[C_p] = [C_p]_{ss} + \left([C_p]_\circ - [C_p]_{ss}\right)e^{-kt} \tag{8.105}$$

where

$$[C_p]_{ss} = \frac{R}{kV_D} \tag{8.106}$$

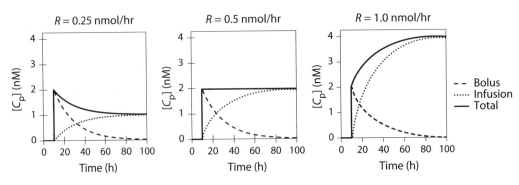

Figure 8.15 The plasma concentration—resulting from an initial 10 nmol bolus, continuous infusion, and summed total, all initiated at $t = 10$ hours—is shown for a 5 L volume of distribution for three infusion rates.

Clearly, the approach to steady state depends on the initial deviation, $[C_p]_o - [C_p]_{ss}$, and on the clearance kinetics. An initial loading dose D_L can be provided to reduce or eliminate this initial deviation:

$$[C_p]_o = \frac{D_L}{V_D} \tag{8.107}$$

If the loading dose is equal to the ratio of the continuous input rate and the clearance rate constant, the plasma concentration will theoretically be steady from the outset:

$$[C_p]\left(D_L = \frac{R}{k}\right) = \frac{R}{kV_D} + \left(\frac{R/k}{V_D} - \frac{R}{kV_D}\right)e^{-kt} = \frac{R}{kV_D} \tag{8.108}$$

Typical $[C_p]$ curves for different infusion rates are shown in figure 8.15.

8.4 Macromolecular Crowding

The cytoplasm is a very dense solution with a concentration of macromolecules of roughly 300–400 mg/mL, or a solute volume percentage of 34–44% [24]. How does such *macromolecular crowding* affect the equilibria and kinetics of biomolecular interactions in the cytoplasm? Primarily in two ways: (1) Crowding raises the apparent concentration of each macromolecule in the equations describing their reactions and (2) crowding slows diffusion.

To a first approximation, weak nonspecific interactions among the wide variety of biomacromolecules can simply be ignored, and we can attempt to account for the geometric effects that result from severely decreasing the space available for a protein of interest to move around in. Figure 8.16 schematically illustrates this effect.

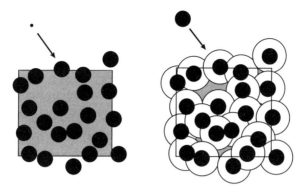

Figure 8.16 Illustration of the effect of macromolecular crowding on the volume accessible to a particle. On the left, an infinitesimally small particle can access any space (shaded gray) not occupied by one of the larger particles. On the right, the larger particle can access a much smaller volume (shaded gray), corresponding to those locations where the center of the probe particle approaches the surface of no other particle to a distance less than its own radius [25].

Because two macromolecules cannot interpenetrate, the larger a macromolecule is, the less space there is to fit it into a suspension of other macromolecules. These effects are often termed *excluded-volume effects*.

In crowded suspensions, we must reframe the concept of concentration on a per-accessible-volume basis rather than on a per-total-volume basis. The space taken up by other macromolecules is, in effect, subtracted from the volume under consideration. This correction is performed in thermodynamic relationships by substitution of a variable called activity a_i for concentration:

$$P + L \rightleftharpoons C$$

$$K_d = \frac{a_P a_L}{a_C} \qquad (8.109)$$

where the activities a_i are related to per-total-volume concentrations by activity coefficients γ_i:

$$a_P \equiv \gamma_P [P] \qquad (8.110)$$

$$a_L \equiv \gamma_L [L] \qquad (8.111)$$

$$a_C \equiv \gamma_C [C] \qquad (8.112)$$

In dilute solutions, $\gamma_P \approx \gamma_L \approx \gamma_C \approx 1$, and all of the relationships we have been using up to this point remain unchanged. In crowded solutions, however, the activity coefficients correct for the perturbations that result from excluded-volume effects. (Activity coefficients can also be used to account for the presence of weak nonspecific interactions among molecules at higher concentrations.) The strictly geometric component of the activity coefficient can be defined straightforwardly as follows:

$$\gamma_i \equiv \frac{\text{Total volume}}{\text{Volume accessible to component } i} \qquad (8.113)$$

Examination of equation 8.113 indicates that the excluded-volume effect causes $\gamma_i > 1$, raising the activity of component i in equilibrium relationships, such as that shown in equation 8.109. Relationships have been derived for suspensions of uniformly sized spherical particles, with the most widely used being the *scaled particle theory* [26], which predicts that

$$\ln \gamma = -\ln(1-\phi) + 7Q + \frac{15}{2}Q^2 + 3Q^3 \qquad (8.114)$$

where ϕ is the fraction of volume directly occupied by the spheres, and $Q \equiv \phi/(1-\phi)$ is the ratio of occupied to unoccupied volume. This relationship is plotted in figure 8.17. Experimental measurements of the activity coefficient γ have been performed for dense protein suspensions, using osmotic pressure, sedimentation, or light scatter as probes. The simplified scaled particle theory does a surprisingly good job of predicting activity coefficients in these experiments [26]. Examination of figure 8.17 indicates that the effective concentration of a protein participating in equilibria inside the cell can be increased by orders of magnitude by the excluded-volume effect. This effect is rapidly amplified as the size of the protein of interest relative to the bulk average molecule size increases, as shown in figure 8.18. The strength of the particle-size dependence of the excluded-volume effect is also illustrated schematically in figure 8.16.

8.4.1 Macromolecular Crowding Effects on Diffusion

By contrast with the fairly simple geometric accessible-volume interpretation of crowding effects on activity coefficients, there is no simple theoretical picture of crowding's effect on diffusion [28]. However, ample experimental evidence exists for the hindering effect of high macromolecular concentrations on diffusion.

There are three general mechanisms by which a protein's diffusive motion through a crowded solution might be impeded: (1) increased fluid viscosity, (2) reversible binding interactions with the crowding macromolecules, and (3) elastic collisions with the crowding molecules. These three mechanisms have been likened to an automotive analogy [29]. The time taken for a car to traverse a given distance is a function of its speed (viscosity), time spent at stop lights (binding), and route (crowding collisions). When studying proteins in their native environment, one should expect some degree of binding interactions, which would confound attempts to measure

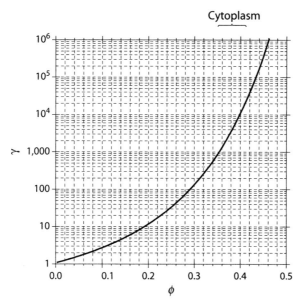

Figure 8.17 Activity coefficient γ predicted by scaled particle theory for suspensions of hard spheres with volume fraction ϕ [26]. The range for cytoplasmic volume fraction was measured for *Escherichia coli* [24].

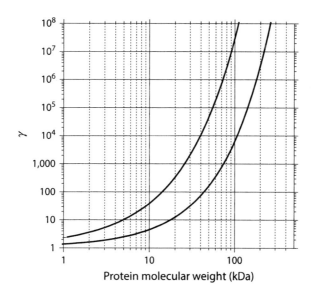

Figure 8.18 Effect of protein size on activity coefficient, as predicted by scaled particle theory (adapted from reference 24, assuming a protein molar volume of ≈ 0.73 g/mL [27]). The two curves represent upper and lower estimates based on different assumptions of average molar volume for the crowding macromolecules. For probes substantially smaller than the average cytoplasmic macromolecule (60 kDa), the excluded-volume effect is relatively weak. However, the activity coefficient rises dramatically as molecular weight increases.

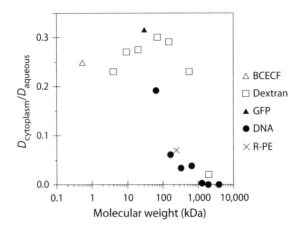

Figure 8.19 Diffusivity of a variety of tracers in the cytoplasm of mammalian cells, measured by FRAP. There appear to be effectively three regimes: <100 kDa, where cytoplasmic diffusivity is ≈0.25 that in free solution; >1,000 kDa, where tracers do not diffuse freely; and an intermediate transition region [3,4,29–32].

just the crowding effects on diffusive transport. Consequently, fluorescence recovery after photobleaching (FRAP, chapter 2) measurements have been made by injection of putatively inert tracer particles, such as dextran, green fluorescent protein (GFP), the phycobiliprotein R-phycoerythrin (R-PE), or the small fluorophore 2′,7′-bis-(2-carboxyethyl)-5-(and-6)-carboxyfluorescein (BCECF), as well as for more densely charged DNA. The results of such studies are summarized in figure 8.19. Surprisingly, for tracer molecules below ≈100 kDa, there is only a ≈3–5 fold decrease in diffusivity in the cytoplasm relative to dilute solution, with no strong molecular weight dependence. Above ≈1,000 kDa, however, FITC-dextran and DNA tracers are found to be essentially immobilized.

8.4.2 Macromolecular Crowding Effects on Binding Equilibria

How does macromolecular crowding alter binding equilibria in vivo? Let's examine equation 8.109, rewriting it to parse out the nonideal effects into a term consisting entirely of activity coefficients:

$$K_d = \frac{\gamma_P [P] \gamma_L [L]}{\gamma_C [C]} \quad (8.115)$$

$$= \left(\frac{\gamma_P \gamma_L}{\gamma_C}\right)\left(\frac{[P][L]}{[C]}\right) \quad (8.116)$$

$$= \frac{1}{\Gamma} \tilde{K}_d \quad (8.117)$$

where

$$\Gamma \equiv \frac{\gamma_C}{\gamma_P \gamma_L} \quad (8.118)$$

is a nonideality correction factor, and

$$\tilde{K}_d \equiv \frac{[P][L]}{[C]} = \Gamma K_d \quad (8.119)$$

is the apparent dissociation equilibrium constant. Crowding generally produces a correction factor Γ that is less than one, because each of the individual activity coefficients is greater than one. Consequently, the nonideal correction factor usually has the effect of decreasing $\tilde{K}_d$, favoring association equilibria. Two examples of this effect are considered below.

1. *Binding to a solid phase.* Consider the case in which a protein binds to something large and effectively external to the solution phase—for example, a site on a membrane surface, a cytoskeletal structure, or a large aggregate. Let $[L]$ refer to the free site concentration and $[C]$ to the concentration of protein bound to the sites. For a large structure, the addition of a single protein has a negligible effect on the activity coefficient (so $\gamma_L = \gamma_C$):

$$\tilde{K}_d = \Gamma K_d = \frac{K_d}{\gamma_P} \quad (8.120)$$

   As we have seen, in vivo, γ_P can be much greater than one (figures 8.17 and 8.18), and so crowding can significantly enhance binding of soluble proteins to solid surfaces. This observation is empirically validated, in particular, for the effects of crowding leading to increased protein aggregate formation in vivo [33–35].

2. *Binding of two molecules in solution.* To estimate effects of crowding on Γ, it is necessary to make some assumptions regarding the shape of the protein-ligand complex. Let's assume that the complex is spherical and has a molar volume that is the sum of the unbound partners' molar volumes. This assumption doesn't affect the overall qualitative conclusions of the analysis [26]. Using the predicted activity coefficient from the lower curve in figure 8.18, one can predict values for Γ upon complex formation, as shown in figure 8.20. Because $\Gamma < 1$, the apparent equilibrium dissociation constant $\tilde{K}_d < K_d$ for all sizes of protein and ligand, and so crowding always favors association at equilibrium. However, the magnitude of this effect is relatively negligible when either the protein or the ligand is less than 1 kDa in size. Crowding dramatically favors binding when both the protein

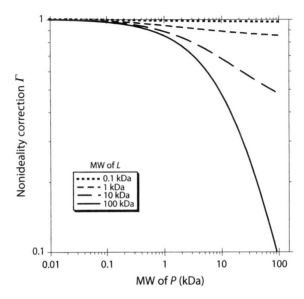

Figure 8.20 Predicted effects of macromolecular crowding on protein-ligand binding equilibria. The nonideality correction is as defined in equation 8.118, and the effect of protein P and ligand L molecular weight on activity coefficients is calculated from scaled particle theory as for the lower curve in figure 8.18.

and ligand are greater than 10 kDa in molecular weight. This observation that crowding favors association reactions has been confirmed experimentally [25].

8.4.3 Macromolecular Crowding Effects on Rates

Macromolecular crowding exerts two opposing effects on rate processes, namely, crowding slows diffusion and thus reduces the encounter frequency of two biomolecular partners in a reaction. Yet crowding favors association at equilibrium (as we have seen in the previous section), and if the transition state intermediate of a reaction is an associated complex, the apparent activation energy will be decreased. How these two effects combine depends on the quantitative magnitudes of the reaction and transport terms (k_1 and k_+, respectively, from equation 8.51):

$$k_{\text{on}} = \frac{k_1 k_+}{k_1 + k_+} \tag{8.121}$$

The acceleration of k_1 by crowding and reduced activation energy is illustrated in the reaction diagram of figure 8.21. The decrease in diffusivity (and consequently, k_+) due to crowding is shown empirically in figure 8.19. Taken together, there is predicted to be an optimum in an overall observed reaction rate k_{on} as a function of crowding, as shown in figure 8.22. Such an optimum has been experimentally

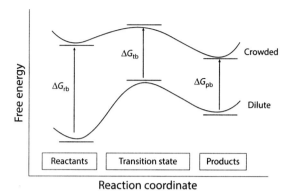

Figure 8.21 Macromolecular crowding can both make an unfavorable reaction thermodynamically favorable and accelerate a reaction by decreasing the apparent activation energy [28].

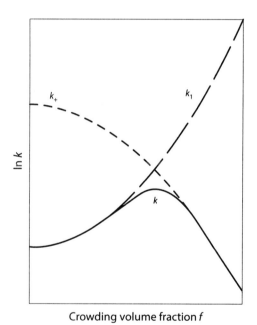

Figure 8.22 The balance between decreased diffusive encounter rate and reduced activation energy theoretically leads to an optimal rate versus crowder concentration [28].

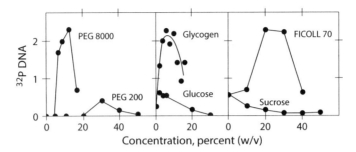

Figure 8.23 Integrated reaction rate for T4 polynucleotide kinase (given as radiolabeled phosphorus incorporation, ^{32}P DNA) as a function of different crowder concentrations. The optimum in reaction rate versus crowding may arise from the trade-off outlined in figure 8.22 [36].

observed previously (figure 8.23) and may be due to the described trade-off between diffusive transport deceleration and reaction acceleration.

Suggestions for Further Reading

H. C. Berg. *Random Walks in Biology*. Princeton University Press, 1993.

W. M. Deen. *Analysis of Transport Phenomena*. Oxford University Press, 2011.

A. J. Grodzinsky. *Fields, Forces, and Flows in Biological Systems*. Garland Science, 2011.

G. A. Truskey, F. Yuan, and D. F. Katz. *Transport Phenomena in Biological Systems*, second edition. Prentice-Hall, 2009.

Problems

8-1 The purpose of this exercise is to examine the effects of immobilization on observed binding and dissociation kinetics. Consider a polystrene bead of diameter 5 μm on which either a protein (150,000 g/mol molecular weight) or its small-molecule ligand (150 g/mol molecular weight) can be immobilized. In either case, the soluble protein or ligand can be labeled so as to track the number of complexes on the bead as a function of time. In solution, the association rate constant for the protein-ligand complex formation is $k_{on} = 10^6$ M^{-1} s^{-1}, and the dissociation rate constant is $k_{off} = 10^{-3}$ s^{-1}. You may assume that the density of the ligand and protein is 1.1 g/cm^3.

  a. Plot the observed association rate constant k_f versus the number of molecules immobilized per bead for both the case of immobilized protein and the case of immobilized ligand.

b. Plot the observed dissociation rate constant k_r versus the number of molecules immobilized per bead for both the case of immobilized protein and the case of immobilized ligand.

c. Under what immobilization conditions (i.e. protein or ligand surface density) will the observed rate constants (k_f, k_r) reflect the intrinsic rate constants (k_{on}, k_{off})?

8-2 An enterprising young faculty member sets out to measure the number of receptors of a certain type on a given cell line. Her method relies on using diffusion-limited characteristics of ligand binding to reveal receptor number.

She has prepared specific cross-linking reagents that bind covalently to a portion of the receptor away from the ligand binding site. With these cross-linkers, she will make separate preparations. One will contain only single cells, another will contain cross-linked pairs of cells, yet another will contain cross-linked triples, and so on for higher-order oligomers.

Assume that the equations derived for spherical single cells also apply to the oligomers created. Using cell-sorting technology, she will be able to get exact counts of numbers of cells.

a. What experiments will this young faculty member carry out to measure the number of receptors per cell?

b. What other measurements will she need to obtain?

c. Are there any critical reasons that this experiment may not work, or at least be extremely difficult?

8-3 You aim to achieve a steady 10 nM plasma concentration of a 75 kDa protein. What dosing strategy should you use to administer the molecule? You have the following data from pilot experiments: a 1 µg tracer dose of protein yields an area under the curve of plasma concentration versus time equal to 10 µg·hr/L; and the volume of distribution is determined to be 5 L.

8-4 You have engineered a drug to combat cardiovascular disease by activating a key target protein inside cells within atherosclerotic plaque. The drug, which is absent from the body initially, is consistently supplied to the bloodstream at R nmol/h. The drug leaves the bloodstream through the kidneys; this process is quantitatively described with first-order rate constant k. It also passively enters the cells of interest. The rate of this process is proportional to the concentration of drug in the bloodstream. There is no mechanism for the drug to leave the cell to re-enter the bloodstream. Instead, the drug can reversibly bind the target protein in a monovalent manner. The number of targets per cell is denoted T and can be assumed to be in significant excess relative to the delivered drug. The drug can also be degraded via a first-order process with rate constant k_{deg}.

Denote bloodstream volume by V_p and intracellular volume of all targeted cells by V_c.

a. Derive an expression for the dynamic concentration of drug in the bloodstream.

b. What is the steady-state concentration of the drug in the bloodstream?

c. What is the steady-state fraction of target proteins that are complexed?

d. Derive an expression for the dynamic concentration of drug-receptor complex in the cells in the atherosclerotic plaque. You can approximate that binding and unbinding are rapidly equilibrated relative to pharmacokinetic transport.

8-5 A cluster of cells in a patient's muscle tissue consistently expresses a cytokine at rate r_p (mol/s). The cytokine is transported between tissue (volume V_t) and bloodstream (volume V_b) as described by first-order rate constants k_{TB} (tissue to blood) and k_{BT} (blood to tissue). In the bloodstream, the cytokine reversibly binds to its receptor (rate constants for association and dissociation are k_{on} and k_{off}). The concentration of total cytokine receptors in the blood is $[R]_o$. The cytokine is cleared through the kidneys with first-order rate constant k_c. Derive expressions for the steady-state concentrations of cytokine in the tissue, in the bloodstream (unbound), and bound to receptor. The expressions should only contain the indicated constants.

8-6 Your team has engineered a drug to combat infectious bacteria that reside in the bloodstream. Your drug reversibly binds to a protein target on the surface of the pathogenic bacterium; the affinity of this interaction is K_d (nM) (association rate constant k_{on} and dissociation rate constant k_{off}). To provide a consistent source of drug, you administer a probiotic that slowly releases the drug into the bloodstream at a rate of R (nmol/h). Unbound drug is cleared from the bloodstream, via the kidneys, as described by rate constant k_c (1/h). The bloodstream volume is V_p (L).

a. What fraction of protein targets are bound when the system reaches a steady state? The answer should only contain constants stated in the problem. Show your derivation.

b. The drug is highly potent. Thus, a single target-drug complex is sufficient to kill a bacterium. If each bacterium has 500 protein targets, what rate of drug release (R) would be needed so that 90% of bacteria have at least one drug molecule bound at steady state?

8-7 You have developed a therapeutic molecule that consists of an antibody non-covalently conjugated with a ^{177}Lu radioisotope. Unfortunately, ^{177}Lu radioisotope can dissociate from the intact antibody-radioisotope complex (dissociation rate constant k_{off}). For multiple reasons, dissociated ^{177}Lu radioisotope does not reassociate with antibody to an appreciable extent. Fortunately, after dissociation, the small ^{177}Lu radioisotope can clear from the body (clearance rate constant k_c). Due to its large size, you may neglect clearance of the antibody-radioisotope complex during the time scale of interest. Derive an expression for the dynamic concentration of dissociated ^{177}Lu radioisotope in the body (that is, uncleared ^{177}Lu radioisotope).

8-8 You inject N moles of a protein drug subcutaneously (under the skin) into a region of volume V_s at time $t = 0$. Preferably, the drug is transported to a nearby tumor of volume V_t with first-order rate constant k_t. Undesirably, the drug can also be transported from the initial subcutaneous region to the bloodstream with first-order rate constant k_b. Both of these transport processes can be considered unidirectional (subcutaneous-to-tumor and subcutaneous-to-blood). The drug cannot reach the tumor from the bloodstream. In the tumor, the drug reversibly binds to receptor (rate constants k_{on} and k_{off}), which is present at concentration $[R]$. The drug can also clear from the tumor to the lymphatic system with rate constant k_c.

a. Derive a dynamic expression for the fraction of receptor bound by drug. Assume that the drug-receptor interaction is essentially in equilibrium compared to the slow transport processes. Also assume that the receptor is dilute relative to the drug concentration.

b. Plot the fraction of bound receptor versus time for the following parameter values:
$k_t = 0.05$ min^{-1}
$k_b = 0.2$ min^{-1}
$k_c = 0.1$ min^{-1}
$K_d = 10$ nM
$V_t = 3$ mL
$N = 1$ nmol

8-9 Your company developed a small-molecule drug that binds—with 12 nM affinity—to a receptor present on a particular subset of neurons. The presence of the drug bound to the receptor activates intracellular signaling that reduces pain. For use in pain management, you would like to administer the drug in such a way that 10% of receptors are bound by drug on initial treatment. For longer-term control, you'd like to maintain at least 2% of receptors in the bound state. You propose to formulate the drug as a pill with two compartments: One that rapidly (assume instantaneously) dissolves on ingestion to release its contents (i.e., the drug), and a second compartment that gradually dissolves to slowly release its contents. You may assume that the drug is at a uniform concentration throughout its 6 L volume of distribution. In an experiment to evaluate clearance from this distribution volume, 90% of a 2 microgram tracer dose was cleared within 12 hours; uniform clearance kinetics were observed for all regions.

For parts a–d, you may approximate that drug-receptor binding/unbinding is rapid relative to pharmacokinetics (i.e., the drug-receptor system is rapidly equilibrated).

a. Calculate the amount of drug (in mg) that should be added to the rapidly dissolving compartment.

b. Calculate the rate of dissolution (in mg/h of drug) that should be designed for the slowly dissolving compartment.

c. If the drug is 2% (by mass) of the formulation, and the pill is limited to 250 mg total mass, how long will the 2% bound design be maintained?

d. Plot the fraction of receptor bound versus time (using the mass-limited case in part c). Provide an analytical description $f(t)$. Show all work.

e. Now consider that receptor binding/unbinding is not rapidly equilibrated relative to pharmacokinetics. Write the differential equation for [drug-receptor complex] in terms of time and constants. The only variables that should appear in the final form are [drug-receptor complex] and t. Also provide the boundary condition(s) needed, but you do not need to solve the equation.

Drug molecular weight: 572 Da
Number of neurons of this type in the body: 10 million
Number of receptors per neuron: 10,000
Valency: monovalent (one drug molecule binds to one receptor; neither the drug nor the receptor dimerizes)

8-10 You are trying to optimize dosing strategy for an antibody-drug conjugate (150 kDa molecular weight). The antibody targets a receptor that is present at 1 million copies per tumor cell but is also present at 20,000 copies per cell in a small subset of healthy cells in the body. The conjugated drug is highly toxic: It only takes 100 receptor-bound antibody-drug conjugates to kill a cell. You have performed several pilot experiments to determine that the antibody's affinity is 0.2 nM and its rate constant for plasma clearance is 0.02 h^{-1}. Because of the large molecular weight, it is reasonable to assume that the antibody remains essentially confined to the plasma (4 L) except when being cleared. You can assume that the numbers of tumor cells and healthy cells are very small.

a. What is the maximum initial dose that can be administered without killing healthy cells expressing 20,000 receptors?

b. How long will this dose maintain efficacy against tumor cells expressing 1 million receptors?

c. On a single graph, plot the number of receptor complexes versus time for tumor cells and healthy cells. Include at least three administration cycles (i.e., doses).

8-11 If the viscosity of water is 0.8937 cP at 298 K and 0.7225 cP at 308 K, what would the activation energy be for a diffusion-limited reaction in water in this temperature range?

References

1. I. G. Zacharia and W. M. Deen. Diffusivity and solubility of nitric oxide in water and saline. *Ann. Biomed. Eng.* 33: 214–222 (2005).

2. R. L. Fournier. *Basic Transport Phenomena in Biomedical Engineering*. CRC Press (2011).

3. M. J. Dayel, E. F. Y. Hom, and A. S. Verkman. Diffusion of green fluorescent protein in the aqueous-phase lumen of endoplasmic reticulum. *Biophys. J.* 76: 2843–2851 (1999).

4. G. L. Lukacs, P. Haggie, O. Seksek, D. Lechardeur, N. Freedman, and A. S. Verkman. Size-dependent DNA mobility in cytoplasm and nucleus. *J. Biol. Chem.* 275: 1625–1629 (2000).

5. E. B. Brown, Y. Boucher, S. Nasser, and R. K. Jain. Measurement of macromolecular diffusion coefficients in human tumors. *Microvasc. Res.* 67: 231–236 (2004).

6. P. Netti, D. A. Berk, M. A. Swartz, A. J. Grodzinsky, and R. K. Jain. Role of extracellular matrix assembly in interstitial transport in solid tumors. *Cancer Res.* 60: 2497–2503 (2000).

7. D. G. Myszka, X. He, M. Dembo, T. A. Morton, and B. Goldstein. Extending the range of rate constants available from BIACORE: Interpreting mass transport-influenced binding data. *Biophys. J.* 75: 583–594 (1998).

8. D. H. Schott, R. N. Collins, and A. Bretscher. Secretory vesicle transport velocity in living cells depends on the myosin-V lever arm length. *J. Cell Biol.* 156: 35–40 (2002).

9. A. J. Grodzinsky. *Fields, Forces, and Flows in Biological Systems*. Garland Science (2011).

10. A. D. Vogt and E. Di Cera. Conformational selection or induced fit? A critical appraisal of the kinetic mechanism. *Biochemistry* 51: 5894–5902 (2012).

11. M. V. Smoluchowski. Versuch einer mathematischen Theorie der Koagulationskinetik kolloider Lösungen. *Z. Phys. Chem.* 92: 129–168 (1917).

12. C. J. Camacho, Z. Weng, S. Vajda, and C. DeLisi. Free energy landscapes of encounter complexes in protein–protein association. *Biophys. J.* 76: 1166–1178 (1999).

13. H. C. Berg and E. M. Purcell. Physics of chemoreception. *Biophys. J.* 20: 193–219 (1977).

14. J. Erickson, B. Goldstein, D. Holowka, and B. Baird. The effect of receptor density on the forward rate constant for binding of ligands to cell surface receptors. *Biophys. J.* 52: 657–662 (1987).

15. S. Y. Shvartsman, H. S. Wiley, W. M. Deen, and D. A. Lauffenburger. Spatial range of autocrine signaling: Modeling and computational analysis. *Biophys. J.* 81: 1854–1867 (2001).

16. E. W. Thiele. Relation between catalytic reactivity and size of particle. *Ind. Eng. Chem.* 31: 916–920 (1939).

17. G. M. Thurber and K. D. Wittrup. Quantitative spatiotemporal analysis of antibody fragment diffusion and endocytic consumption in tumor spheroids. *Cancer Res.* 68: 3334–3341 (2008).

18. L. A. Goentoro, G. T. Reeves, C. P. Kowal, L. Martinelli, T. Schupbach, and S. Y. Shvartsman. Quantifying the Gurken morphogen gradient in *Drosophila* oogenesis. *Dev. Cell* 11: 263–272 (2006).

19. D. J. Fink, T.-Y. Na, and J. S. Schultz. Effectiveness factor calculations for immobilized enzyme catalysts. *Biotechnol. Bioeng.* 15: 879–888 (1973).

20. E. S. Avgoustiniatos and C. K. Colton. Effect of external oxygen mass transfer resistances on viability of immunoisolated tissue. *Ann. N. Y. Acad. Sci.* 831: 145–167 (1997).

21. J. J. Rhoden and K. D. Wittrup. Dose dependence of intratumoral perivascular distribution of monoclonal antibodies. *J. Pharm. Sci.* 101: 860 (2012).

22. D. M. Benson, C. C. Hofmeister, S. Padmanabhan, A. Suvannasankha, S. Jagannath, R. Abonour, C. Bakan, P. Andre, Y. Efebera, J. Tiollier, M. A. Caligiuiri, and S. S. Farag. A phase 1 trial of the anti-KIR antibody IPH2101 in patients with relapsed/refractory multiple myeloma. *Blood* 120: 4324–4333 (2012).

23. M. L. Reitman, X. Chu, X. Cai, J. Yabut, R. Venkatasubramanian, S. Zajic, J. A. Stone, Y. Ding, R. Witter, C. Gibson, K. Roupe, R. Evers, J. A. Wagner, and A. Stoch. Rifampin's acute inhibitory and chronic inductive drug interactions: Experimental and model-based approaches to drug-drug interaction trial design. *Clin. Pharmacol. Ther.* 89: 234–242 (2011).

24. S. B. Zimmerman and S. O. Trach. Estimation of macromolecule concentrations and excluded volume effects for the cytoplasm of *Escherichia coli*. *J. Mol. Biol.* 222: 599–620 (1991).

25. A. P. Minton. The influence of macromolecular crowding and macromolecular confinement on biochemical reactions in physiological media. *J. Biol. Chem.* 276: 10,577–10,580 (2001).

26. A. P. Minton. Molecular crowding: Analysis of effects of high concentrations of inert cosolutes on biochemical equilibria and rates in terms of volume exclusion. *Methods Enzymol.* 295: 127–149 (1998).

27. T. V. Chalikian, M. Totrov, R. Abagyan, and K. J. Breslauer. The hydration of globular proteins as derived from volume and compressibility measurements: Cross correlating thermodynamic and structural data. *J. Mol. Biol.* 260: 588–603 (1996).

28. S. B. Zimmerman and A. P. Minton. Macromolecular crowding: Biochemical, biophysical, and physiological consequences. *Annu. Rev. Biophys. Biomol. Struct.* 22: 27–65 (1993).

29. A. S. Verkman. Solute and macromolecule diffusion in cellular aqueous compartments. *Trends Biochem. Sci.* 27: 27–33 (2002).

30. H. P. Kao, J. R. Abney, and A. S. Verkman. Determinants of the translational mobility of a small solute in cell cytoplasm. *J. Cell Biol.* 120: 175–184 (1993).

31. O. Seksek, J. Biwersi, and A. S. Verkman. Translational diffusion of macromolecule-sized solutes in cytoplasm and nucleus. *J. Cell Biol.* 138: 131–142 (1997).

32. M. Goulian and S. M. Simon. Tracking single proteins within cells. *Biophys. J.* 79: 2188–2198 (2000).

33. D. M. Hatters, A. P. Minton, and G. J. Howlett. Macromolecular crowding accelerates amyloid formation by human apolipoprotein C-II. *J. Biol. Chem.* 277: 7824–7830 (2002).

34. M. D. Shtilerman, T. T. Ding, and P. T. Lansbury, Jr. Molecular crowding accelerates fibrillization of α-synuclein: Could an increase in the cytoplasmic protein concentration induce Parkinson's disease? *Biochemistry* 41: 3855–3860 (2002).

35. L. A. Munishkina, E. M. Cooper, V. N. Uversky, and A. L. Fink. The effect of macromolecular crowding on protein aggregation and amyloid fibril formation. *J. Mol. Recog.* 17: 456–464 (2004).

36. B. Harrison and S. B. Zimmerman. Stabilization of T4 polynucleotide kinase by macromolecular crowding. *Nucleic Acids Res.* 14: 1863–1870 (1986).

9

Discrete Stochastic Processes

Never think that lack of variability is stability. Don't confuse lack of volatility with stability, ever.
—Nassim Nicholas Taleb

The world is noisy and messy. You need to deal with the noise and uncertainty.
—Daphne Koller

In the rate equations written so far in this book, we have used concentrations (such as $[P]$ and $[L]$) that vary continuously, even though any given volume has a discretely countable number of molecules. This *continuum approximation* is quite accurate for large numbers of molecules, in the same way that one need not perform quantum mechanical calculations to accurately simulate the lift and drag forces on an airplane. However, the continuum approximation breaks down for small numbers of molecules in a given control volume (e.g., inside a single cell). For example, if there are one or two copies of the DNA encoding a gene, it is clearly inappropriate to describe this quantity as a smoothly varying "gene concentration." A handful of molecules can nevertheless exist at biologically significant concentrations in small volumes: For an *Escherichia coli* cell with a typical intracellular volume of ≈ 1 fL, a 1 nM concentration corresponds to just ≈ 0.6 molecules. Thus, it is not possible in *E. coli* to smoothly vary the concentration of a biomolecule in the low nanomolar range, because each biomolecule added or subtracted results in a discrete change of ≈ 1.7 nM in concentration. From this simple calculation, it should be evident that it is also impossible to achieve a (nonzero) subnanomolar concentration inside a single *E. coli* cell, because this would require less than one copy of the biomolecule.

Table 9.1 shows that small numbers of molecules can still be present at significant concentrations in organisms ranging from bacteria to humans. Additionally, larger mammalian cells contain numerous membrane-bounded subcellular compartments (nucleus, Golgi, endosomes, mitochondria, etc.) in which small numbers of molecules can exert significant effects due to their high concentrations in these restricted volumes.

Table 9.1 Small numbers of molecules in individual cells.

Cell type	Intracellular volume (fL)	Molecules corresponding to 1 nM concentration
Escherichia coli	≈1	≈0.6
Saccharomyces cerevisiae	≈40	≈24
Human erythrocyte	≈100	≈60

As discussed in chapter 8, biomolecules randomly bump into one another as the result of thermal collisions. Given that the impact of this randomness is magnified when dealing with small numbers of biomolecules, how do living cells carry out reproducible programs of gene expression, growth, movement, differentiation, and development? Before we consider biological strategies for dealing with or even harnessing randomness, let's first discuss how to quantitatively describe noisy, random processes involving small numbers of biomolecules.

9.1 Poisson Statistics

Many processes involving small numbers of molecules are well described by Poisson statistics. Let's derive the probability distribution for a Poisson random variable, building up from the probability of single binary events to describe the statistics for larger numbers of events. The simplest event to consider is one that is binary—either a success or a failure. For example, either a molecule is present in a given volume or it is not; or a protein molecule binds to a ligand or it does not. A simple yes-no experiment of this kind is called a Bernoulli trial, and the number of successes in a series of independent Bernoulli trials is a binomial random number. For n Bernoulli trials, the probability of k successes $P(k)$ is

$$P(k) = \binom{n}{k} p^k (1-p)^{n-k} \quad (9.1)$$

where p is the probability of success in any individual Bernoulli trial, and $\binom{n}{k} = \frac{n!}{k!(n-k)!}$ is the number of different ways to pick k items from a pool of n (see chapter 3). Let's now consider a series of Bernoulli trials in space (e.g., presence or absence of a molecule in a volume) or time (e.g., occurrence or nonoccurrence of a reaction in a period of time). Let's make the following three assumptions that define a Poisson process:

1. In a small enough increment in space or time, only zero or one event will occur (i.e., a single Bernoulli trial).

2. Events in each increment of space or time are independent of events in every other increment.
3. The probability of success of a Bernoulli trial is proportional to the size of the increment (e.g., $p = \lambda V/n$ or $p = \lambda t/n$, where V/n and t/n are increments of volume and time, respectively, and λ is a proportionality constant).

It is intuitively clear that many biomolecular processes are consistent with these assumptions (with the independence of subsequent events as the assumption most likely to be violated). For simplicity, let's continue the derivation of what we will now call the Poisson probability (P_P) in terms of volume V/n for the moment, recognizing that the mathematics are identical for a Poisson process occurring in time rather than in space. For a volume of interest V subdivided into n equal and independent volumes, we have

$$P_P(k) = \binom{n}{k} \left(\frac{\lambda V}{n}\right)^k \left(1 - \frac{\lambda V}{n}\right)^{n-k} \tag{9.2}$$

$$= \frac{n!}{k!(n-k)!} \left(\frac{\lambda V}{n}\right)^k \left(1 - \frac{\lambda V}{n}\right)^{n-k} \tag{9.3}$$

To satisfy the first assumption for a Poisson process, let's shrink the volume increment ($n \to \infty$) such that only one or zero molecules can be found in it:

$$\lim_{n \to \infty} P_P(k) = \lim_{n \to \infty} \frac{n!}{k!(n-k)!} \left(\frac{\lambda V}{n}\right)^k \left(1 - \frac{\lambda V}{n}\right)^{n-k} \tag{9.4}$$

Collecting constant terms outside the limit gives

$$\lim_{n \to \infty} P_P(k) = \frac{(\lambda V)^k}{k!} \lim_{n \to \infty} \frac{n!}{(n-k)!} \left(\frac{1}{n}\right)^k \left(1 - \frac{\lambda V}{n}\right)^{n-k} \tag{9.5}$$

The first two terms in the limit can be restated as a product of fractions:

$$\frac{n!}{(n-k)!} \left(\frac{1}{n}\right)^k = 1 \left(1 - \frac{1}{n}\right) \left(1 - \frac{2}{n}\right) \cdots \left(1 - \frac{k-1}{n}\right) \tag{9.6}$$

Consequently,

$$\lim_{n \to \infty} \frac{n!}{(n-k)!} \left(\frac{1}{n}\right)^k = \lim_{n \to \infty} \left(1 - \frac{1}{n}\right) \left(1 - \frac{2}{n}\right) \cdots \left(1 - \frac{k-1}{n}\right) \tag{9.7}$$

$$= 1 \tag{9.8}$$

And so equation 9.5 can be simplified as follows:

$$\lim_{n\to\infty} P_P(k) = \frac{(\lambda V)^k}{k!} \lim_{n\to\infty} \left(1 - \frac{\lambda V}{n}\right)^{n-k} \qquad (9.9)$$

$$= \frac{(\lambda V)^k}{k!} \lim_{n\to\infty} \left(1 - \frac{\lambda V}{n}\right)^{n} \qquad (9.10)$$

$$= \frac{(\lambda V)^k}{k!} e^{-\lambda V} \qquad (9.11)$$

The mean (average) μ of a discrete random number with probability distribution $P(k)$ is equal to its expectation, $\mathbb{E}(k)$, which is defined as

$$\mathbb{E}(k) = \mu = \sum_{k=0}^{\infty} kP(k) \qquad (9.12)$$

For a Poisson random number, this is

$$\mu = \sum_{k=0}^{\infty} k \frac{(\lambda V)^k}{k!} e^{-\lambda V} \qquad (9.13)$$

$$= \lambda V \qquad (9.14)$$

The probability distribution for a Poisson random number can then be rewritten as

$$P_P(k) = \frac{\mu^k}{k!} e^{-\mu} \qquad (9.15)$$

The shape of the Poisson probability distribution for increasing values of the average μ is shown in figure 9.1. Of particular note is the asymmetry of the distribution at low values; the probability of zero events $P_P(0) = e^{-1} \approx 0.37$ when $\mu = 1$.

The primary reason to apply a discrete probabilistic description of molecular events is to account for the random fluctuations intrinsic to systems of small numbers of molecules. Examination of figure 9.2 shows that, for small values of μ, the probability that $k \neq \mu$ can be significant. These fluctuations correspond to the width of the probability distribution around μ and can be quantified in terms of variance σ^2, or standard deviation σ. The variance of a random number is defined as

$$\sigma^2 = \mathbb{E}\left[(k - \mu)^2\right] \qquad (9.16)$$

Discrete Stochastic Processes

$$= \mathbb{E}(k^2) - [\mathbb{E}(k)]^2 \tag{9.17}$$

$$= \sum_{k=0}^{\infty} k^2 P(k) - \mu^2 \tag{9.18}$$

where $\mathbb{E}(k)$ is the expectation value of the variable k as defined above. For the Poisson distribution, the variance is

$$\sigma^2 = \mu \tag{9.19}$$

so a Poisson random number has a mean and variance that are equal. A more intuitive measure of peak width is the coefficient of variation (CV), which equals σ/μ. Thus, for a Poisson random number,

$$\text{CV} = \frac{1}{\sqrt{\mu}} \tag{9.20}$$

As a result, the fluctuations in a Poisson random number fall as the reciprocal of the square root of the mean.

(As an aside, it can be shown that for large μ, a Poisson distribution tends to a Gaussian distribution. Thus, the theories for small-number systems and large-number systems are not fundamentally different. However, depending on the system properties, multiple measurements on a small-number system may not necessarily converge to the equivalent large-number result.)

In addition to knowing the average and fluctuations of a Poisson process in the time domain, it is often of interest to know the distribution of waiting times between events. The probability density function for the waiting time between events in a Poisson process is given by a random number with probability density function

$$p(t) = \lambda e^{-\lambda t} \tag{9.21}$$

with mean $\mu = 1/\lambda$ and variance $\sigma^2 = 1/\lambda^2$.

The probability that the next event in a Poisson process will occur at $t \geq T$ is

$$P(t \geq T) = \int_T^{\infty} p(t) dt \tag{9.22}$$

$$= \int_T^{\infty} \lambda e^{-\lambda t} dt \tag{9.23}$$

$$= e^{-\lambda T} \tag{9.24}$$

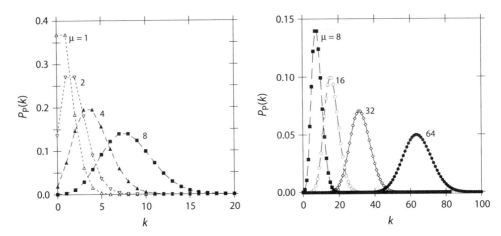

Figure 9.1 The Poisson probability distribution for means $\mu = 1, 2, 4, 8, 16, 32$, and 64. The probability $P_P(k)$ is defined only for integer values of k, represented by the points. The lines connecting the points are present only as a convenience.

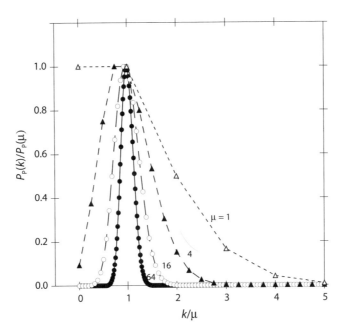

Figure 9.2 As μ increases, the CV of the Poisson probability distribution drops as the reciprocal of $\sqrt{\mu}$. For values of $\mu \geq 100$, fluctuations from the mean can be expected to be $\leq 10\%$.

There is a useful relationship between the first-order rate constant in a chemical process and the exponential random number distribution. The rate constant k is effectively the Poisson parameter λ, and so the average time τ between reaction events for a single molecule is

$$\tau = \frac{1}{k} \qquad (9.25)$$

Example 9-1 The average number of mRNA molecules for a particular regulatory gene is four in a given cell type. What is the fraction of cells that have no copies of the message? What fraction has the average? What is the fraction of cells with six or more copies?

Solution The number of copies of a particular mRNA molecule in a cell is a Poisson random variable, because by parsing out the volume of the cell into small enough increments, there will be either zero or one copy of the message in each volume increment. We imagine that all cells are equivalent in size and that the occurrence of the message in an increment is independent of its occurrence in other increments.

Using equation 9.15 and the observed $\mu = 4$ copies per cell, we can compute the probability of finding a cell with any individual number of copies. The table below is for $k = 0, 1, 2, 3, 4$, and 5, as well as the cumulative sum:

k	$P_P(k)$
0	$\frac{4^0}{0!} e^{-4} = 0.018$
1	$\frac{4^1}{1!} e^{-4} = 0.073$
2	$\frac{4^2}{2!} e^{-4} = 0.147$
3	$\frac{4^3}{3!} e^{-4} = 0.195$
4	$\frac{4^4}{4!} e^{-4} = 0.195$
5	$\frac{4^5}{5!} e^{-4} = 0.156$
	Sum $= 0.785$

Thus, the fraction of cells with no copies of the message is 0.018, and the fraction with the average (four) is 0.195. The fraction with six or more is one minus the fractions with zero, one, two, three, four, or five: $1 - 0.018 - 0.073 - 0.147 - 0.195 - 0.195 - 0.156 = 0.215$.

Example 9-2 In a particular experiment, it is desired to place an average of 1,000 cells into each well of a microtiter plate. Assuming that the cells are placed using a Poisson process, what is the expected coefficient of variation across the wells?

Solution For a Poisson process, the mean and the variance are equal, and the coefficient of variation is CV $= \sigma/\mu = 1/\sqrt{\mu}$. In this case, $\mu = 1,000$, so CV $= 1/\sqrt{1,000} = 0.03$, which is unitless.

Example 9-3 Assume that the population of the United States is 300,000,000 people and that 1,300,000 cases of cancer are diagnosed per year. In a small town of 500 people, it is found that five cases are diagnosed in a particular year. Because this is significantly higher than the expected number for a town of that size, one might suspect that this could be a *cancer cluster*, in which some local environmental agent, for example, causes a locally increased rate of cancer incidence. To evaluate that possibility, compute the probability that this number of cancer cases would appear in a town of this size in a given year by chance. If there are 1,000 towns with a population of 500 people, what is the probability that no such town will have five or more cases in a particular year?

Solution We assume that the occurrence of cancer is a Poisson process, which means that cases are independent (i.e., nonclustered). The rate of occurrence of cancer in the United States in this example is $(1.3 \times 10^6 \text{ yr}^{-1})/(3.0 \times 10^8) = 4.333 \times 10^{-3}$ person^{-1} yr^{-1}, so the average number of cases expected in a town of 500 people is $\mu = 500 \times 4.333 \times 10^{-3} = 2.167$ yr^{-1}. The probability that five or more cases would occur in a town of this size in one year is

$$P_P(k \geq 5) = 1 - [P_P(0) + P_P(1) + P_P(2) + P_P(3) + P_P(4)] \quad (9.26)$$

$$= 1 - e^{-2.167}\left[\frac{2.167^0}{0!} + \frac{2.167^1}{1!} + \frac{2.167^2}{2!} + \frac{2.167^3}{3!} + \frac{2.167^4}{4!}\right] \quad (9.27)$$

$$= 0.069 \quad (9.28)$$

Thus, it is unusual for a town this size to have so many cases by chance—but not unprecedented, because there is nearly a 7% chance of it happening in any given town. Thus, these statistics alone do not provide evidence for a cancer cluster due to a specific cause. With 1,000 towns of this size, the average number of towns with five or more cancer cases would be about 69. The probability that no towns of this size would have five or more cases is given by

$$P_P(0) = \frac{69^0}{0!} e^{-69} = 1.08 \times 10^{-30} \qquad (9.29)$$

which is extremely unlikely. Thus, it would be unusual *not* to find towns with substantially more cancer cases than average, even if all cancer cases were independent and there were no local reasons that would produce multiple cases. For this reason, it is important to understand the mechanistic causes of events such as environmentally-influenced cancer cases, rather than rely solely on statistical analyses. Note that for a cluster with a specific cause, one would expect elevated rates in this town year after year. If instead these are the expected statistics for a set of random, independent events, elevated rates should not repeat regularly.

Case Study 9-1   S. E. Luria and M. Delbrück. Mutations of bacteria from virus sensitivity to virus resistance. *Genetics* 28: 491–511 (1943) [1].

In 1943, Luria and Delbrück published a classic paper in which they distinguished two fundamentally different hypotheses for the emergence of genetic variation [1]. By studying the statistical distribution of the occurrence of phage-resistant bacteria in cultures grown from a phage-sensitive strain of *E. coli*, they were able to definitively demonstrate that heritable mutations occurred randomly and spontaneously prior to the application of selective pressure, rather than being a result of induction by selective pressure. Of particular importance in this study was examination of statistical distributions of the numbers of resistant variants across a number of identically prepared cultures; the spontaneity hypothesis predicted large fluctuations across the cultures, whereas the induction hypothesis predicted small fluctuations.

Consider a simple scenario in which a culture is inoculated (at generation $g=0$) with a single sensitive bacterium that undergoes exponential growth. The number of bacteria in generation g will be

$$N_g = 2^g \qquad (9.30)$$

According to the induction mechanism (or Lamarckian mechanism), all bacteria in the culture will be phage sensitive at the final generation, just before they are plated onto phage-containing plates. If the probability that any given bacterium will be induced to become resistant when challenged with phage is I, then a Poisson distribution will describe the expected distribution of phage-resistant bacteria in the culture at generation g. The average number of resistant bacteria in the distribution will be

$$\mu_I = N_g I = 2^g I \qquad (9.31)$$

which gives a standard deviation of

$$\sigma_I = \sqrt{\mu_I} = \sqrt{2^g I} \qquad (9.32)$$

and a coefficient of variation of

$$CV_I = \frac{1}{\sqrt{\mu_I}} = \frac{1}{\sqrt{2^g I}} \qquad (9.33)$$

The coefficient of variation decreases exponentially with the number of generations for resistance that is induced by exposure to phage. Moreover, because the distribution of resistant bacteria is Poisson in the final exposed generation, the variance is equal to the mean of the distribution:

$$\sigma_I^2 = \mu_I \qquad (9.34)$$

For mutations arising spontaneously (or via a Darwinian mechanism), we assume that the mutation rate M is constant per bacterium per generation. The average number of new resistant variants arising in generation g, given by $\mu_{R',g}$, is

$$\mu_{R',g} = N_g M \qquad (9.35)$$

The total number of resistant variants in generation g is the number of new resistant variants created in the last cell division plus reproduction of previously created variants:

$$\mu_{R,g} = N_g M + 2\mu_{R,g-1} \qquad (9.36)$$

Applying the initial condition that the inoculum in generation $g=0$ contains a single sensitive bacterium,

$$\mu_{R,0} = 0 \qquad (9.37)$$

produces an inductive set of equations for which the solution is

$$\mu_{R,g} = g 2^g M \qquad (9.38)$$

We are assuming that resistant and sensitive bacteria reproduce at the same rate, that the total number of resistant bacteria is so small that the total number of bacteria is nearly the same as the number of sensitive bacteria, and that resistant bacteria do not mutate back to be sensitive. Also, for convenience, the discussion has been phrased in terms of synchronized generations. The argument is essentially the same if the discrete generation g is replaced by the continuous time t.

The population of mutants introduced at a particular generation g_o is expected to be Poisson distributed when it is generated. What will the distribution be when the culture

Discrete Stochastic Processes 537

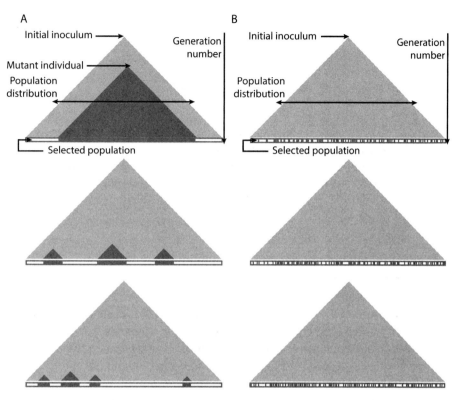

Figure 9.3 Luria-Delbrück [1] experiment of distribution of mutant populations. (A) In the Darwinian mechanism, in the absence of selective pressure, mutants appear randomly throughout the growth of the culture. Each gray triangle schematically represents the growth of a population from an initial inoculum (topmost point) through increasing generations. Light gray indicates wild-type individuals; dark gray indicates mutant individuals. All progeny resulting from a mutant individual are also mutant. Three different trials of the identical experiment are illustrated. Depending on the stochastic behavior of individual mutant appearance timing in a given trial, the number of mutants in the final generation of the culture can be highly variable. (B) In the Lamarckian mechanism, no mutants occur until the selection is applied to the individual in the final generation. In this model, which particular individuals develop adaptive mutations in response to selective pressure is determined by stochastic processes; nevertheless, the population is sufficiently large at the time of selection that the distribution is very similar in independent trials, three of which are illustrated here.

finishes growing and is plated at generation g? The result is obtained by propagating the appropriate cell divisions of the resistant bacteria:

$$\mu_{R',g,g_\circ} = N_{g_\circ} M 2^{g-g_\circ} = N_g M \qquad (9.39)$$

$$\sigma^2_{R',g,g_\circ} = N_{g_\circ} M 2^{g-g_\circ} 2^{g-g_\circ} = N_g M 2^{g-g_\circ} \qquad (9.40)$$

A superposition of all generations produces the overall average and variance

$$\mu_{R,g} = \sum_{g_\circ=1}^{g} \mu_{R',g,g_\circ} = gN_g M = g2^g M \qquad (9.41)$$

$$\sigma_{R,g}^2 = \sum_{g_\circ=1}^{g} \sigma_{R',g,g_\circ}^2 = N_g M 2^g \sum_{g_\circ=1}^{g} 2^{-g_\circ} = M 2^{2g} \qquad (9.42)$$

where the final evaluation of the sum in the variance calculation assumes a reasonably large number of generations. Notice that the variance is significantly larger than the mean for spontaneously arising mutations, with a ratio of $2^g/g$ for a large number of generations, which contrasts sharply with the model for phage-induced variation, in which the ratio was unity. Luria and Delbrück carried out careful experiments and observed the large fluctuations across multiple cultures characteristic of the model for spontaneous variation. In this manner, they firmly established a Darwinian mechanism of evolution over a Lamarckian one (figure 9.3).

9.2 Stochastic Simulations

The models considered so far in this text have been almost exclusively constructed using ordinary differential equations (ODEs). In their simplest form, they represent the behavior of the contents of a single well-mixed vessel containing a number N of different molecular species. Each type of molecule—ligand, receptor, complex, metabolite, and so forth—is a separate species, and there may be many copies of each species in the solution. These molecular species react according to a given set of chemical reactions. It is useful to number the reactions, with the forward and reverse directions of a reversible reaction counted separately. Let M be the total number of reaction *channels*.

We have expressed such models as a series of ODEs in which changes in the concentration of species i, $[X]_i$, with respect to time are given by functions of the instantaneous concentrations through the relevant reaction channels:

$$\frac{d[X]_1}{dt} = f_1([X]_1, [X]_2, \ldots, [X]_N) \qquad (9.43)$$

$$\frac{d[X]_2}{dt} = f_2([X]_1, [X]_2, \ldots, [X]_N) \qquad (9.44)$$

$$\vdots$$

$$\frac{d[X]_N}{dt} = f_N([X]_1, [X]_2, \ldots, [X]_N) \qquad (9.45)$$

Given a set of initial conditions, these equations can be integrated to produce a trajectory of how the concentrations vary with time. The solutions have some properties that we may be comfortable with but that should still be questioned to understand their range of validity. The concentration of each species is a continuous function of time, and the same result is obtained every time the simulation is run from the same initial conditions.

However, concentrations are reflective of the number of molecules per unit volume. Because the number of molecules must be an integer (there are no fractional molecules), the concentration should actually move in discrete steps in time that represent the gain or loss of an integral number of molecules. For large numbers of molecules, this is of little concern (for example, 1 pM protein in a microliter of volume still corresponds to more than 6×10^5 molecules, and the addition of one molecule changes the concentration by less than 10^{-5} pM). However, for small numbers of molecules, the discrete picture can be very different from the continuous. For example, some transcription factors appear to be present at less than 10 copies per cell.

In a somewhat related fashion, the pure repeatability of ODE simulations is comforting but not reflective of all realities. Stochastic processes are those involving discrete events; stochastic processes with small counts often lead to large fluctuations between repeats. A common example is the flipping of a fair coin. A large number of flips results in a head:tail ratio very nearly equal to unity. In contrast, a small number of flips leads to a high chance that a significantly different ratio will be observed; two coin flips will result in either all heads or all tails 50% of the time. Likewise, chemical reactions involving large numbers of molecules tend to produce the same result in repeated trials. Similar reactions with much smaller numbers of molecules tend to produce a much wider range of results. These differences among results, which we term *stochastic fluctuations*, are often described as *noisy*, which should not be taken as an indication that the variation is due to inaccuracies in measurement. Instead they are a natural, intrinsic result of the processes that involve small numbers of counts.

9.2.1 Stochastic Gillespie Simulations

A framework for thinking about the dynamics of stochastic reactions and simulating their behavior has been developed by Daniel T. Gillespie [2,3]. Imagine that we track not the concentrations of species $[X]_i$, but rather the number of molecules of each species Y_i, given some set of concentrations at time t. Let these molecule counts be $\bar{Y} = \{Y_1, Y_2, \cdots, Y_N\}$. With some thought, we realize that in the near future, the system will evolve by first staying the same for some period of time τ and then changing by a single chemical reaction occurring along channel R_μ. Reactions fire at discrete times and infinitely quickly; if we divide time into thin enough slices, in most time slices, no reaction occurs, and in no slice does more than one reaction occur. We need to somehow gain an understanding of what affects the time between reactions and which reaction occurs next. However, because we do not track the positions and momenta of all the constituent atoms of all molecules in the system (including solvent), it is not possible to compute with high certainty the waiting time between reactions or the next reaction to occur. Instead, we only track the number of molecules of each type, and so the knowledge that we can possess about the future evolution of the reacting system can be only probabilistic in nature.

The general formalism for treating systems of this type is the so-called chemical master equation, which will be briefly outlined here, following the presentation of Gillespie [2]. Our probabilistic knowledge of the system can be represented by the probability density $P(\vec{Y};t)$, which is the probability that the number of molecules of each type at time t is given by the vector $\vec{Y}$. The chemical master equation describes how $P(\vec{Y};t)$ evolves in time:

$$P(\vec{Y};t+dt) = P(\vec{Y};t)\left[1 - \sum_{\mu=1}^{M} a_\mu dt\right] + \sum_{\mu=1}^{M} B_\mu dt \qquad (9.46)$$

where the first term on the right-hand side gives the probability that the system was in the final state $\vec{Y}$ at time t and had no reactions occur in the time interval t to $t+dt$, and the second term represents the probability that at time t the system was in some state one reaction away from the final state and that exactly one reaction occurred in the interval t to $t+dt$. The term B_μ includes the probability of the system being in the state one reaction away from final $\times$ the probability for undergoing the required reaction. It is useful to express that reaction probability a_μ as the product of an intrinsic reaction propensity c_μ and the number of combinations of molecules capable of undergoing the reaction h_μ ($a_\mu = h_\mu c_\mu$).

Rearrangement of equation 9.46 leads to the chemical master equation itself:

$$\frac{\partial P(\vec{Y};t)}{\partial t} = \sum_{\mu=1}^{M} [B_\mu - a_\mu P(\vec{Y};t)] \qquad (9.47)$$

Example 9-4 Analytically solve the chemical master equation for a single first-order degradation reaction $C \rightarrow \phi$ starting with $n = n_o$ molecules.

Solution Consider the starting situation of n_o molecules. How could one arrive at that state? One way would be to come from this state and have no reactions occur (essentially "staying in the state"). Are there other ways?

There will never be more than n_o molecules (i.e., $P_n = 0$ for $n > n_o$), so we can't arrive at this state from the state with $n_o + 1$ molecules. Likewise, no reaction will increase the number of molecules C from $n_o - 1$ to n_o, so we can't arrive from there. Thus, all $B_\mu = 0$ for reactions μ that end with n_o copies of molecule C, and equation 9.47 reduces to

$$\frac{dP_{n_o}}{dt} = -a_\mu P_{n_o} = -h_\mu c_\mu P_{n_o} = -n_o c_\mu P_{n_o} \tag{9.48}$$

with c_μ corresponding to the intrinsic reaction propensity for the degradation of C. Thus,

$$P_{n_o}(t) = e^{-n_o c_\mu t} \tag{9.49}$$

Repeating this for P_{n_o-1} results in

$$\frac{dP_{n_o-1}}{dt} = n_o c_\mu P_{n_o} - (n_o - 1) c_\mu P_{n_o-1} \tag{9.50}$$

Thus,

$$P_{n_o-1}(t) = e^{-(n_o-1)c_\mu t} n_o \left(1 - e^{-c_\mu t}\right) \tag{9.51}$$

Continuing this approach, it can be shown by induction that the general solution is

$$P_n(t) = \binom{n_o}{n} e^{-nc_\mu t} \left(1 - e^{-c_\mu t}\right)^{n_o - n} \tag{9.52}$$

This solution is simply the probability density function for a binomial random variable (equation 9.1) with $p = e^{-c_\mu t}$. (Note that this probability p is mathematically equivalent to the fraction of molecules calculated from the deterministic solution for this first-order degradation reaction: $[C]/[C]_o = e^{-k_{\text{deg}} t}$, where k_{deg} is the intrinsic reaction propensity.)

Although the chemical master equation is not difficult to write for many systems and can be analytically solved for trivial problems, as in the previous example, it has proven exceedingly difficult to solve analytically or simulate numerically for most systems of interest.

The Gillespie approach develops a treatment for these simulations as follows. Let $P(\tau, \mu)d\tau$ be the probability that a system in state $\vec{Y}$ at time t will evolve such that the next reaction will occur in the time interval $(t + \tau, t + \tau + d\tau)$ and be along reaction channel R_μ. This is termed the *reaction probability density function* and is assumed to have the form

$$P(\tau,\mu)d\tau = P_0(\tau) \cdot a_\mu d\tau \qquad (9.53)$$

where $P_0(\tau)$ is the probability that the state does not react for a time interval of τ, and $a_\mu d\tau$ is the probability that reaction R_μ occurs in the interval $(t+\tau, t+\tau+d\tau)$. The assumption inherent in equation 9.53 is that the waiting-time probability density function does not depend explicitly on which reaction occurs next. $P_0(\tau)$ can be obtained by noting from equation 9.46 that $[1 - \sum_{\mu=1}^{M} a_\mu dt]$ is the probability that no reaction occurs in time dt, so

$$P_0(\tau + d\tau) = P_0(\tau) \left[1 - \sum_{\mu=1}^{M} a_\mu d\tau \right] \qquad (9.54)$$

whose solution is

$$P_0(\tau) = e^{-\sum_{\mu=1}^{M} a_\mu \tau} = e^{-a_0 \tau} \qquad (9.55)$$

where $a_0 \equiv \sum_{\mu=1}^{M} a_\mu$. Thus, the waiting time between reactions follows an exponential decay, the same as in Poisson processes discussed earlier in this chapter. Substituting equation 9.55 into equation 9.53 provides the key equation for simulating stochastic dynamics:

$$P(\tau,\mu) = a_\mu \cdot e^{-a_0 \tau} \qquad (9.56)$$

Is this result sensible? It says that the probability that the next reaction will be reaction μ at a time τ in the future is the product of one term that depends only on the reaction μ (through a_μ), and another that depends on the time τ and all reactions (through a_0). Moreover, it says that if we focus on a given time τ_1, the probability for the next reaction is just proportional to a_μ. Likewise, if we focus on a given reaction whose probability is a_1, then the waiting time is an exponential that decays faster for increasing a_1, everything else remaining the same. This sounds consistent with the same framework that we used to describe the Poisson process at the start of this chapter. We can make very thin time slices, t/n, into the future and imagine that up to one reaction can occur in each time slice. For the case of only one reaction channel, equation 9.56 simplifies to $P(\tau) = a_1 e^{-a_1 \tau}$, which matches exactly the Poisson waiting-time expression in equation 9.21.

To run dynamic simulations, we need a mechanism for generating random numbers for τ and μ from the probability density function given by equation 9.56. This can be achieved by selecting two random numbers r_1 and r_2 uniformly distributed

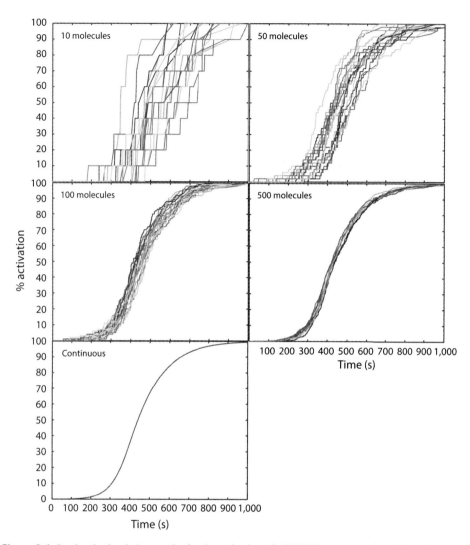

Figure 9.4 Stochastic simulation results for the activation of a MAP kinase network model [4] for initial conditions consisting of 10, 50, 100, and 500 copies of each enzyme, compared to the continuous results from an ODE model of the same network. The stochastic results from multiple independent simulations are plotted on the same graph. The results differ more from one another for small numbers of molecules as initial conditions than for larger numbers. Stochastic simulations with large numbers of molecules behave similarly to the continuous model.

between zero and one and using them to create values for τ from

$$\tau = \left(\frac{-1}{a_0}\right)\ln(r_1) \tag{9.57}$$

and μ as the integer for which

$$\sum_{v=1}^{\mu-1} a_v < r_2 a_0 \leq \sum_{v=1}^{\mu} a_v \tag{9.58}$$

This provides a fundamental methodology for running stochastic dynamic simulations of biological systems.

For simulations of large numbers of molecules, we expect the stochastic result from independent simulations to be similar to one another and to be similar to the continuous result obtained from simulations of ODEs. For simulations where only a small number of reacting molecules are present, however, we expect the results of individual stochastic simulations to differ somewhat from one another, although frequently the average is similar to the continuous result obtained with the corresponding ODEs. This behavior is illustrated in figure 9.4.

Case Study 9-2   T. S. Gardner, C. R. Cantor, and J. J. Collins. Construction of a genetic toggle switch in *Escherichia coli*. *Nature* 403: 339–342 (2000) [5].

In a foundational paper, Gardner, Cantor, and Collins designed, built, and tested a small genetic network that could exist in one of two stable states. Left alone, it tended to remain in its current state indefinitely; alternatively, it could be manipulated through the application of inducers to flip to the other state and remain there. This engineered *genetic switch* could be *flipped* from one state to another at will, apparently indefinitely. Their design is shown in figure 9.5. The protein encoded by the gene for Repressor 1 binds at Promoter 1 and represses transcription of the constitutively expressed gene for Repressor 2, corresponding to a state with high concentration of the protein for Repressor 1 (as well as an adjacently encoded Reporter protein, GFP) and low concentration of the protein for Repressor 2 (which the study's authors call State 2 or the *high* state due to the Reporter expression). Likewise, Repressor 2 binds at Promoter 2 to repress transcription of the constitutively expressed gene for Repressor 1, corresponding to a state with high concentration of the protein for Repressor 2 and low concentration of the protein for Repressor 1 and Reporter (called State 1 or the *low* state). It turns out that both these states are stable, as shown by the authors in an analysis of the stationary states of the ODEs that model the system, outlined below (see section 6.1 on nonlinear dynamics). Switching between these two stable states was accomplished by transient application of inducers (molecules or conditions) that inactivated Repressor 1 or Repressor 2.

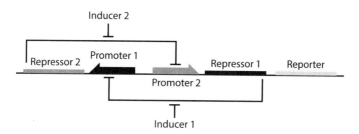

Figure 9.5 A schematic diagram of the genetic toggle switch designed and constructed by Gardner, Cantor, and Collins [5]. The protein encoded by the gene for Repressor 1 binds at Promoter 1 to repress transcription of the gene for Repressor 2; Repressor 2 binds at Promoter 2 to repress Repressor 1. A fluorescent reporter protein is expressed together with Repressor 1 for readout purposes. Small molecules or other changes of conditions act as inducers that interfere with the repression activity of the proteins of Repressor 1 or Repressor 2.

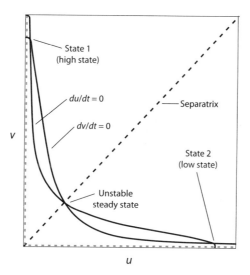

Figure 9.6 Analysis of the steady-state behavior of a genetic toggle switch for a case of symmetric parameterization. The two solid lines represent the two nullclines, one for $du/dt = 0$ and the other for $dv/dt = 0$, which intersect at three stationary points. Two stable stationary states (labeled State 1 and State 2) surround an unstable one, which defines a separatrix. Data from reference 5.

An *E. coli* plasmid encoded the genetic components. Lactose (Lac) repressor (*lacI*), the Ptrc-2 promoter, and isopropyl-β-D-1-thiogalactopyranoside (IPTG) served as Repressor 2, Promoter 2, and Inducer 2, respectively. In some plasmids, a temperature-sensitive variant of the λ repressor (*cIts*), the P_Ls1con promoter, and temporary inactivation by incubation at a nonpermissive temperature were used as Repressor 1, Promoter 1, and Inducer 1, respectively. In others, the tetracycline (Tet) repressor (*tetR*), the $P_L$tetO-1, and anhydrotetracycline (aTc), respectively, were used.

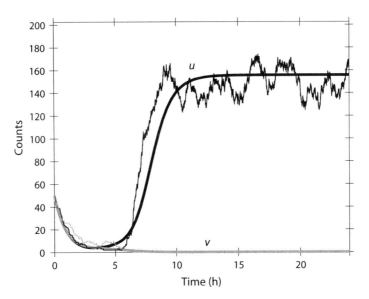

Figure 9.7 Trajectories from simulations of dimensionalized versions of equations 9.59 and 9.60. The amount of u is plotted as a function of time in black, and v in gray. The two smoother curves represent deterministic simulations, and the other two curves are for stochastic simulations. Both simulations start with similar values of u and v, so both repressors are at similar numbers of molecules (concentrations) and mutually drive down each other's expression and concentration. Near the trough for u, in each simulation u eventually takes over and dominates expression of v going forward. Interestingly, before this point in the stochastic simulation, v dominates u for much of the time. Simulations run by K. Shi.

A simple nondimensionalized model describing the system used u to represent the concentration of Repressor 1 protein, v for the concentration of Repressor 2 protein, and the following equations to describe the dynamics of the system [5]:

$$\frac{du}{dt} = \frac{\alpha_1}{1+v^\beta} - u \quad (9.59)$$

$$\frac{dv}{dt} = \frac{\alpha_2}{1+u^\gamma} - v \quad (9.60)$$

Here α_1 and α_2 are the constitutive expression rates of synthesis of u and v, respectively, and β and γ represent the cooperativity of repression of expression of their respective gene products. Often this cooperativity is associated with multimerization of the DNA-binding molecules constituting the repressors (chapter 6). In the absence of the other repressor, each repressor is expressed at a constant rate independent of its concentration and is degraded with first-order kinetics. The presence of the other repressor diminishes expression in a nonlinear fashion for $\beta, \gamma > 1$. A nonlinear dynamics analysis of the steady states of this system, as described in chapter 6, is shown in figure 9.6. The two nullclines representing steady-state levels of u and v intersect in three stationary points. Two are stable, representing State 1 (the low state) and State 2 (the high state), whereas the third

Discrete Stochastic Processes

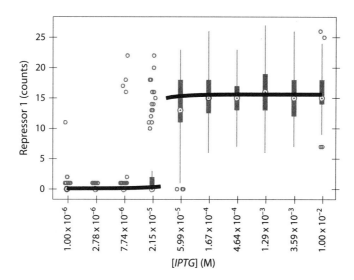

Figure 9.8 Results of simulations of equations 9.61 and 9.62 using parameters from the caption of figure 5 of Gardner et al. [5] with initial conditions of zero counts for both repressor species. For each concentration of inducer IPTG, 100 stochastic simulations were run to steady state, and a gray box-and-whiskers plot shows the median (circle with centered dot), first and third quartiles (ends of box), min and max (ends of whiskers), and outliers (gray circles). The corresponding deterministic simulation steady states (using ODEs) are shown as solid black lines with a clear break at intermediate inducer concentration. The cooperative switch from the low Repressor 1 state to the high Repressor 1 state in response to an increase in IPTG concentration, in both the stochastic and deterministic simulations, represents a bifurcation. Simulations run by K. Shi.

is unstable and is on a separatrix that separates the two stable basins of attraction. Under sufficiently strong inducer conditions, the unstable stationary state and one of the stable ones coalesce and disappear, leaving only the remaining stable stationary state. This can be used to set the system in one or the other basin of attraction; removal of the inducer conditions returns the system to a three-stationary-state (stable-unstable-stable) phase diagram shown in figure 9.6. This changeover between three stationary states and one stationary state represents a bifurcation of the phase diagram (see chapter 6).

In their paper, Gardner et al. [5] go on to demonstrate experimentally that the system behaves consistently with the model. Using the inducers, the system can be deterministically toggled between the high and low states. In the absence of inducers, the system tends to remain in the state in which it was placed. It is a tremendously interesting set of results, and we recommend it to the curious reader. The current chapter is about discrete stochastic processes, and for the rest of this case study, we describe the results of stochastic simulations of the dimensionalized versions of equations 9.59 and 9.60, parameterized (and augmented) as described in the caption to figure 5 of Gardner et al. [5], using the Gillespie algorithm.

Equations 9.59 and 9.60 represent a symmetric system with respect to u and v, which is reflected in the symmetry of the stationary points and nullclines in figure 9.6. However, the symmetry axis $u = v$, with equal concentrations of the two repressors, is a separatrix between two basins of attraction. Any particular simulation at long time should settle into one or the other basin of attraction, most likely remaining in the basin it started in when molecular counts are sufficiently high. Figure 9.7 shows this behavior for two simulations, one stochastic and one deterministic, that end in a state with high *u* and low *v*.

To simulate the effect of the inducer IPTG, the model from equations 9.59 and 9.60 was augmented to give

$$\frac{du}{dt} = \frac{\alpha_1}{1+v^\beta} - u \qquad (9.61)$$

$$\frac{dv}{dt} = \frac{\alpha_2}{1+\left(\frac{u}{(1+[IPTG]/K)^\eta}\right)^\gamma} - v \qquad (9.62)$$

where [*IPTG*] is the concentration of the inducer IPTG, *K* is its dissociation constant from the lactose repressor (*u*), and η is the cooperativity of IPTG binding [5]. The results of stochastic simulations carried out with increasing concentrations of IPTG are shown in figure 9.8. At low IPTG concentration, nearly all independent stochastic simulations end with very low counts of λ repressor (*v*). As the IPTG concentration grows, the set of simulations splits, with some simulations ending in the low *v* state and an increasing fraction ending in the high *v* state. Finally, at higher IPTG concentrations, the large majority of simulations end in the high *v* state. The abruptness of the transition from low- to high-*v* near an IPTG concentration of 3×10^{-5} M demonstrates the cooperativity of the transition reflected by a bifurcation.

9.2.2 Approximate Stochastic Methods: A Bridge between Exact Stochastic Methods and Deterministic Formalisms

The original Gillespie method, also known as the stochastic simulation algorithm (SSA), is exact in its implementation of the chemical master equation, but it can run very slowly when the time steps between reactions are short. Some improvements have been made to the SSA, but any method that still explicitly simulates every individual reaction event will be impractical for many applications of interest. Therefore, approximate stochastic methods have been developed that tolerate some loss in exactness of the SSA in order to reduce the required computing time. One such approximate method that retains the discrete and stochastic nature of the SSA is known as the *tau-leaping* method.

Tau-leaping entails preselecting a value of τ such that no propensity function a_μ is likely to change significantly during the time interval $[t, t+\tau]$. When this condition is satisfied, each a_μ is approximately constant, so the number of times that a reaction channel R_μ fires in the time period τ is a Poisson random variable with mean and variance $a_\mu \tau$. The tau-leaping approximation therefore allows us to leap the system forward as follows:

$$Y_i(t+\tau) = Y_i(t) + \sum_{j=1}^{M} P_j(a_j\tau)v_j \qquad (9.63)$$

where P_j is an independent Poisson random variable (with mean and variance both equal to $a_j\tau$), and v_j is the state-change vector describing the change in number of molecules due to reaction R_j. This simplifying approximation in the tau-leaping method requires less frequent updating of the rates, thus accelerating the simulation in many cases.

When the tau-leaping criterion is satisfied and it is also true that all $a_\mu\tau$ values are much greater than 1, not only do the propensity functions remain relatively constant, but also every reaction channel fires many more times than once during the time period τ. Mathematically, this additional constraint ($a_\mu\tau \gg 1$) allows Poisson random variables to be approximated by normal random variables with the same mean and variance. This is notable, because we replace a discrete probability distribution (Poisson) with a continuous probability density (normal). The resulting continuous and stochastic formalism is known as the chemical Langevin equation (CLE):

$$Y_i(t+\tau) = Y_i(t) + \sum_{j=1}^{M} N_j(a_j\tau, a_j\tau)v_j \qquad (9.64)$$

where $N_j(m, \sigma^2)$ is an independent normal random variable with mean m and variance σ^2. Furthermore, using the property of normal random variables that $N(m, \sigma^2) = m + \sigma N(0, 1)$, the CLE can be rewritten as

$$Y_i(t+\tau) = Y_i(t) + \sum_{j=1}^{M} v_j a_j \tau + \sum_{j=1}^{M} v_j \sqrt{a_j\tau} N_j(0, 1) \qquad (9.65)$$

or equivalently,

$$\frac{dY_i}{dt} = \sum_{j=1}^{M} v_j a_j + \sum_{j=1}^{M} v_j \sqrt{a_j}\, \Gamma_j(t) \qquad (9.66)$$

where Γ_j is an independent Gaussian white-noise process. Thus, the right-hand side of this equation is the sum of a deterministic term based on the underlying reaction rates (recall that a_j is a function of $Y_i(t)$) and a stochastic *noise* term. This is called a stochastic differential equation.

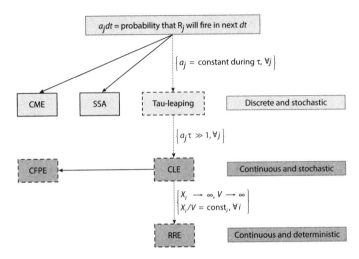

Figure 9.9 The relationships between the chemical master equation, the SSA, the tau-leaping method, the CLE, and the RRE. CME = chemical master equation; SSA = stochastic simulation algorithm; CLE = chemical Langevin equation; CFPE = chemical Fokker-Planck equation; RRE = reaction rate equation. Schematic adapted from reference 6.

Finally, in the limit that the number of molecules of each species (Y_i) and the reaction volume (V) tend to infinity while maintaining constant concentration of each species ($Y_i/V = $ constant), the noise term in the CLE becomes vanishingly small. Therefore, the result reduces to the continuous and deterministic reaction rate equations (RRE)

$$\frac{dY_i}{dt} = \sum_{j=1}^{M} v_j a_j \qquad (9.67)$$

which are the classical ODEs from chemical kinetics that we have used prior to this chapter (figure 9.9).

9.3 Stochastic Gene Expression

Cell-to-cell variability in gene expression in a monoculture (i.e., a single species of microbe or a monoclonal cell line) can arise from two general sources: intrinsic or extrinsic noise [7]. The latter arises from cell-to-cell differences in synthetic capacity, for reasons such as size inhomogeneity. Cell size distributions are generally log normal, with a severalfold range in single-cell mass well represented in the population. Naturally, larger cells will possess more ribosomes and will therefore have a greater capacity for protein synthesis per molecule of mRNA. However, if one analyzes a cohort of cells of equivalent size, there remain fluctuations in the copy number of a reporter protein that are intrinsic to the noise resulting from a small discrete number of transcription and translation events. Parsing these two sources of variation, intrinsic and extrinsic, sheds light on the stochastic variability built into the central dogma of molecular biology at the single-cell level [8].

Transcription is found to be *bursty* in both prokaryotes and eukaryotes: A given promoter fluctuates between an *on* and *off* state, during which initiation and progression of transcription are Poisson processes while transcription is *on*. The number of protein molecules made during a burst is found experimentally to approximate a geometric or exponential distribution, and convoluting this with the statistics of burst occurrence leads to a predicted gamma distribution for the number of protein molecules n:

$$p(n) = \frac{n^{a-1} e^{-n/b}}{b^a \Gamma(a)} \tag{9.68}$$

where a is the burst frequency, and b is the average number of molecules produced per burst [8]. Experimental measurements of single-cell protein copy number distribution are consistent with this broad gamma distribution.

9.4 Stochasticity in Phage Infection

Human-made systems tend to produce the same outcome when subjected to the same conditions. If you purchase a dozen of the same model calculator and type the same arithmetic expression into each, every one of them will produce the same answer. Of course, this should be true even if the calculators are different models from different manufacturers. Biological systems, even when they look the same, often behave differently. Imagine a population of growing cells that is plated out and exposed to a uniform dose of ionizing radiation. Given an appropriate dose, some of the cells will survive, and others will not. Is this difference in behavior due to random chance, or is there some inherent difference among the cells when first plated out? We can imagine the stochasticity of DNA damage events may lead some cells to receive more damage than others, and stochasticity in the sensing, repair, signaling, and cell-cycle control processes can also lead to nondeterministic outcomes. However, it is also possible that preexisting differences among the cells, such as cell-cycle stage and current number of DNA repair proteins, might account for the cells that survive versus those that do not. Indeed, variability of genotype and phenotype is an important hallmark of biological systems and is the fuel for evolution. Variability of behavior of cells with the same genotype is an extension that permits even a genetically identical population to survive under conditions in which individuals may perish. However, stochasticity can also be used to model things that we don't yet understand or haven't yet measured. As a simple analogy, one may use probability and statistics to guide betting in a game of blackjack, because the order of the cards in the deck is unknown. By counting cards as they are dealt from the top of the deck, one may improve one's bets and increase the expected return. These bets are placed assuming cards come off the deck randomly, which they don't—the order is predetermined once the deck has been shuffled and cut—but because the order is not known, the best one can do is rely on a stochastic model. Thus, stochasticity introduces probabilistic behavior that could represent randomness or merely the lack of detailed knowledge.

Case Study 9-3 A. Arkin, J. Ross, and H. H. McAdams. Stochastic kinetic analysis of developmental pathway bifurcation in phage λ-infected *Escherichia coli* cells. *Genetics* 149: 1633–1648 (1998). F. St-Pierre and D. Endy. Determination of cell fate selection during phage lambda infection. *Proc. Natl. Acad. Sci. U.S.A.* 105: 20,705–20,710 (2008) [9, 10].

Arkin, Ross, and McAdams constructed a stochastic model of the initial stages of phage λ infection of *E. coli*, parameterized it by comparing it to experimental results, and analyzed its behavior by conducting simulations. The full model is quite detailed, and we will focus here only on its key features. Careful and elegant experimentation has led to a mechanistic understanding of the underlying molecular events. Interestingly, a population of infected bacteria splits, with some members of the population producing large numbers of progeny phage before ultimately undergoing cell lysis (this pathway is *lytic* growth) and others integrating the phage genome but otherwise seemingly unaffected (the *lysogenic* pathway). The key switch controlling this cell fate decision rests with a pair of mutually repressing transcription factors, CI and Cro. Soon after infection, a "race" occurs whereby if the concentration of CI is high compared to that of Cro, CI represses Cro expression and directs progression along the lysogenic pathway. If the race turns out the other way, Cro represses CI expression, and the cell undergoes phage reproduction and lysis. Arkin, Ross, and McAdams modeled the race as controlled by stochastic processes of transcription and translation, with fluctuations in expression in different cells responsible for different outcomes. Encouragingly, this model produced results that compared favorably with experiments for how the probability of lytic versus lysogenic growth varied with respect to the average number of phage infecting each cell (the so-called multiplicity of infection), with greater multiplicity of infection leading to lower rates of lysis. Taken together, the results demonstrate that a stochastic model is capable of accounting for the available data, but do they prove that random fluctuations in gene expression are responsible for the lysis-lysogeny decision? One theme of this text has been that consistency with experiments does not necessarily imply correctness for a model, and so agreement between experiment and modeling here is not proof that they operate by the same mechanism. It could be that other factors are strongly determining of which cells will have a higher propensity to undergo lytic or lysogenic growth.

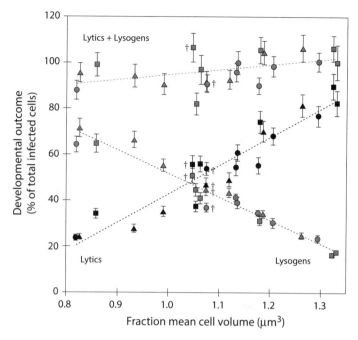

Figure 9.10 Developmental outcomes as a function of cell size in the λ phage infection of *E. coli*. Data from reference 10.

Indeed, 10 years after the Arkin, Ross, and McAdams paper was published, St-Pierre and Endy published a study [10] in which they examined cells for preexisting properties that might determine their propensity to follow the lytic versus lysogenic pathway after phage infection. Interestingly, when they took account of *E. coli* cell size at the time of infection, they found that smaller cells were significantly more likely to undergo lysogenic growth, whereas larger cells had a much greater propensity to undergo lytic growth (figure 9.10). Cell size alone was not a perfect predictor of cell fate, which indicated that some combination of stochasticity and other preexisting differences is responsible for the remaining variability.

9.5 Diffusion and Random Walks

In chapter 8, we invoked Fick's first law to describe diffusion of a molecule to its binding site. Here we derive that relationship for the one-dimensional case from a simple model of diffusion based on a random walk made up of discrete steps. Similar to the relationship between discrete stochastic simulation algorithms and continuous reaction rate equations, we will see that taking the continuous limit of the discrete random walk model leads to Fick's first law.

The one-dimensional form of this relationship is

$$J_x = -D\frac{\partial [C]}{\partial x} \tag{9.69}$$

where J is the flux of material expressed as a quantity per unit area per unit time (such as mol m^{-2} s^{-1}), the diffusivity D is a coefficient that characterizes the process (units of m^2 s^{-1}), and $[C]$ is the concentration of solute.

The movement of such molecules can also be described by Brownian motion, which is named after the botanist Robert Brown who, in 1827, observed that particles of pollen suspended in water appeared to move randomly (but Brown himself did not develop mathematical descriptions of such movement) [11]. We shall show that Fick's first law can be derived from a highly simplified description of Brownian motion, a one-dimensional *random walk* [12]. The assumptions for a one-dimensional random walk are:

1. *Discrete time and length steps.* During every time increment τ, each particle moves a distance δ to the right or to the left.

2. *Unbiased walk.* The probability of moving to the right is equal to the probability of moving to the left.

3. *Independent particles.* The particle concentration is considered to be so dilute that a particle's motion is unaffected by its neighbors.

Clearly, this is a highly oversimplified model of the process, and yet it captures the gross features of molecular diffusion remarkably well. Let's consider a suspension of particles undergoing a one-dimensional random walk: At distance x there are $N(x)$ particles, and at distance δ to the right, there are $N(x+\delta)$ particles. At a given time step τ, on average half the particles at x will step to the right, and half the particles at $x+\delta$ will step to the left. Consequently, the net number of particles crossing from left to right between these two points will be

$$-\frac{1}{2}(N(x+\delta) - N(x)) \tag{9.70}$$

This number can be converted to flux J_x by dividing by the cross-sectional area A and the time step τ:

$$J_x = -\frac{1}{2A\tau}(N(x+\delta) - N(x)) \tag{9.71}$$

Multiplying both the numerator and the denominator by δ^2 and rearranging terms yields

Discrete Stochastic Processes

$$J_x = -\frac{\delta^2}{2\tau}\frac{1}{\delta}\left(\frac{N(x+\delta)}{A\delta} - \frac{N(x)}{A\delta}\right) \tag{9.72}$$

Making the substitutions $D \equiv \frac{\delta^2}{2\tau}$ to define the diffusivity, and $\frac{N(x)}{A\delta} \equiv [C](x)$ as the concentration of particles, gives

$$J_x = -D\frac{1}{\delta}\left([C](x+\delta) - [C](x)\right) \tag{9.73}$$

In the limit as $\delta \to 0$, this becomes Fick's first law (matching equation 9.69):

$$J_x = -D\frac{\partial [C]}{\partial x}$$

So it can be seen that Fick's first law is a simple, direct consequence of an unbiased random walk: Probabilistically, particles will move from regions of high concentration to regions of lower concentration.

9.6 Combinatorial Bioinformatics

Genetic sequences and their resultant mRNA and polypeptide products comprise discrete monomers in a precise order. As there are four possible nitrogenous bases in DNA, a gene of n nucleotides has 4^n possible sequences. Such a gene would have $n/3$ codons, each encoding for one of the 20 natural amino acids, resulting in $20^{n/3}$ possible polypeptides. Evaluation of the likelihood of particular scenarios—from random evolution to focused manipulation of genetic sequences to de novo emergence of new function—requires simultaneous consideration of the multitude of these discrete probabilities.

Evolution functions by the accumulation of beneficial mutations via selective retention. Mutations are sampled nearly randomly, with rare beneficial mutations. For example, only 5–9% of single amino acid mutations improve stability [13], which motivates sampling numerous mutations to achieve a reasonable likelihood of identifying a benefit. Consider N mutations, each with an equal probability (p_b) of benefiting function. The likelihood of at least one of the mutations providing benefit is the complement of the probability of none of the mutations providing benefit:

$$p_{\geq 1 \text{ benefit}} = 1 - p_{0 \text{ benefits in } N \text{ mutations}} \tag{9.74}$$

Because each mutation is independent, the probability of N mutations all lacking benefit is the product of all N nonbeneficial probabilities:

$$p_{0 \text{ benefits in } N \text{ mutations}} = p_{\text{nonbenefit}}^{N} = (1-p_b)^N \quad (9.75)$$

where the probability of lacking benefit is the complement of the probability of providing benefit. Thus we have

$$p_{\geq 1 \text{ benefit}} = 1 - (1-p_b)^N \quad (9.76)$$

The final solution uses a pair of complementary probabilities to avoid potential complications with double counting a set of mutations with multiple benefits.

Here we consider another explanation for arriving at the same result, which will prove useful in the remainder of this section. Consideration of each mutation as a Bernoulli trial yields an equivalently valid solution. Applying equation 9.1 to compute the probability of $k \geq 1$ *successes* (beneficial mutations) in N *trials* (mutations) yields

$$p(k \geq 1) = \sum_{k=1}^{N} \binom{N}{k} p_b^k (1-p_b)^{N-k} \quad (9.77)$$

which can be stated from a complementary perspective as

$$p(k \geq 1) = 1 - p(k=0) = 1 - \binom{N}{0} p_b^0 (1-p_b)^{N-0} = 1 - (1-p_b)^N \quad (9.78)$$

Simultaneous incorporation of multiple mutations—naturally in a single organism or synthetically in directed evolution of a single protein—increases search breadth but is countered by the fact that protein mutations are more frequently detrimental than beneficial. Approximately 50–65% of single amino acid mutations hinder stability [13]. Moreover, many mutations are severely detrimental, to the extent of eliminating any protein function via drastic destabilization or elimination of a key functional moiety. We can add such function-eliminating mutations as a layer of complexity to the model. To yield an improved organism or protein, a collection of mutations must now contain at least one beneficial mutation and lack an extensively detrimental mutation. The probability that a random collection of N mutations will include i beneficial ones and the remainder being neutral is $p_b^i (1-p_b-p_e)^{N-i}$, where p_e is the probability of a function-eliminating mutation. Thus, the probability of the collection of N mutations yielding net benefit considers all possible combinations of at least one benefit:

$$p_{\text{improved}} = \sum_{i=1}^{N} \binom{N}{i} p_b^i (1-p_b-p_e)^{N-i} \quad (9.79)$$

which is a close analog of summing over the successes of Bernoulli trials (equation 9.1), with the exception of function-eliminating mutations. Because of this difference, we can not just take the complement of the $i = 0$ term, as we had done previously.

We can further expand the model to include mildly detrimental mutations. To provide an improved phenotype, a collection of mutations must have more beneficial mutations than hindering mutations and lack function-eliminating mutations. We again consider all possible combinations of $i = 1$ to N beneficial mutations, while now also considering up to $h = i - 1$ hindering mutations (but at most $N - i$ hindrances, because there are only N mutations). These h hindrances may occur in $\binom{N-i}{h}$ combinations:

$$p_{\text{improved}} = \sum_{i=1}^{N} \binom{N}{i} \sum_{h=0}^{\min(i-1,N-i)} \binom{N-i}{h} p_b^i p_h^h p_n^{N-i-h} \quad (9.80)$$

where p_n is the probability of a neutral mutation and is equal to $1 - p_b - p_e - p_h$.

The balance between a more extensive sampling of mutational space and an increased likelihood of ruinous mutation is shown in figure 9.11. This model is consistent with the stated characteristics of useful models, in that it is a simplification that omits certain complexities for the sake of tractability and insight.

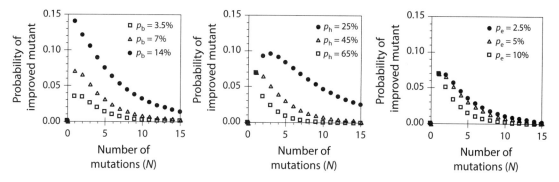

Figure 9.11 The probability of a mutant organism or protein exhibiting improved function is dependent on the number of mutations (N) and the likelihoods of each mutation being individually beneficial (p_b), hindering (p_h), or function eliminating (p_e). Plots are generated with $p_b = 7\%$, $p_h = 45\%$, and $p_e = 5\%$ unless otherwise indicated.

Example 9-5 You have created a combinatorial library of ligands in which 10 codons within a peptide loop were diversified using NNN degenerate codons (equal probability of each nucleotide at all three codon positions). After testing 10,000 proteins from the library, you found a high-affinity ligand with a serine-histidine-proline-glutamine (SHPQ) tetrapeptide motif. On further testing, you determined that presence of this tetrapeptide motif is sufficient to provide binding affinity regardless of its location in the ligand loop; that is, any nontruncated peptide loop with SHPQ will bind to the target of interest. (Note that the HPQ motif, in a broad array of contexts, enables streptavidin binding [14].) What is the likelihood that the library contains a functional ligand with the SHPQ motif?

Solution Because all codon sequences are equally probable, the probability of any particular amino acid being encoded by an NNN codon at a given site is the number of different codon sequences that encode for that amino acid divided by the total number of possible codon sequences:

$$p(\text{amino acid } i) = \frac{\text{Codons encoding for amino acid } i}{\text{Number of possible codons}} \quad (9.81)$$

A quick look at the genetic code reveals that there are six, two, four, and two different codons that encode for S, H, P, and Q, respectively:

$$p(S) = \frac{6}{64} \quad (9.82)$$

$$p(H) = \frac{2}{64} \quad (9.83)$$

$$p(P) = \frac{4}{64} \quad (9.84)$$

$$p(Q) = \frac{2}{64} \quad (9.85)$$

Because each site's codon is assumed to be independent of the other sites, the probability of all four correct amino acids being encoded at four particular sites (in this case, sequentially) is the product of the individual probabilities:

$$p(\text{SHPQ at a particular set of sites}) = \frac{6}{64} \cdot \frac{2}{64} \cdot \frac{4}{64} \cdot \frac{2}{64} = 5.7 \times 10^{-6} \quad (9.86)$$

which is equivalent to computing the probability as the ratio of the 96 possible sequences ($6 \times 2 \times 4 \times 2$) encoding for SHPQ to the nearly 17 million possible four-codon sequences (64^4).

In addition to these four sites encoding SHPQ, the remaining six sites must not encode for a stop codon:

$$p(\text{no stop at 6 sites}) = \left(\frac{61}{64}\right)^6 = 0.75 \tag{9.87}$$

Thus, the probability that a peptide loop has SHPQ at a particular set of four sites and the remaining sites lack a stop codon is

$$p(\text{valid sequence (SHPQ at particular sites and no stops))}$$

$$= (5.7 \times 10^{-6})(0.75) \tag{9.88}$$

$$= 4.3 \times 10^{-6} \tag{9.89}$$

The proper SHPQ motif may occur at any location within the 10-mer randomized region, that is, it can occur at sites 1–4, 2–5, ..., or 7–10. Given the very low frequency of the motif, it is a good approximation to assume the occurrence of a motif to be exclusive, thereby allowing for summation of the probabilities for all seven sites:

$$p(\text{valid sequence with SHPQ at any set of sites})$$

$$\approx \sum_{j=1}^{7 \text{ sets}} 4.3 \times 10^{-6} \tag{9.90}$$

$$\approx (4.3 \times 10^{-6})(7) \tag{9.91}$$

$$\approx 3.0 \times 10^{-5} \tag{9.92}$$

Yet this overestimates the actual probability by double-counting rare sequences with multiple SHPQ motifs. To avoid double-counting, reframe "at least one" as the complement of none:

$$p(\text{at least one SHPQ at 7 sets of sites}) = 1 - p(\text{no SHPQ at any set}) \tag{9.93}$$

One must also account for the lack of stop codons elsewhere. Yet the lack of stop codons and the SHPQ probability are conditionally dependent. Note that the probability of A and B is the product of the probability of A and the probability of B given A:

$$p(A \text{ and } B) = p(A)p(B|A) = p(B)p(A|B) \qquad (9.94)$$

Although this simplifies to the product of the individual A and B probabilities for independent events, the conditional probability is an important consideration in many scenarios, including the current problem. The generality of the choice for A and B enables selection of the more mathematically convenient pair of probabilities in equation 9.94:

$$p(\text{proper sequence}) = p(\text{no stops})p(\text{at least 1 SHPQ} \mid \text{no stops}) \qquad (9.95)$$

Applying equation 9.93, the independence of non-SHPQ sets, and a version of equation 9.86 modified for the condition of no stop codons, yields

$p(\text{proper sequence})$

$$= p(\text{no stops})[1 - p(\text{no SHPQ in 7 sets} \mid \text{no stops})] \qquad (9.96)$$

$$= p(\text{no stops})[1 - p(\text{no SHPQ at a set} \mid \text{no stops})^7] \qquad (9.97)$$

$$= p(\text{no stops})[1 - \{1 - p(\text{SHPQ at a set} \mid \text{no stops})\}^7] \qquad (9.98)$$

$$= \left(\frac{61}{64}\right)^{10}\left[1 - \left\{1 - \frac{6 \times 2 \times 4 \times 2}{61^4}\right\}^7\right] \qquad (9.99)$$

$$= 3.0 \times 10^{-5} \qquad (9.100)$$

Note that, for this rare event, the result is not appreciably different from the previous approximation (equation 9.92). The difference is only 0.002%.

To calculate the probability of the library having at least one of these peptides among the 10,000 tested, note that this is simply the complement of the library having no full-length SHPQ peptides; in other words, it's the complement of the probability that all peptides lack SHPQ or are truncated:

$$p(\text{in library}) = 1 - (1 - p(\text{valid}))^{10,000} \qquad (9.101)$$

$$= 1 - (1 - 3.0 \times 10^{-5})^{10,000} \qquad (9.102)$$

$$= 0.26 \qquad (9.103)$$

There is a 26% likelihood of finding at least one full-length peptide with an SHPQ motif.

Example 9-6 One approach to comparing different sources of DNA is DNA fingerprinting. Enzymes called restriction endonucleases cut DNA at a specific sequence. The resulting pattern of DNA fragments, which is dependent on the sequence, can be compared between different sources. For example, the enzyme NheI (named for its origins from the *Neisseria mucosa heidelbergensis* bacterium) cuts at the sequence 5'-GCTAGC-3'. Note that this is a palindromic sequence, so if it exists on one strand, it will simultaneously exist on the reverse complement:

$$5' \text{-} G\,C\,T\,A\,G\,C \text{-} 3'$$
$$3' \text{-} C\,G\,A\,T\,C\,G \text{-} 5'$$

(a) What is the expected number of cuts that will result from treating plasmid DNA (15,000 base pairs) with the NheI enzyme?

(b) A sample is evaluated by this technique and is observed to have no cuts. The lab technician is concerned that the experiment was faulty, perhaps due to denatured enzyme. What is the probability that a random 15,000 base-pair sequence would lack the 5'-GCTAGC-3' sequence?

Solution The probability that a random 6-base section has the exact sequence is

$$p = \prod_{i=1}^{6} p(\text{site } i \text{ is correct base}) \tag{9.104}$$

$$= \left(\frac{1}{4}\right)^{6} \tag{9.105}$$

The expected number of cuts is effectively approximated, for this case of a rare event, by the sum of probabilities over each possible set of 6-base sequences:

$$E(\text{number of cuts}) = \left(\frac{1}{4}\right)^{6} (15{,}000) = 3.7 \tag{9.106}$$

The probability of lacking this particular sequence in 15,000 circularized base pairs is the product of 15,000 equivalent probabilities of missing the sequence:

$$p = \prod_{i=1}^{15{,}000} p(\text{no cutting at a random 6-base section}) \tag{9.107}$$

$$= p(\text{no cutting at a random 6-base section})^{15,000} \tag{9.108}$$

$$= \left(1 - \left(\frac{1}{4}\right)^6\right)^{15,000} \tag{9.109}$$

$$= 0.026 \tag{9.110}$$

The result—a 2.6% probability of lacking the recognition sequence—is also obtained from equation 9.15 by using the approximation of the expectation value.

Case Study 9-4 T. A. Hopf, J. B. Ingraham, F. J. Poelwijk, C. P. I. Schärfe, M. Springer, C. Sander, and D. S. Marks. Mutation effects predicted from sequence co-variation. *Nat. Biotechnol.* 35: 128–135 (2017) [15].

In the post-genomic era, abundant sequences are known from various species for many protein homologs. Despite the high likelihood of functional hindrance of a single amino acid mutation, the enormous size of the potential sequence space enables natural protein evolution to achieve high sequence variance across functional protein homologs. Sitewise amino acid frequencies, across evolution, can be used to predict the phenotypic impact of mutation. In slightly more detail, under the hypothesis that species survival—and, thus, the existence of that sequence in the evolutionary dataset—correlates with protein function, one can compute a predicted phenotypic impact of mutation from the frequency of that amino acid at that site across species. Although this hypothesis is limited by mismatch between evolutionary pressures and the particular phenotypic trait of the particular protein of interest (as well as by biases and incompleteness of sequencing data), the approach has proven useful. In this paper, Marks and colleagues demonstrate that this approach can be markedly improved by consideration of pairwise, or epistatic, interactions.

The general strategy to predict the functional impact of a set of mutations is as follows: (1) identify sequence homologs using a hidden Markov model search of a sequence database; (2) weight sequences based on identity to others in the homolog list to reduce biases from database deposition (e.g., based on heightened study of model organisms) or evolutionary age (i.e., greater sampling in the vicinity of older sequences); (3) compute sitewise (h_i) and pairwise (J_{ij}) parameters that are consistent with the sequence data on the basis of a Potts model:

$$P(\sigma) = \frac{1}{Z} \exp\left(\sum_i h_i(\sigma_i) + \sum_{i<j} J_{ij}(\sigma_i, \sigma_j)\right) \tag{9.111}$$

in which $P(\sigma)$ is the probability of observing sequence σ, Z is a factor to normalize the summed probabilities to unity, and the exponential argument is the statistical energy (using a convention of positive energy changes as beneficial); (4) use the resultant probability distribution model to compute the likelihood of observation (and, thus, presumed quality of function) of any protein mutant.

Explicit computation of the set of all h_i and J_{ij} values that maximize the likelihood of observing the weighted sequence data is intractable because of the 20^N possible sequences (where $N =$ number of sites in the protein sequence) to compute for normalization. Thus, a pseudo-likelihood maximization is used to infer the parameter values. In addition, the number of parameters ($\sim(20N)^2/2$) presents the potential challenge of overfitting, even for large protein families. Thus, regularization—in particular, l_2 regularization—is used to restrict model complexity. The resultant calculation is done to identify h_i and J_{ij} values that maximize the objective:

$$\left\{\hat{h},\hat{J}\right\} = \operatorname*{argmax}_{h,J} \sum_{s,i} \pi_s \ln P_i^s(\sigma) - \lambda_h \sum_{i,a} h_i(a)^2 - \frac{\lambda_J}{2} \sum_{i,j,a,b} J_{ij}(a,b)^2 \qquad (9.112)$$

where i and j index the sites, a and b index the amino acids, π_s is the weight of sequence s from step 2, P_i^s is the conditional likelihood of sequence s at site i, and λ values are regularization parameters to penalize complexity.

The approach, termed *EVmutation*, yields mutational energy predictions that correlate (Spearman's $\rho = 0.4$–0.7) with deep experimental data across 34 data sets of 21 proteins and a tRNA gene. These correlations are superior to multiple alternative techniques, including an implementation of EVmutation lacking pairwise coupling (i.e., no J terms). The approach benefits not only from the consideration of pairwise coupling but also from the global fitting of these coupling parameters, which avoids the error of linking covarying sites that are only transitively linked via a third interactor rather than directly coupled.

This conceptual approach of sitewise and pairwise statistical analysis—via global inference—of naturally evolved sequences has also empowered accurate de novo structure prediction [16, 17].

Suggestions for Further Reading

H. C. Berg, *Random Walks in Biology*, Princeton University Press, 1993.

A. J. Grodzinsky, *Fields, Forces, and Flows in Biological Systems*, Garland Science, 2011.

Problems

9-1 Derive equation 9.21 for the waiting time of a Poisson process from the expressions leading up to it in this chapter.

9-2 You are interested in a cellular activity that requires that four or more intact copies of a particular mRNA be present. Given that the average number of intact copies is one, what is the number of cells in a population of 1 million cells that are active?

9-3 You are planning an accelerated evolution experiment with a mutation rate of 10^{-2} mutations per base pair per generation, in which a gene requires five specific, individual DNA mutations for its protein product to acquire a particular desired change of function.

   a. What is the expected proportion of the population that will have acquired all five desired mutations after 20 generations?

   b. If the gene has 1,005 nucleotides, what is the average number of additional DNA mutations beyond the desired five that will have accumulated?

9-4 a. Show that equation 9.56 is an appropriate probability expression by integrating over all future times $\tau \in (0, \infty]$ and summing over all reactions μ from 1 to M, for which the total probability should be unity.

   b. Imagine a reaction vessel containing reactants for two independent reactions, μ_1 and μ_2. If each reaction were propagated in separate vessels, the probability distributions would be $a_1 e^{-a_1 \tau}$ for the first and $a_2 e^{-a_2 \tau}$ for the second. Show that the probability distribution for the first reaction in the combined reaction vessel is $a_1 e^{-(a_1+a_2)\tau}$.

   c. Now generalize your result from part b for a single reaction vessel containing reactants for M reactions that are not necessarily independent, showing that the probability distribution for the first reaction is $a_1 e^{-a_0 \tau}$.

9-5 Tuftsin is a polypeptide with sequence Thr-Lys-Pro-Arg (TKPR) that stimulates many immune cells.

   a. How many different DNA sequences could encode for tuftsin?

   b. You generate a combinatorial library to randomize the three amino acids aside from the terminal Arg (i.e., the sites that are TKP in the natural peptide). You use NNN codons, which provide an equal frequency of A, C, G, and T nucleotides at these sites. You test 1,000 variants randomly selected from this library. Unbeknownst to you, to demonstrate functionality, the protein must have (1) a neutral hydroxyl-containing side chain at the first mutated site; (2) a positively charged side chain (K or R, not H) at the second site; and (3) a proline at the third site. How many of the 1,000 variants do you expect to be (neutral -OH)—(positively charged)—(proline)?

9-6 You are performing mutagenic polymerase chain reaction on the gene encoding for a protein with 50 amino acids. You use a very high salt concentration, resulting in 2% errors in base pairing. You produce 100 proteins from the library of mutated genes and test their activities. What is the probability that you find at least one improved protein? *Important data:* Pilot experiments with this protein indicate that only 3% of amino acid mutations yield an improved protein, whereas 20% of amino acid mutations hinder function and 25% destroy protein function. (*Hint*: Remember to account for silent mutations.)

9-7 You design a DNA primer, of length 18 nucleotides, to anneal to the start of a specific gene.

   a. What is the probability that the primer will anneal to only its intended location in the human genome?

   b. What is the probability of the primer matching 17 of 18 nucleotides to an unintended sequence somewhere in the genome?

9-8 One approach to DNA sequencing, called Sanger sequencing, is to perform unidirectional polymerase chain reaction with one modification: the inclusion of $2'$, $3'$-dideoxynucleotide triphosphates (ddNTPs). These are analogs of the usual deoxynucleotide triphosphates used for DNA replication with the exception that the $3'$ hydroxyl is replaced by hydrogen. All four ddNTPs (A, C, G, and T nitrogenous bases) are added in low concentrations in addition to the standard dNTPs. When a ddNTP is incorporated in a "growing" strand of DNA, additional extension of that particular strand is stopped. Incorporation of ddNTPs and dNTPs is based on their relative concentrations. Thus, different growing strands of DNA will be stopped at different lengths. Each of the ddNTPs is labeled with a different fluorescent moiety, which enables determination of the final nucleotide that was added for a PCR product of a particular length. For example,

   ACTCAGCTGACGAATGCGAGCATTACGAG**C**
   ACTCAGCTGACGAATGCGAGCATTACGAGC**T**
   ACTCAGCTGACGAATGCGAGCATT**A**
   ACTCAGCTGACGAATGCGAGCATTACG**A**
   ACTCAGCTGACGAATGCGAGCATTA**C**
   ACTCAGCTGACGAATGCGAGCATTACGA**G**

If strands of many lengths are synthesized and evaluated, the sequence can be assembled from knowledge of these terminal nucleotides. If the ddNTP concentration is high, the strands will terminate rapidly, and it will be difficult to generate sequence information far from the primer. If the ddNTP concentration is too low, there will be too few short strands and a resultant lack of sequencing data in the region near the primer.

   a. Why is further replication on that strand not possible?

   b. Plot a histogram of DNA product length for dNTP/ddNTP ratios of 10, 100, 1,000, and 10,000.

c. What dNTP/ddNTP ratio (values aside from the four plotted are allowable) provides the longest detectable products and, thus, the most sequencing information? Assume that the reaction is performed with 10 picomoles of primer and that at least 3 femtomoles of a particular product are needed to detect the fluorescent signal. You may assume that sufficient cycles are performed to amplify all primers.

9-9 You are encapsulating cells into spherical droplets (70 μm in diameter) for phenotypic sorting. What cell concentration will maximize the fraction of droplets that contain exactly one cell?

References

1. S. E. Luria and M. Delbrück. Mutations of bacteria from virus sensitivity to virus resistance. *Genetics* 28: 491–511 (1943).
2. D. T. Gillespie. Exact stochastic simulation of coupled chemical reactions. *J. Phys. Chem.* 81: 2340–2361 (1977).
3. D. T. Gillespie. Approximate accelerated stochastic simulation of chemically reacting systems. *J. Chem. Phys.* 115: 1716–1733 (2001).
4. C. Y. F. Huang and J. E. Ferrell. Ultrasensitivity in the mitogen-activated protein kinase cascade. *Proc. Natl. Acad. Sci. U.S.A.* 93: 10,078–10,083 (1996).
5. T. S. Gardner, C. R. Cantor, and J. J. Collins. Construction of a genetic toggle switch in *Escherichia coli*. *Nature* 403: 339–342 (2000).
6. D. T. Gillespie. Stochastic simulation of chemical kinetics. *Annu. Rev. Phys. Chem.* 58:35–55 (2007).
7. N. Maheshri and E. K. O'Shea. Living with noisy genes: How cells function reliably with inherent variability in gene expression. *Annu. Rev. Biophys. Biomol. Struct.* 36: 413–434 (2007).
8. G. W. Li and X. S. Xie. Central dogma at the single-molecule level in living cells. *Nature* 475: 308–315 (2011).
9. A. Arkin, J. Ross, and H. H. McAdams. Stochastic kinetic analysis of developmental pathway bifurcation in phage lambda-infected *Escherichia coli* cells. *Genetics* 149: 1633–1648 (1998).
10. F. St-Pierre and D. Endy. Determination of cell fate selection during phage lambda infection. *Proc. Natl. Acad. Sci. U.S.A.* 105: 20,705–20,710 (2008).
11. R. Brown. A brief account of microscopical observations made in the months of June, July, and August 1827, on the particles contained in the pollen of plants; and on the general existence of active molecules in organic and inorganic bodies. *Philos. Mag.* 4: 161–173 (1828).
12. H. C. Berg. *Random Walks in Biology*. Princeton University Press (1993).
13. N. Tokuriki, F. Stricher, J. Schymkowitz, L. Serrano, and D. Tawfik. The stability effects of protein mutations appear to be universally distributed. *J. Mol. Biol.* 369: 1318–1332 (2007).
14. K. S. Lam, S. E. Salmon, E. M. Hersh, V. J. Hruby, W. M. Kazmierski, and R. J. Knapp. A new type of synthetic peptide library for identifying ligand-binding activity. *Nature* 354: 82–84 (1991).
15. T. A. Hopf, J. B. Ingraham, F. J. Poelwijk, C. P. I. Schärfe, M. Springer, C. Sander, and D. S. Marks. Mutation effects predicted from sequence co-variation. *Nat. Biotechnol.* 35: 128–135 (2017).
16. D. S. Marks, L. J. Colwell, R. Sheridan, T. A. Hopf, A. Pagnani, R. Zecchina, and C. Sander. Protein 3D structure computed from evolutionary sequence variation. *PLoS One* 6: e28766 (2011).
17. T. A. Hopf, L. J. Colwell, R. Sheridan, B. Rost, C. Sander, and D. Marks. Three-dimensional structures of membrane proteins from genomic sequencing. *Cell* 149: 1607–1621 (2012).

Index

ΔG, 34
ΔH, 34
ΔS, 34
ε, 79
ϕ^2, 500–503

absorbance, 77
activated complex, 68
activity coefficient, 513
actuarial tables, 418
adaptation, 351
Airy disk, 97
alanine scanning mutagenesis, 51
Alexa Fluor dyes, 93
allometric scaling, 381
allophycocyanins, 94
Alon, U., 355
amensalism, 452
analytical ultracentrifugation, 178
antithetic integral feedback, 359
apoptosis, *see* cell death
approximate stochastic methods, 548
area under the curve, 510, 521
Arkin, A., 552
Arrhenius relationship, 64, 234
association rate constant, 31
AUC, *see* area under the curve
Aurelius, Marcus, 373
avidity, *see* multivalent binding, 128, 138, 143, 197

Bailey, J. E., 295, 296
Ball, Lucille, 477
Baltimore, D., 363
Beer-Lambert law, 79
bell-shaped response, 139, 232, 394
Bernoulli trial, 528
Berra, Yogi, 19
bifurcation, 332, 368, 369, 464, 547
binding interactions, 29
binding isotherm, 112, 126, 142, 191, 196, 197
biolayer interferometry, 166

bioreactors, 404
bistability, 3, 331, 345, 348, 544
bisubstrate kinetics, 227
blood coagulation, 261
Bohr, Niels, 19
Box, George, 16
Briggs-Haldane assumption, 215, 217
Brownian motion, 554
bursty transcription, 551

C_p, 37
Callahan, Harry, 477
cancer cluster, 534
Carroll, Lewis, 16
caspases, 260
cell cycle, 365, 431
cell death, 413, 431
cell signaling, 6, 251
cellular biomass, 397
cellular maintenance, 408
center [nonlinear dynamics], 331
CFSE, 433
charge-charge interactions, 43
chemical Langevin equation, 549
chemical master equation, 540
chemostat, 403, 462, 465, 467–469
chemotaxis, 355
Cheng-Prusoff equation, 245
chi squared, 20, 180
chromatography, 178
circadian rhythms, 365
Claude, Albert, 281
Collins, J. J., 4, 544
combinatorial bioinformatics, 555
commensalism, 452
compartmental growth models, 431
compartmental models, 297, 431, 504, 505, 508
compensation, enthalpy-entropy, 38
competitive inhibition, 235, 252
confocal microscopy, 99
conformational selection, 486

constraint-based models, 258
continuum approximation, 45, 61, 289, 527
cooperative enzymes, 231
cooperativity, 24
Coulomb's law, 43
cows, 424
crosslinked receptors, 138
crowding, molecular, 29, 512

D, 479, 481, 505, 516, 554
Dall'Acqua, W. F., 143
data collection, 179
death
 by injury, 413
 by senescence, 417
 organismal, 413
Debye-Hückel model, 73
Debye-Hückel theory, 69
deconvolution microscopy, 99
deformation energy, 50
degradation, 281
degree of reductance balance, 398
Delbrück, M., 535
Di Cera, E., 486
diffraction, 77
diffusion, 478, 514, 553
diffusion-limited binding, 86, 478, 489, 496, 502, 521
diffusivity, 479, 481, 505, 516, 554
dilution rate, 407
directed evolution, 164
dissociation rate constant, 32, 36
DNA fingerprinting, 561
double-jump experiment, 151
double-mutant cycle, 59, 104, 105
doubling time, 27, 284, 298, 375, 377, 393, 462, 463
Doyle, J., 355
Drosophila, 251, 349, 418
dynamic rate processes, 1

EC_{50}, 125, 243, 311, 318, 338
Eastwood, Clint, 477
effective concentration, 131, 156, 198, 199, 243, 256, 508, 514
Eigen, M., 23, 149
Einstein, Albert, 16
electromagnetic radiation, 75
electrophoresis, 178
electrostatic energy, 44
electrostatic interactions, 43
electrostatic steering, 69
enthalpy, 34
enthalpy-entropy compensation, 38
entropy, 34
Enzyme Commission, 211
enzyme kinetics, 209
enzyme-linked immunosorbent assay, 169
epidemic, 443

equilibration time, 153
equilibrium, 22, 31, 35, 111, 136, 191, 193–198
equilibrium constant, 15, 32, 36, 111, 192, 194
equilibrium screening, 164
exponential growth, 375
extinction coefficient, 79
extreme value theory, 418
Eyring equation, 68

fed-batch fermentors, 409
feedforward, 358
Fermi, Enrico, 20
Fersht, A. R., 59, 75, 190
Fick's first law, 479, 553
Fick's second law, 481
fixed point, 327, 368, 455, 465
flow cytometry, 100, 434
fluctuations, 289, 362, 407, 447, 530, 535
fluorescence, 81
fluorescence microscopy, 97
fluorescence recovery after photobleaching, 83, 96
fluorescence resonance energy transfer, 87
flux-control coefficient, 257
fractional saturation, 112, 124, 175, 308
FRAP, 83, 96
Frechet distribution, 420
free energy, 34
frequency modulation, 362
FRET, 87
Furano, Hokkaido, 465
futile cycle, 342
Förster resonance energy transfer, 87

G protein-coupled receptors, 256
Gaddum equation, 243
Galilei, Galileo, 152
Gard, Sven, 281
Geisel, Theodor, 453
gene expression, 3, 10, 281, 296, 550
generalized growth equations, 388
genetic toggle switch, 4, 544
GFP, *see* green flourescent protein
Gibbs free energy, 34
Gibbs-Helmholtz equation, 37
Gillespie simulations, 539
Gillespie, D. T., 539
Glacken, M. W., 392
global parameter fitting, 190
Goldbeter, A., 342
Gompertz death kinetics, 418
Gompertz growth kinetics, 18, 380
Gompertz, B., 417
green fluorescent protein, 89
growth
 curves, 374
 parameter estimates, 403
 phases, 374

growth-associated products, 411
growth-factor dose responses, 391

Hall, J., 366
Hartwell, L., 365
Haugh, J. M., 310
Hayflick limit, 425
Hayflick, L., 425
heat capacity, 37
Henri, V., 213
Hill coefficient, 126, 127, 231, 326, 337
Hill plot, 126, 231, 324
Hinshelwood, Cyril, 111, 390
HIV, 7, 235, 248, 435
  dynamics, 447
homeostasis, 351
Hood, Leroy, 323
hot spot, 52
human growth hormone, 139, 310, 394
Hunt, T., 365
hybridoma, 412
hydrogen bond, 45
hydrolytic regulatory enzymes, 260
hydrophobic effect, 49
hypothesis testing, 9
hysteresis, 334

IC_{50}, 243, 247, 273, 274, 395
incoherent feedforward, 358
index of refraction, 76
induced fit, 486
infection, 7, 248, 443, 551
inhibition, 230, 235
  competitive, 235, 252
  noncompetitive, 235, 238
  uncompetitive, 235, 240
integral feedback control, 351, 358
interface properties, 51
internalization, 5, 10, 307
ionic screening, 69
ionic strength, 69
ionizing radiation, 415, 551
isotherm, *see* binding isotherm
isothermal titration calorimetry, 38, 176

K_d, 32, 36
K_I, 237
K_m, 217, 219
κ, 68
k_{cat}, 215, 220
k_{off}, 32, 116
k_{on}, 31, 116
Kelvin, Lord, 152
kinetic screening, 164
kinetics
  association, 117
  bisubstrate, 227
  dissociation, 117

Koller, D., 527
Koshland, D. E., 232, 342
Koshland-Némethy-Filmer theory, 232

labeling artifacts, 158
Langevin equation, 549
Langmuir, 112, 390
Lauffenburger, D. A., 497
Leibler, S., 355
Lennard-Jones potential, 47
ligand depletion, 113, 160
light, 75
light-matter interactions, 75
limit cycle, 331, 365
limiting substrate, 389
linear-quadratic models, 415
Lineweaver-Burk plot, 217, 246
Lippincott-Schwartz, J., 315
local depletion, 495
logistic equation, 378
Lotka-Volterra model, 454
low-pass filter, 361
luciferase, 271
Luria, S. E., 535

macromolecular crowding, 29, 512
Malthus, Thomas Robert, 373, 377
Mann, K. G., 262
Marks, D. S., 562
Martin, L., 432
mass-action kinetics, 13, 24, 31, 281
master equation, 540
mathematical modeling, 9
Mendeleev, Dmitri, 111
Menten, M. L., 213
metabolic control analysis, 257
metabolic heat generation, 401
metabolism, 257
metalloproteases, 269
Michaelis, L., 213
Michaelis-Menten
  equation, 24, 212
  mechanism, 217
  range of validity, 221
microbial bioreactors, 404
microfluidic assays, 170
mixed populations, 451
molecular crowding, 29, 512
molecular titration, 340
monoclonal antibody production
  kinetics, 412
Monod growth form, 18, 381
Monod, J., 390
Monod-Wyman-Changeux theory, 232
Morrison equation, 248, 253
multimeric macromolecules, 121, 192, 231,
  337, 546
multiphoton microscopy, 99

multiple binding sites, 121
  cooperativity, 125, 194, 196, 197, 231
  independent sites, 121
multistability, 331, 345
multivalent binding, 121, 128, 139
mutiple-hit models, 415
mutualism, 452

negative feedback, 5, 266, 297, 352, 362, 369
network dynamics, 3, 323
networks, biomolecular, 319, 323
noncompetitive inhibition, 235, 238
noncovalent binding interactions, 29
nondimensionalization, 119, 225, 326, 343, 382, 383, 406, 445, 500, 501, 546
nonlinear dynamics, 324, 455
nonlinear least squares, 180, 190
nonspecific binding, 162
nonspecific protein adsorption, 157
Norrish, R. G. W., 23, 149
Northrop, John, 209
nullclines, 327, 368, 546
Nurse, P., 365
Nutches, 454

Occam, William of, 16
optical spectroscopy, 175
optimal dilution rate, 407
organismal death, 413
oscillations, 359, 365, 456

Pe, 482
pK, 233
Péclet number, 482
parameter estimation, 20, 159, 179, 180, 185, 186, 190, 285, 315
  confidence intervals, 186
  sensitivity analysis, 20
Pearl, R., 378
Perelson, A. S., 435, 447
perfect adaptation, 351
pH, 273, 276
pH effects, 232
pH-dependent binding, 55
pharmacodynamics, 508
pharmacokinetics, 7, 10, 508
pharmacological inhibition, 243
phase portrait, 326, 368, 369
Phillpotts, Eden, 1
photoactivatable fluorophores, 95
photobleaching, 83
phycobiliproteins, 94
phycocyanins, 94
phycoerythrins, 94
Plato, 29
Poisson statistics, 528
Poisson-Boltzmann equation, 70

Popper, Karl, 17, 19
populations, 7
Porter, G, 23, 149
positive feedback, 262, 345
positive-feedback, 326
precision of adaptation, 353
predation, 454
product-formation kinetics, 411
protein trafficking, *see* trafficking
pseudo-first-order binding, 106, 116, 197, 203
pseudochemical reaction, 377, 396
pulse-chase labeling, 24, 294, 433, 435

quasi-steady-state approximation, 135, 215, 217, 227, 233, 275, 484
Quastler, H., 441

R_o, 446
R_μ, 542
Re, 505
radiationless transition, 82
rate constant
  association, 31
  crowding effects on, 518
  dissociation, 32
  typical range, 33
Rayleigh limit, 98
reaction channel, 538, 542
reaction probability density function, 542
recapture, 495
receptor downregulation, 12, 308
Reed, L. J., 378
regulatory enzymes, 256, 260
relaxation kinetics, 150
replicative senescence, 425
reproductive ratio, 446
respiratory quotient, 397
Reynolds number, 505
Rosbash, M., 366
Rosenblueth, Arturo, 16
RQ, 397

Sc, 505
Sh, 505
saddle point, 330, 458
salt-bridge energetics, 61
Sarkar, C. A., 348
Savageau, M., 389
scaling analysis, 500
Scatchard plot, 115
scattering, 76
Schild analysis, 246
Schmidt number, 505
secondary metabolites, 411
senescence, replicative, 425
sensitivity analysis, 257, 265
separatrix, 331

sequence co-variation, 562
Seuss, Dr., 453
Shakespeare, William, 31, 323
Sherman, F. G., 441
Sherwood number, 505
signaling pathways, *see* cell signaling
single-target kinetics, 413
SIR models, 444
small numbers of molecules, 527
Smith, J. A., 432
Smoluchowski diffusion limit, 489
Socrates, 29
sorting, *see* trafficking
specific growth rate, 19, 66, 283
spiral [nonlinear dynamics], 331
stable node, 331
stable spiral, 331
standard state, 36
state functions, 34
stationary point, 327, 368, 455, 465
steady state, 22
stem cell dynamics, 439
Stern, L. J., 139
stochasticity, 3, 289, 359, 551
  fluctuations, 539
  gene expression, 550
  Gillespie simulations, 539
  simulations, 538
Stokes shift, 82
Stokes-Einstein equation, 479
stopped-flow spectrometry, 97
substrate inhibition, 230
super-resolution microscopy, 100
surface plasmon resonance, 1, 166
switchlike behavior, 3, 24, 126, 127, 231, 337, 345, 348, 393, 544

τ_D, 375, 377, 393, 462, 463
Taleb, Nassim Nicholas, 527
tau-leaping method, 548
telomere end-replication problem, 428
telomeres, 426
temperature-dependent binding, 64
thermodynamic cycle, 54, 59, 132, 238, 273
thermodynamic pathway, 130
thermodynamics, 34
Thiele modulus, 500, 503
Thomas, Lewis, 209
threshold theorem, 446
Tidor, B., 61
time delay, 285, 287, 294, 296, 305, 362, 432, 461
tissue penetration, 500
toxic metabolite accumulation, 394
trafficking, 5, 10, 281, 297, 306
trajectory of concentrations, 325
transcription, 281, 289, 297, 544, 551
  bursty, 551

transition state, 66, 67
transition state theory, 67, 73
translation, 3, 66, 281, 297
transmission coefficient, 68
transport effects on binding, 477
trimeric G proteins, 256
tryptophan fluorescence, 175
turnover number, 220
Twain, Mark, 19
two-compartment model, 504, 505, 508
two-photon microscopy, 99
two-step binding, 484
Tyrannosaurus rex, 462

μ, 19, 66, 283
U.S. population, 379
ultrasensitivity, 127, 337, 340, 342, 348
uncompetitive inhibition, 235, 240
unimolecular binding constant, 130
unstable node, 331
unstable spiral, 331

V_D, 511
v_{max}, 217
van der Waals interactions, 46
van't Hoff equation, 37, 150
Verhulst, P.-F., 17, 378
Volterra, V., 452
volume of distribution, 511
von Bartalanffy rate form, 382

waiting time, 447, 531
washout, 406
Weibull distribution, 420
West-Brown-Enquist rate law, 382
Wiener, Norbert, 16
William of Occam, 16
Wittrup, K. D., 164, 503
wormlike chain model, 133

χ^2, 20, 180

yield coefficient, 398
Young, M., 366

zero-order ultrasensitivity, 342